Radioactivity and Pollution in the Nordic Seas and Arctic Region

Observations, Modeling, and Simulations

Ola M. Johannessen, Vladimir A. Volkov,
Lasse H. Pettersson, Vladimir S. Maderich,
Mark J. Zheleznyak, Yongqi Gao,
Leonid P. Bobylev, Andrey V. Stepanov,
Ivan A. Neelov, Viktor P. Tishkov and
Sven P. Nielsen

Radioactivity and Pollution in the Nordic Seas and Arctic Region

Observations, Modeling, and Simulations

Springer

Published in association with
Praxis Publishing
Chichester, UK

PRAXIS

Professor Ola M. Johannessen
Mohn–Sverdrup Center and Nansen Environmental and Remote Sensing Center
Bergen, Norway; and
Geophysical Institute, University of Bergen,
Bergen, Norway

Mr Lasse H. Pettersson
Mohn–Sverdrup Center and Nansen Environmental and Remote Sensing Center
Bergen, Norway

Dr Mark J. Zheleznyak
Institute of Mathematical Machines and System Problems
Kiev, Ukraine

Dr Leonid P. Bobylev
Nansen International Environmental and Remote Sensing Center
St Petersburg, Russia

Dr Ivan A. Neelov
Nansen International Environmental and Remote Sensing Center
St Petersburg, Russia; and
Arctic and Antarctic Research Institute
St Petersburg, Russia

Dr Vladimir A. Volkov
Nansen International Environmental and Remote Sensing Center
St Petersburg, Russia

Dr Vladimir S. Maderich
Institute of Mathematical Machines and System Problems
Kiev, Ukraine

Dr Yongqi Gao
Mohn–Sverdrup Center and Nansen Environmental and Remote Sensing Center
Bergen, Norway

Mr Andrey V. Stepanov
V. G. Khlopin Radium Institute
St Petersburg, Russia

Dr Viktor P. Tishkov
V. G. Khlopin Radium Institute
St Petersburg, Russia

Dr Sven P. Nielsen
Risø National Laboratory for Sustainable Energy
Technical University of Denmark
Roskilde, Denmark

SPRINGER–PRAXIS BOOKS IN GEOPHYSICAL SCIENCES
SUBJECT *ADVISORY EDITOR*: Philippe Blondel, C.Geol., F.G.S., Ph.D., M.Sc., Senior Scientist, Department of Physics, University of Bath, Bath, UK
Published in association with the Nansen Centers in Bergen and St Petersburg
Nansen Center's Polar Series No. 5

ISBN 978-3-642-26273-9 Springer-Verlag Berlin Heidelberg New York
Springer is part of Springer-Science + Business Media (springer.com)

Softcover reprint of the hardcover 1st edition 2010

Cover design: Jim Wilkie
Project management: OPS Ltd, Gt Yarmouth, Norfolk, UK

Contents

APPENDICES

Preface

This book is the result of a decade of studies dedicated to the study of the spread of radioactive pollution in the Nordic Seas and Arctic, focused on the many land-based industrial sources in Siberia, as well as pollution from European nuclear fuel–processing plants in Sellafield (into the Irish Sea) and La Hague (into the English Channel).

The main aim of investigations has been to assess the spread of radioactive pollution under both present-day natural conditions and the future potential scenarios of accidents. The results of our studies are also intended to support decision making in risk prevention, forecasting, and preparedness. This study is in line with recommendations from the Arctic Monitoring and Assessment Program's (AMAP) *Arctic Pollution 2002* report (AMAP, 2002), which also concludes that: "The major sources of radioactive contamination of the Arctic environment remain fallout from atmospheric nuclear weapons testing in the period 1945 to 1980, discharges from European spent nuclear fuel reprocessing plants, and fallout from the 1986 Chernobyl nuclear power plant in the Ukraine."

The work was carried out on the basis of the publicly available radionuclide observations published in the open literature. These observations were integrated into our new Generic Model System (GMS), which includes four interconnected numerical models: (1) a model of radionuclide spread in river systems, (2) a model of radionuclide spread in the stratified water of estuaries, (3) a model for the Kara Sea, and (4) a North Atlantic and Arctic Ocean model. Simulations were made for 20th-century climate conditions as well as for 21st-century global-warming conditions reflecting a doubling of atmospheric CO_2 concentrations.

The study was carried out within the framework of the EU INCO-COPERNICUS-II project "Simulation Scenarios for Potential Radioactive Spreading in the 21st Century from Rivers and External Sources in the Russian Arctic Coastal Zone—RADARC" (ICA2-CT-2000-10037), 2001–2003, which included the results of the following projects funded by the International Association for the Promotion of Co-operation with Scientists from the New Independent States of the Former Soviet Union (INTAS): (1) "Observations and Modeling of Transport and Dilution of Radioactive Waste and Dissolved Pollutants in the Kara Sea" (INTAS 93-814), 1994–1997, (2) "Study of Influence of Land-based Sources of Radionuclides on Radioactive Contamination of the Kara Sea through Ob′ and Yenisei River Systems" (INTAS 97-31278), 1999–2001; (3) "Detection and Modelling of Greenhouse Warming in the Arctic and Sub-Arctic" (INTAS 97-1277), 1998–2000; and (4) the ARC project funded by the Research Council of Norway. Preparation of the book was undertaken under the EC FP6 project "Generic Model Simulations (GMS) of spreading of marine pollutants in the Arctic environment during the 21st Century", 2005–2006 and was also supported by the Co-operation Program with Central and Eastern Europe funded by the Research Council of Norway and the Norwegian Ministry of Foreign Affairs. These projects were coordinated by Ola M. Johannessen, Lasse H. Pettersson, Vladimir A. Volkov, and Leonid P. Bobylev.

Work on the book was led by the Mohn–Sverdrup Center for Global Ocean Studies and Operational Oceanography (MSC)/Nansen Environmental and Remote Sensing Center (NERSC), Bergen, Norway, and the Nansen International Environmental and Remote Sensing Center (NIERSC), St. Petersburg, Russia. There were major contributions from the Institute of Mathematical Machines and System Problems, Kiev, Ukraine, the V. G. Khlopin Radium Institute, St. Petersburg, Russia, and the Risø National Laboratory for Sustainable Energy at the Technical University of Denmark, Roskilde, Denmark.

Acknowledgments

The authors acknowledge support from the several projects and agencies mentioned in the Preface. Support for the completion and editing of this book was provided by the Mohn–Sverdrup Center for Global Ocean Studies and Operational Oceanography (MSC), Bergen, Norway, through a generous donation to the MSC by Trond Mohn on behalf of Frank Mohn ASA, Bergen, Norway. We also acknowledge contributions to Chapter 2 from Dr. T. A. Akimenko and Dr. E. A. Zakharova, Moscow State University, and to Chapters 1 and 2 from Kirill Bobylev and Vadim Plekhov, V. G. Khlopin Radium Institute/NIERSC. The authors also acknowledge contributions from Morten Sickel and Per Strand, Norwegian Radiation Protection Authority, and Lars Otto Reiersen, Arctic Monitoring and Assessment Program (AMAP), for contributions through their participation in the INCO-COPERNICUS-II project "Simulation Scenarios for Potential Radioactive Spreading in the 21st century from Rivers and External Sources in the Russian Arctic coastal zone—RADARC". The authors also acknowledge Tor Olausson, NERSC, Bergen, Norway, for improving many of the figures.

Figures

Tables

Abreviations and acronyms

AARI	Arctic and Antarctic Research Institute
AMAP	Arctic Monitoring and Assessment Program
APC	Approximately Permissible Concentration
AW	Atlantic Water
BCM	Bergen Climate Model
Bq	becquerel
CBDW	Canadian Basin Deep Water
Cd	cadmium
CHARIMA	Charriages de Rivières Maillés
Ci	curie
CO_2	carbon dioxide
CPI	Carbon Preference Index
Cr	chromium
^{137}Cs	cesium-137
Cu	copper
DBE	Dnieper–Bug Estuary
DCW	Digital Chart of the World
DDD	1,1-dichloro-2,2-bis(*p*-chlorophenyl) ethane
DDE	1,1-dichloro-2,2-bis(*p*-chlorophenyl) ethylene
DDT	1,1-trichloro-2,2-bis(*p*-chlorophenyl) ethane
DEM	Digital Elevation Model
DKWAV	Distributed Kinematic Wave model for channel flows
DMA	Defense Mapping Agency
DOC	Dissolved Organic Carbon
EBDW	Eurasian Basin Deep Water
EC	European Council
ECHAM	Hamburg climate model based on ECMWF model
ECMWF	European Center for Medium-Range Weather Forecasts

EGC	East Greenland Current
EPA	Environmental Protection Agency
ESP	Elementary Synoptic Process
ESRI	Environmental Systems Research Institute
FANTOM	Fate and Transport Ocean Model
FLESQOT	Flow, Energy, Salinity, Sediment, and Containment Transport model
FUIC	Factor of Using the Installed Capacity
GIS	Geographic Information System
GMS	Generic Model System
GSAIW	Greenland Sea Arctic Intermediate Water
GSDW	Greenland Sea Deep Water
HAMSOM	HAMburg Shelf Ocean Model
HCB	hexachlorobenzene
HCH	hexachlorocyclohexane
HCP	Homogeneous Circulation Period
HEC-2	Hydrologic Engineering Center model of river hydraulics
HEU	Highly Enriched Uranium
Hg	mercury
HLW	High-Level nuclear Waste
HM	Heavy Metal
HPP	Hydroelectric Power Plant
IAEA	International Atomic Energy Agency
IBRAE RAS	Nuclear Safety Institute, Russian Academy of Sciences
ICRP	International Commission on Radiological Protection
ILW	Intermediate-Level nuclear Waste
IMMSP	Institute of Mathematical Machine and System Problems
INTAS	International Association for the Promotion of Co-operation with Scientists from the New Independent States of the Former Soviet Union
ISAIW	Iceland Sea Arctic Intermediate Water
ISDW	Iceland Sea Deep Water
ISW	Irminger Sea Water
JS NOGC	Joint Stock National Oil and Gas Company
KAREX (-94, -95)	Kara Sea expeditions
KMCC	Krasnoyarsk Mining and Chemical Combine
KRI MINATOM	Khlopin Radium Institute, Ministry of Atomic Energy
KTE	Kinetic Turbulent Energy
LEU	Lowly Enriched Uranium
LLW	Low-Level radioactive Waste
LRW	Liquid Radioactive Waste
MCC	Mining and Chemical Combine (Krasnoyarsk)
MCi	megacurie

MEIW	Modified East Icelandic Water
MICOM	Miami Isopycnal Coordinate Ocean Model
ML	Mixed Layer
MMBI	Murmansk Marine Biological Institute
MNAW	Modified North Atlantic Water
MOX	mixed oxides (uranium, plutonium)
MPC	Maximum Permissible Concentration
MW	megawatt
NAC	Norwegian Atlantic Current
NAO	North Atlantic Oscillation
NATO	North Atlantic Treaty Organization
NCAR	National Center for Atmospheric Research
NCC	Norwegian Coast Current
NCCP	Novosibirsk Chemical Concentrate Plant
NCEP	National Centers for Environmental Prediction
NEAW	North East Atlantic Water
NERSC	Nansen Environmental and Remote Sensing Center
Ni	nickel
NIERSC	Nansen International Environmental and Remote Sensing Center
NPP	Nuclear Power Plant
NSAIW	Norwegian Sea Arctic Intermediate Water
NSDW	Norwegian Sea Deep Water
NwAC	Norwegian Atlantic Current
OA	Objective Analysis
OC	Organochlorine Compound
OGCM	Ocean General Circulation Model
OGP	Oil–Gas Province
OILTOX	Oil transport and fate model
ONC	Operational Navigation Chart
PA	Production Association
PAH	Polycyclic Aromatic Hydrocarbon
Pb	lead
PBq	petabecquerel
PCB	polychlorinated biphenyl
PGF	Pressure Gradient Force
PINRO	Knipovich Polar Research Institute of Marine Fisheries and Oceanography
PIW	Polar Intermediate Water
POC	Particulate Organic Carbon
POM	Princeton Ocean Model
POP	Persistent Organic Pollutant
^{239}Pu	plutonium-39
QN	Quasi-Normal (control run)
RADARC	Simulation scenarios for potential radioactive spread

	in the 21st century from rivers and external sources in the Russian Arctic coastal zone
RAW	Recirculating Atlantic Water
RBMK	Reaktor Bolshoy Moshchnosti Kanalniy (high-power channel-type reactor)
RIVTOX	Transport of radionuclides in a river system
RW	Radioactive Waste
SAR	Synthetic Aperture Radar
SAT	Surface Air Temperature
SC	Special Combine
SCC	Siberian Chemical Combine (Tomsk)
SFT	*Statens ForurensningsTilsyn*
SIGRID	Sea Ice Grid (a standard for archival and exchange of data in digital form)
SMPDBK	SiMPlified Dam Break Code (National Weather Service)
SNF	Spent Nuclear Fuel
SNOP/RUSNOP	Soviet/Russian Norwegian Oceanographic Program
^{90}Sr	strontium-90
SRW	Solid Radioactive Waste
SRWB	Station of Radioactive Waste Burial
SSS	Sea Surface Salinity
SST	Sea Surface Temperature
TBq	terabecquerel
THORP	THermal Oxide Reprocessing Plant
THREETOX	Three-dimensional dispersion simulation of radionuclides in stratified water bodies
TODAM	One-dimensional channel model
TOH	Total Oil Hydrocarbons
TVD	Total Variation Diminishing model
^{235}U	uranium-235
UHMI	Ukrainian HydroMeteorological Institute
VPF	Vector Product Format
VVER	Vodo–Vodyanoi Energetichesky Reaktor (water-cooled, water-moderated energy reactor)
WASP/TOXI	Water quality Analysis Simulation Program/Toxic chemical model
WOCE	World Ocean Circulation Experiment
WSC	West Spitsbergen Current
Zn	zinc

Contributors

NATALYA N. DZIUBA
Institute of Mathematical Machines and System Problems
Glushkov Pr. 42, 03187 Kiev, Ukraine

GENNADY V. ILYIN
Murmansk Marine Biological Institute
Vladimirskaya Str. 17, 183010 Murmansk, Russia

VLADIMIR I. KOSHEBUTSKY
Institute of Mathematical Machines and System Problems
Glushkov Pr. 42, 03187 Kiev, Ukraine

ALEXEI V. KOURAEV
Nansen International Environmental and Remote Sensing Center
14th Line Str. 7A, Office 49, Vasilievsky Island, 199034 St. Petersburg, Russia
Moscow State University, Russia
Moscow State University, Faculty of Geography, 119899 Vorobyevy Gory, MGU, Moscow, Russia
Current affiliation: OMP-LEGOS (Observatoire Midi-Pyrénées, Laboratoire d'études en géophysique et océanographie spatiale), avenue Edouard Belin 18, 31401 Toulouse cédex 9, France

MARTIN W. MILES
Environmental Systems Analysis Research Center
4450 Arapahoe Ave., Suite 100, Boulder, CO 80303, U.S.A.

STEINAR ORRE
G. C. Rieber Climate Institute, Nansen Environmental and Remote Sensing Center
Thormøhlensgate 47, 5006 Bergen, Norway

Authors

OLA M. JOHANNESSEN
Mohn–Sverdrup Center and Nansen Environmental and Remote Sensing Center
Thormøhlensgate 47, 5006 Bergen, Norway
Geophysical Institute, University of Bergen
Allégaten 70, 5007 Bergen, Norway
Email: *ola.johannessen@nersc.no*

VLADIMIR A. VOLKOV
Nansen International Environmental and Remote Sensing Center
14th Line Str. 7A, Office 49, Vasilievsky Island, 199034 St. Petersburg, Russia
Email: *vladimir.volkov@niersc.spb.ru*

LASSE H. PETTERSSON
Mohn–Sverdrup Center and Nansen Environmental and Remote Sensing Center
Thormøhlensgate 47, 5006 Bergen, Norway
Email: *lasse.pettersson@nersc.no*

VLADIMIR S. MADERICH
Institute of Mathematical Machines and System Problems
Glushkov Pr. 42, 03187 Kiev, Ukraine
Email: *vlad@immsp.kiev.ua*

MARK J. ZHELEZNYAK
Institute of Mathematical Machines and System Problems
Glushkov Pr. 42, 03187 Kiev, Ukraine
Email: *mark@immsp.kiev.ua*

YONGQI GAO
Mohn–Sverdrup Center and Nansen Environmental and Remote Sensing Center
Thormøhlensgate 47, 5006 Bergen, Norway
Email: *yongqi.gao@nersc.no*

LEONID P. BOBYLEV
Nansen International Environmental and Remote Sensing Center
14th Line Str. 7A, Office 49, Vasilievsky Island, 199034 St. Petersburg, Russia
Email: *leonid.bobylev@niersc.spb.ru*

ANDREY V. STEPANOV
V. G. Khlopin Radium Institute
2nd Murinski Av. 28, 194021 St. Petersburg, Russia
Email: *stepanov@khlopin.ru*

IVAN A. NEELOV
Nansen International Environmental and Remote Sensing Center
14th Line Str. 7A, Office 49, Vasilievsky Island, 199034 St. Petersburg, Russia
Arctic and Antarctic Research Institute
Bering Str. 38, 199397 St. Petersburg, Russia
Email: *neelov@aari.nw.ru*

VIKTOR P. TISHKOV
V. G. Khlopin Radium Institute
2nd Murinski Av. 28, 194021 St. Petersburg, Russia
Email: *tishkov@khlopin.ru*

SVEN P. NIELSEN
Risø National Laboratory for Sustainable Energy, Technical University of Denmark
Frederiksborgvej 399, Postboks 49, 4000 Roskilde, Denmark
Email: *spni@risoe.dtu.dk*

Introduction

This book covers a range of important topics related to the sources and spread of radioactive pollution in the Nordic Seas and the Arctic, including the basins of the Ob′ and Yenisei Rivers that feed into Russia's Arctic seas (see figure overleaf). The book is intended to provide a relatively detailed and comprehensive—though not exhaustive—treatment of some of the major aspects of both the past spread and potential spread of radionuclides in the environment. We also present the same for non-radioactive toxic pollutants such as persistent organic pollutants (POPs). The book represents an integrated and thoroughly referenced compilation of material from international and Russian sources, put together in this way for the first time. We combine results from observations and new numerical modeling of the spread of contamination, as well as conducting a risk assessment.

Chapter 1 describes the range of sources of radioactive and non-radioactive contamination in the Arctic and Nordic Seas. These include primary sources such as nuclear weapons testing, emissions from nuclear power stations, nuclear reactors installed on nuclear-powered submarines and icebreakers, and intentional discharges of radioactive waste into the environment from nuclear reprocessing plants. Here we focus on point sources including the Russian nuclear industry enterprises along the Ob′ and Yenisei Rivers, as well as other sources in Russia and Europe. The chapter concludes with a description of the main sources of non-radioactive pollution focused on the Norwegian–Russian Arctic and sub-Arctic marine environment, namely the Nordic, Barents, White, Kara, and Laptev Seas.

Chapter 2 describes the geography of the study region and the observational data used in this study. The geographical and oceanographic features of the study region are described in the first section, which is organized in three subsections: the Ob′ and Yenisei river basins, the Kara Sea region, and the Nordic Seas and adjacent seas. The second section presents an overview of the environmental observations that are used in this study, organized in two subsections: natural environmental (e.g., hydrological, oceanographic, and geophysical) data and radioactive pollution data.

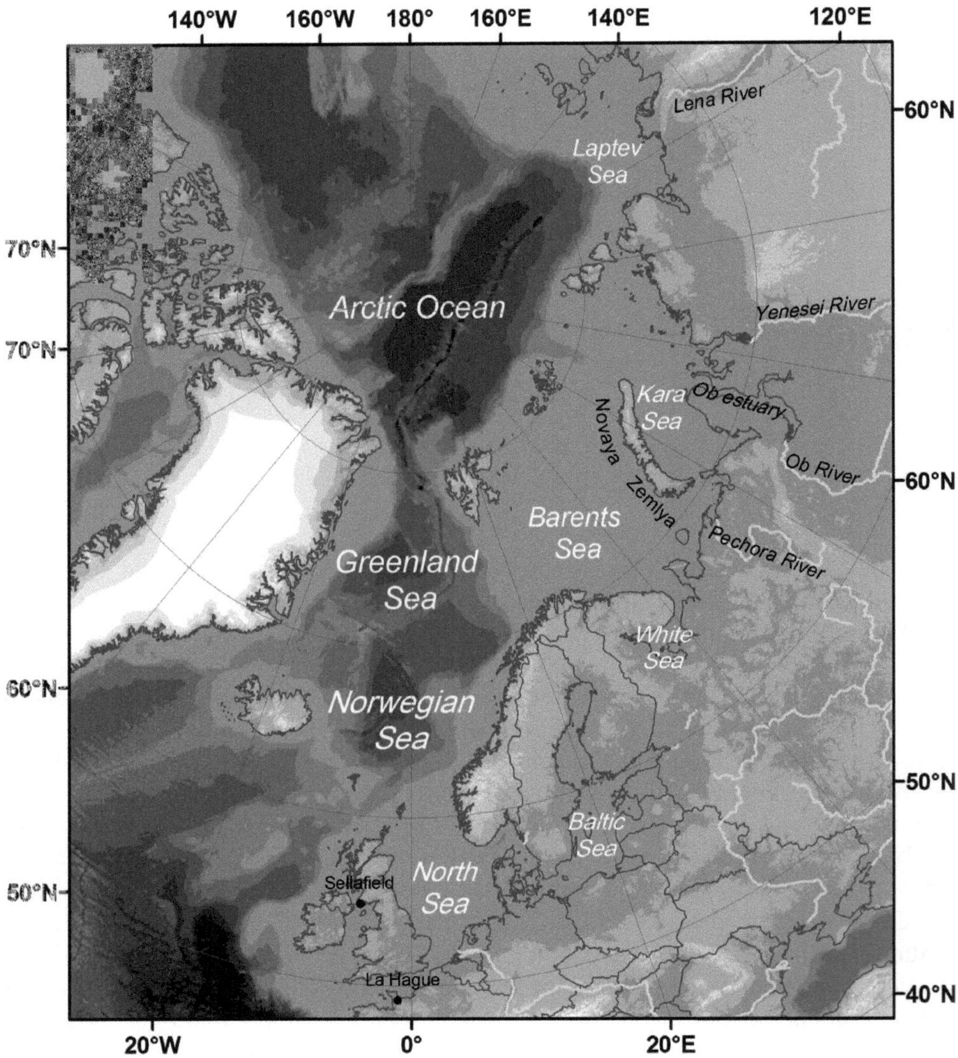

The Nordic Seas and Arctic study region. Map generated from *ETOPOI* (Amante and Eakins, 2009).

Chapter 3 presents a set of numerical modeling techniques to simulate the spread of radioactivity in the aquatic environment, in both marine and inland waters. The first section describes the concept and structure of the Generic Modeling System (GMS). We then present a model for the Atlantic and Arctic Oceans, followed by a shelf sea model for the Kara Sea. Then we describe river and estuary models for the Ob′ and Yenisei Rivers. These comprise a one-dimensional model to simulate the transport of radionuclides in the river system, and a numerical model for three-dimensional dispersion simulation of radionuclides in stratified water bodies. For each model, we validate the models against observational data and knowledge.

Chapter 4 is dedicated to studies of the spread of radioactivity in the Arctic using the GMS. Two sets of numerical experiments are presented: (1) simulations or hindcasts of past contamination by anthropogenic radionuclides originating from various sources: nuclear weapons tests, atmospheric fallout from Chernobyl, the Sellafield (U.K.) reprocessing plant, and Russian nuclear enterprises located along rivers; (2) simulations of contamination as a result of potential accidents at nuclear plants and submarines.

The first series of experiments allowed reconstruction of past radionuclide pollution of the Ob′ and Yenisei Rivers and the flux of radioactivity to the Kara Sea from nuclear plants. The GMS was applied to simulate the pathways of radioactivity from different sources in the Arctic and North Atlantic. The simulation demonstrates the distribution and evolution of radionuclides over the integration time in detail. The model chain was validated against the data available in this period.

In the second series of experiments, the scenarios of accidental release of radionuclides influencing the Arctic environment and their transport from external sources to the Arctic marine environment are considered. We evaluate various realistic scenarios for numerical modeling of hypothetical situations at selected Russian industrial plants and onboard submarines resulting in potential radioactivity release into the environment. The scenarios include accidental radionuclide release into the environment from three major Russian radiochemical plants. Several different situations of radionuclide pollution to rivers are simulated using the GMS to predict their spread in the Arctic under present-day and CO_2-doubling global-warming scenarios. In addition, two scenarios based on nuclear submarine accidents are simulated to show the capability of using the GMS as an operational tool in decision support systems.

Chapter 5 describes the spread of non-radioactive toxic pollution, including POPs, petroleum hydrocarbons, and heavy metals. Here we demonstrate that the GMS can be modified to simulate the spread of POPs, hydrocarbons, and other toxic non-radioactive pollutants. The first section describes the approach used to model non-radioactive pollutant transport. The next section presents an example of GMS simulation of non-radioactive pollutant transport in the Arctic, focused on POPs. The chapter also describes some aspects of modeling the spread of petroleum hydrocarbons in the marine environment.

Chapter 6 assesses the risk of the radiological impact on humans resulting from the potential release of radioactive material. Because it is clearly beyond the scope of the chapter to provide a comprehensive risk assessment of all potential environmental and human impacts from all scenarios of radioactive releases in the Arctic marine and terrestrial realms, we target one important set of risks: risks to humans associated with potential releases along the Ob′ and Yenisei Rivers, including an assessment of how anthropogenic climate change may affect the consequences. Risk assessment is introduced and various scenarios are described. The formulation and implementation of dose models are then described. The results of risk assessment modeling are provided, followed by a summary and conclusions.

Appendices of tables and figures follow Chapter 6 and provide support for and details of the results.

The Afterword presents the key points and considers the present status and future prospects of the spread of radioactivity and pollution in the Nordic Seas and the Arctic.

1

Sources of anthropogenic pollution in the Nordic Seas and Arctic

This chapter describes the sources of radioactive and non-radioactive contamination in the Arctic and Nordic Seas. The primary and secondary sources of radioactivity in the study region are enumerated and described in Section 1.1. Section 1.2 focuses on a detailed description of three major Russian nuclear industries: (1) the Mayak Production Association, Chelyabinsk; (2) the Siberian Chemical Combine, Tomsk-7; and (3) the Mining and Chemical Combine, Krasnoyarsk-26. Section 1.3 provides a substantial description of non-radioactive pollution, including its sources and spread in the marine environment of the Barents, White, Kara, and Laptev Seas. The descriptions given here have been prepared from information from a range of open-literature material, including a wealth of Russian material and more widely known publications; for example, Arctic Monitoring and Assessment Program (AMAP) reports. An effort has been made to provide the most recent information, although the ever-changing status of pollution sources challenges an evaluation of the present situation.

1.1 RADIOACTIVE CONTAMINATION: CLASSIFICATION AND DESCRIPTION OF SOURCES

1.1.1 Classification of sources

1.1.1.1 Primary and secondary sources

The problem of radioactive contamination of the environment has arisen relatively recently, having a less than 70–year history, beginning from the moment of the first nuclear weapon tests in the atmosphere and underwater, including the only wartime use of nuclear weapons over Hiroshima and Nagasaki in 1945. Radioactive contamination sources that are direct and essentially instantaneous can be considered *primary* sources, whereas those that are manifested through environmental interactions are classified as *secondary* sources. Here the primary and secondary sources of radioactive contamination are classified and ranked by significance.

The most significant primary source (hereafter "P1") of radioactivity in the environment is probably nuclear weapon testing. In total, about 1,500 nuclear explosions have been carried out, more than 90% of which have been in the Northern Hemisphere, mainly in the U.S.A. and U.S.S.R. (Matishov and Matishov, 2001, 2004). Underground nuclear explosions can also be considered in this category, although these are less environmentally hazardous.

The second primary source (P2) is accidental emissions from nuclear power stations, the Chernobyl accident leaving the largest imprint. The same class applies to non-controllable emissions from any nuclear reactors (e.g., mobile reactors onboard nuclear-powered submarines and icebreakers, as well as stationary ones in industrial centers and research institutes).

The third primary source (P3) is a separate "potentially hazardous" class of nuclear reactors installed on nuclear-powered submarines and icebreakers, which create a particular background in the waters of the Barents and Kara Seas—about 180 such submarines and vessels have been equipped with such reactors (Nilsen *et al.*, 1996).

The fourth primary source (P4) of radioactive contamination is the intentional discharge of radioactive waste to the environment—namely, discharge from industrial plants producing and processing radioactive materials, including the non-controllable leakage of radioactive components (P4). Deliberate discharges of radioactive waste were made to artificial and natural water bodies both in areas near the location of the plants themselves and in remote, sparsely populated regions. For example, during the second half of the 20th century, periodic dumping and disposal of liquid and solid anthropogenic isotopes into the Russian Arctic Seas was commonplace.

The fifth primary source (P5) is waste disposed in specially sealed containers. Such disposals are also referred to as "potentially hazardous" primary sources. Sunken submarines with nuclear reactors onboard are included to this group.

It is important to emphasize that all primary sources are local in character; however, by getting into the environment the contaminating radioactive elements become entrained into air and water masses on a regional to global scale. They can undergo modification and form secondary radioactive contamination sources, and the contamination scale can significantly increase up to global.

Secondary sources are formed under the influence of external environmental factors. The spatial and temporal scales of primary and secondary sources can differ significantly, with the scale of secondary contamination being generally greater. These aspects underscore the importance of investigating the role of atmospheric circulation, oceanic currents, and mass exchanges within the environment when estimating the consequences of radioactive contamination. It should be noted that primary and secondary sources differ in the compositions of contaminants.

The most significant—in terms of the spatial scale—secondary source (S1) of radioactive contamination is the radioactive components suspended in the atmosphere that form the global fallout of radioactive substances from the atmosphere.

The present flows of global fallout of radioactive substances to the ground and sea surface from the atmosphere have their origins mainly before the 1990s. This source of radioactive contamination is significant not only because of its large scale, but also due to the fact that its formation is influenced by all sources of primary contamination as well as secondary sources as a result of the natural processes of weathering and washout.

The other classes of secondary sources of radioactivity include contaminants contained in the floodplain and bottom sediment in areas where liquid waste is openly stored, including solid radioactive contamination components at the Earth's surface where contaminants from primary sources were discharged (S2); floodplain sediment of natural water bodies (S3); and a class of potentially hazardous secondary sources. These are regular repositories of radioactive contaminants of different types (S4): underground disposal, pool accumulators, radioactive waste pools, radioactive waste repositories, and spent fuel repositories. The fifth group is the organisms that through biotic activity have accumulated contaminants in concentrations hazardous for health. A block diagram of the proposed classification is presented in Figure 1.1.

All sources of radioactivity in the environment interact between each other under the influence of the physical, chemical, and biological processes occurring in the

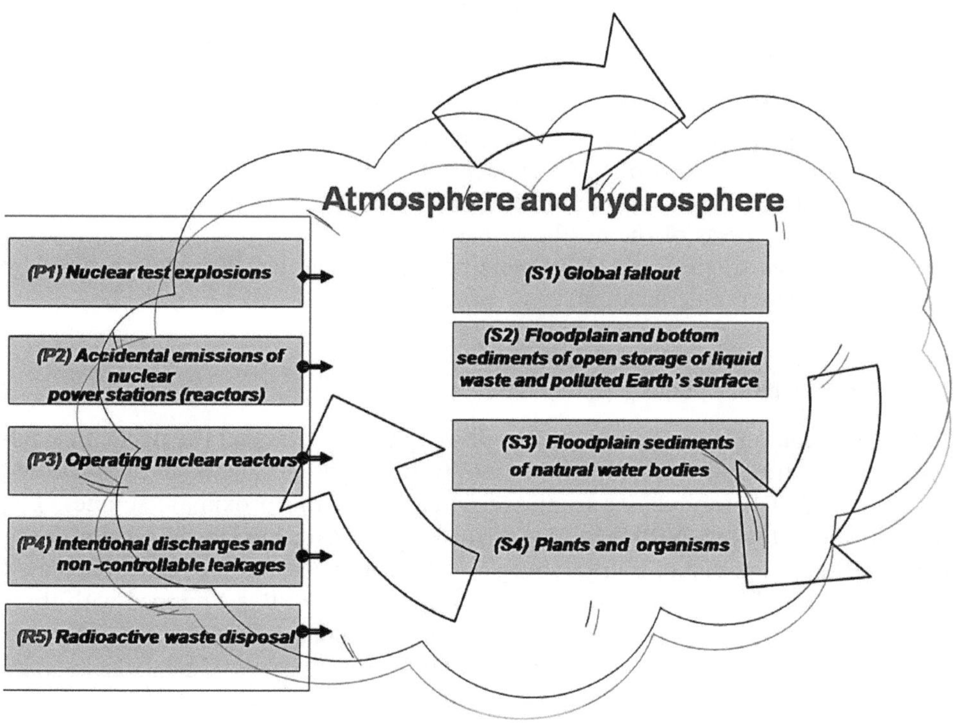

Figure 1.1. Primary (P) and secondary (S) sources of artificial radionuclide contamination of the environment.

Earth's biosphere. Thus, in studying and modeling the spread of radioactivity in the environment, one has to take into account both primary and secondary sources. If we consider the spread of radioactive pollution over large water areas, the types of pollution that cover wide areas are of primary significance. On the other hand, even local primary sources can play a significant role under natural conditions in the source area for transfer of pollution. Such conditions are determined by the dynamics of atmospheric processes in the area of the sources, the presence of rivers and groundwater, local relief, type of soil, etc. In the presence of such conditions, even the point source and local primary sources can be a great hazard to the environment and biosphere.

1.1.1.2 Global and point sources

The scale of radioactive pollution can be subdivided into global scale and point sources. Global depositions of radionuclides involve sources of anthropogenic environmental contamination from the atmosphere, as a cumulative consequence of nuclear tests conducted from 1945 to the 1990s, reaching a maximum in 1963. Estimations of the total amounts thrown into the atmosphere are 910×10^{15} Bq of ^{137}Cs, 600×10^{15} Bq of ^{90}Sr, and 240×10^{18} Bq of ^{3}H (UNSCEAR, 1982). From 1945 to 1992, it is estimated that atmospheric depositions amounting to 1.4×10^{15} Bq of ^{137}Cs and 1.0×10^{15} Bq of ^{90}Sr were deposited on Kara Sea. An example of the ground deposition of ^{138}C from nuclear weapon fallout is presented in Figure 1.2.

Sources of contamination on the global scale include radioactive releases as a consequence of the explosion at the Chernobyl nuclear power plant (NPP). The accident at Chernobyl in 1986 increased the content of cesium in the atmosphere by approximately 100 PBq of ^{137}Cs (Izrael, 1998; Strand, 1998). Chernobyl radioactive depositions still contribute to contamination of the ecosystems of the Arctic coast and harbor areas of the northern seas (Figure 1.3).

Point sources are the focus of the remainder of the chapter; they are organized by category and geographic location.

1.1.2 Nuclear power plants (NPPs)

Radioactive contamination from nuclear power plants can impact the Arctic through long-distance transport of contamination from major accidents in mid-latitude locations (e.g., Chernobyl), or from accidents from more proximate sources. For example, the Beloyarsk NPP is located in the town of Zarechnyi in the Ekaterinburg region, 40 km to the east of Ekaterinburg on the eastern coast of the reservoir created as a cooling pond on the Pyshma River. The Beloyarsk NPP is equipped with three power installations: two thermal neutron reactors and one fast reactor. The first power installation, with reactor AMB-100 and a capacity of 100 MW, was stopped in 1981. The second power installation, with reactor AMB-200 and a capacity of 200 MW, was stopped in 1989. The third power installation, with a reactor BN-600 (Table 1.1) that started up in April 1980 and became the world's first industrial power unit with the fast neutron reactor, is presently still operating. The possibility of

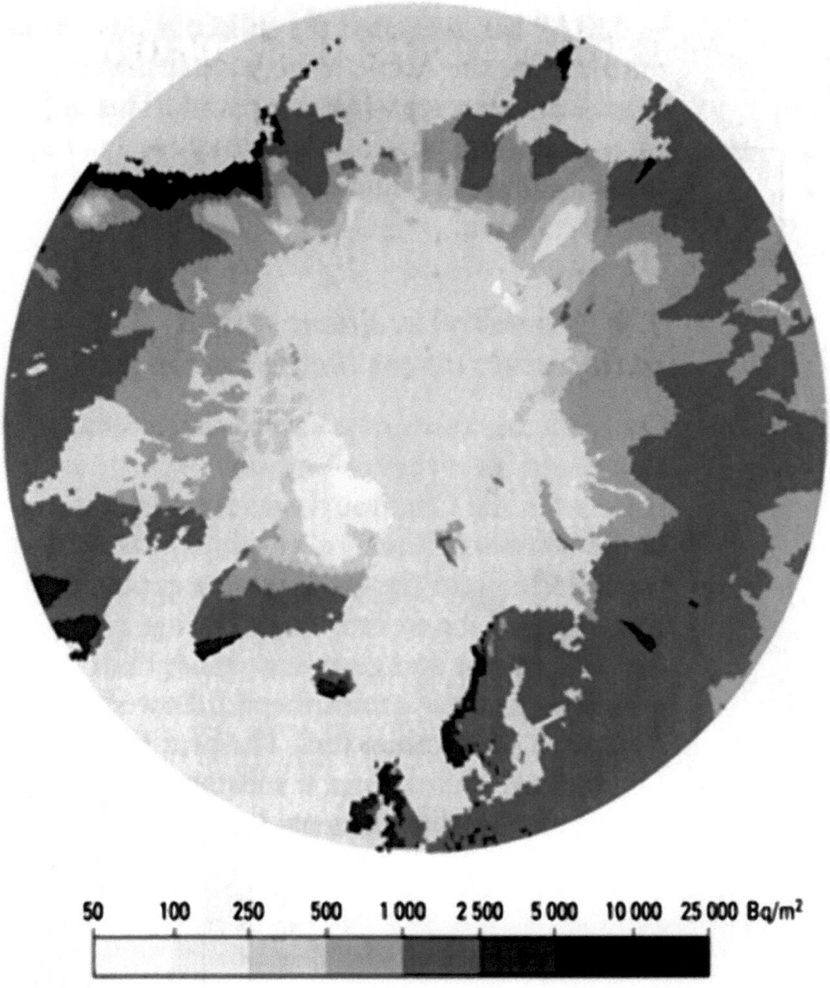

Figure 1.2. Estimated ground deposition of ^{137}Cs from nuclear weapon fallout (AMAP, 1997, 1998).

recommencement of construction of the fourth power unit (BN-800), frozen after the accident at Chernobyl NPP, should also be considered.

The spent nuclear fuel of the first power unit (AMB-100) of the Beloyarsk NPP has been completely removed from the reactor, and the second power unit contains a few fuel assemblies that remain in the reactor. There are 4,990 fuel assemblies stored in the cooling ponds of the NPP, some of which are no longer leak-proof, resulting in increased levels of radioactivity (by a factor of 1.6 over 3 years) due to the direct contact of uranium with reservoir water. Despite the fact that the first power unit was stopped, decommissioning operations are still not complete.

During the operational history of the Beloyarsk NPP, there were a few accidents. Two accidents occurred with two old reactors. In 1977 half of the heating elements

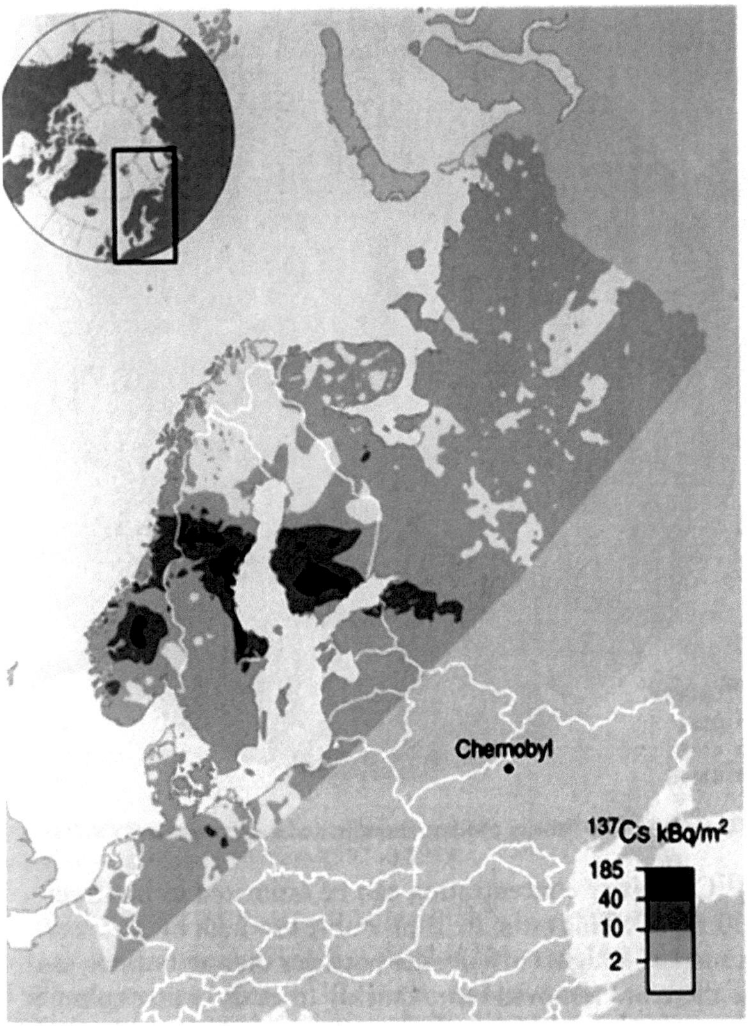

Figure 1.3. Ground deposition of ^{137}Cs from the Chernobyl accident (AMAP, 1997, 1998).

fused in reactor AMB-100. In December 1978 the same reactor burst into flames when part of the roof collapsed on the fuel reservoir of the turbine. In May 1994 there was a fire in the third power unit, caused by leakage of the liquid sodium used as the heat carrier (*http://decomatom.org.ru/?q=node/42*).

1.1.3 Nuclear industry enterprises

1.1.3.1 European reprocessing plants

There are three main European spent nuclear fuel (SNF) reprocessing plants that contribute to Nordic Sea/Arctic Ocean pollution: two are in the U.K. (Sellafield and Dounreay) and one in France (Cap de la Hague).

Table 1.1. General characteristics of the Beloyarsk NPP Unit 3, 1997.

Type of reactor	BN-600
Power (MW)	600
Entry into the network	April 8, 1980
Produced electrical energy (million kWh)	3,835.4
Own needs (%)	7.55
Runtime of a reactor (days)	278
Runtime of a reactor (effective days)	266
FUIC* (%)	72.99*

* FUIC = factor of using the installed capacity. This index describes usage efficiency in the generation of power.

Sellafield plant

The Sellafield (formerly Windscale) plant is located near Whitehaven in Cumbria, U.K. Besides facilities to reprocess SNF, the Sellafield plant also has six reactors for the production of weapons-grade plutonium (NATO, 1995a, b). In 1994 the U.K. had 63 tons of plutonium and over the next 25 years it was planned to produce 170 tons more (Bohmer, 1994). The plant has the capacity to process 1,500 tons of SNF from the MAGNOX reactor per year. The liquid radioactive waste is discharged into the Irish Sea. Discharges began in 1952, with the highest values observed in the 1970s (Figure 1.4). Total discharge from Sellafield is estimated as 160 PBq from 1952 to 1986. In 1986, an ion exchange system started to operate at Sellafield (Gladkov and Sivinzev, 1994) and since then discharges have become stable and are at low levels. The total estimated Sr and Cs discharge (NATO, 1995a, b) is 6.2 PBq of ^{90}Sr and 41 PBq of ^{137}Cs, while the radioactivity in 1995 is estimated at 3.8 PBq of ^{90}Sr and 27 PBq of ^{137}Cs. The total "present" radioactivity is 51 PBq.

Most of the radionuclides in sedimentation stay at the region of discharge, whereas cesium and strontium contribute to the pollution of the Arctic through water mass transport. Dilution greatly reduces radioactive contamination, such that levels in the Norwegian Sea are 1,000 times less than in the Irish Sea. The Barents Sea receives about 20% of the ^{137}Cs and 30% of ^{90}Sr (Nikitin *et al.*, 1991). The Joint Russian–Norwegian Expert Group (1992) estimated that 2% of the North Cape branch of the North Atlantic Current (NAC) reaches the Kara Sea and the sea should receive 150 TBq ^{137}Cs and 33 TBq of ^{90}Sr for the whole period of Sellafield discharges up to that time.

Figure 1.4 shows the concentration of ^{137}Cs in the Barents Sea. It is estimated that it takes between 5–7 years (Aarkrog, 1993) or 4–6 years (NATO, 1995a, b) for water from Sellafield to reach the Arctic Ocean. The estimation of 4–6 years seems

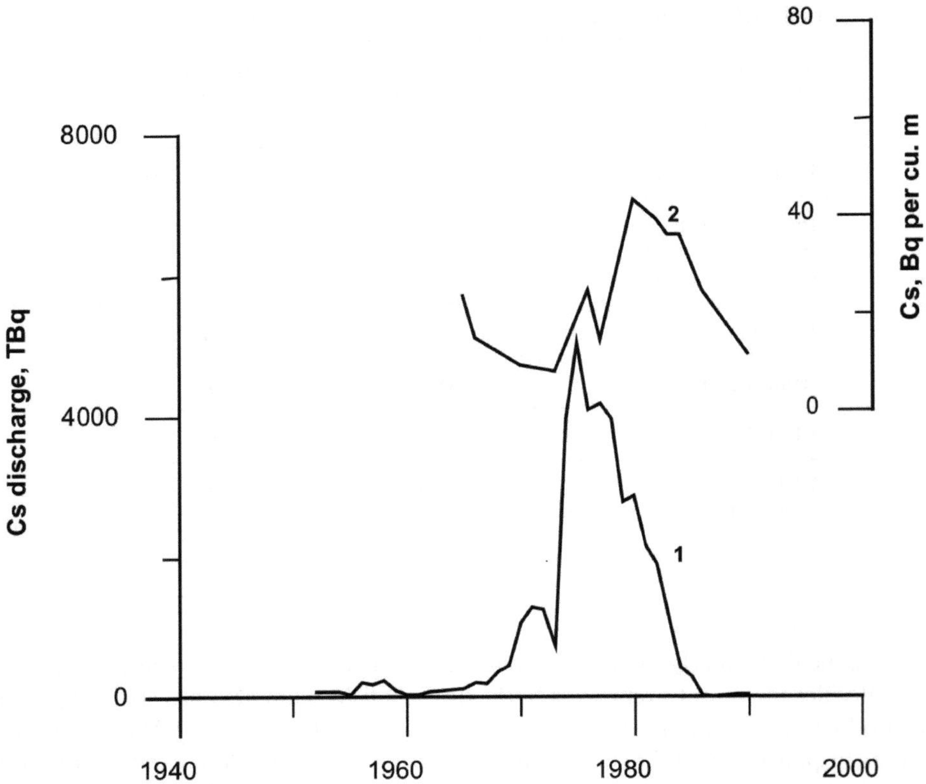

Figure 1.4. Liquid discharge of ^{137}Cs from Sellafield (1) and ^{137}Cs concentration in the Barents Sea (2).

to be more reasonable (Figure 1.5). According to observations in the Kara Sea (1978–1987) there is an inflow of waters with elevated ^{137}Cs concentrations through Karskiye Vorota Strait (Nilsen, 1993), which is believed to be due to transport of radioactive waste from Sellafield.

A new plant called THORP (Thermal Oxide Reprocessing Plant) was constructed near Sellafield and began operation in the 1990s (NATO, 1995a, b) to process SNF from Japan, Germany, and France and, moreover, to produce 7 tons of plutonium per year (Bohmer, 1994). The beginning of operations at THORP was expected to decrease the radioactive discharge into the environment by 2000% (Bohmer, 1994).

Dounreay plant

The Dounreay Nuclear Power Development Establishment is located in Scotland near the town of Thurso. From 1958 this plant began discharging liquid radioactive waste through a pipeline to the Pentland Firth. The capacity of Dounreay is much

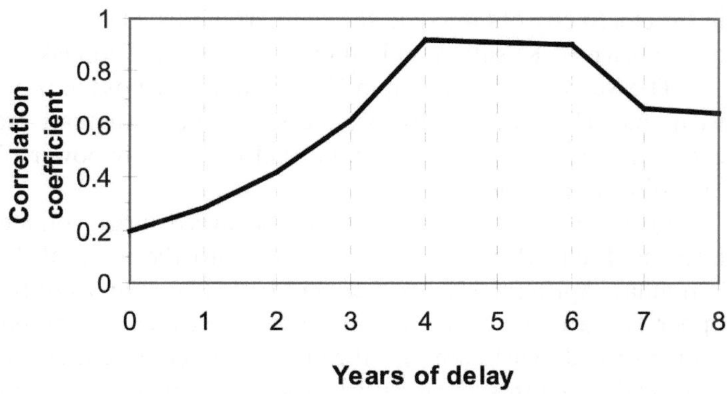

Figure 1.5. Dependence between ^{137}Cs discharge from Sellafield and ^{137}Cs concentration in the Barents Sea with different delay time.

smaller than that of Sellafield—it can process about 1 ton of SNF per year. In 1986 the total release of radioactive waste from Dounreay was 11 PBq (NATO, 1995a, b). More recently, the amount of liquid discharged seems to have stabilized and even reduced (NATO, 1995a, b) but the planned operations included growth of discharge by 1,000% in the next 3 years (Bohmer, 1994). This was due to plans to reprocess 30 tons of SNF from the closed former reactor at Dounreay. According to the capacity of the plant, there are doubts if this processing could be done in such a short time; more realistic estimates indicate that it would take more than 5 years (Bohmer, 1994).

Cap de la Hague

The Cap de la Hague plant is located near Cherbourg, France. Reprocessing began in 1966. Values of liquid discharges from this plant are less precise and detailed than those from the British plants. Due to the use of filtering installations from the beginning of operation, discharges from Cap de la Hague were lower than from Sellafield (Gladkov and Sivinzev, 1994). The plant capacity is about 800 tons of SNF per year and total discharges are estimated until 1986 to be 23 PBq (NATO, 1998). As in the case of Sellafield and Dounreay, the most hazardous elements for the Barents and Kara Seas are ^{137}Cs and ^{90}Sr.

1.1.3.2 Russian nuclear industry enterprises

There are several notable nuclear industry enterprises in Russia that should be considered as sources of (potential) radioactive contamination reaching the Arctic. The first three summarized here are the most significant and are detailed in Section 1.2.

Mayak Production Association (Mayak PA), Chelyabinsk-65

The Mayak PA was the first enterprise in the former Soviet Union created for production of nuclear materials for military purposes. The Mayak PA is located in the closed city of Ozersk in the southern Urals. Ozersk is approximately 15 km east of the town of Kyshtym and 70 km north of Chelyabinsk.

The Mayak PA was built in 1948 with the purpose of producing ^{239}Pu for nuclear weapons. The Mayak PA produces isotopes, weapons-grade plutonium, radiochemicals, enrichment of uranium, fuel for nuclear power plants, and reprocessing of radioactive waste.

From 1949 to 1956, the plant discharged large volumes of radioactive waste into the Techa River, and (from 1951) into the internal drainage system of Lake Karachay. Spring floods have contaminated large areas of the Asanov swamp in the upper Techa River. Some radionuclides, especially such mobile isotopes as ^{90}Sr, were transported downstream and through the river system eventually reached the Kara Sea. Aside from this, the territory (about 20,000 km^2) surrounding the enterprise was contaminated as a result of the Kyshtym accident in 1957, in which the tank for storage of high-level waste exploded.

During the period from 1951 to 1966, a system of dams was constructed in the upper Techa River with the purpose of holding nuclear waste. The risk of potential radioactive contamination of the environment is from possible leakages of waste from the gathering ponds. One extreme radioactive contamination scenario is a complete break in a dam resulting in radionuclide inflow into the Asanov swamp downstream, then into the Techa River, and finally into the Ob′ River. Other large-scale releases could occur in the event of the Asanov swamp drying out, resulting in the high-water washout of radionuclides into the Ob′ River. Further, there is the possibility of a continuous leakage of radioactive waste into the groundwater from one of the dams through its bottom segment.

The Karachay Lake has no external drainage system, but it caused surrounding areas to be polluted in the drought period of 1967 when lake bottom sediments became exposed and were scattered by the wind. Now the bottom of the lake is paved with concrete blocks to prevent a repeat. However, radionuclides may move from the bottom sediments into the groundwater, and further into the Techa River. The annual planned industrial release of radioactive waste into Lake Karachay is about 20,000 m^3/yr with a total radioactivity of 370 MBq/yr to 746 MBq/yr (*Atom without "Classified" Stamp*, 1992; Cochrane *et al.*, 1995).

In recent years, releases from Mayak PA have considerably decreased because the last of the reactors producing weapons-grade plutonium was stopped in 1990.

Siberian Chemical Combine (SCC), Tomsk-7

The Siberian Chemical Combine (SCC) is located near the town of Seversk (formerly Tomsk-7), 20 km northwest from Tomsk, on the Tom′ River. The city was established in 1949 for the production of fission materials for nuclear weapons. The population of Tomsk-7 is about 119,000, 15,000 of whom are employed at the combine. The health protection zone of SCC is 192 km^2.

SCC activities consist in producing weapons-grade plutonium, enriching uranium hexafluoride, manufacturing uranium protoxide–monoxide and uranium hexafluoride, reprocessing irradiated basic blocks in order to produce and refine uranium and plutonium salts, producing nuclear materials, temporarily storing spent nuclear fuel, and disposing of waste.

The SCC infrastructure consists of a reactor plant (for plutonium production), isotope separation plants (centrifugal method), a sublimate plant (for production of uranium protoxide–monoxide and uranium hexafluoride), a radiochemical plant (for reprocessing of irradiated blocks), and a chemical–metallurgical plant (for production of weapons-grade plutonium).

The SCC is presently the largest complex in the world for plutonium, uranium, and transuranium production. The combine has a complete cycle of reprocessing: primary enrichment, plutonium production and its separation from high-level decay products, as well as the secondary reprocessing of spent uranium using fluoridation technology (Cochrane *et al.*, 1989, 1995).

Mining and Chemical Combine, Krasnoyarsk-26

Zhelenogorsk (formerly Krasnoyarsk-26) was founded in 1950 for production of weapons-grade plutonium. Krasnoyarsk-26 is located on the right bank of the Yenisei River approximately 50 km to the northeast of Krasnoyarsk. The population of Zhelenogorsk is about 100,000, 8,000 of whom are employed at the combine. The site's industrial area is 17 km^2 and the health protection area is 131 km^2.

The basic activities of the Krasnoyarsk Mining and Chemical Combine (MCC) are reprocessing of the spent nuclear fuel from nuclear power plants (three reactors), production of weapons-grade plutonium, and temporary storage of spent nuclear fuel (industrial accumulators and regional burial grounds).

The MCC (known as Combine 815 in its early years) was set up as an enterprise for production of weapons-grade plutonium. The main production activity of the MCC is concentrated 10 km from the city of Zhelenogorsk on the right bank of the Yenisei River in a mine inside the rocky massif stacked by Archaean–Proterozoic crystalline rocks on the western slope of the southern extremity of the Yenisei mountain ridge. In 1958 the first direct flow uranium graphite reactor, AD, started up. The second direct flow reactor, ADE-1, started in 1961 and the third, ADE-2, started in 1964. These entities formed the underground reactor plant (Plant A). At the same time a radiochemical plant (Plant B) was set up, being the second part of a technological chain (production string) of the main production, which started in 1964.

From the inception of the MCC enterprises, radioactivity safety measures were put in place and by 1958 there were treatment facilities to refine liquid low-level waste, as well as to utilize and dispose of radioactive waste. The underground liquid low-level and high-level storage facility Severnyi, which started up in 1967, belongs to the last-mentioned class. Since the start-up of main production, the MCC has produced weapons-grade plutonium and delivered it (together with spent refined uranium) to other enterprises of the Russian Nuclear Complex. The MCC has never produced nuclear weapon charges nor fuel elements for nuclear power plants.

In 1992 two of the three industrial direct flow reactors—AD and AD-1—were disabled as a result of the MCC's activities being converted towards production of different materials (polycrystalline silicon, optical quartz glass, quartz crucibles, germanium tetrachloride, and particularly pure germanium). The radiochemical specialization and the applicable professional skills of the staff built up before the changeover led to plans for a future nuclear enterprise (the RT-2 plant), whose purpose would be the reprocessing of spent nuclear fuel from nuclear power plants. However, due to financial difficulties, the building of the RT-2 plant was suspended (*Atom without "Classified" Stamp*, 1992; Cochrane *et al.*, 1995).

Krasnoyarsk-45, the Electromechanical Plant

As already mentioned, Zhelenogorsk (formerly Krasnoyarsk-45) was founded at the beginning of the 1950s to produce enriched uranium for the Soviet nuclear weapons program. The city is located on the River Kan approximately 70 km east of Krasnoyarsk. The population is approximately 67,000, 10,000 of whom are employed at the complex.

The design of a gas diffusion plant was believed to have started in the early 1960s, and the plant began to produce enriched uranium in 1964. In parallel, a hydroelectric power station (GRES-2) was built to supply the complex with heat and electrical energy. In the 1960s the U.S.S.R. began to replace gas diffusion installations with centrifuges. The last gas diffusion cascade in Krasnoyarsk-45 was stopped in 1990.

The Electromechanical Plant in Krasnoyarsk-45 now constitutes up to 29% of the concentrating capacity of Russia. In addition to uranium extraction, the complex also extracts isotopes of tungsten, molybdenum, krypton, xenon, germanium, iron, sulfur, oxygen, and carbon.

The Electromechanical Plant produces enriched uranium for Russian nuclear reactors and for export. Using a similar baseline technology as that optimized for lighter elements, the Electromechanical Plant produces a wide range of stable isotopes for sale worldwide.

Since 1997 the complex has been involved in reprocessing highly enriched uranium from disassembled warheads in accordance with the Russian–American agreement (*Atom without "Classified" Stamp*, 1992; Cochrane *et al.*, 1995).

Zlatoust-36 (Chelyabinsk area)

The town of Trekhgornyi (formerly Zlatoust-36) was established as a machine-building plant for assembly/disassembly of nuclear warheads. The population is about 33,000, 10,000 of whom are employed at the plant. The assembly plant equips warheads using parts produced by other enterprises. The factory makes use of two sites to store warheads some 10 km to the east of the plant (*Atom without "Classified" Stamp*, 1992; Cochrane *et al.*, 1995).

Sverdlovsk–44, the Ural Electrochemical Plant

The town of Novoural′sk (formerly Sverdlovsk-44) is located in the Sverdlovsk region, approximately 40 km northwest of Ekaterinburg. The population of the town is 96,000, 15,000 of whom are employed at the combine.

Sverdlovsk-44 was founded in 1945 for the production of highly enriched uranium for nuclear ammunition. The Ural Electrochemical Plant (originally Combine 813) is the oldest and largest producer of enriched uranium in Russia.

The enrichment factory was built close to Verkhnevinsk to take advantage of nearby railroads, power lines, and two artificial lakes (the Demidov ponds) for the supply of cooling water. The construction of the gas diffusion plant, D-1, began in January 1946. Its first line was placed in operation in 1948, but because of technical difficulties the production of 90% uranium at the D-1 plant did not start until 1950. The factory operated until 1955 and then was stopped and disassembled because of low efficiency. Three newer and bigger gas diffusion plants (D-3, D-4, and D-5) were started in 1953.

Sverdlovsk-44 was also the pioneer in the U.S.S.R in the technology behind centrifugal enrichment. The experimental centrifugal factory started its operation in October 1957. The industrial centrifugal plant was built in three stages from 1962 to 1964 (*Atom without "Classified" Stamp*, 1992; Cochrane *et al.*, 1995).

Sverdlovsk-45, Plant 418/ElectroChimPribor Combine

The closed town of Lesnoy (formerly Sverdlovsk-45) is located on the banks of the Nizhnaya Tura River, about 160 km to the north of Ekaterinburg. The population is about 58,000, 10,000 of whom are employed at the combine.

Sverdlovsk-45 was founded in 1947 with the purpose of building Plant 418, which was intended for electromagnetic separation in the production of highly enriched uranium (HEU). The construction of a complex for industrial separation, SU-20, was completed at Plant 418 at the same time as the first Soviet plant on gas diffusion, D-1, in Sverdlovsk-44. Originally, the SU-20 complex was used to raise the enrichment quality level of the uranium received from plant D-1 from 70% up to 90%. Improvements in the technology of gas diffusion later reduced the necessity for electromagnetic separation of uranium isotopes, and therefore the SU-20 complex was reoriented towards the enrichment of non-uranium isotopes.

In the second half of the 1950s, a part of Plant 418 was re-equipped for the assembly and disassembly of nuclear warheads and became known as the ElectroChimPribor Combine. The combine was founded to double in size the electro-mechanical plant in Arzamas-16 for the assembly of physical packages and nuclear warheads. It eventually became the largest in Russia for the assembly of warheads. The plant for the assembly and disassembly of warheads is located northeast of the inhabited area. The production complex has two large storehouses for nuclear ammunition, the nearest of which is located 10 km west of the basic production area (*Atom without "Classified" Stamp*, 1992; Cochrane *et al.*, 1995).

Novosibirsk Chemical Concentrate Plant

The joint stock company Novosibirsk Chemical Concentrate Plant (NCCP) is located north of Novosibirsk. The NCCP worked initially on reprocessing uranium concentrates. It now produces fuel elements for power engineering. The tailing dump for radioactive waste burial including uranium radionuclides, lithium, and mercury is located 4 km from the NCCP on the route from Novosibirsk to Kemerovo. The waste products are transported to storehouses as a pulp by sludge line directly from the plant. The volume of the tailing dump is 384,000 m^3. The pulp moves at a flow rate of 800 m^3 per day with a 1:60 solid-to-liquid ratio, and solid flow up to 15 tons per day. The radioactive component is uranium (solid 0.05%, liquid 0.05 mg/L). Such a waste product delivery scheme was not available in the early years, when delivery was carried out by motor transport and resulted in environmental contamination. Radioactive contamination can only occur now from accidents along the sludge line, which probably has happened, given the local presence of radioactive contamination under the sludge line. The level of contamination is local in character.

The existing tailing dump has operated since 1949 and consists of three sections (section 1 was suspended in 1963, section 2 has operated since 1964 and is nearly filled, and section 3 still operates). Neither the first nor second sections were equipped with devices to prevent soil and groundwater pollution; furthermore, their burial capacities are exhausted.

The NCCP is thus a potential source of radioactive contamination of the environment. First, natural and enriched uranium may pollute the air at the main site of the plant through windborne losses from the factories. Second, pollution of the soil could occur from accidents and depressurization of the sludge line. Third, there could be uranium-containing waste from the NCCP after refining and additional extraction of uranium at the tailing dump when disposed of in the Ob′ River.

In 1994 some 364,000 m^3 of waste was discharged into the Ob′ River, but the concentration of radionuclides was lower than the detection threshold. Only 0.012 Ci of natural and 0.01 Ci of enriched uranium were ejected into the atmosphere; this constitutes about 5% of the allowed ejection (*Atom without "Classified" Stamp*, 1992; Cochrane *et al.*, 1995).

Angarsk Electrolysis Chemical Combine

The Angarsk Electrolysis Chemical Combine was founded at the end of the 1950s with the purpose of producing enriched uranium for the Soviet nuclear program. The combine is in the open city of Angarsk 30 km northwest of Irkutsk.

The gas diffusion plant achieved its greatest capacity in 1964 and apparently was at that time the most effective concentrate enterprise. The concentrate plant in Angarsk never produced HEU, and its partially enriched uranium was shipped to other concentrate combines of the U.S.S.R. for production of HEU. The combine comprised about 8% of the concentrate capacity of Russia. Moreover, this complex is one of two Russian combines that produce uranium hexafluoride (UF6), which is raw material for the NCCP (*Atom without "Classified" Stamp*, 1992; Cochrane *et al.*, 1995).

1.1.4 Scientific and research reactors and laboratories

Chelyabinsk-70, Scientific and Research Institute of Technical Physics

The Scientific and Research Institute of Technical Physics (Russian Federal Nuclear

Center) is located in the city of Snezhinsk (formerly Chelyabinsk-70) and is one of two Russian centers for the design of nuclear weapons.

Chelyabinsk-70 was founded in 1955 near Lake Sinara 80 km south of Yekaterinburg and 20 km north of Kasly. In 1998 its population was 48,000, 11,000 of whom work at the institute. In the early years of the Cold War, this number was a quarter higher.

The primary goal of the institute is the design of nuclear weapons and bringing about scientific and technical support for nuclear weapons during their lifetime. The institute has wide theoretical and experimental expertise in the engineering and non-nuclear testing of nuclear weapons. Plants 1 and 2 are capable of producing experimental and pre-production models of nuclear weapons (*Atom without "Classified" Stamp*, 1992; Cochrane *et al.*, 1995).

1.1.5 Special combines

Sverdlovsk Special Combine (SC)

The Sverdlovsk Special Combine serves the Sverdlovsk, Tyumen, and Perm regions. It is responsible for burial of solid radioactive waste (SRW) and spent sources of ionizing radiation. Burial is carried out in accordance with both container and containerless technology. As of 1995, 135.8×10^3 Ci of waste products had been buried. Information on the sources of radioactive contamination is not available.

Chelyabinsk SC

The Station of Radioactive Waste Burial (SRWB) of the Chelyabinsk RADON SC is located 55 km to the north of Chelyabinsk and 6 km from Chishma. The burial station serves the Chelyabinsk and Kurgan regions and buries SRW and spent sources of ionizing radiation; liquid radioactive waste (LRW) is absent. By January 1995 SRWB had buried 32.4×10^3 Ci of waste products.

Novosibirsk PA

The Novosibirsk Production Association RADON is located 2.5 km from Chik (Kochenevsky district, Novosibirsk region) and 30 km west of Novosibirsk. The RADON PA is responsible for the burial of radioactive waste and serves the Novosibirsk, Tomsk, Kemerovo, and Omsk regions, the Krasnoyarsk and Altai territories, and Kazakhstan (until 1992). Reprocessing of radioactive waste is not done here. Solid waste is stored in reinforced concrete buried tanks (at a depth of 3–6 m). The radioisotopic composition of waste products includes both short-lived and long-lived radionuclides, such as ^{137}Cs, ^{90}Sr, and ^{241}Am. The radioactivity of buried waste products at the end of 1993 was about 2.3×10^5 Ci.

Irkutsk SC

This SRWB is located 36 km along the Aleksandrovskiy road, 15 km from the Angara River. It has operated since 1963 and serves the Irkutsk and Chita regions, and the Buryat, Tuva, and Sakha-Yakutia republics.

Tanks for the storage of closed sources of ionizing radiation at this SRWB are stored indoors of the "Module" type. The filled tanks are concreted. The tank for storage of low-level radioactive waste is outdoors; it is made of concrete and the working compartment is closed by a metallic cover that weighs 3 tons.

The SRWB only disposes of SRW, closed sources of ionizing radiation, and materials from radioactive contamination locations after deactivation. The building of a new tank for the storage of SRW has been planned.

The SRWB covers 20 ha. Geologically, it is above sedimentary rock with weakly sintered sandstones 200 m thick. The groundwater level is at a depth of 112 m.

Institute of Nuclear Physics of the Kazakhstan Academy of Sciences

The institute is involved in the collection, reprocessing, and burial of radioactive waste in Kazakhstan; it also serves the Alma Ata, Djambul, Chimkent, and east Kazakhstan regions.

1.1.6 Nuclear weapons tests and "peaceful" nuclear explosions

The former Soviet Union used nuclear explosives for weapons testing as well as for mining and construction work. Hundreds of such explosions were carried out in different parts of the former U.S.S.R. (Table 1.2).

Table 1.2. Soviet nuclear weapons tests and "peaceful" nuclear explosions by location.

Sites of nuclear weapons tests	*Number of tests*
Semipalatinsk Test Site	456
Northern Test Site (Novaya Zemlya)	130
Total at nuclear test sites	*586*
Russian Federation European part—59 Asian part —32	91
Ukraine	2
Kazakhstan	33
Uzbekistan	2
Turkmenistan	1
Total outside nuclear test sites	*129*
Grand total	***715***

1.1.6.1 Nuclear military test explosions

Past atmospheric tests of nuclear weapons remain by far the largest global source of radioactive contamination. Most atmospheric tests were done before 1962, and atmospheric testing ceased entirely in 1980. About 12% of the fallout from the tests has normally been deposited close to the test site, and about 10% has ended up in a band around the same latitude as the test site. The remaining 78% is global fallout (see above), most of which ends up in the same hemisphere as the tests (AMAP, 1997, 1998).

Nuclear weapons testing began in 1945 when the U.S.A. tested the first fission weapon in New Mexico. Up to 1952 nuclear tests were done with fission weapons and the fallout was deposited mainly at test sites and in the latitude layers corresponding to the test sites (Aarkrog, 1993). The main test areas were Bikini Atoll, Nevada, and Semipalatinsk. In 1952 the U.S.A. tested its first thermonuclear bomb at Eniwetok Atoll. After that a series of thermonuclear explosions were carried out. The estimates of global fallout (Aarkrog, 1993 according to UNSCEAR, 1982) show that fallout was minimal in the polar regions. Comparison of ^{90}Sr fallout shows that in regions between 60°N and 90°N the density of fallout was 3.1 times less than in regions between 30°N and 60°N (Gladkov and Sivinzev, 1994). The Kara Sea is located between 70°N and 80°N, a latitudinal range that should have contamination of ^{90}Sr of about 0.68 kBq/m^2 (Gladkov and Sivinzev, 1994). Estimates of the total amount of radioactivity for the Arctic in 1993 are 9 PBq of ^{90}Sr and 15 PBq of ^{137}Cs (Aarkrog, 1993).

Semipalatinsk-16 (Kurchatov)

The largest test site is located near Semipalatinsk, Kazakhstan. The Semipalatinsk test site was founded in 1948. The first test was conducted in 1949 and the last in 1989. Of the 467 explosions conducted here, 346 were underground (first test in 1961), 87 atmospheric, and 26 surface explosions; some tests included multiple explosions.

During the first tests, a number of incidents took place; for example, after the first explosion in 1949 about 25,000 people were affected by fission products, mainly iodine-131 (*Nuclear Tests ...*, 1997; *Sources Contributing to Radioactive Contamination ...*, 1997).

Novaya Zemlya

The Russian Arctic islands of Novaya Zemlya were the only site for atmospheric nuclear weapons tests in the Arctic. The adjacent Barents and Kara Seas should be expected to be much more affected by the nuclear tests at Novaya Zemlya than any other seas.

The Soviet Union carried out a series of different types of tests (underwater, underground, and atmospheric). Nuclear explosions on Novaya Zemlya were carried out at three sites (Figure 1.6). The tests are listed in Tables 1.3 and 1.4 (Matuschenko *et al.*, 1994). Proper atmospheric explosions were done exclusively at point C from 1957 to 1962 until the moratorium on atmospheric nuclear tests—no such tests have been conducted there since 1962. At the beginning of the 1960s, three explosions were

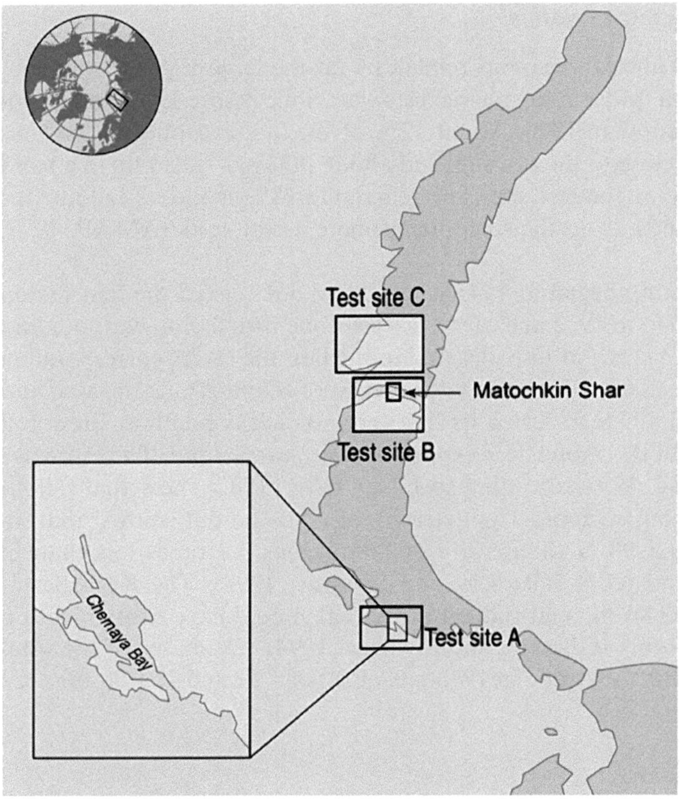

Figure 1.6. Nuclear weapon test sites on Novaya Zemlya (AMAP, 1998).

made over the sea surface at Chernaya Bay, in order to study ship destruction in closed bays as a result of such explosions (Yablokov *et al.*, 1993); there were also three underwater explosions at the same place. The total number of explosions in the atmosphere above Novaya Zemlya is 87 (83 purely atmospheric, 3 over the water surface, and 1 from a tower).

After an underwater explosion, all radioactivity should remain in the water and in sediments. There will be high radioactive contamination in the sediments around the test area. In Chernaya Bay, where three underwater explosions were carried out, the concentration of ^{137}Cs in bottom sediments is 1.44 kBq/kg; for example, standards for the Murmansk region are 0.6 kBq/kg of ^{137}Cs (Semenov, 1993), 5.5 kBq/kg of $^{239,240}Pu$, and 0.395 kBq/kg of ^{60}Co (Matishov *et al.*, 1994). The total release from these three tests is estimated as 150 TBq of ^{137}Cs, 100 TBq of ^{90}Sr, and 60 TBq of ^{239}Pu (NATO, 1995a, b).

Underground explosions took place between 1972 and 1990 primarily at point B (Matochkin Shar Strait) and secondarily at point A (Chernaya Bay). From 1976, the yield of explosions was limited to 150 kt in accordance with the Russian–American treaty known as the Threshold Test Ban Treaty, signed in Moscow on July 3, 1974 (NATO, 1995a, b). Despite underground tests being done at great depths and meas-

Table 1.3. Atmospheric explosions on Novaya Zemlya (Matuschenko *et al.*, 1994) (all explosions took place in region C—see Figure 1.6).

Date	*Power* (kt)	*Date*	*Power* (kt)	*Date*	*Power* (kt)
24-Sep-57	150–1,500	14-Sep-61	150–1,500	2-Sep-62	20–150
6-Oct-57	1,500–5,000	16-Sep-61	150–1,500	8-Sep-62	1,500–5,000
23-Feb-58	150–1,500	18-Sep-61	150–1,500	15-Sep-62	1,500–5,000
27-Feb-58	150–1,500	20-Sep-61	150–1,500	16-Sep-62	1,500–5,000
27-Feb-58	150–1,500	22-Sep-61	150–1,500	18-Sep-62	150–1,500
14-Mar-58	20–150	2-Oct-61	150–1,500	19-Sep-62	>10,000
21-Mar-58	150–1,500	4-Oct-61	1,500–5,000	21-Sep-62	1,500–5,000
30-Sep-58	150–1,500	6-Oct-61	1,500–5,000	25-Sep-62	>10,000
30-Sep-58	150–1,500	8-Oct-61	<20	27-Sep-62	>10,000
2-Oct-58	20–150	20-Oct-61	150–1,500	7-Oct-62	150–1,500
2-Oct-58	150–1,500	23-Oct-61	>10,000	9-Oct-62	<20
4-Oct-58	<20	25-Oct-61	150–1,500	22-Oct-62	>10,000
5-Oct-58	<20	30-Oct-61	50,000	27-Oct-62	1,500–5,000
6-Oct-58	<20	31-Oct-61	150–1,500	29-Oct-62	1,500–5,000
10-Oct-58	20–150	31-Oct-61	1,500–5,000	30-Oct-62	1,500–5,000
12-Oct-58	150–1,500	2-Nov-61	20–150	1-Nov-62	1,500–5,000
15-Oct-58	150–1,500	2-Nov-61	150–1,500	3-Nov-62	20–150
18-Oct-58	1,500–5,000	2-Nov-61	1,500–5,000	3-Nov-62	1,500–5,000
19-Oct-58	20–150	4-Nov-61	<20	18-Dec-62	20–150
20-Oct-58	150–1,500	4-Nov-61	<20	18-Dec-62	20–150
21-Oct-58	<20	4-Nov-61	150–1,500	20-Dec-62	<20
22-Oct-58	1500–5,000	5-Aug-62	>10,000	22-Dec-62	<20
24-Oct-58	150–1,500	10-Aug-62	150–1,500	23-Dec-62	<20
25-Oct-58	150–1,500	20-Aug-62	1,500–5,000	23-Dec-62	<20
10-Sep-61	<20	22-Aug-62	1,500–5,000	23-Dec-62	150–1,500
10-Sep-61	1,500–5,000	25-Aug-62	1,500–5,000	24-Dec-62	150–1,500
12-Sep-61	150–1,500	27-Aug-62	1,500–5,000	24-Dec-62 25-Dec-62 25-Dec-62	>10,000 <20 1,500–5,000
Total number of explosions: 83					

Table 1.4. Non-atmospheric nuclear explosions at Novaya Zemlya (Matuschenko *et al.*, 1994).

Date and time	*Type*	*Power* (kt)	*Region*	*Date and time*	*Type*	*Power* (kt)	*Region*
13-Sep-61	OW	< 20	A	10-21-75 15:00	UG	150–1500	B
27-Oct-61	OW	< 20	A	9-29-76 6:00	UG	20–150	B
22-Aug-62	OW	< 20	A	10-20-76 11:00	UG	1–20	B
7-Sep-57	Surface (tower)	< 20	A	9-1-77 6:00	UG	20–150	B
7-27-72 13:00	UG	Extremely small	A	10-9-77 14:00	UG	1–20	B
9-27-73 10:00	UG	20–150	A	8-10-78 11:00	UG	20–150	B
10-27-73 10:00	UG	Megaton class	A	9-27-78 5:05	UG	20–150	B
11-2-74 8:00	UG	Megaton class	A	9-24-79 6:30	UG	20–150	B
10-18-75 12:00	UG	150–1,500	A	10-18-79 10:10	UG	20–150	B
10-18-75 12:00	UG	150–1,500	A	10-11-80 10:10	UG	1–20	B
9-18-64 11:00	UG	Small power	B	10-11-80 10:10	UG	20–150	B
10-25-64 11:00	UG	Small power	B	10-1-81 15:15	UG	20–150	B
10-27-66 9:00	UG	Megaton class	B	10-11-82 10:15	UG	20–150	B
10-27-66 9:00	UG	Megaton class	B	8-18-83 20:10	UG	20–150	B
10-21-67 8:00	UG	150	B	9-25-83 17:10	UG	20–150	B
10-21-67 8:00	UG	150	B	8-26-84 7:30	UG	1–20	B
11-7-68 13:02	UG	150–1,500	B	10-25-84 9:30	UG	20–150	B
10-14-69 10:00	UG	150–1,500	B	8-2-87 6:00	UG	< 1	B
10-14-69 10:00	UG	150–1,500	B	5-8-88 2:50	UG	1–20	B
10-14-70 9:00	UG	Megaton class	B	12-4-88 8:20	UG	1–20	B
9-27-71 9:00	UG	Megaton class	B	10-24-90 17:58	UG	1–20	B
8-28-72 9:00	UG	150–1,500	B	21-Sep-55	UW	< 20	A
9-12-73 10:00	UG	100	B	10-Oct-57	UW	< 20	A
8-29-74 13:00	UG	150–1,500	B	23-Oct-61	UW	< 20	A
8-23-75 12:00	UG	150–1,500	B				

OW = over the water, UG = underground, UW = underwater nuclear explosions. Total number of explosions: 3 over water, 1 on the surface, 42 underground, and 3 underwater.

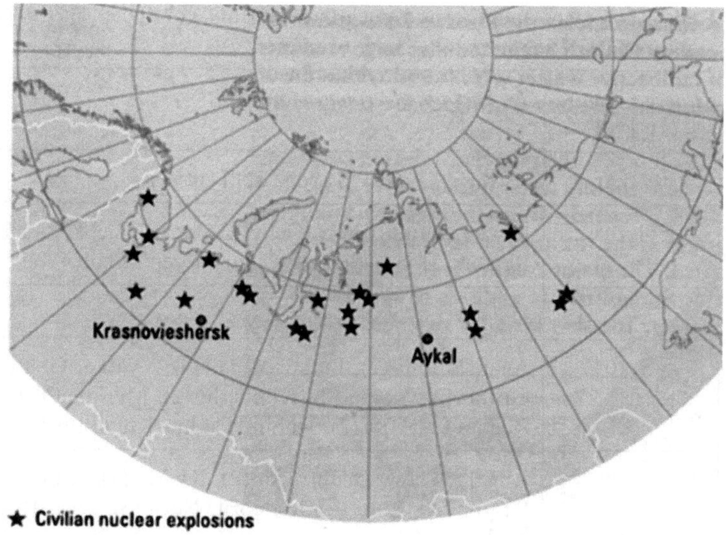

Figure 1.7. Civilian nuclear explosions (AMAP, 1997, 1998).

ures taken to prevent the escape of radioactive elements (Dumik and Evseev, 1993), in some cases there were accidental releases of radioactive gases to the Earth's surface (Matuschenko *et al.*, 1994). As a result there are some hotspots that are highly radioactive around the test areas (Aibulatov, 1993), but the level decreases significantly with distance from the source. There is no information about the effect of underground explosions on marine pollution. The total yield of all explosions carried out on Novaya Zemlya is about 357.5 Mt (Gladkov and Sivinzev, 1994).

1.1.6.2 Underground civilian ("peaceful") nuclear explosions

The civilian nuclear explosions that took place in Russia during 1967–1988 should have some contribution to overall radioactive pollution (Figure 1.7, Table 1.5). These explosions were usually done for seismic surveys, geological surveys, ease of mineral resource exploration, and to stop blow-outs at gas wells. Some elevated levels of radioactive elements around some the explosion sites have been noted, but there is incomplete information about how these sources affect the Arctic Ocean through atmospheric transport or river runoff. Nonetheless, it is known that at three of the sites accidents led to significant local contamination (*Nuclear Tests . . .*, 1997; AMAP, 1997, 1998).

The nuclear explosion Taiga in March 1971 was the first of 250 planned underground charges to create a canal in an attempt to redirect some northern Russian rivers. The explosion contaminated a 700-m long trench 100 km from Krasnovieshersk. Radiation levels were still 0.01 mSv per hour in the most contaminated areas 15 years after the explosion.

A nuclear explosion was detonated on October 2, 1974 to construct a dam about 90 km from Aykhal. The detonation Kristall did not go according to plan, and an array of radionuclides was released into the nearby environment. The contamination

Table 1.5. Soviet underground nuclear explosions conducted in the interests of the national economy, 1965–1988.

Energy release (kt TNT)	*Technological ground name*	*Total*
Less than 1	Klivazh, Lazurit, Tavda, Tel′kem (2)	5
1–3	Benzol, Butan, Vega, Geliy, Globus, Dniepr, Kristall, Neva, Region (2), Sovkhoznoye, Sary-Uzen′, Galit	24
3–5	Dniepr, Fakel	2
5–10	Agat, Batolit (2), Butan, Vega (13), Vyatka, Galit, Gorizont (4), Grifon (2), Kama (2), Kimberlit (2), Meridian (3), Meteorit (3), Region (3), Rift (2), Rubin, Takhta-Kugul′tinskoye, Shpat	43
10–20	Angara, Galit, Dedurovka (2), Kvarts (3), Kimberlit, Krater, Kraton (4), Lura (6), Meteorit, Neva (3), Oka, Rift, Rubin, Sheksna	27
20–50	Galit (2), Pamuk, Pirit, Say-Utyos, Taiga, Urta′Burlak	7
50–100	Galit (5), Say-Utyos (2)	7
100–150	Chagan	1
	Total	*116*

was not measured until 1990–1993, and there is not enough information to estimate the amount of radioactivity that was released. The report mentions a dead forest but not whether this was the result of radioactive fallout.

The detonation Kraton-3 took place on August 24, 1978 about 120 km from Aykhal at a depth of 577 m. Its purpose was seismic sounding of the Earth's crust. A few seconds after the explosion, instruments recorded a radioactive release, probably from an incompletely sealed well. A radioactive cloud passed over a campsite and exposed about 80 people. In 1981 there was a large-scale attempt to decontaminate the area. The reports also mention a dead forest at this site (AMAP, 1997, 1998).

1.1.7 Military bases, nuclear icebreakers, and submarines

Several countries—including the U.S., U.K., France, China, and Russia—have nuclear-powered military vessels that can transit the Arctic. The ships in Russia's Northern Fleet, stationed at bases on the Kola Peninsula, contain approximately 150 nuclear reactors, most of them in submarines. There are thus potential radioactive pollution risks associated with bases for nuclear submarines and icebreakers, and with facilities for the storage of SNF on the Kola Peninsula (Figure 1.8, Table 1.6).

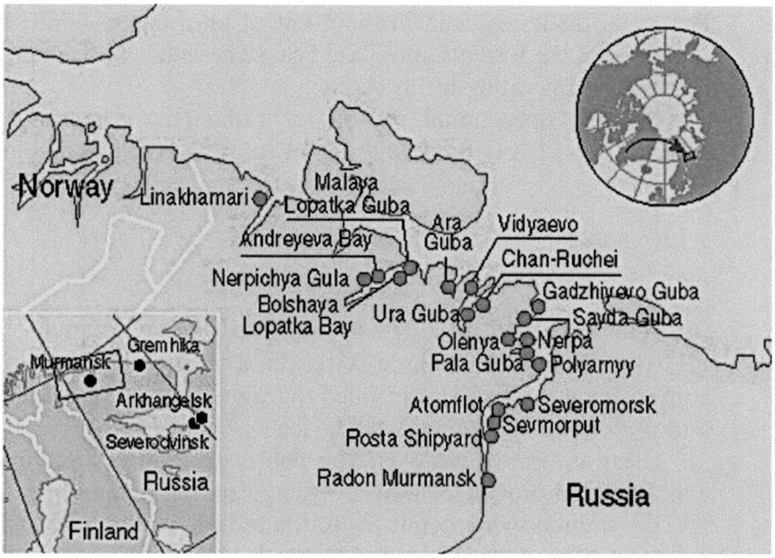

Figure 1.8. Major Russian naval bases along the Kola Peninsula and White Sea (AMAP, 1998).

Table 1.6. Details of radioactive wastes stored at Navy facilities and at RF Ministry of Economy enterprises.

Enterprise	*Quantity of LRW* (thousand m^3)	*Total radioactivity of LRW* (Bq)	*Quantity of SRW* (thousand m^3)	*Total radioactivity of SRW* (Bq)
Russian Navy				
Northern Fleet	8.695	3.05×10^{12}	5.863	5.41×10^{13}
Pacific Fleet	5.767	1.36×10^{12}	1.7029	6.29×10^{14}
Ministry of Economy enterprises				
JC Amur Shipbuilding	1.0	0.5×10^{9}	0.012	No data
Nerpa Shiprepairing	0.105	1.11×10^{10}	0.5	No data
Zvezda Far East	0.95	1.78×10^{11}	1.99	4.7×10^{12}
PE Sever	1.033	No data	5.76	5.44×10^{12}
PE Sevmashpred-priyatie	0.123	No data	1.8	No data
Ship Equipment Enterprise	0.02	3.7×10^{5}	0.0015	0.67×10^{10}
Total	*17.693*	4.6×10^{12}	*17.629*	6.91×10^{14}

Several accidents have been reported. For example, a 1982 accident at an SNF storage facility led to the release of 37 TBq (1,000 Ci) through an escape of cooling water onto the soil around the facility. In 1989, during anchorage of a damaged submarine, 74 TBq (2,000 Ci) of liquid RW was released into the environment (Petrov, 1993). Though these accidents had no negative impact on the marine environment in the region, the increasing amount of SNF that will be stored at Kola

Peninsula facilities could in the event of an accident contribute to the radioactive pollution of the Barents and Kara Seas. The same applies to nuclear submarines and icebreakers operating in the Arctic.

Routine operational releases of radioactive material from nuclear-powered vessels are probably small, but documentation is not available (AMAP, 1997, 1998).

1.1.8 Miscellaneous accidents

Thule

An American B-52 aircraft carrying four nuclear weapons crashed on the sea ice in Bylot Sound near Thule, Greenland in January 1968. The impact triggered conventional explosives, which led to fragmentation of the nuclear weapons onboard and plutonium spreading over the ice.

Clean-up crews removed the debris from the crash and the upper layer of contaminated snow. However, it was agreed that not all the plutonium was recovered and some unknown amount fell to the bottom of Bylot Sound. Subsequent environmental studies carried out by Denmark have concluded that approximately half a kilogram of plutonium sank to the underlying sediment (the results of the Danish study have not officially been endorsed by the U.S. government).

Immediately after the accident, the seawater in Bylot Sound had slightly higher levels of plutonium than other places along the Greenland coast. By 1970 the levels were down to background, with the most likely explanation being that the plutonium adhered tightly to sediment particles. Figure 1.9 is a map of radioactivity in the sediment (AMAP, 1997, 1998).

Komsomolets and Kursk

The nuclear submarine *Komsomolets* sank in the Barents Sea on April 7, 1989 as a result of an onboard fire. The submarine is lying at a depth of 1,680 m at 73°43.27′N and 13°15.87′E (Yablokov *et al.*, 1993). The *Komsomolets* was equipped with one reactor of the water-cooled, water-moderated energy (VVER) type (Kuznetsov *et al.*, 1993). At the moment of the accident, its reactor held 3.1 PBq (83 kCi) of ^{137}Cs, 2.8 PBq (75 kCi) of ^{90}Sr, 4.4 TBq (118 Ci) of ^{239}Pu, and 1.7 TBq (46 Ci) of ^{240}Pu, according to Gladkov *et al.* (1994), although other estimates (Yablokov *et al.*, 1993) are 42 kCi of ^{90}Sr and 55 kCi of ^{137}Cs. In addition, the submarine has aboard two nuclear warheads with radioactivity 400 Ci (6.4 kg of ^{239}Pu) (Gladkov *et al.*, 1994). Total ^{239,240}Pu radioactivity is 564 Ci and the total radioactivity overall is 722 Ci (2.6 PBq).

The submarine has a titanium hull which allowed it to descend to such a great depth. The same feature brings about very rapid corrosion in the presence of different parts that are made from steel and other metals. Diving expeditions to evaluate the possible outcome of radionuclides from the submarine showed that, though the hull of the *Komsomolets* is not hermetically sealed, there is a very weak release of radioactivity. The concentration of ^{137}Cs was less than 1×10^{-11} Ci/L (0.37 Bq/L) (Yablokov *et al.*, 1993).

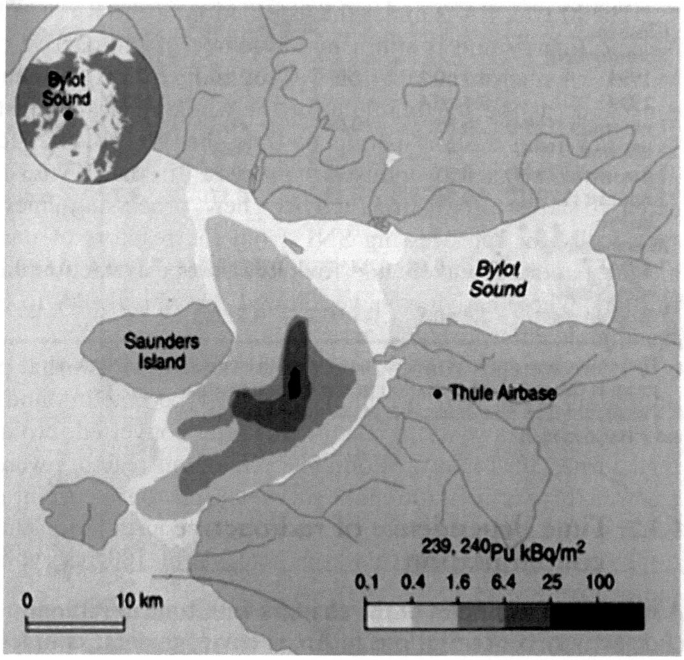

Figure 1.9. Radioctivity concentrations of $^{239,240}Pu$ in sediments near Thule, Greenland (AMAP, 1998).

Should the *Komsomolets* be raised, the risks would be much greater than to leave it undisturbed (NATO, 1995a, b). In any case, in the highly improbable event that all of the radioactivity from the *Komsomolets* is released to the environment, it would be minor compared with the discharges of liquid radioactive waste from European reprocessing plants (Kouraev, 1996).

Another accident onboard the Russian nuclear submarine *Kursk* (K-159) happened on August 12, 2000 at a depth of 240 m north of the Kola Peninsula, where the submarine sank (see also Section 4.2.5). Thus, this is another potential danger of radioactive leakage—considered in scenarios elsewhere in this book (Section 4.5).

1.2 RADIOACTIVE POLLUTION: MAJOR RUSSIAN NUCLEAR INDUSTRIES

1.2.1 The Mayak Production Association, Chelyabinsk

General information

The Mayak Production Association (Mayak PA) is located in the northern part of the Chelyabinsk region near Kyshtym and Kasli (*Sources Contributing to Radioactive Contamination* ..., 1997). The Chelyabinsk region is where nuclear reactors were used for industrial production of weapons-grade plutonium for defense purposes (Kruglov, 1994).

The Mayak PA is under the jurisdiction of the Russian Federation's Ministry of Atomic Energy and is still a major source of radioactive contamination. The first Mayak PA nuclear installation was set in operation on June 18, 1948, which may be considered as the starting date for the whole nuclear industry of the former Soviet Union. Within a short period of time the Mayak PA succeeded in developing a high-capacity nuclear industrial process. In the 1970s, together with the basic defense-related production activity, new processing lines were set in operation specifically for reprocessing SNF from the reactors of nuclear power plants, submarines, and icebreakers and for radioisotope extraction and production of isotope emitters. These activities have allowed the Mayak PA to become an economically sustainable enterprise and an exporter.

There were six running reactors at the Mayak PA that produced weapons-grade plutonium. Five were uranium graphite–type reactors and one was a heavy-water reactor. The heavy-water reactor was later converted into a light-water reactor and then turned on for operation. The remaining reactors were stopped. Besides these reactors there is an additional light-water reactor in current use for the production of isotopes for civil purposes. There is also one installation that verifies the concentrations of liquid radioactive wastes and about 100 reservoirs with high-level liquid radioactive wastes.

Geographical location

The activities of the Mayak PA have affected areas in three Ural regions: the northern part of the Chelyabinsk region, the southern part of the Sverdlovsk region, and a small part of the Kurgan region along the Techa and Iset′ Rivers. The affected territory is situated on the eastern slopes of the central and southern Urals and the western Siberian lowlands. The southwestern and central parts of this territory are located within an abrasion–erosion platform 220 m to 250 m above sea level. The northeastern part is situated in an erosion-accumulative plain (Kruglov, 1994). There are more than 3,000 lakes in the region and the major rivers have drainages that are controllable.

The Chelyabinsk region has an area of 88,500 km^2, with a population of 3,636,800, 82% of whom live in cities. The region is rich in mineral deposits, particularly iron ore, magnetite, and gold. The Chelyabinsk region is the home of Russia's sixth-largest industrial production.

The enterprises of the Mayak PA (55°44′N, 60°54′E) are located in the closed town of Ozersk at the interfluves of the Techa and Misheluak Rivers, in the northern part of the Chelyabinsk region in the vicinity of the old southern Ural towns of Kyshtym and Kasli within a protected territory (Figure 1.10). The Mayak PA occupies a total area of 90 km^2. Ozersk is situated 15 km to the east of Kyshtym and 70 km to the north of Chelyabinsk. Ozersk's population is about 85,000, about 17,000 of whom are employed at the Mayak PA.

The enterprise was established at the end of the 1940s to produce and reprocess weapons-grade plutonium. The first industrial "A-reactor" was started up in June 1948. Four more reactors were started up between 1950 and 1952. Although the

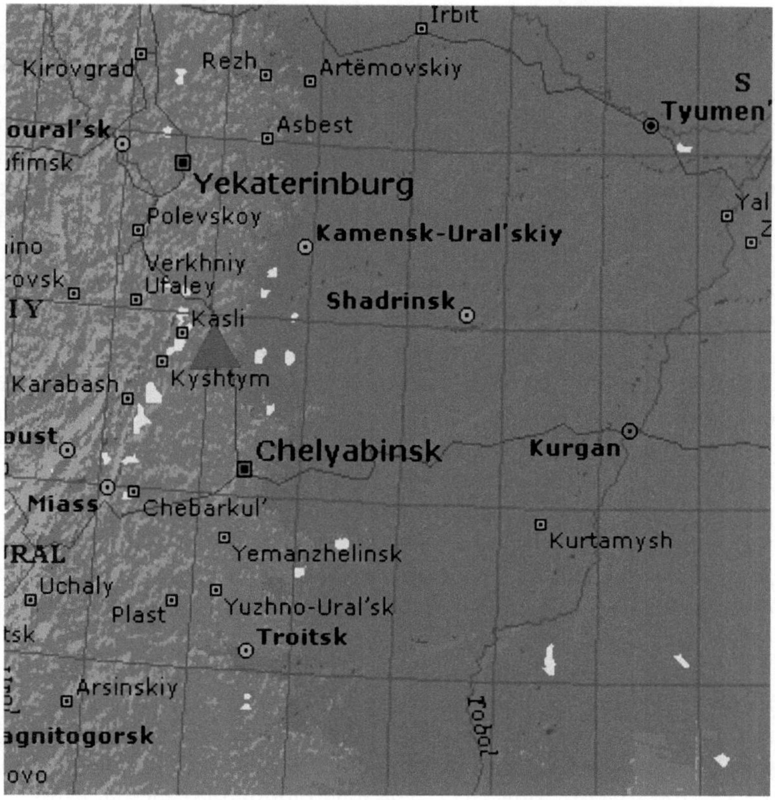

Figure 1.10. The geographical position of the Mayak PA (© 2000 Microsoft Corp. All rights reserved).

Mayak PA is still engaged in military work, the production of weapons-grade plutonium ceased in 1987 and the enterprise has increasingly directed its technical potential towards non-military applications.

The geographic features of the Mayak PA location (geology, hydrogeology, climate, etc.) influence the atmospheric dissipation of radioactive discharges, and thus play a role in defining the radiation control area.

The source of the River Techa is in Lake Irtash; further downstream, the river flows through Lake Kyzyltash. The Techa is a part of the Ob′ river system and is a tributary of the Iset' River. The river system Techa–Iset–Tool–Irtash–Ob′ discharges into the Kara Sea. Hence, it is an important transport route for radioactivity from the Urals to the Arctic Seas (Kruglov, 1994).

The Techa River is considered to begin from the outflow of dam 11 of the Mayak PA's system of dams and bypass channels. The catchment area of the Techa and Iset′ Rivers is situated between the Ural Mountains and the Tool River valley. The length of the Techa River up to its confluence with the Iset′ is 187 km, whereas its natural length is 243 km. The length of the Iset′ is 606 km, 364 km of which lie downstream of its confluence with the Techa. Both rivers have average mineralization and moderate carbonate hardness of water.

Climatic conditions

The region of the Mayak PA enjoys a continental climate with a long cold winter and a warm summer. The territory is exposed to the northeast, east, and southeast allowing the entry of different continental air masses from Asia. The winter is long (170–174 days) and cold and under the dominant influence of the Siberian High. In the summer dry continental airmasses from the south and southeast predominate above Central Asia and Kazakhstan. Therefore, summer air temperatures are higher than on the same latitude of the European part of the country. In autumn, cold Arctic air extends from the north to the south, accompanied by a fall in air temperature and early light frosts (AMAP, 1997).

Westerly winds predominate throughout the whole year, although there is seasonal variability in wind directions and speeds. In the winter the meteorological conditions of the southern Urals are influenced by the high-pressure ridge of the Asian Anticyclone, and westerly and southwesterly winds with mean velocities of 5 m/s prevail. In the summer, the Azores High causes an increase of atmospheric pressure to the west of the southern Urals, and as a consequence the 4 m/s northerly/northwesterly winds bringing rain predominate at this time. In some years during May–June tornado-like windstorms with wind velocities of 25 m/s to 30 m/s are observed. The mean annual wind velocity is 4.1 m/s. Maximum wind velocity may reach 24 m/s (AMAP, 1997).

Air temperature data indicate that the cold period begins in October and ends in May. The mean temperature of the coldest month (January) is −14.7°C. The minimum air temperature of −43.2°C was observed in December 1955. The warmest month is July, when mean air temperature is 18.7°C, with an absolute maximum of 37.3°C observed in July 1952. The mean annual air temperature is 2.6°C. Daily air temperature fluctuations are strongly pronounced in all seasons of the year. The earliest frosts are usually observed by September 15, with September 3 being an early date and October 5 a late date.

The temperature of the subsurface ground layer at a depth of 0.5 m as a rule coincides with air temperature, but this value delays for 1 month at a depth of more than 1 m. At a depth of 0.2 m the highest mean soil temperature reaches 15.8°C, while the lowest temperature is 4.2°C.

The Mayak PA is located in an arid region (Figure 1.11). The precipitation maximum occurs in the summer months. Mean annual precipitation water equivalent is equal to 429 mm. The first blanket of snow occurs in the middle of October. The permanent snow cover is in place at the beginning of November. The duration of the snow cover varies from 150 to 170 days. The depth of snow may reach 30 cm to 35 cm in open areas and 45 cm to 50 cm in wooded areas. The snow begins to melt in the second half of March and takes 15–20 days to completely disappear. Precipitation during the warm season comprises 75% to 77% of annual precipitation, falling as rain showers and continuous drizzle.

Structure of the Mayak PA

The structure of the Mayak PA consists of plants for production of weapons-grade

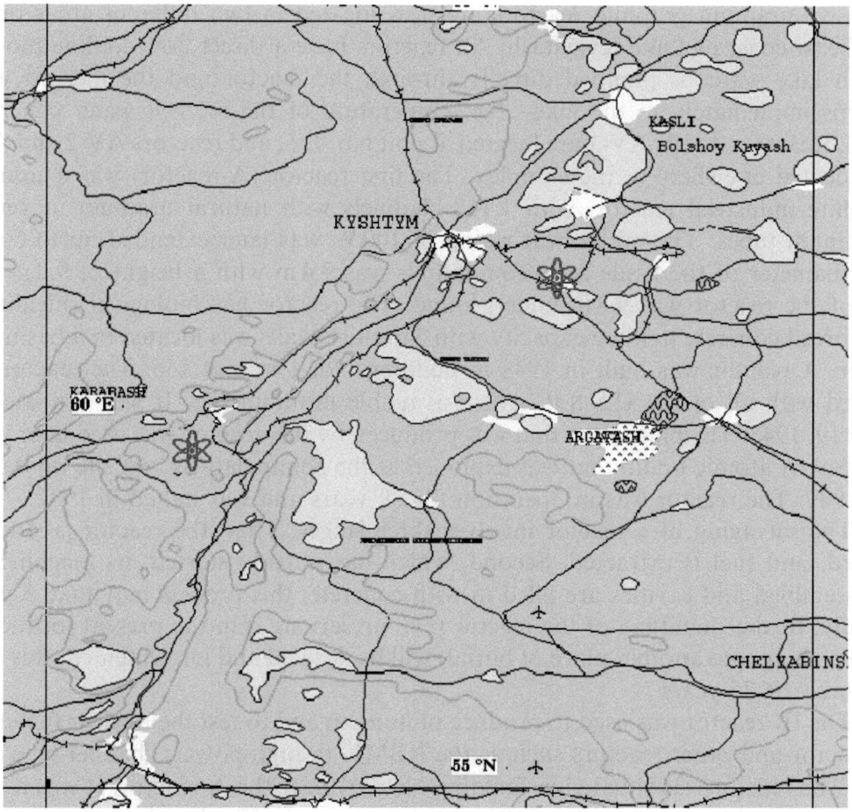

Figure 1.11. The territory surrounding the Mayak PA.

plutonium; plant RT-1 for reprocessing spent nuclear fuel from nuclear power plants and Navy and research reactors; a radioisotope plant; complex for curing and storage of high-level waste products in a vitrified state; spent nuclear fuel and regenerated plutonium and uranium storage; and nuclear waste stores and repositories (*Nuclear tests of the USSR* ..., 1997).

At present the Mayak PA is a large, relatively modern nuclear industry enterprise that is successful in a range of non-nuclear spheres of activity stemming from its fundamental defense purpose. Production of weapons-grade plutonium at the radiochemical plant of the Mayak PA was ceased in 1987 when five production reactors were halted in response to initiatives on the reduction of strategic arms.

Types of reactor installations

Five of seven military uranium graphite reactors were halted during 1987–1991, and the two that survived produce a wide range of radioisotopes for military and civil purposes, including ^{14}C and ^{60}Co (with a specific radioactivity up to 10 TBq/g) and ^{192}Ir (up to 30 TBq/g).

Five uranium graphite reactors are now located in two different areas on the southeast coast of Lake Kyzyltash. All reactors have a direct flow cooling mode in which lake water is pumped directly through the reactor and the merged water returns immediately to the lake—the temperature of the merged water can reach 70°C. Reactors A, IR, AV-1 are located at unit No. 156, and reactors AV-2 and AV-3 are located elsewhere in the complex. The first reactor (A-reactor) was a uranium graphite industrial reactor, with 1,168 channels with natural uranium in vertical aluminum tubes. The initial heat rating of 100 W was later extended up to 500 W. The diameter of the fissile zone compounds was 9.4 m with a height of 9.2 m. The top of the reactor was 9.3 m underground. The reactor has biological shielding: a reinforced concrete holding capacity with 3 m thick walls was located in a basin with water. A reactor was built in 1948 in building No. 1 in unit 156. The reactor was loaded with all of the U.S.S.R.'s then-available uranium-235. It was activated on June 19, 1948. The plutonium that was produced by this reactor was used later in the first Soviet atomic bomb that was exploded at the Semipalatinsk test site on August 29, 1949. The reactor was in operation for 39 years and was halted in 1987.

The salvaging of a reactor involves three stages. First, the reactor is stopped, cooled, and fuel is extracted. Second, which has already started, its machinery is disassembled and cavities are filled in with concrete; this process may take 5 years. Finally, further handling of the reactor (i.e., preserving it in the present location or transporting it to another place of burial) will be decided and implemented; this stage can take 20–25 years.

The IR reactor was used to produce plutonium and to test the nuclear fuel of the A-reactor and other reactors such as the RBMK (a high-power, channel-type reactor). The reactor was situated in building No. 701 near the A-reactor. It was a small uranium graphite reactor of 65 MW with 248 channels. The building of the reactor was initiated on August 18, 1950. The IR reactor was started up on December 22, 1951 and stopped on May 24, 1987 after 36 years of operation.

The AV-1, AV-2, AV-3 reactors were large uranium graphite reactors, which most probably had a common design, each having 2001 channels. Public data are available only on reactor AV-2. The active zone of AV-2 was a vertical barrel with a diameter of 11.8 m and height of 7.6 m. The biological shielding of the active zone consisted of three layers: the first layer combined a 1.5 m thick water-and-sand coating and a 2 m thick concrete wall; the second layer was a mixture of sand and bauxite ore 1.5 m thick and a 3 m thick concrete coating; the third layer was the 1.5 m thick water basin.

AV-1 was started up in 1955 and stopped on August 12, 1989; AV-2 started up in April 1951 and stopped in July 1990; AV-3 started up on September 15, 1952 and stopped on November 1, 1990. AV-3, located in building No. 501 in unit No. 156, was the last operating uranium graphite reactor.

The heavy-water reactor Ruslan was the next reactor to be started up at the Mayak PA, presumably between 1948 and 1951; it operated until 1980. At the end of the 1980s it was converted to a 1,000 MW light-water reactor. The reactor is used for production of tritium for H-bombs and the special isotope ^{238}Pu. Lyudmila is another operational light-water reactor of the Mayak PA, with a capacity of

1,000 MW. The reactor is also used for production of tritium and different isotopes, including ^{238}Pu.

The five Mayak PA uranium graphite reactors in total have produced 58.3 tons of ^{238}Pu. According to the available data, the two remaining reactors should have produced 14.7 tons of plutonium up to 1992; if so, the total production of ^{238}Pu at Mayak PA can be estimated as 73 tons.

Reprocessing Plant RT-1

Reprocessing Plant RT-1 was started up in 1956. The plant was initially constructed to reprocess weapons-grade plutonium from spent nuclear fuel from five breeder reactors. In 1976, the plant was modified for reprocessing of civilian purpose plutonium received from nuclear power plants, reactors of atomic submarines and icebreakers, research reactors and fast reactors (BN-350 and BN-600). Reconstruction of the plant was aimed primarily at reprocessing fuel in stainless steel or zirconium envelopes, which are used in the aforementioned types of reactors.

The reprocessing plant capacity comprises 600 tons of spent nuclear fuel per year (*Nuclear tests of the USSR* ..., 1997). Until recently the plant reprocessed spent nuclear fuel from 23 VVER-440 reactors including those from countries in Eastern and Central Europe (15 reactors). Furthermore, countries in Eastern and Central Europe and Finland also send spent nuclear fuel from their own Soviet Russian–built pressurized water reactors to this facility.

The 300 tons of spent nuclear fuel annually used by plant RT-1 have a radioactivity of 180 MCi, including up to 90 MCi per year of long-lived fission fragments; the plant produces about 3 tons of plutonium annually.

During its operating history, plant RT-1 reprocessed a large quantity of spent nuclear fuel from VVER-440 and other types of reactors. As a result of this activity, over 25 tons of plutonium with a total radioactivity of 350 MCi was accumulated in temporary surface reservoirs. Between 1976 and 1991 about 200 tons of spent fuel were reprocessed per year on average. Since 1991 the amount of reprocessed spent fuel has decreased, in part due to the difficulties in importing spent fuel from the CIS countries Azerbaijan and Ukraine, and from the Central European countries Bulgaria, Hungary, and the Czech Republic. In 1995, Finland ceased forwarding its spent fuel from the Lovissa power plant for reprocessing. In 1992, Finland was reported to have had 120 tons of fuel reprocessed, corresponding to an annual production of 1 ton of reactor-grade plutonium.

The plant has a spent nuclear fuel accumulating pond, three disassembly sectors and a fuel extraction installation. For several years the plant was engaged in analyzing techniques for fractionating high-level waste products and encapsulating medium-level waste in cement or concrete shapes. RT-1 is capable of reprocessing 400 tons of spent nuclear fuel per year, or 300–900 fuel assemblies per year. The technique uses tributyl phosphate, through which 99% of the uranium and plutonium contained in spent fuel is extracted. Reprocessing of 1 ton of spent fuel results in 45 m^3 of high-level waste, 150 m^3 of mid-level waste, 2,000 m^3 of low-level waste, and 7,500 kg of solid nuclear waste.

Radioisotope Plant

The Radioisotope Plant is one of the world's largest suppliers of radionuclides, emitting sources, and radionuclide medicaments. Well-known companies in the U.K., France, Germany, and the U.S.A. are among its customers. It has a wide spectrum of production: alpha-emitting and beta-emitting sources, gamma and X-radiation sources, fast neutron sources, thermal sources (based on ^{90}Sr and ^{238}Pu), and radioactive isotopes (^{14}C, ^{137}Cs, ^{90}Sr, ^{241}Am, ^{238}Pu, ^{237}Np, ^{147}Pm).

Vitrification Facility

The Vitrification Facility for liquid radioactive waste has been in operation since 1987. This facility has a capacity of 500 L/h and transforms radioactive waste into phosphate glass. The first ceramic smelter ran for only 13 months before the electrodes were destroyed by their own very high electrical current (2,000 A). Vitrification of liquid radioactive waste was resumed on June 25, 1991, after a new smelter had been built.

The processed glass is placed in stainless steel containers (0.6 m in diameter and 3.4 m in height) which are stored in groups of three in surface storage facilities containing air-cooling systems. At the present time, there are nearly 4,000 such containers stored at the Mayak Chemical Combine. The containers will be stored for 20 to 30 years until a permanent underground site has been built for them.

By the beginning of 1995, 8,100 PBq (218 MCi) had been vitrified, extracted from liquid waste measuring 8,500 m^3 in volume. The vitrified waste weighed approximately 1,600 tons. The facility vitrified on average approximately 1,850 PBq/yr (50 MCi/yr) during its first 5 years of operation. Now all high-level liquid waste is vitrified. The liquid waste that undergoes vitrification has a concentration of 10 TBq/L (400 Ci/L).

MOX production

There used to be a number of production facilities at Mayak PA for the fabrication of MOX fuel (a mix of uranium and plutonium). Of the five MOX fabrication facilities, two have been shut down, two continue to operate, and construction of the fifth has been halted, at least temporarily.

The first pilot facility operated during the 1960s and 1970s, and used approximately a ton of weapons-grade plutonium to produce test fuel assemblies for fast research reactors. Between 1986 and 1987, there was a small facility in operation to produce MOX fuel for fast reactors of the BN type. It had a capacity of 35 kg of weapons-grade plutonium per year (five assemblies per year).

Since 1988 another facility has been producing MOX fuel for trial in fast reactors. Its capacity is 70 kg to 80 kg of weapons-grade plutonium a year, or 10 fuel assemblies per year. In 1993, the Atomic Research Institute in Obninsk (just outside Moscow) carried out a trial using 150 kg of MOX fuel in a fast reactor.

Construction had also begun on a MOX fuel fabrication facility, but work was suspended when the plant was 50% to 70% complete. The plant was intended to produce fuel for three planned breeder reactors in the South Urals Project at a capacity of 5 t to 6 t of plutonium per year. Plans for this facility also included the production of MOX fuel assemblies for VVER reactors.

Experimental Research and Development Station

The Experimental Research and Development Station was established in 1958—just after the 1957 Kyshtym accident—in order to work on problems connected with large-scale radioactive contamination. The priority activities of the station are related to radio-agrochemistry, agricultural radioecology, wood radioecology, aqueous radioecology, ecological monitoring, radiation physics and chemistry, the biological effects of ionizing radiation, assessment of radiation doses, and radioecological and agricultural rehabilitation.

The emphasis was on the study of the main mechanisms of the behavior and migration of radionuclides in different components of the environment, with consequent actions for radioecological rehabilitation of contaminated lands and programs on their return to practical use. The station is a highly developed institute with powerful experimental capabilities. Due to the results attained, more than 80% of "eliminated" lands of the eastern Urals were returned to economical and agricultural use.

Cascade of reservoirs and system of bypass channels (Techa cascade of reservoirs)

From 1949 to 1951, discharges of low-level and medium-level waste products were made into the open hydrographic system of the Ob′ Basin through the Techa River. As a result, 76 million m^3 of waste with a total radioactivity of 2.8 MCi (beta activity) containing 10% long-lived fission products were discharged during that period.

Later, the Techa River was dammed up by a system of reservoir dams used for handling low-radioactive and medium-radioactive waste products (Figure 1.12). In 1951–1964, a cascade of artificial reservoirs and a system of bypass channels in the upper Techa was constructed, essentially confining the inflow of radionuclides to downstream of the Techa River. Reservoirs 2, 3, 4, 10, and 11 are part of this system (*Nuclear Tests of the USSR* ..., 1997).

Reservoir 2 was used as a cooling pool. The series of natural lakes and ponds have been used up to the present as pools for the discharge of nuclear wastes, including Lake Karachay (Reservoir 9) and Lake Kyzyltash (Reservoir 2). Reservoirs 3, 4, 10, 11, and 17 are now mainly used as industrial pools.

Low-level solutions are treated for desalinization and clearing of radionuclides in ion exchange installations. Decontaminated waters are routed to a pool of reverse water supply and reclaimed waste is discharged into the Techa reservoir cascade.

The inflow of long-lived radionuclides amounts to 4 kCi per year into Reservoir 2 and more than 2 kCi per year into Reservoir 3. At present, about 200 kCi of long-lived radionuclides are deposited in reservoirs.

The high levels of radioactive contamination of the Mayak PA region have necessitated the creation of a health protection zone covering an area approximately 350 km^2 around the enterprise. Agricultural activity and human dwellings in this zone

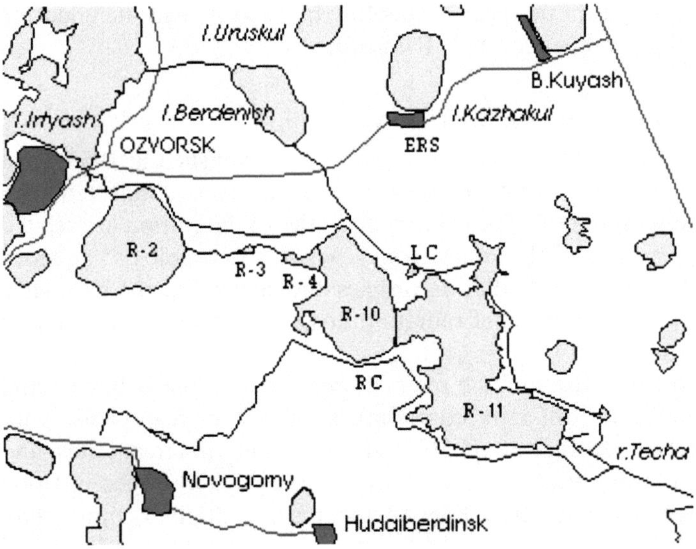

Figure 1.12. The system of pools and dams in the Mayak PA territory.

are prohibited. There is also a wider band, known as the observation zone, where regular monitoring is carried out; agricultural activity in the observation zone is allowed.

Radioactive waste management

More than 40 years of activity at the Mayak PA have resulted in local accumulation of an extremely large quantity of radionuclides and considerable contamination of the region (Chelyabinsk, Ekaterinburg, Kurgan, and Tyumen regions). An anomalously high radiation environment is observed in the health protection zone at the interfluve of the Techa and Myshelyak Rivers with an area of 270 km^2. Waste products of radiochemical production with a total value of 1 billion Ci are concentrated here. A considerable proportion of these waste products (more than 120 MCi) is deposited in the open environment: reservoir water, bottom sediment, etc. Contaminated ground water with an area about 10 km^2 and a total value of 6 kCi (beta activity) can be found in the disposal areas. The greatest potential hazard from the Mayak PA is represented by nuclear waste deposited in the open environment. High-level liquid and solid nuclear wastes are the next greatest (*Nuclear Tests of the USSR* ..., 1997).

According to the results of an inventory estimation in 1990, the Mayak PA has accumulated 30,000 PBq (800 MCi) of liquid and solid radioactive waste including waste products concentrated in the industrial reservoirs. Bulk amounts of spent nuclear fuel from nuclear power plants received for reprocessing have been reduced in recent years. Also, owing to the reprocessing and vitrification of high-level and medium-level waste, the amount of liquid waste contained in the reservoirs has steadily decreased.

Table 1.7. Total quantity of high-level nuclear waste (HLW) stored at the Mayak PA, March 1, 1995.

Type of waste	*Quantity of waste* (m^3)	*Total activity* (PBq–MCi)
Suspensions	19,000	5,000–135
Nitric acid solutions	11,800	9,800–264
Vitrified waste	8,673	8,700–235

High-level nuclear waste (HLW)

Each year the Radiochemical Plant creates 2,000 m^3 to 3,000 m^3 of liquid HLW—with a total radioactivity of up to 2,200 PBq (60 MCi)—from reprocessing spent nuclear fuel from nuclear power reactors and naval power reactors. As of March 1, 1995, a total of 30,800 m^3 of liquid HLW had been accumulated, including suspensions, with a total radioactivity of 15,000 PBq (Table 1.7).

Spent nuclear fuel arriving at the Radiochemical Plant is transferred to the storage ponds. After the decay of short-lived fission products, the fuel is sectioned and dissolved in nitric acid. Uranium and plutonium are extracted for further use in RBMK reactors and fast-breeder reactors (Kruglov, 1994).

After solvent extraction, raffinates are concentrated to various degrees by evaporation and the residues are blended with previously accumulated liquid HLW. After evaporation, waste from civil fuel reprocessing has a radioactivity concentration of up to 2 TBq/L (60 Ci/L) with the following chemical composition: nitric acid (up to 4 M) containing aluminum (1–3 g/L), sodium (0.3–2.5 g/L), iron (1–10 g/L), nickel (1–5 g/L), chromium (0.3–2.0 g/L), calcium (0.2–2.0 g/L), SO (1–3 g/L), and platinoids and rare earth elements. Reprocessing of spent highly enriched fuel from naval power units gives solutions containing up to 200 g/L salts, with a radioactivity concentration of up to 1 TBq/L (30 Ci/L). The chemical and radionuclide composition of the suspensions is quite complex, since solutions from different types of irradiated fuel elements have been concentrated by sedimentation (Kruglov, 1994).

High-level nitric acid solutions are stored in cylindrical tanks with stainless steel covers and bottoms. Each tank is installed inside a separate cell lined with stainless steel. The cells have sumps with instruments to indicate liquid levels and pipes to collect any leakage.

Nitric acid is also extracted during the evaporation of high-level solutions; condensates from this process are discharged into Reservoir 17 without any additional purification. A fractionation unit known as the UE-35 has been set up at the enterprise to radiochemically separate transuranium elements, and to extract strontium and cesium from the waste using a system based on chlorinated cobalt dicarbollid.

There are 18 tanks containing waste from the former military program, each is 9 m in diameter and 5 m deep, with an effective volume of 285 m^3. For heat removal, the tanks are equipped with three-section coils each with a cooling surface of 60 m^2. Solutions from civil reprocessing are kept in three tanks of diameter 22 m, depth 4.25 m, and an effective volume of 1,500 m^3 each. The tanks are equipped with 16 evenly distributed sectional coils, with a total heat removal surface of 300 m^2. Civil purpose tanks are made of stainless steel and military waste tanks are made of chromium–nickel–titanium stainless steel.

Suspensions are stored in 20 concrete storage tanks. The tanks are $19.5 \times 9.5 \times 7$ m in size, lined with stainless steel, and have an effective volume of 1,170 m^3 each. Twelve of the tanks have an internal cooling system; the other eight do not need internal cooling. All storage tanks are equipped with level, pressure, temperature, overfill, and gas emission monitoring systems. All high-level waste stores are also equipped with control systems for corrosion. After the Kyshtym accident of 1957, the construction of facilities for high-level liquid waste and storage control systems were substantially improved.

Intermediate-level nuclear waste (ILW)

The Mayak PA produces 16,000 m^3 to 20,000 m^3 of liquid ILW per year, with a total annual radioactivity of less than 30 PBq (0.8 MCi), according to recent estimates. The specific radioactivity of the waste is 370 MBq/L to 750 MBq/L (20 mCi/L) and the average salt content is 12 g/L to 15 g/L. The source of the ILW is drainage, desorption solutions, solutions from washing and decontaminating equipment, extraction agent and extractor flushing solutions, cladding and container decontamination solutions, and special gas scrubber condensates. Liquid ILW disposal takes place mainly into Lake Karachay (Reservoir 9) and Reservoir 17. In Reservoir 17 the ^{90}Sr concentration was 300 kBq/L and the ^{137}Cs concentration 150 kBq/L. Radioactivity concentrations in reservoir waters remained relatively constant during the previous 3–4 years leading up to the report by Kruglov (1994). Between 1951 and 1995, some 3.7 million m^3 of ILW were discharged into Lake Karachay, and 4,400 PBq (120 MCi) have accumulated in the reservoir. In 1993 the ^{90}Sr concentration in Lake Karachay was 70 MBq/L and the ^{137}Cs concentration 100 MBq/L. Lake Karachay is considered the most radioactively contaminated stretch of water on the planet.

Liquid low-level radioactive waste (LLW)

The liquid low-level waste at the Mayak PA comprises trap water from the facility's sewage system, secondary circuit cooling water from the two reactors, as well as cooling water from the high-level waste storage cells. Raw sewage water (0.7–1.0 million m^3/yr) has a maximum specific radioactivity of 0.56 MBq/L (15 μCi/L) and an average salt content of up to 1 g/L. Salts and radionuclides are removed at the purification plant, using quartz filters for mechanical purification and cation (H^+) and anion (OH^-) exchange filters connected in pairs. The purified water, with a maximum radioactivity of 7 kBq/L (0.2 μCi/L), is discharged into water supply circulation Reservoir 2 (Lake Kyzyltash). Over the period 1956–1990, it has been estimated that about 6.7 PBq (0.18 MCi) of long-lived radionuclides have entered the

Table 1.8. Inventory data on burial grounds and waste products accumulated at the Mayak PA.

Waste product category	*Number of burial sites*	*Volume of waste products* (thousand m^3)	*Total radioactivity of waste products* (Ci)	*Burial site area* (ha)
Low-level and medium-level	203	685.1	31×10^3	20.2
High-level	24	41.3	12×10^6	1.1

reservoirs as liquid LLW, the largest part being retained in Reservoir 10 (Kruglov, 1994).

Solid radioactive waste

There are about 500,000 tons of solid nuclear waste accumulated at the Mayak PA, 30% of which is metallic. Table 1.8 gives inventory data on burial grounds and the waste products accumulated in them.

The accumulation rate of solid waste was about 10,000 t/yr until 1981, but then it decreased due to a setback in production activity. The accumulation rate more recently has been 2,000 t/yr to 2,500 t/yr.

The disposal sites for low-level and medium-level solid nuclear waste are trenches dug in the ground. After being filled by waste, the trenches are covered by waterproof soil. Ground disposal sites as a rule are located where groundwater levels are deep enough that the distance between the bottom of the disposal site and the groundwater is at least 4 m. Waterproofing of the bottom and walls is ensured by a layer of clay. The leakage of radionuclides from disposal sites can be through migration with infiltrated precipitation (during burial at the disposal site before the waste is covered), migrations in water-bearing horizons, and diffusions in the wet soils.

High-level solid waste is disposed of in permanent ferro-concrete buildings with a multiple waterproof system (bitumen, stainless steel, and concrete). An additional barrier against radionuclide leakage is the silt soil covering the floors and walls of the permanent building. Only high-level waste disposals are equipped with control-measuring techniques and a warning system. There are no control and measuring systems for trench-type waste disposal.

Almost all solid waste generated at the enterprise is buried without reprocessing because of the country's lack of specially designed installations for incineration, decontamination, and melting. The existence of the great number of waste disposal sites (occupying an area of 30 ha) is due to the fact that every department has its own waste disposal sites for all types of waste.

Accidents and discharges of radioactivity at the Mayak PA

Three principal accidents have caused major radioactive contamination of the environment in the vicinity of the Mayak PA:

(1) direct discharges of liquid low-level radioactive waste into the Techa River (1949–1956);
(2) the Kyshtym accident of 1957;
(3) wind transport of contaminated bottom sediment from Lake Karachay during the drought in the spring of 1967.

Direct discharges of liquid middle-level waste into the Techa River were made between 1949 and 1956. During this time more than 100 PBq of radioactive material was discharged into Reservoir 3, which together with Reservoir 4 was used as a core pool before release into the Techa River. The isotopes of ruthenium (^{103}Ru, ^{106}Ru) and rare earth elements comprised more than 50% of the radioactivity. It has been estimated that approximately 12 PBq of ^{90}Sr, 13 PBq of ^{137}Cs, and 2 TBq of plutonium and americium isotopes have been discharged into the reservoir (*Sources Contributing to Radioactive Contamination* ..., 1997). Approximately 98% of the total value was discharged between December 1949 and November 1951. It has been stated that during the period of the most intense discharges (1951) lowering of the specific radioactivity in the vicinity of Metlino (Figure 1.13) reached 68% for ^{90}Sr and 89% for ^{137}Cs, which demonstrates the important role of the reservoirs as core pools for radionuclide adsorption. However, the levels of radioactive matter that were not accumulated in reservoirs were still sufficiently high to cause serious radioactive contamination of the whole Techa River area. About 7,500 inhabitants were evacuated from 20 settlements on the banks of the Techa River with external doses between 35 mSv and 1,700 mSv. Discharges of ^{90}Sr and ^{137}Cs between 1949 and 1957 caused radioactive contamination of 240 km^2 of floodplain soils, within which an area of 80 km^2 had a contamination level of 3.7×10^{10} Bq km^{-2}. Contamination of floodplain soils took place by two principal means: deposition of radionuclides from contaminated water and transport of contaminated sediment during high waters. The Sanov swamp contained approximately 40 TBq of ^{90}Sr and

Figure 1.13. Location of settlements on the Techa River.

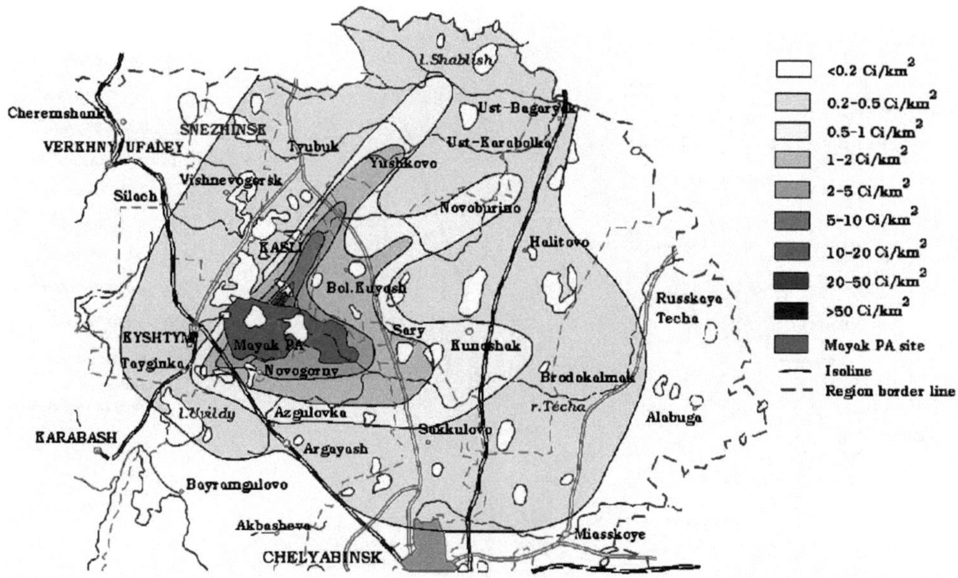

Figure 1.14. ^{137}Cs contamination of the Chelyabinsk region after the Kyshtym accident of 1957.

210 TBq of ^{137}Cs, as of 1992. It is believed that the swamp lost the majority of its accumulated ^{90}Sr during the periods of the most intense discharges (1962–1992); this assumption is based on the low proportion of $^{90}Sr/^{137}Cs$ in the swamp.

The Kyshtym accident of 1957. A tank with liquid radioactive weapons-grade plutonium waste exploded on September 29, 1957 in the grounds of the Mayak PA. As a result of this explosion, surface soils of Chelyabinsk, Sverdlovsk and Tyumen regions were significantly contaminated. About 90% of the nuclear waste from the tank (20 MCi) was retained in the explosion area and the rest (2 MCi) was scattered by wind, resulting in significant radioactive contamination of a vast territory of 23,000 km^2 home to a population of 250,000 inhabitants (Figures 1.14 and 1.15). The radioactive contamination was termed the "East Ural radioactive trace". About 10,800 inhabitants were evacuated from contaminated territories after this accident. Later, after practical rehabilitation, about 82% of the area covered by the East Ural radioactive trace was returned to economic use. However, agricultural usage of the most contaminated central part (180 km^2) was limited.

Wind transport of contaminated bottom sediment from Lake Karachay. Between April 10 to May 15, 1967 the region surrounding the Mayak PA became radioactively contaminated. The contamination was caused by wind transport of bottom sediment from uncovered Karachay Lake sites. Since 1951 the lake was used as a storage reservoir for medium-level waste. A drop in water level caused

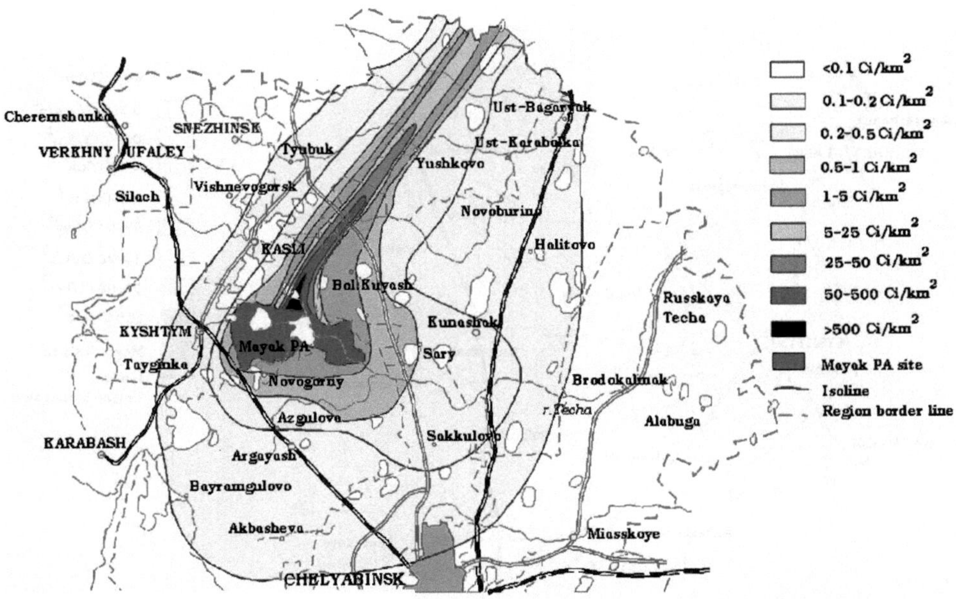

Figure 1.15. ^{90}Sr contamination of the Chelyabinsk region after the Kyshtym accident of 1957.

by low snowfall in the winter led to the exposure of littoral bottom sediment in the following spring. Early spring was dry with strong gusty winds in April and May. As a result, most of the radioactive matter, 22 TBq (600 Ci), combined with fine bottom sediment was scattered over the territory occupied by the Mayak PA. The density of the distribution and boundaries of the contaminated territory was complicated due to the different wind directions.

1.2.2 Siberian Chemical Combine, Tomsk-7

General information

The Siberian Chemical Combine (SCC) is located near Seversk (formerly Tomsk-7) 20 km northwest of Tomsk on the River Tom′ (Figure 1.16). Tomsk-7 was founded in 1949 for the production of fission materials for nuclear weapons. Its population is about 119,000, 15,000 of whom work at the combine.

The enterprise consists of a reactor plant (reactors for production of plutonium), plants for isotope separation by means of centrifuges, a sublimate plant (for production of uranium protoxide–monoxide and uranium hexafluoride), a radiochemical plant (for reprocessing irradiated blocks), and a metallurgical–chemical plant (for production of weapons-grade plutonium).

The SCC is presently the largest enterprise for the production of plutonium, uranium, and transuranic elements in the world. The combine has a complete cycle of reprocessing: primary enrichment, accumulation of plutonium, separation of plutonium from highly radioactive decay products, secondary processing of spent

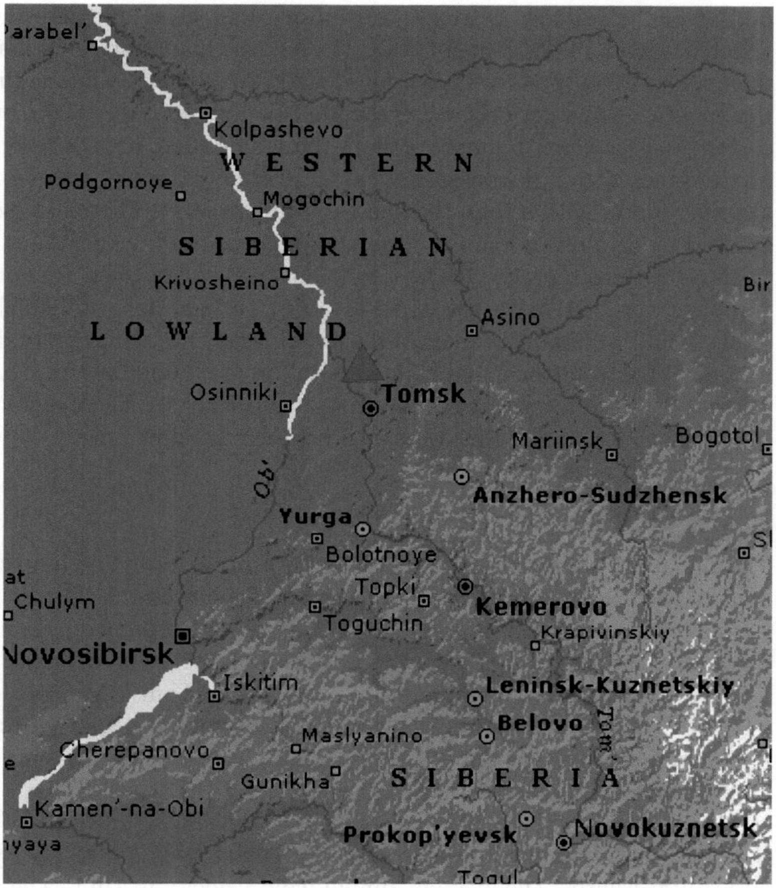

Figure 1.16. Location of the Tomsk SCC (© 2000 Microsoft Corp. All rights reserved).

uranium by means of fluoridation technology. The size of the radioactivity control zone around SCC is 192 km^2.

General geographical characteristics of the area around the SCC

The combine occupies an area on the right bank of the River Tom′ as viewed from the south. The surface of this area is weakly undulating and is covered by mixed forests and discontiguous swamps. The general incline of the region points toward the valley of the River Tom′ and decreases from 160 m in the northeast to 70 m in the southwest.

The elevation of the watershed area to the east of the industrial site is 170 m to 180 m. The western side of the watershed area has a number of deep gullies some 13 m deep. The surface of the gullies is turfed and further covered by mixed forests and bushes. The northern part of the River Tom′ is a flat plain with a small incline toward the Tom′.

The industrial combine is bordered to the south and west by the river. The industrial site occupies territory mainly on the third terrace and partly on the western border of the watershed area and a hillside that gives onto the third terrace. The watershed area of the plateau consists of Late Quaternary clay loams with a total thickness of 20 m to 30 m. The surface loams are covered by fine-grained sands whose thickness at the western border of the plateau does not exceed 2 m, but reaches at the eastern border 6.0 m or more. Below the loams there are Tertiary deposits: white quartz sands with a thickness of 15 m to 20 m underlain by green–brown–blue clays and clay loams with a total thickness more than 40 m. The hillside of the plateau is covered by de-alluvial loams and sandy loams whose thickness does not exceed 7 m. Within the third terrace, Quaternary loams of the watershed area of the plateau are completely eroded, as are the upper layers of Tertiary deposits. Thick alluvial deposits are found within the borders of the present-day river valley up to the bottom of the hillside of the watershed area plateau. Swamps are found in several isolated low-lying parts of the third terrace (especially in its northern part). These regions are turfed and the thickness of turf in the swamps is between 5 m and 7 m.

Climatic conditions

The climate of the area is continental with warm summers and long cold winters. The area is open to the north, west, and south. Penetrating Arctic air masses result in sudden changes in air temperature. Air masses from the west result in higher temperatures and thaws in the winter, but leads to storms and drizzle in the summer.

The average annual temperature of air at the Tomsk Meteorological Station is +0.6°C. The highest monthly average temperature (+18.1°C) occurs in July; the lowest monthly average temperature (−19.5°C) occurs in January. The most extreme temperatures recorded in Tomsk are +36°C and −55°C, which is a 91°C temperature range.

The prevailing wind directions in the Tomsk area are southerly, southeasterly, and southwesterly. The average annual wind velocity is 4 m/s. The highest monthly average wind velocity is observed in March (6 m/s), the lowest in July with a velocity about 3 m/s. The maximum wind velocity in the area can reach 30 m/s with southwesterly winds. Seasonally, the highest velocity is noted in the spring and autumn, rarely in the summer. In the winter (December–February) the maximum wind velocity does not exceed 20 m/s.

Mean annual precipitation during the 60-year period of observations in Tomsk is 548 mm, with a minimum of 359 mm and maximum of 769 mm. For the warm period of the year (April–October) it is 372 mm (68%), for the cold period (November–March) it is 176 mm (32%). The minimum precipitation (3 mm) falls in February, the maximum in July (172 mm).

Hydrography of the Tom′ River

The Tom′ River is the right tributary of the River Ob′ and originates on the western hillside of Abakan Ridge at a height of 800 m to the east of Tashalyk. The length of the River Tom′ is 839 km. The pear-shaped catchment area is 61,240 km^2; the width of the top part of the catchment area is about 160 km, the middle part is 40 km to

50 km, and the lower part about 80 km. In the lower part the river divides into two arms, the length of the right arm is 10 km and that of the left arm 5 km.

From the industrial site of the SCC the Tom′ flows basically northwest. The combine is located on the right shore of the river. Section lines for industrial water intake are situated about 50 km to 52 km from the river mouth.

The Tom′ river basin passes through four geomorphological zones, one of which—the alluvial plain—is pertinent to the combine site. The Tom′ is part of the alluvial plain from Beloborodovo up to the river mouth, as are the small Malaya Chernaya, Bolshaya Chernaya, Um, Chernaya Kislovka, Poros, and Kamenushka rivers. These rivers have lengths between 30 km and 50 km and catchment areas between 80 km^2 and 100 km^2. This area is part of the west Siberian lowland where elevations do not exceed 200 m a.s.l. The area is a flat plain, with a gradual incline directed toward the Tom′. The southern part of the plain has a broad dune landscape covered by pine forests and small rivers. In the floodplain of the Tom′, the small rivers flow only as small branches. In the floodplain there are numerous lakes and swamps. The banks are low and overgrown with trees and bushes. The River Tom′ in this area can be categorized as a lowland river. Below Beloborodovo the shores are lower and woody vegetation creeps mangrove-like into the riverbed.

The Tom′ becomes a tributary of the Ob′ about 65 km south of Tomsk. In its upper part—before the inflow of tributaries (Us, Mras-Su, and Kondom)—the Tom′ is a fast mountain river coursing down a deep, narrow valley. At 140 km downstream of its source, the width of the river reaches 200 m. Then the valley widens forming a floodplain between 2 km and 3 km wide.

Near Tomsk the Tom′ regains some of its character as a mountain river with a width of 500 m in low water and a mean incline from 0.00025 m to 0.00018 m. South of Tomsk the depth of the river does not fall below 1.25 m. The lower part of the river (from Beloborodovo to the mouth) flows through a plain, which has a wide floodplain, low shores, and small inclines. In this part there are many islands, rifts, and spits. The floodplain is between 5 km and 10 km wide and during spring high waters it is rarely fully inundated. Near the SCC the Bol′shaya Kirgizka and Romashka (both small rivers) flow into the Tom′—the drainage canal for the combine's waste waters was built in its riverbed.

Water level regime of the Tom′ River

The Tom′ River according to its water level and flood regime is related to the group of rivers with Altai-type spring high water distinguished by low-level, high-water broad-crested floods, high water flows in summer, and low water flows in winter.

The level of the Tom′ is characterized by three phases: (1) spring high water, during which maximum levels are achieved; (2) summer/fall low water broken by rainfall-induced high waters; and (3) winter low water with the minimal water levels.

Spring high water. As a result of the overlying snow cover thawing at the end of March/beginning of April, the flood begins. In some years the beginning of the flood takes place as late as the middle/end of April. Spring high water is forceful, with water levels reaching 7.5 m to 9.5 m above winter levels.

After the maximum level has been achieved, high levels remain for 3–4 days and then the flood begins to subside: after a sudden fall, there are often slight increases in water level forming "crests" before dropping once more. The lowering of the water level proceeds in this manner for 40–50 days.

Summer/fall low water. Summer low water begins from the middle/end of July, when it is broken by rainfall-induced high waters which can raise the river level by 0.5 m to 3.0 m. However, the average level decreases gradually throughout this period.

Winter low water. Winter low water comes about with the formation of ice on the river and is characteristic of the lowest levels observed during the year. The average duration of winter low water is 189 days, with the minimum being 159 days and the maximum 243 days.

Water supply and its fluctuations. The catchment area of the River Tom′ near Tomsk is 57,300 km^2, or 58,300 km^2 at the water level measuring post near Beloborodovo. Long-term mean water discharge (for the period of observations up to 1959) is equal to 1,110 m^3/s. Average long-term water supply volume is 35.6 billion m^3/yr.

Soil structure of the SCC observation area

The topsoil of the territory consists of turf-podzolic, gray woodland, alluvial turfen, and swamp low-lying soils. The basic structure (50–55% of the observed area) of topsoil is determined by turf-podzolic soils in combination with light-gray woodland soils on light loams. These soils are widely distributed in the northern, northwestern, central, and eastern parts of the territory.

Light-gray and gray woodland soils on heavy and medium loams are distributed in central, eastern, and southeastern parts of the observed territory and occupy 35% to 40% of this territory. The floodplain of the River Tom′ (5–7% of the area) consists of alluvial-turfen acid soils. Swampy, low-lying turf soils on medium turfs are localized in a small territory in the northern part (3–5% of the area).

The SCC structure

The SCC consists of the following industrial units: a reactor plant where carbon–uranium reactors (ADE-4.5) are used for production of weapons-grade plutonium and for generation of electricity; a plant for isotope separation using centrifuges for the production of enriched uranium hexafluoride; a sublimate plant for production of uranium protoxide–monoxide and uranium hexafluoride; a radiochemical plant for reprocessing irradiated blocks to generate and refine uranium and plutonium salts; a metallurgical–chemical plant for generation of nuclear materials; and an experimental physical department (former reactor plant) with three uranium graphite industrial reactors, now decommissioned and the equipment removed.

The first reactor was put into operation in September 1958, the second in December 1959, and the next three reactors became operational at 1-year intervals (according to other data, 2-years intervals).

The change in direction of the defense industry in the 1990s led to the beginning of a shutdown of reactor operations at the SCC. By 1994, three of the five reactors were out of use. Two reactors remained in operation generating heat and electrical power for the nuclear complex, for Seversk, and for the nearest oil and chemical complexes. Direct flow reactors are basically responsible for the content of radionuclides in the combine's waste waters. Irradiated reactor fuel is recycled at the Radiochemical Plant which was put into operation in 1956.

Until recently, plutonium was transported to the Metallurgical–Chemical Plant where it was recycled in metal for further weapon production. Since October 1994, plutonium has been recycled as plutonium dioxide and stored.

The Metallurgical–Chemical Plant was also designed for the production of HEU for nuclear weapons. In 1994 the plant began to recycle HEU in its oxide form, which is then recycled into lowly enriched uranium (LEU) for use as the fuel for uranium reactors in accordance with the U.S.A.–Russia agreement on HEU. In 1996 production began on the fluoridation and dilution of HEU. HEU oxide is sent to electrochemical plants in Novouralsk and Zelenogorsk where it is reprocessed into LEU for civil reactors.

The Tomsk plant for enrichment was built and put into operation in 1953. Recent estimates are that the plant represents about 14% of Russia's enrichment capacity. It is also involved in recycling HEU under the U.S.A.–Russia agreement. Furthermore, the SCC is one of the largest Russian producers of uranium hexafluoride (UF6), the raw material for enrichment plants.

Waste management

Production at the SCC involves a great volume of radioactive effluents: liquid and solid radioactive waste, and gaseous waste including inert radioactive gases (in particular, uranium and plutonium aerosols). Hence, it is necessary to capture and remove it to burial sites. According to SCC data, gaseous aerosol waste products are purified before discarding (depending on their aggregation state and a type) and after purification at an efficiency equal to 80% to 99.9% the gas–air mixture is vented from the plant chimney into the atmosphere.

As a byproduct of SCC activity, various technological and non-technological waste solutions including radioactive elements are formed. Some waste solutions are directed toward refining structures, and after multi-step purification (settling, mechanical refinement, ion exchange resin purification), such that concentrations meet permissible sanitary levels, they are discarded into the River Tom′. Some of the waste products are directed toward open storages. As a result of long-term storage, silty bottom sediments form with high uranium and plutonium content, which clearly is the source of radioactive contamination of the environment. Industrial waste from the combine enters the River Tom′ near Chernilschikovo via the waste channel through the River Romashka.

Large-scale studies on the design of the optimum technologies for the destruction of open radioactive waste storage are carried out at the SCC. The most radioactive solutions are injected into an underground accumulative layer. The cumulative radioactivity of radioactive waste stored in underground layers is evaluated to be about 4×10^8 Ci, and about 1.25×10^8 Ci in open stores. The most dangerous—from the point of view of radioactive contamination of the environment—are open stores of radioactive waste (basins, pulp storage, water storage basins) and underground storage, if active exchanges with surface water takes place and in case of natural disasters.

Moreover, the municipal water intake of Tomsk is located 12 km from underground radioactive waste stores. This is a cause of concern because of the danger of radionuclide penetration into potable water.

Solid radioactive waste formed during production activity at the combine is buried in special concrete stores located in the grounds of the various plants or in the centralized store built to handle the combine's waste products.

According to information from the combine, the basic substances entering the atmosphere along with discarded air are alpha/beta-emitting nuclide aerosols, short-lived radioactive noble gases, iodine-131, strontium-90, nitrogen oxides, and fluoric compounds.

In the waste water discarded from the combine into the River Tom′, only radionuclides of induced radioactivity (sodium-24, manganese-56, phosphorus-32) are registered. The cooled water is directed to the water storage basin and after a calculated decay time, it is released into the Tom′. During the decay time in the water storage basin, all short-lived radionuclides of induced radioactivity decay down to a level below sanitary norms, except for sodium-24 and phosphorus-32, whose content in the release waters exceeds sanitary norms.

Data from the SCC would appear to suggest that the situation within its territory and within the sanitary protective zone is safe (i.e., there is practically no contamination of SCC territory by long-lived radionuclides and the only concern may be short-lived radionuclides). However, given the magnitude of production at the Mayak PA, Chelyabinsk-7 (Section 1.2.1) and the levels of contamination in its territory, it is difficult to believe that such a benign situation exists at the SCC, Tomsk-7. Indeed, some publications mention the presence of significant contamination levels at different sites of SCC territory. The radioactivity level in soils near the water channel going from Tomsk-7 is equal to 300 μR per hour. Analysis of the soil demonstrates abnormal quantities of cobalt-58, chromium-51, and zinc-65, indicating significant leakage due to the corrosion and depreciation of the reactor body as a result of its 30–40-year operation period. Testing has also shown the abnormal content of manganese-54, iron-59, and scandium-46 in the soil. Nonetheless, the most important thing is the absence of plutonium leakage.

Accidents

There was an accident at the SCC on April 6, 1993. The release contaminated a territory of 100 km^2. The primary radionuclide in the contaminated area is ruthenium-106 reaching a level of 8 Ci/km^2 in territory in close vicinity to the combine in 1993. Furthermore, after the accident short-lived niobium-95 (up to 3.5 Ci/km^2) and zirconium-95 (up to 1.5 Ci/km^2) were observed in the contaminated area. During procedures to correct the radioactive situation (e.g., gamma-shooting of the territory around the SCC), areas contaminated by cesium-137 at a density between 1 Ci/km^2 and 2 Ci/km^2 were also observed. The contamination areas take the form of a wide fan emanating from the SCC: 10 km to the north, 2 km to 3 km to the west and east, and also along the banks of the Tom′. Studies of the cesium contamination show that it is not connected with the accident of 1993. Apparently, it is simply the result of long-term activity at the combine.

1.2.3 Mining Chemical Combine. Krasnoyarsk-26

The description below was prepared mainly on the basis of several well-known publications (Russian Federal Program, 1995; AMAP, 1997, 1998; Mining Chemical Combine, 2000).

General information

Krasnoyarsk Mining Chemical Combine (MCC) was set up to generate weapons-grade plutonium. Building of the MCC and town for its personnel began in 1952. Main production is situated several kilometers from the town (Figure 1.17) on the right shore of the Yenisei, in an excavation inside a rocky massif made of Archean Proterozoic crystal soils on the western decline of the southern side of the Yenisei Range. The first direct flow uranium graphite nuclear power reactor (AD) was put into operation in 1958. The second direct flow nuclear power reactor (ADE-1) was put into operation in 1961, and the third (ADE-2) in 1964. These reactors comprise the underground reactor plant (plant A). Simultaneously with the building of the reactors a radiochemical plant (plant B) was constructed, the second part of the main production line; this plant started in 1964.

During the design and building of the MCC, measures for radioecological safety were accepted. The waste disposal plant for purification of low-level liquid and technological gaseous releases and for waste recovery and radioactive waste disposal had become operational by 1958. The Severnyi polygon, an underground disposal place for low-level and middle-level liquid wastes, was put into operation in 1967. Since the beginning of the main production and up to the middle of 1992, the MCC generated weapons-grade plutonium and delivered it (together with spent purified uranium) to other enterprises. Nuclear warheads and fuel assemblies for the atomic power station were not produced at the combine.

In 1992 two industrial direct-flow reactors (AD and AD-1) were put out of operation because of a change of direction (conversion) for the main production at the MCC. Within the conversion framework, the MCC complex began to

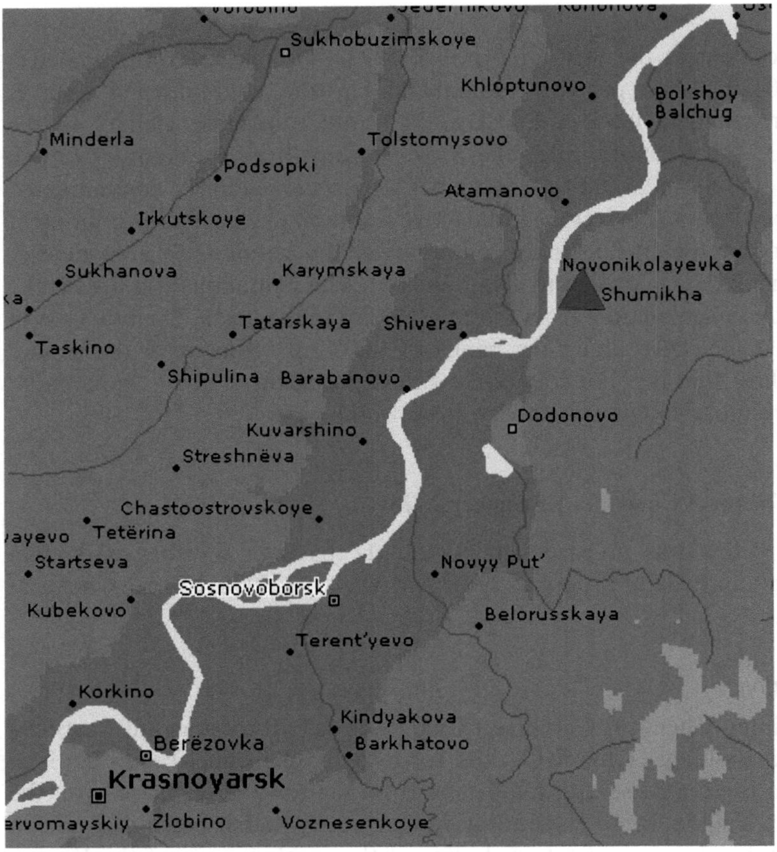

Figure 1.17. Location of the Krasnoyarsk MCC (© 2000 Microsoft Corp. All rights reserved).

manufacture other kinds of materials (polycrystalline silicon, optical quartz glass, quartz crucibles, germanium tetrachloride, and especially pure germanium). The radiochemical orientation of MCC production and the proficiency of its personnel enabled the MCC to plan the future creation and start-up of a nuclear enterprise (plant RT-2) for reprocessing the spent fuel of atomic power plants. However, as a result of financial difficulties, the building of RT-2 was stopped.

The long-term operational activity of MCC enterprises was accompanied by releases of controlled quantities of radioactive and harmful chemical agents into the atmosphere and to the aquatic environment. A radioactivity-monitoring system for the area around the MCC has been in place since the MCC's inception. Observational results concerning radioactivity in the environment are registered in journals and summarized in reports.

The geographical position

The territory occupied by the facility is about $360\,km^2$. It covers a 15 km wide band along the right shore of the Yenisei River. The sizes and boundary of the control area were designed keeping prevailing westerly and southwesterly winds in mind. The control area is elongated from the southwest to northeast with a length of 17 km and its boundary on the opposite side of Atamanovo (a village) runs along the Yenisei.

The climate of the region is continental with long cold winters and short, dry summers, late springs, and rainy falls. The mean air temperature of the coldest month (January) is about $-18°C$ and mean air temperature of the warmest month (July) is about $+19°C$. The mean air humidity of the coldest month is 83%, and 75% during the warmest. Mean wind velocity is 2.4 m/s and has a prevailing southwesterly direction. Mean annual precipitation is 490 mm. Snow cover is established in October, while the spring thaw/snowmelt season occurs in April–May. The prevailing wind direction is southwesterly.

The Yenisei River runs through the area occupied by the combine, and as its water flow is regulated it does not freeze in winter. Mean water temperature is 1.7°C, stream flow speed is 1.7 m/s, average depth is 2 m, and average width is about 1,000 m. According to long-term observations the average flow rate is $\sim$2,800 m^3/s.

Industrial facilities

The main industrial facilities whose operational activities may result in radioactive contamination of the environment are

1. *Carbon–uranium reactors* (AD and ADE-1). These reactors used metallic natural uranium, which were loaded as blocks covered by aluminum. For reactor cooling, water from the Yenisei River was used but only after being refined to remove mechanical impurities. After cooling of the reactor active zone the water was discarded back to the Yenisei. Technological gases and ventilating air were specially purified and then released into the atmosphere. The reactors (AD and ADE-1) were put out of operation in 1992.

   Reactor ADE-2 has a closed circuit with chemically refined water. In addition to generating plutonium it works as the power reactor. Since 1965 ADE-2 has provided thermal energy for the enterprise and the population of Krasnoyarsk-26. The underground nuclear heat-and-power plant operates on this basis.
2. *Underground radiochemical plant* (plant B). This plant recycles the blocks of irradiated uranium that ultimately generate the uranium and weapons-grade plutonium. High-level waste formed during radiochemical reprocessing is stored in special stainless steel reservoirs located in mine workings inside the rocky massif; in the future this waste is to be recycled into solid radioactive waste. Liquid mid-level and low-level wastes are delivered for underground injection after preliminary special treatment. Air–gas and gas–aerosol releases as a result of careful multi-step purification are removed via the high chimney (150 m) into the atmosphere.
3. *Spent nuclear fuel storage department*. This is located in a building at the RT-2 Plant. This department was put into operation in 1985. Here the assemblies of

VVER-1000 reactors from Russia and the Ukraine are delivered. The radioactive waste of spent fuel from the nuclear power plants of other countries is also received and reprocessed. Spent fuel assemblies are delivered by railway in special transport systems (TK-10 and TK-13). The reloading of spent nuclear fuel and its further storage are carried out in the cooling pond with the water purified of chemical and mechanical impurities.

4. *Radioactive waste–reprocessing department.* This department opened in 1958 and is now part of the RT-2 Plant. It was built to refine liquid low-level and technological gaseous releases as a result of the main production of the combine. Here the re-use of all kinds of wastes and their preparation for disposal is made. Waste water is discarded to the Yenisei River after being refined to meet sanitary norms. Mid-level and low-level liquid wastes are delivered to the Severnyi polygon for underground injection.
5. *The polygon Severnyi.* This area is located on the right bank of the River Yenisei (within the watershed area of the Yenisei and Bolshoi Tel) 55 km to 60 km from Krasnoyarsk downstream of Balchug (a village) and 10 km from the center of underground facilities to the northeast. The total area is approximately 45 km^2. Liquid radioactive waste is delivered to the polygon via the main pipeline that has a length of 15 km and is then injected under the ground in water-bearing sandy clay layers at depths of 130 m to 220 m and 400 m to 500 m. Simultaneously, underground water is pumped down and its purity controlled. Medium-level waste is injected at a depth of 400 m to 500 m (horizon 1); low-level waste is injected in the above-lying horizon 2, isolated from horizon 1 and from the upper layer by watertight soil at a depth of 180 m to 220 m. The store is used as the disposal place for liquid radioactive waste of nuclear reactors and radiochemical production and already contains millions of cubic meters of radioactive waste with a total radioactivity of approximately 108 Ci. In the future, in the event the RT-2 Plant is relocated here, a disposal place for tritium is planned.

   The now-suspended surface water basins were earlier used for liquid low-level waste releases. The problem was that the shores of these basins used to get dry and create the conditions for wind transfer and dispersion over significant territories of radionuclide-contaminated bottom sediment. In addition, liquid from the basins was probably trickling out drop by drop. Operations on liquidation of these basins are being carried out.

Radioactive releases and discharges

The sources of the environmental contamination in the MCC region are gas–aerosol releases into the atmosphere from chimneys and uncontrolled main devices, liquid discharges consisting of cooling water, technologically contaminated and waste water, and wind transfer of radioactive solutions from storage.

Radioactive releases. Radioactive matter at the atmospheric boundary layer is caused by the release of technological gases and ventilating air from seven air

Table 1.9. Sources of radionuclide release at the MCC.

Sources	*Description*
Source 1	Technological release from the Radiochemical Plant after purification at the plant's gas purification station
Source 2	Technological release from the Reactor Plant after purification at the gas purification station of the Isotope Separation Plant near the release source
Source 3	Ventilating releases from the second-zone rooms of the reactor and the Radiochemical Plant
Sources 6, 7	Ventilating releases from the Central Laboratory which is charged with analytical control and carries out experimental work for the Radiochemical Plant; and ventilating releases from the second-zone rooms of the reactor and the Radiochemical Plant
Source 10	Ventilating air from rooms inside the Nuclear Power Plant
Source 11	Steam and gas released from the waste channel of cooling water of the bores of the reactor plant handling protection system

chimneys. Entry into the atmosphere from organized radionuclide releases at the enterprise is through the following sources (Table 1.9).

Radioactive discharges. The MCC's waste water is discharged to the Yenisei River at five points (below). Waste water discharge containing radionuclides to the Yenisei is made by means of releases 2–5.

Discharge 1. Cooling water for the turbines and coolers of the combined heat-and-power plants of Plant A and other virtually non-radioactive water.
Discharge 2. It is here that the main quantity of radionuclides are released into the Yenisei River.
Discharge 3. Utility water for fuel assembly storage after biological purification.
Discharge 4. Water for special laundry, decontamination water for rooms, and transport of RT-2 water purified up to sanitary norms.
Discharge 5. Utility water for the RT-2 Plant.

1.3 NON-RADIOACTIVE POLLUTION

1.3.1 Main sources of marine pollution in the Russian Arctic

Except for the Barents and White Seas, economic activity in the areas occupied by Russia's Arctic seas is so small that the seas have a generally low baseline level of anthropogenic pollution. The main sources of pollutants are freshwater runoff from the vast watersheds and trans-boundary transport of pollutants by sea currents and winds. Conversely, in the area of the Barents and White Seas, there is intensive economic activity mainly connected with fishery, merchant and naval shipping, and oil–gas exploration operations. There are about 200 fishing vessels from Russian fishery companies that ply their trade in the Barents Sea. Oil and gas fields on the shelf can serve as local natural sources of oil hydrocarbons and PAHs for the basins of the Barents and Kara Seas.

The role of continental runoff as a source of pollutants has been widely studied, whereas the contribution of other sources is very poorly covered.

The polluting effect of discharge from coastal sources is clearly observed in the coastal zone and is commensurate with the level of industrial–social development of the economical centers located along the shore. Bays and gulfs where there are ports become local accumulators of anthropogenic pollution.

1.3.1.1 Barents Sea

The Barents Sea has an open exchange of water with the Norwegian and Greenland Seas coupled with a system of warm Atlantic currents that determines the hydro-physical and ecological situation in the southern Barents Sea. The long-distance, trans-boundary transport of pollutants by oceanic currents and the proximity of North European industrial centers means that pollution may be more significant for the Barents Sea ecosystem than the other seas. Micro-elements, heavy metals, and organochlorine pesticides are carried to the basin with North Atlantic water. The role of warm Atlantic currents in the transfer of arsenic is especially notable (*Biotesting* ..., 2003).

The atmospheric transfer of polluting aerosols from Northern Europe and the Kola Peninsula, especially in winter, is also significant (Golubeva, 2002; Ilyin *et al.*, 2004; Matishov *et al.*, 2004a, b). Metal aerosols migrating in the atmosphere above the Barents Sea sub-region are supplemented to a large extent by heavy metals from regional sources, which puts this group of elements (Ni, Cu, Zn, Cd, Pb, Cr, and Hg) among the primary pollutants. In the Murmansk region, among the most significant sources of emissions of pollutants to the atmosphere—including metals and benzo(*a*)pyrene—are the metallurgy combines Severonikel (Monchegorsk) and Pechenganikel (Zapolyarny and Nikel), thermoelectric power stations (Murmansk, Kola, Apatity, Kirovsk), and the enterprises of OJSC Apatit (Apatity, Kirovsk). Dust emissions from metallurgy enterprises are enriched by a number of heavy metals, predominantly Ni, Zn, and Cu. For example, the emissions of these metals from the Severonikel combine comprise 3,720 t, 3,060 t, and 2,995 t per year, respectively, in addition to 10.4 t per year of Cd.

During the year the concentrations of Pb, Cd, Ni, and Cu in atmospheric precipitation are constantly or periodically higher than in seawater. This shows that atmospheric precipitation is a source of water pollution in the Barents Sea (Matishov *et al.*, 2004a, b).

The mass of heavy metals annually reaching the Barents Sea by means of precipitation is much lower than the mass of heavy metals already in the seawater.

In the geochemically active marine environment, especially in the open regions of the Barents Sea, heavy metals brought by precipitation do not lead to significant changes in seawater composition. However, short-term local effects from the atmospheric fallout of heavy metals to the water are possible in the polar front zone in a narrow strip of melting sea ice that has accumulated atmospheric fallout during the wintertime.

The role played by other local pollution sources—merchant shipping, fishing vessels, and natural filtration of hydrocarbons from sediments—remains poorly investigated (*Chemical Processes* ..., 1997).

In coastal regions the most significant pollution sources in terms of ecology are the advection of pollutants (e.g., metals, oil hydrocarbons, organochlorine compounds or OCs, surfactants) with water of the Norwegian Coastal Current, atmospheric fallout (e.g., sulfate ions, heavy metals, OCs), and discharge of pollutants across Kola Bay with effluents from the Murmansk industrial complex. About 84 million m^3 of effluents per year are discharged to the Barents Sea basin. About 50 million m^3 to 70 million m^3 per year are discharged into Kola Bay, of which 80% are discharged without treatment. Between 2004 and 2006, 20 t to 28 t of oil products and 25 t to 38 t of metals and other pollutants were discharged to the bay (*Quality of Sea Water* ..., 2005; *Report on Environmental Protection* ..., 2005, 2006, 2007).

The local sources of these discharges include the port, industrial, and communal complexes of Kola Bay (oil hydrocarbons, heavy metals, polychlorinated biphenyls or PCBs, surfactants, etc.); merchant shipping, fishing vessels, and Navy ships (oil hydrocarbons, economic and domestic discharges); and offshore oil–gas production (potential danger of accidental spillages and leakages of oil hydrocarbons and drilling muds).

In Varangerfjord, pollution can be traced to continental runoff (Pechenga River) and discharges from the coastal infrastructure of Pechenga Bay and the port of Liinahamari.

About 3.5 million m^3 of sewage water containing about 0.6 t of oil products is discharged into Motovsky Bay from infrastructure facilities (*Report on Environmental Protection* ..., 2005, 2006, 2007; *Environmental Status* ...2003).

The Barents Sea watershed has large quantities of iron, copper–nickel, and rare earth ores, as a result of which the natural background concentrations of copper, manganese, nickel, iron, and aluminum in freshwater runoff are elevated (Dauwalter, 1997, 1998). Acidification of atmospheric precipitation by discharges from industrial enterprises leads additionally to intensive leaching of metals from ore bodies and increases their concentration in surface waters, especially in the basins of the Pechenga, Kola, and Patsoyoki Rivers. In their rivermouth areas the concentrations of Fe, Mn, Zn, and Cu are known to exceed between two and six times the maximum permissible concentration (MPC) for fisheries in 60% of the observations (*Review of Pollution*, 2001; *State of the Environment* ..., 2001; *Report on Environmental Protection* ..., 2007). In the eastern regions of the Barents Sea the water quality is characterized by a high level of zinc (up to 3.5 MPC), copper (1–7 MPC), and iron (2–8 MPC). River runoff leads to local areas of relatively elevated concentrations of metals in the water and sediment in the rivermouth areas of the coast, all varying with runoff volume. The eastern regions of the Kola coast are sparsely populated and there is no industrial infrastructure.

In the Pechora Sea the sources of anthropogenic pollution are the Pechora River and the port and economic complex of Varandey. The Pechora River annually discharges to the sea about 6,000 t of oil hydrocarbons and 0.4 t of detergents. The river water is polluted with compounds of zinc, copper, iron, and phenols in which metabolites of DDT have been detected (maximum concentration of 0.009 μg/L) (*Quality of Sea Water ...*, 2004). Pollution of the open sea area also occurs to some degree as a result of water exchange with bays and gulfs polluted by continental runoff.

1.3.1.2 White Sea

Pollutants in the White Sea are brought primarily with river runoff, which transports pollutants from the pulp and paper industry, energy enterprises, the communal and domestic economy, and the river fleet. Through rivermouth areas about 250.5 million m^3 of sewage water is discharged to the sea. Industrial effluents contribute up to 6.18 t of oil and oil products and up to 0.354 t of phenols. River runoff (Severnaya Dvina and Onega Rivers) brings about 3,840 t of oil and oil products and about 167 t of phenols. More than 90% of the discharges come through Dvina Bay.

Nine enterprises divert effluents to Kandalaksha Bay, the largest being the OJSC Kandalakshsky Aluminum Enterprise SAUL, the CJSC Belomorskaya Oil Base, the State Unitary Enterprise Kandalakshavodokanal, and the State Unitary Enterprise Apatityvodokanal. About 7.7 million m^3 of untreated water is discharged to the bay. In general, up to 125 t of suspended matter, 4.5 t to 6.6 t of oil, about 2 t of dissolved iron as well as phenols and detergents are annually discharged from infrastructure facilities to the water of the bay (*Quality of Sea Water ...*, 2005; *Report on Environmental Protection ...*, 2005, 2007). The discharge of heavy metals with surface and communal effluents forms stable elevated concentrations of iron and molybdenum (up to 2 MPC) and episodically of copper in the bay.

The concentration of aerosols in the atmosphere over the White Sea is practically the same as over the Barents Sea: 0.34 $\mu g\, m^3$ (Shevchenko, 2006). However, unlike the Barents Sea, most of the sea ice cover that has accumulated aerosol fallouts is exported in spring from the basin to the Barents Sea, so the role played by the atmospheric transfer of pollutants in the White Sea decreases.

1.3.1.3 Kara Sea

Pollutants in the Kara Sea are brought with the abundant runoff of the Ob′ and Yenisei Rivers, which have extensive economically developed watersheds. Surface runoff water is taken far out to sea in summer due to a system of currents. Data limitations make quantitative estimation of the role of river runoff impossible, although there were observations between 1995 and 2002. During this period the Ob′, Nadym, Taz, and Yenisei Rivers transported to the Kara Sea, respectively, from

535,000 t to 470,000 t of oil and oil products; 83.5 t to 3 t of HCH (α-isomers and γ-isomers); about 9.5 t of DDT and DDE; and salts of heavy metals and phenols (*Yearbook of Surface Water Quality* ..., 1997; *Quality of Surface Water* ..., 2004).

Water exchange with the Barents Sea is a source of pollutants. The most significant is a deep flow of Atlantic water from the Barents Sea through St. Anna Strait along the Novozemelsky trough. In the central part of the sea, a significant component of oil pollution is the invasion of hydrocarbons through the sedimentary mantle (Ilyin, 2005). There are no modern quantitative data on the discharge of pollutants to the Kara Sea from Barents Sea water.

Transported as aerosols, heavy metals and organic pollutants are brought to the sea. The levels of aerosols in the near-water air layer over the Kara Sea (0.18 $\mu g/m^3$) are half as large as those over the seas of the European part of the Arctic (i.e., the Barents and White Seas). At the same time, Pb, Ni, Cu, Cr, Hg, Cd, Co, Mn, and the organochlorine pesticides HCH and DDT are present in the spectrum of pollutants. Some pollutants are contained in aerosols in greater concentrations than in the actual water of the Pechora Sea; in particular, pesticides, chromium, cadmium, cobalt, and manganese. The influence of local eolian transfer is primarily reflected in pollution of the ice cover in coastal water areas (Matishov *et al.*, 1998).

1.3.1.4 Laptev Sea

The sources of pollution of Laptev Sea waters are little studied. The main source of pollutants is runoff from the Lena River, and other significant sources of pollutants are the Anabar, Olenek, and Yana Rivers. A relatively large industrial source of pollutants in the basin is Tiksi, a port at the western periphery of the sea. The most typical pollutants are oil products, phenols, and compounds of iron, copper, and lead. The highest recurrence of pollution (excesses of MPC) can be blamed on oil products and phenols. The mean annual concentration of the majority of pollutants in rivers is between 1 MPC and no more than 3 MPC. The transport of oil products to the coastal zone along different rivers varies, but may well total 390,000 tons. More than 98% of this volume of oil products passes along the Lena River.

No discharge of OCs into river water has been recorded (*Yearbook of Surface Water Quality* ..., 1997; *Quality of Surface Water* ..., 2004). Pollution of the Laptev Sea environment by these compounds is connected with trans-boundary transfer by sea currents and airflows. For example, the role of Atlantic water entering from the Barents Sea and spreading in the intermediate water layer is clearly manifested in the spread of PAHs in the Laptev Sea (Ilyin *et al.*, 1998a, b). Nonetheless, the discharge of OCs and PCBs into surface water does appear to occur seasonally at the time of snowmelt and as a result of aerosol fallout.

In general, aerosol fallout to the Laptev Sea area is smaller than that to the Barents Sea. The concentration of aerosol particles in the atmospheric boundary layer is practically the same as in the Kara Sea (Shevchenko, 2006).

1.3.2 Distribution of pollution in the Russian Arctic Seas and coastal areas

Comprehensive assessment of the relative roles played by local sources and transboundary transfer of pollutants for all marine marginal regions of the Russian Arctic is clearly beyond the current level of knowledge. Nevertheless, one can note a west-to-east decrease in the volumes of pollutants brought by ocean currents. Local sources of pollution centered along sea coasts and in the watershed basins mainly have a local impact within the areas of surface water spread in the coastal zone. Coastal economic activities are sources of local areas of pollution of coastal water, predominantly by oil products (*Yearbook of Sea Water Quality* ..., 1997). The role of river runoff as a source of pollutants is more expressed in seas in the eastern sector of the Russian Arctic due to the large size of watershed territories and annual runoff volume compared with river runoff in the western sector of the Arctic coast (Kola Peninsula). However, the lack of development of industrial and economic complexes in the northern territories equates in general to a low level of anthropogenic pollution in the Russian sector of the Arctic. As far as the marginal seas are concerned, the dominating role of atmospheric fallout and trans-ocean currents as a source of persistent OCs in the open sea, especially DDT and HCH, is evident, which is clearly observed, for example, by the ingress of these compounds to the Barents Sea.

The coastal areas of the Russian Arctic Seas have long been endangered by a range of dangerous pollutants such as stable OCs (pesticides like DDT and HCH and polychlorinated biphenyls), oil products, PAHs, and heavy metals. The spread of these pollutants and their levels of accumulation have been investigated to a greater extent than the spread of other pollutants. However, the spatiotemporal patterns of pollution of most Russian Arctic Seas have been sparsely studied up to the present.

In Arctic regions where pollution sources are the consequence of local or remote economic activity, river and atmospheric transfer become the major mechanisms contributing to the buildup of organic and non-organic pollutants. The only exceptions are the Barents and Kara Seas where the role of sea currents in transporting pollutants from outside is more significant.

The volume of river runoff and the extent of river water pollution in Arctic Seas differ significantly (Figure 1.18, see color section). Annual discharges of untreated or poorly treated water into the basins of rivers of the Barents–White Sea region are assessed as 800 million m^3, whereas in the Ob′ and Yenisei basins the discharges are an order of magnitude greater (3,300 and 2,000 million m^3). The most common pollutants are oil products, phenols, heavy metals, and OC pesticides (*Quality of Surface Water* ..., 2004).

The transport of anthropogenic aerosols into the Arctic atmosphere is most intensive in winter and spring, when winds from lower latitudes predominate. The background pollution of air masses forms in the industrial regions of Europe and southern Siberia. At the same time, local sources of atmospheric pollution are also of great importance, with the Norilsk and Kola mining complexes being major players (Table 1.10) (Matishov *et al.*, 1998b; Shevchenko, 2006). By means of smoke emissions, between 1% and 10% of the annual production of nickel, copper, and cobalt are discharged to the atmosphere (Doncheva, 1975). The introduction of environ-

Table 1.10. Fluxes of the water-insoluble fraction of aerosols (larger than 1 μm) to the surface of Arctic Seas as a result of dry deposition (adopted from Shevchenko, 2006).

Localization	*Concentration of aerosols* (μg/m^3)	*Water area* (thousand km^2)	*Flux to the sea surface* (thousand t/yr)
Barents Sea	0.23	1,424	224
Northern part	0.34	356	54
Western boundary	0.17	356	64
Eastern part	0.29	356	65
Southern part	0.19	356	40
White Sea	0.37	90	2.1
Kara Sea	0.18	883	108
Laptev Sea	0.15	662	96

mental protection technologies at enterprises in the Kola mining–metallurgic complex between 1995 and 1999 made it possible to reduce the emissions of metals in the region almost twofold relative to 1990. This has partly diminished the environmental load in the region.

Knowledge of marginal water areas is greater in some areas than others. Economically developed areas, primarily in the Barents Sea, have been investigated to a greater extent.

1.3.2.1 Barents Sea

Taking into account the proximity of European industrial centers, water exchanges with the Nordic and White Seas, and the prevailing wind directions in the region, it should be expected that a larger number of various pollutants are discharged to the Barents Sea than to other seas of the Russian Arctic. The main agents of the spread of pollutants in the sea include a system of warm Atlantic currents and coastal currents, which transport continental runoff. The main areas of concentration of anthropogenic pollutants are parts of the frontal zones in the sea and depressed seabed relief forms, bays, and gulfs used by business.

Aquatic environment

Data over several years indicate that total oil hydrocarbons (TOH) in water vary in different parts of the water area between 0 mg/L and 0.11 mg/L. The mean annual concentration of oil products in water comprises about 0.01 mg/L (*Chemical Processes* ..., 1997; Ilyin *et al.*, 2004; *Quality of Sea Water* ..., 2005).

Non-uniformity in the concentration of oil products in open sea regions is determined to a greater extent by hydrological processes such as currents and hydrological fronts. In the coastal zone the non-uniformity of pollution is determined by the influence of water polluted by industrial and domestic effluents on the coastal

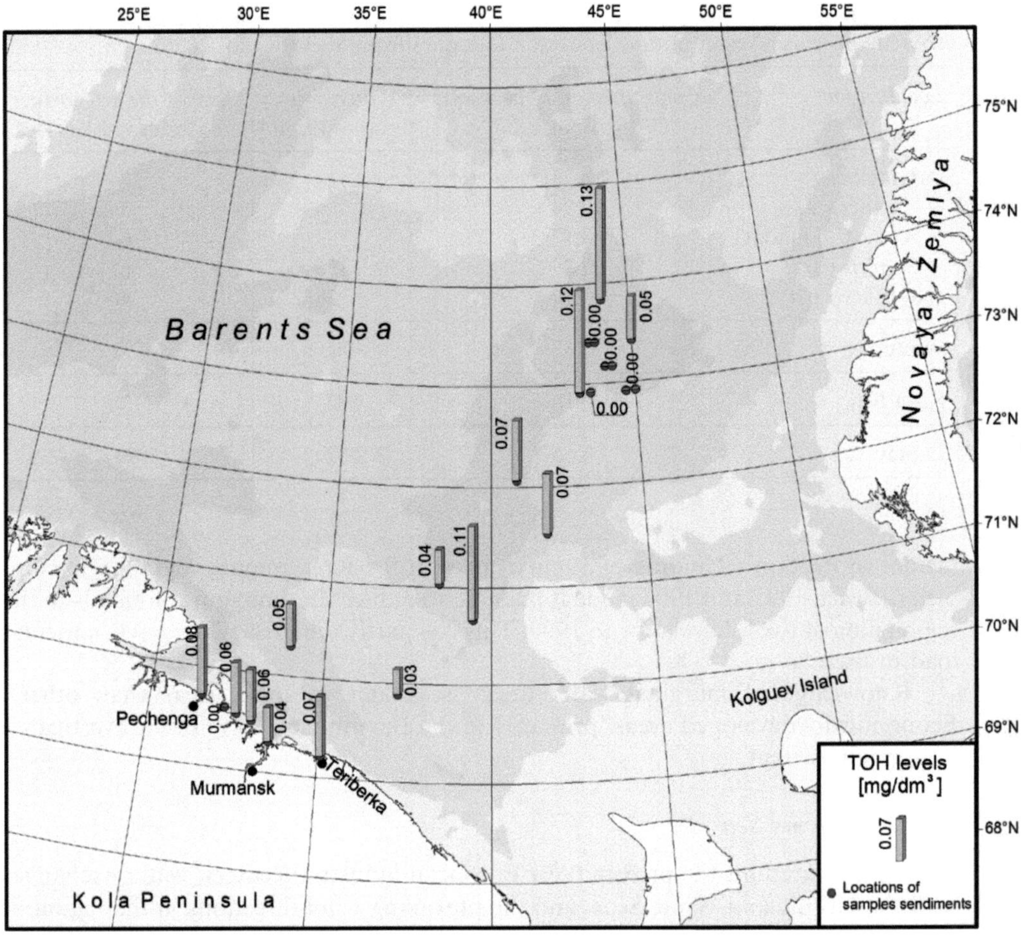

Figure 1.19. Distribution of total oil hydrocarbons in water along a transect across the southern part of the Barents Sea, 2003.

current. Increased concentrations of oil products exceeding the MPC (0.05 mg/L) are generally observed in coastal regions and sometimes in the zones of warm fronts in central sea regions (Figure 1.19).

In the pycnocline layer and in the near-bottom water layer, oil products were either not detected or present in insignificant quantities (Table 1.11).

In areas with economically developed gulfs and inlets, such as Kola and Motovsky Bays and Teriberskaya Gulf, the total concentrations of oil products vary from 0.02 mg/L to 0.11 mg/L. However, in individual cases much higher concentrations (more than 5 MPC) were observed.

The Pechora Sea is an extensive shallow-water region at the southeastern periphery of the Barents Sea. Several oil fields have been discovered here (e.g.,

Table 1.11. Concentration and characteristics of the composition of oil products in water from the surface layer of the pipeline area, 2001–2004.

Indicators	*Localization region*		
	Central trough	*Murmansk shallow, Rybachiya banka*	*Coastal zone*
$\sum$ TOH (mg/L)	0.00–0.11	0.00–0.12	0.02–0.11
$\sum$ paraffins (μg/L)	1.1–9.0	1.2–90.0	2.2–50.0
Pristane–phytane	0.8–13.2	1.4–13.2	9.0–14.0
CPI coefficient	0.3–3.9	0.6–1.1	0.8–0.2
Dominating paraffins	C_{14}–C_{15}; C_{22}–C_{25}	C_{10}–C_{12}; C_{14}; C_{15}	C_{10}–C_{12}; C_{14}–C_{17}

Medynskoye, Prirazlomnoye, and Varandey Sea). Marine trans-shipment of oil produced at onshore fields (Varandey and Peschanoozerskoye) is carried out year-round (Figures 1.20 and 1.21). In the water of the Pechora Sea the levels of TOH vary between 0.0 mg/L and 0.09 mg/L, but average out at 0.02 mg/L (*Chemical Processes* ..., 1997; Ilyin *et al.*, 2004; *Quality of Sea Water* ..., 2005).

Figure 1.20. Peschanoozersky terminal, Kolguev Island (© B. Frantsen) (Frantsen and Bambulyak, 2003).

Figure 1.21. The tanker *Saratov* being loaded at Varandey terminal (Ban'ko, 2002).

TOH is a complex conglomerate of oil hydrocarbons mainly comprised by paraffins. The spectrum of natural paraffins in the water of the Barents Sea is represented by compounds from C_{10} to C_{30}. Their total concentration ranges from 1.1 μg/L to 20.0 μg/L. In some areas, concentrations up to 90 μg/L are recorded (Plotitsyna *et al.*, 2002; Ilyin *et al.*, 2004). In the structural series of paraffins, the short-chain C_{12}–C_{22} alkanes prevail. Hydrocarbons of plant and bacterial origin (C_{20}–C_{25}) comprise between 30% and 33%. Unlike the open sea area, there is an increased ratio between the isoprenoids pristane and phytane, and there is a low-CPI (carbon preference index) value in the structure of alkanes in coastal waters, especially in Motovsky and Kola Bays, which points to relatively elevated background values for hydrocarbons of oil origin. The CPI reflects a ratio between the sum of alkanes with an odd number of carbon atoms and the sum of alkanes with an even number of atoms. The mixture of biogenic paraffins is characterized by an increased index value ($CPI > 1$).

Polyaromatic hydrocarbons are characteristic of both oil and technogenic environmental pollution. Aromatic compounds are contained in crude oil, in products from the burning of organic fuel, and in products of diagenetic transformation of organic matter in soil and sediment. The group of polycyclic aromatic compounds (PAHs) includes not only multi-nuclear compounds but also bi-nuclear and tri-nuclear compounds such as naphthalene, anthracene, and phenanthrene (Rovinsky *et al.*, 1995). Most multi-nuclear aromatic compounds (arenes) belong to the class of carcinogenic compounds. The most widespread compound of this group is benzo(*a*)pyrene. It is considered an indicator of the carcinogenicity of PAHs. The concentration of PAHs and individual arenes in water is not normalized by any indicators.

In the water of the Barents Sea the total concentration of PAHs is low and the composition of aromatic hydrocarbons is poor. In the southern part of the sea, the concentration of PAHs varies from 12 ng/L to 79 ng/L. Of 14 analyzed compounds, the concentrations of perylene, pyrene, phenanthrene, fluorene, fluoranthene, and benzo($b + k$)fluoranthene are reliably determined. The concentration of carcinogenic polyaromatic compounds, including benzo(*a*)pyrene, was below the detection limit.

The products of combustion and incineration are discharged to the atmosphere by industrial centers, automobile transport, and engines of marine ships; they are then transported by air flows. Therefore, the spread of PAHs in open areas of the sea is dictated by atmospheric fallout and ocean currents. The arenes of perylene and benzo($b + k$)fluoranthene (an indicator of technogenic emissions resulting from the pyrolysis of organic fuel) predominate. The numerous fishing vessels operating in the Barents Sea can also be a source of some PAHs in the surface water layer (Ilyin *et al.*, 2004).

The concentration of PAHs increases towards the coast. Near the western sea margin, concentrations of 79 ng/L have been observed (Plotitsyna *et al.*, 2002). In the southeastern part (i.e., the Pechora Sea) the concentrations of PAHs change from 15 ng/L to 67 ng/L (*Quality of Sea Water* ..., 2005). According to some data, the concentration of PAHs in the near-bottom layer in the coastal zone increases in places to 420 ng/L and even to 500 ng/L (Ivanov, 2002).

In the near-bottom water layer the concentration of PAHs increases in the proximity of oil fields and gas fields. According to observations by the Murmansk Marine Biological Institute (MMBI) within the deep-water central trough, where the Shtokman gas condensate field is located, the concentration of PAHs can comprise between 260 ng/L and 330 ng/L (Ilyin *et al.*, 1996a, b). Near the Prirazlomnoye field in the shallow Pechora Sea, the total concentration of PAHs increases to 120 ng/L in the near-bottom water layer.

Despite the Barents Sea's remoteness from agricultural regions, industrial complexes, and large settlements, many pesticides that degrade stably were detected in water by using modern study methods (Table 1.12).

Dichlorodiphenyltrichloroethane (DDT) metabolites are the most widespread. In water of the open sea, five metabolites have been detected: *p′p*-DDE, *p′p*-DDD,

Table 1.12. Concentrations of organochlorine pesticides and polychlorinated biphenyls (PCBs) in water of the southern Barents Sea (ng/L) (*MMBI Achived Materials*, 2001–2004).

Ingredient	$\sum$ *HCH*	$\sum$ *DDT*	$\sum$ *chlordanes*	*Aldrin*	*Dieldrin*	*HCBs*	$\sum$ *PCBs*
Range of concentrations	0.55–6.13	1.8–8.5	0.13–3.42	0.02–0.35	0.00–0.30	0.00–0.04	0.5–7.0
MPC for fisheries	10	10	10	10	10	10	10

o′p-DDD, *o′p*-DDT, and *p′p*-DDT. The total amount of DDT is generally not large and is comparable with other relatively clean sea areas.

As a rule, the total concentrations of DDT range between 1.84 ng/L and 3.25 ng/L. Higher values are episodically recorded locally in coastal and frontal zones. In general, the level of concentration of DDT metabolites does not reach MPC values. The highest DDT concentrations have been recorded in water of the coastal zone of the western Murman (Ivanov, 2002; Plotitsyna, 2002). In parts of Kola and Motovsky Bays, DDT concentrations of 2.5 ng/L in the surface layer and 0.2 ng/L in the near-bottom layer have been recorded (*Status Report* ..., 2003).

The differences in concentrations between surface and near-bottom water layers are low. DDT in the Barents Sea is mainly discharged from relatively remote sources brought by North Atlantic currents and the Norwegian coastal current.

The ratio of DDT to DDE metabolites in water varies around a value of 2. This value indicates a small degree of transformation of metabolites and advection of a comparatively "fresh" pollutant to the Barents Sea environment from outside.

DDT metabolites are carried to the peripheral area of the Pechora Sea by sea currents. The total concentration of DDT in this area decreases and varies from 0.04 ng/L to 0.50 ng/L. The composition of metabolites is depleted (*Quality of Sea Water* ..., 2005).

Another widespread complex pesticide—hexachlorocyclohexane (HCH)—is also detected over the entire southern area of the sea, having however half the concentrations of DDT. HCH is comprised of α-isomers, β-isomers, and γ- isomers. The ratio between α-HCH and γ-HCH in the water of the Barents Sea is biased towards the γ-isomer. This is testament to a periodic influx of relatively little transformed HCH chemicals carried by Atlantic waters to the study area. In the coastal zone, the concentration of HCH varies from 0.69 ng/L to 1.51 ng/L. Only the γ-HCH isomer is observed. In the western part of the coastal zone in the upper water layer of Motovsky Bay, the concentration of HCH is 1.7 ng/L.

In the near-bottom water layer, no HCHs were detected (*Status Report* ..., 2003). In water of the coastal areas of the Pechora Sea, the concentration of α-CH changes from 0.25 ng/L to 1.06 ng/L, γ-HCH from 0.05 ng/L to 0.36 ng/L, and β-HCH from 0.05 ng/L to 0.57 ng/L.

In addition to the abovementioned pesticides, which are the most widespread, *cis*-chlordane, hexachlorobenzene, aldrin, and dieldrin have been detected in the open water areas of the sea except for the Pechora Sea (Table 1.12). In the Pechora Sea only chlorobenzenes (0.05–0.13 ng/L) have been recorded. The presence of these pesticides is a result of the global spread of OCs by warm currents.

Polychlorinated biphenyls (PCBs) present a very widespread industrial pollutant, as they are widely used as a plasticizer in most industries from paintwork to energy. The composition of PCBs consists of several high-molecular congeners—#28, #31, #52, #99, #101, #105, #118, #126, #138, #153, #156, #180, and #187—some of which (e.g., #118), because of their toxic properties, are referred to as dioxin-like compounds.

Modern observations of the levels of PCBs in water are not numerous. Investigations have been carried out mainly by MMBI and partly by the Knipovich

Polar Research Institute of Marine Fisheries and Oceanography (PINRO). According to these observations, the concentration of PCBs in the water of open sea areas is very small and can only be detected locally. Higher concentrations were noted in the frontal zones of Barents Sea warm-jet currents and in the coastal zone. The concentration of these compounds is about 4.5 ng/L on average, which is much less than the MPC of 10 ng/L (*List of Fishery* ..., 1999). In the area of the central trough, low-molecular congeners #28, #31, #52, #101 prevail in the composition of PCBs (60–80%), and in the coastal waters high-molecular compounds #99, #118, #138, #153, #180 prevail (60–78% of the entire mass of PCBs) (Plotitsyna, 2002: Ilyin *et al.*, 2004).

Pollution of the aquatic environment of the Pechora Sea by PCBs is between 0.05 ng/L and 1.8 ng/L. The water of Motovsky Bay contains about 2.5 ng/L of PCBs in the surface layer and about 1.2 ng/L in the near-bottom layer (*Status Report* ..., 2003). In the near-bottom layer of open sea areas, the concentrations of PCBs are less than 0.5 ng/l, increasing only in the coastal zone to between 0.7 ng/L and 0.9 ng/L (Ivanov, 2002).

No detergents have been detected in the open water area. In water of the coastal zone, the concentration of surfactants is insignificant, being episodically noted in the waters of Motovsky Bay and near the Kola coast (up to 0.05–0.07 mg/L) (*Yearbook* ..., 1991, 1992, 1993; *Status Report* ..., 2003; Ilyin *et al.*, 2004). Domestic effluents are the main source of surfactants.

Heavy metals (those metals with a specific weight of more than 4.5 g/cm^3) and microelements are normal environmental components and are necessary in low concentrations for biochemical processes in ecosystems. However, if threshold concentrations are exceeded the majority of heavy metals become toxic for organisms. Especially toxic are cadmium, mercury, and some others. The natural background of some of them (as a rule non-toxic metals) is determined by concentrations of tens and even hundreds of micrograms in one liter of seawater (e.g., iron, manganese, and zinc), the normal concentration of other metals comprises tenths and hundredths of micrograms (cadmium, cobalt, mercury, etc.). Heavy metals are carried to the marine environment by sewage water and atmospheric fallout.

The background of heavy metals in the Barents Sea is significantly formed by the atmospheric transfer of pollutants, namely the transport of aerosols from the industrial centers of Western Europe and the Kola Peninsula. Atmospheric emissions from regional sources—such as the industrial complexes Severonikel and Pechenganikel, OJSC Apatit, and thermal electric power stations—are enriched with metals: Ni, Cu, Zn, Cd, Pb, Cr, Hg, and benzo(*a*)pyrene (Table 1.13) (Golubeva, 2007).

Table 1.13. Estimates of the mass of heavy metals from rain reaching the Barents Sea annually.

Ingredients	*Elements*						
	Pb	*Cd*	*Cu*	*Zn*	*Ni*	*Cr*	*Hg*
Mass of HM (thousand tons)	1.17	0.036	3.48	17.7	0.500	0.71	0.004

The composition of river runoff, due to the geochemical characteristics of the province and anthropogenic impact, is a natural source of the increased concentration of metals found in local coastal areas. The territory of the Kola Peninsula has abundant iron ore bodies, which is reflected in the runoff that accumulates in the coastal zone. Metals are leached from the ore bodies by soil waters, especially during the fallout of acid rain. The atmospheric fallout of heavy metals in the watershed territory also accumulates in surface waterways and is carried to the marine environment (Ilyin, 2007a).

The concentration of metals in the water layers of open regions is characterized by low-level concentrations, although their distribution is non-uniform (Figure 1.22, see color section).

Nickel, lead, chromium, cadmium, tin, and arsenic form concentrations that are one to two orders of magnitude below the MPC for fisheries (*List of Fishery* ..., 1999). The more toxic metals, mercury and cobalt, have only trace concentrations or are absent over most of the water area. Only iron and copper have concentrations close to the MPC in various parts of the sea.

The concentrations of copper and nickel (and iron) are large due to their elevated levels in continental runoff and in atmospheric fallout (Ilyin, 2004). Nickel is characterized by a regular decrease in concentrations from the coastal zone to water of the central trough. The distribution of copper in the water area has a patchy character. Relatively increased concentrations are noted in a zone of the polar front and in the coastal zone.

The spread of lead, tin, and cadmium is probably determined to a great extent by the advection of these metals with warm currents. In their spatial distribution, the areas of increased and decreased concentrations are fragmented, which reflects a non-uniform episodic advection of Pb, Sn, and Cd to the Barents Sea.

The increased concentration of metals in the upper water layer of the polar front can be determined by the winter accumulation of aerosol fallout on the sea ice cover and release of pollutants at icemelt in the front zone.

Different tendencies are noted in the distribution of other metals. Manganese is distributed relatively uniformly in the water area. The concentrations of zinc, chromium, and arsenic are observed to increase from the coast to the central deep-water regions and to the central trough.

Zones of relatively elevated concentrations are noted in the water of the warm currents in the vicinity of Rybachiya Banka and Murmansk shallow (Pb up to 1.2 μg/L; Sn up to 3.5–4.0 μg/L; Cd 0.1–0.6 μg/L), and also in the polar front zone in the area of the central trough (Sn up to 6.0 μg/L; Cd about 0.4 μg/L; Pb >6 μg/L) (Ilyin, 2004, 2007a).

Although arsenic is not a metal, it is traditionally analyzed in ecological studies along with heavy metals as a widespread toxic metalloid. It is supposed that arsenic comes to the Barents Sea mainly with the water of warm currents (Ilyin, 2004, 2007a). Therefore, local zones with the highest concentrations of arsenic (up to 4 μg/L) are episodically recorded in the western part of the sea and in the area of the polar frontal zone (the central trough) in the upper and near-bottom layer.

In general, the concentrations of many metals in coastal water are higher than in the water of open areas such as Murmansk shallow and Rybachiya Banka. Localization of areas of increased and decreased concentrations of metals in the coastal zone especially in bays and gulfs depends on the intensity and location of water from continental runoff. For example, copper and arsenic in the coastal zone form the second local zone of increased concentrations (*Biotesting* . . ., 2003).

Bottom sediments

Marine bottom sediments accumulate pollutants according to their concentrations in water and according to the mineralogical and grain-size composition of the sediments themselves. Bottom sediments at a specific stage of biogeochemical processes in the sea deposit pollutants due to sorption from water. A reverse process can occur under changed conditions: this process involves desorption and ingress of pollutants from sediments to the aquatic environment, so-called "secondary pollution". So, in addition to the peculiarities of pollution entering the marine environment, the geochemical properties of sediments themselves increase the variability of their accumulation in deposits.

For example, the accumulation of oil products in bottom sediments of the open areas of the Barents Sea is rather heterogeneous, and their concentration varies from trace values to 80 μg/g dry weight. The highest concentrations are noted in sediments of the central trough and coastal zone. The ranges of TOH concentrations in different morphological zones of the sea are shown in Table 1.14 (*Chemical Processes* . . ., 1997; Ilyin *et al.*, 2004). The heterogeneous character of TOH distribution in sediments is determined by hydrodynamic and lithological factors. In sediments of the coastal zone of Murmansk, especially near such water bodies as Kola and Motovsky Bays, the concentration of oil products is elevated. In different years in this coastal zone,

Table 1.14. Concentration of oil hydrocarbons and PAHs and structural characteristics of the composition of paraffins in bottom sediments in the southern Barents Sea.

Parameters	*Area of localization*			
	Central trough	*Murmansk shallow, Rybachiya Banka*	*Coastal zone*	*Gulfs*
$\sum$ TOH (μg/g dry weight)	0.0–80.0	0.0–60.0	10–700	8–120
$\sum$ n-C_{10}–C_{34} (μg/g dry weight)	1.7–2.0	1.6–2.3	1.6–20.9	2.0–8.0
CPI	0.69–0.97	0.41–1.11	0.53–0.93	1.2
$\frac{n\text{-}C_{12}-C_{22}}{n\text{-}C_{23}-C_{34}}$	1.24–2.16	0.14–6.73	1.16–3.61	2.5
$\sum$ PAHs (ng/g dry weight)	20–399	20–188	38–197	25–2,060

areas have been noted where the accumulation of oil products reaches between 120 μg/g to 700 μg/g dry weight. Beyond coastal areas used for business, the TOH concentrations in sediments decrease.

The structural composition of paraffins in bottom sediments is wider than that of water. The concentrations and characteristics of the spectrum of paraffins are given in Table 1.14. In the composition of aliphatic compounds, alkanes of different structural groups such as C_{12}–C_{17} and C_{18}–C_{24} are predominant. Short-chain compounds such as C_{10}–C_{14} are detected over practically the entire water area. The low-boiling point compounds of C_{10} are mainly spread in deposits in the northern periphery of the central trough, which can be caused by the elimination of light hydrocarbons from the sedimentary mantle.

The values of organic matter transformation indices at the slopes of the Murmansk shallow are the most non-uniform, which reflects the mixing of hydrocarbons of oil and biogenic genesis (Romankevich *et al.*, 1982). In the sediments of this area, two groups of paraffins dominate: C_{12}–C_{17} and C_{18}–C_{24} (30% and 45%, respectively, of the total quantity of paraffins). An increased concentration of biogenic paraffins C_{20}–C_{21} and C_{25}–C_{26} is recorded during the intensive development of micro-algae.

In sediments of the coastal zone the fraction of light compounds C_{12}–C_{17} is much less (25%) than sediments in the open parts of the sea. However, in the coastal zone, the fraction of long-chain compounds C_{25}–C_{28} is increased (up to 11%), with their ingress to sediments being related to terrestrial biocenoses.

Although the distribution of PAHs in bottom sediments is non-uniform—ranging from 20 ng/g to 400 ng/g dry weight—the maximum accumulation occurs in deposits of the Murmansk coastal gulfs (where business enterprises are located) and in the central trough. The composition of polyaromatic hydrocarbons in bottom sediments also has qualitative differences compared with water.

Over the entire water area, pyrogenic compounds prevail in the composition of PAHs—pyrene, benzo-anthracene, dibenzo-anthracene, fluoranthene, etc.—which possess carcinogenic properties. The concentration of benzo(*a*)pyrene varies from 0 ng/g to 14 ng/g dry weight. The maximum level of the accumulation of carcinogens is also typical of the deposits of the frontal zone.

A group of typically oily PAHs—represented by diatomic naphthalene, tri-nuclear anthracene, and indene—comprises an insignificant fraction of the composition in all seabed areas (Ilyin *et al.*, 1996a, b, 2004; *Scientific–Methodological Approaches* ..., 1997). The bottom sediments of open regions are also characterized by the spread of natural PAHs such as chrysene and phenanthrene. These arenes are the result of transformations (metamorphization) of organic matter and come to the environment as a result of being leached from ancient soils and bottom sediments (Rovinsky *et al.* 1988). Some PAHs such as fluorene, benzo(*g*, *h*, *i*)perylene, di-benzo(*a*, *h*)anthracene, anthracene, acenaphthene, and benzo(*a*)anthracene have a restricted spread in sediments and are detected only in some local zones (Ilyin *et al.*, 1996a, 2004; *Scientific–Methodological Approaches* ..., 1997).

In sediments of the coastal zone and especially in economically developed zones, the prevailing compounds are diatomic naphthalene and anthracene, multi-nuclear fluoranthene, and phenanthrene. In the eastern areas of the coastal zone (Pechora Sea), the concentration of PAHs varies from 5 ng/g to 80 ng/g dry weight. In the proximity of oil and gas fields, the total concentration of PAHs is elevated. The main arenes are naphthalene (5–40 ng/g dry weight) and fluoranthene (0.2–0.4 ng/g dry weight), both compounds of oil and technogenic genesis. The concentration of benzo(*a*)pyrene in this region is very low between 0.0 ng/g and 5.2 ng/g dry weight. The low background of carcinogens is probably determined by that area's remoteness from densely populated and industrial northern European and Eurasian cities (Ilyin *et al.*, 1996b, 2004; *Quality of Sea Water* ..., 2005).

In western coastal areas the spectrum of PAHs significantly increases. Near the mouth of Kola Bay, in Motovsky Bay, and in Varanger Fjord, the oil-borne compounds naphthalene and fluoranthene become the main aromatic compounds in sediments, testament to the long-term anthropogenic pressure on the coastal zone. The total concentration of PAHs is generally not high—from 7 ng/g to 147 ng/g dry weight—much less than the pollution threshold of the Norwegian *Statens forurensningstilsyn* (SFT) classification. A tendency for increasing concentrations from Varanger Fjord to Kola Bay is noted. Benzo(*a*)pyrene is also present in sediment of the western sector of the coastal zone (0–3 ng/g dry weight) but has a patchy distribution, similar to other carcinogenic PAHs (*Environmental Status*..., 2003; Ilyin *et al.*, 2004). Sediments in the Pechenga and Ura Gulfs and in the southern and middle bends of Kola Bay are much more intensively polluted than the open areas of Varanger Fjord, Motovsky Bay, and Kola Bay, respectively. In some localities of these gulfs the concentration of PAHs occasionally increases to the weakly polluted level (Molvaer *et al.*, 1997).

The accumulation of polychlorinated biphenyls in bottom sediments is very low. In 35% of samples of bottom sediments, the concentration of these compounds was below the detection threshold (*Scientific–Methodological Approaches* ..., 1997; Ivanov, 2002).

The concentration of PCBs in the sediment from open regions comprises on average 0.3 ng/g dry weight. A relatively increased level of PCBs is detected in the sediment of the central trough (1.0–1.5 ng/g dry weight) and in the coastal area (in some places up to 2.5 ng/g). The composition of PCBs in the bottom sediment is greater than in water and is represented by such compounds as #28, #31, #52, #101, #105, #118, #138, #153, and #180. The main mass fraction in the sum of compounds is comprised by such congeners as #52, #118, #138, and #153. In the sediment of the Pechora Sea the concentration of PCBs is practically the same as that in the southern and central sea areas ranging from 0.05 ng/g to 2.3 ng/g dry weight (*Scientific–Methodological Approaches* ..., 1997).

In the sediment of the coastal zone there is a tendency toward increasing concentrations of PCBs westward. In some parts of Motovsky Bay, PCB concentrations of more than 2.5 ng/g have accumulated in the sediment. In the sediment of Varanger Fjord the concentration of PCBs increases in some places to 4.0 ng/g dry weight. The composition of PCBs and the ratio of their congeners at the western periphery of the sea are identical to the composition of PCBs in other parts of the sea. In general, the concentration of PCBs in the sediment of the Barents Sea can be qualified as background and the sediment as "non-polluted" by PCBs (by the SFT qualification: $\sum$ PCB $< 5\,\mu g/g$ dry weight) (Molvaer *et al.*, 1997). The highest concentrations of these pollutants were observed in the sediment of Pechenga Gulf (4–19 ng/g dry weight) due to the amount of shipping (*Environmental Status* ..., 2003).

It has been suggested that the most significant sources of PCBs in the marine environment are fishing fleets, merchant shipping, and Navy fleets. The water and sediment in port areas and adjoining regions are particularly polluted. The main flow of pollutants in the Barents Sea comes from the Norwegian coast, which includes the many large and small fisheries in Varanger Fjord. The PCBs are transferred by the Norwegian Coastal Current.

Another group of xenobiotics (organochlorine pesticides) that reach the Barents Sea are predominantly brought by warm Atlantic currents and by atmospheric fall-out, all of which determines their concentrations. Although these substances are not synthesized in nature and have an exclusively anthropogenic origin, their levels in the sediment are usually normalized in some countries including Norway (Molvaer *et al.*, 1997). The total accumulation of determined stable organochlorine pesticides in the sediment of the Barents Sea is incomparably less than the approximate levels for non-polluted sediment (<500 ng/g dry weight by SFT).

The total concentration of DDT ranges from 0.4 ng/g to 15.8 ng/g dry weight, comprising 3.3 ng/g on average. Maximum concentrations are typical of the deposits of the central trough. The main fraction of DDT (about 55%, on average) is comprised of $p'p$-DDT (*Scientific–Methodological Approaches* ..., 1997). The concentration of HCH in the sediment of the open sea is between 1.53 ng/g and 5.18 ng/g, on average 3.06 ng/g. The maximum concentration is found in a zone of the Murmansk shallow. Practically everywhere the level of the unstable γ-isomer of HCH (lindane) exceeds 50% of the sum of HCH (*Scientific–Methodological Approaches* ..., 1997; Plotitsyna *et al.*, 2002). Such a ratio between isomers indicates the deposition of little-changed pollutant from warm sea currents.

Chlordane and hexachlorobenzene (HCB) also accumulate in the sediment. The accumulation level of chlordane varies from 0 ng/g to 0.74 ng/g dry weight. The maximum level was noted in deposits in the central trough. The main portion of chlordane (72%) is comprised of its metabolite *trans*-chlordane, the presence of which is by and large only found in deposits in the central trough. Other isomers of chlordane such as *cis*-chlordane and *cis*-non-chlordane have also been recorded in the sediment of areas in the central trough and Murmansk shallow. The concentrations of HCB in the bottom sediment of the southern sea area are no greater than 0.23 ng/g (Savinova, 1994; *Scientific–Methodological Approaches* ..., 1997).

In the sediment of the coastal zone, the concentration of DDT is lower than in open regions (0.5–1.5 ng/g dry weight). In one area of Motovsky Bay a concentration of 0.5 ng/g is recorded and in some areas of Varanger Fjord it increases to 1.3 ng/g dry weight. Congeners of DD, DDD, and DDT have been detected. The DDT metabolite comprises 75% of the total DDT level. Similar to the aquatic environment, this ratio indicates the ingress of largely untransformed substances via oceanic currents and atmospheric transfer. The concentration of HCH in the sediment in the bay is about

0.65 ng/g dry weight, with the γ-isomer comprising 73% of the total HCH (*Environmental Status* ..., 2003).

Most heavy metals and microelements do not form significant concentrations in sediments except for iron and manganese (Table 1.15). The spatial distribution of lead, copper, cadmium, nickel, chromium, iron, manganese, and zinc is characterized by a general tendency toward increasing concentrations from coastal areas to the deep-water part of the central trough (Figure 1.23, see color section). This tendency is most expressed in the distribution of lead, cadmium, manganese, and iron (Ilyin and Petrov, 1994b; *Scientific–Methodological Approaches* ..., 1997; Ilyin *et al.*, 2004). Arsenic has the same pattern of accumulation.

The accumulation of these elements is apparently largely down to transboundary transfer by currents, atmospheric aerosols, and ice drifting to the area of the polar frontal zone, with subsequent accumulation in the finely dispersed sediment of the central trough, where the downwelling circulation of water and suspended matter occurs during winter. The accumulation level of most metals changes abruptly at the boundaries of the spread zones of different types of sediments. In open sea regions the patchy character of their distribution depends in many respects on the ratio between the pelite and silty fractions of bottom sediments.

In narrow coastal zones, especially in coastal gulfs, such elements as copper, cobalt, zinc, nickel, and chromium have a second concentration maximum (Figure 1.24). The formation of this maximum is determined by continental runoff and deposition of metals in the geochemical barrier zone (the zone in which water mixes).

The sediment in the Pechora Sea is characterized by a lower level of accumulation of heavy metals and microelements than all other areas of the Barents Sea (*Quality of Sea Water* ..., 2005).

1.3.2.2 White Sea

The spread of pollutants in the White Sea mainly occurs by means of runoff currents from bays that receive runoff from the large rivers. Pollutant accumulation zones include Onega, Dvina, Kandalaksha, and Mezen′ Bays, and the central deepwater area of the sea basin with its closed-water circulation system. This can be clearly seen in the distribution of oil products, OCs, and phenols. Pollutants are partly exported from the White Sea basin through Gorlo and Voronka to the Pechora Sea. Pollutants transported by waters of the Barents Sea are probably insignificant, but this ecological problem has so far been insufficiently investigated. Among the traditional sources of pollution that have long existed in the basin, marine oil transportation has been on the increase in recent years (Figure 1.25, see color section) (Ilyin and Shavykin, 2006). This commercial activity can significantly influence the load of oil hydrocarbons in this semi-closed marine basin.

Aquatic environment

In contrast to the Barents Sea proper, the White Sea receives a much greater continental runoff from the large rivers flowing through the northwestern part of

Table 1.15. Concentration of heavy metals and microelements in the sediment of the Barents Sea (μg/g dry weight).

Area of localization	*Pb*	*Co*	*Cu*	*Ni*	*Cd*	*Fe*[a]	*Cr*	*Mn*	*Zn*	*As*	*Hg*	*Sn*
Central trough	7–18	1–5	12–35	20–50	0.0–0.3	38–60	32–54	160–470	70–85	7–67	0.03–0.07	1–7
Southern part of the sea	2–18	1–3	10–20	18–31	0.0–0.10	14–30	18–38	151–280	40–70	2–24	0.0–0.07	0.7–3.0
Pechora Sea	1–6	0.4–1.5	3–8	1–3	0.0–0.1	—	1.4–4.0	21–115	4–10	0.4–1.2	0.01–0.02	0.0
Coastal zone and gulfs	5–10	3–30	4–80	12–24	0.0–0.15	9–17	18–57	128–240	26–150	1–15	0.03–0.10	1–4
APC_0[b]	<30	—	<35	<30	<0.25	—	<70	—	<150	<20	<0.15	—

[a] Iron concentrations are given in mg/g dry weight.
[b] APC_0 = approximately permissible concentrations for non-polluted sediments by SFT (Molvaer *et al.*, 1997).

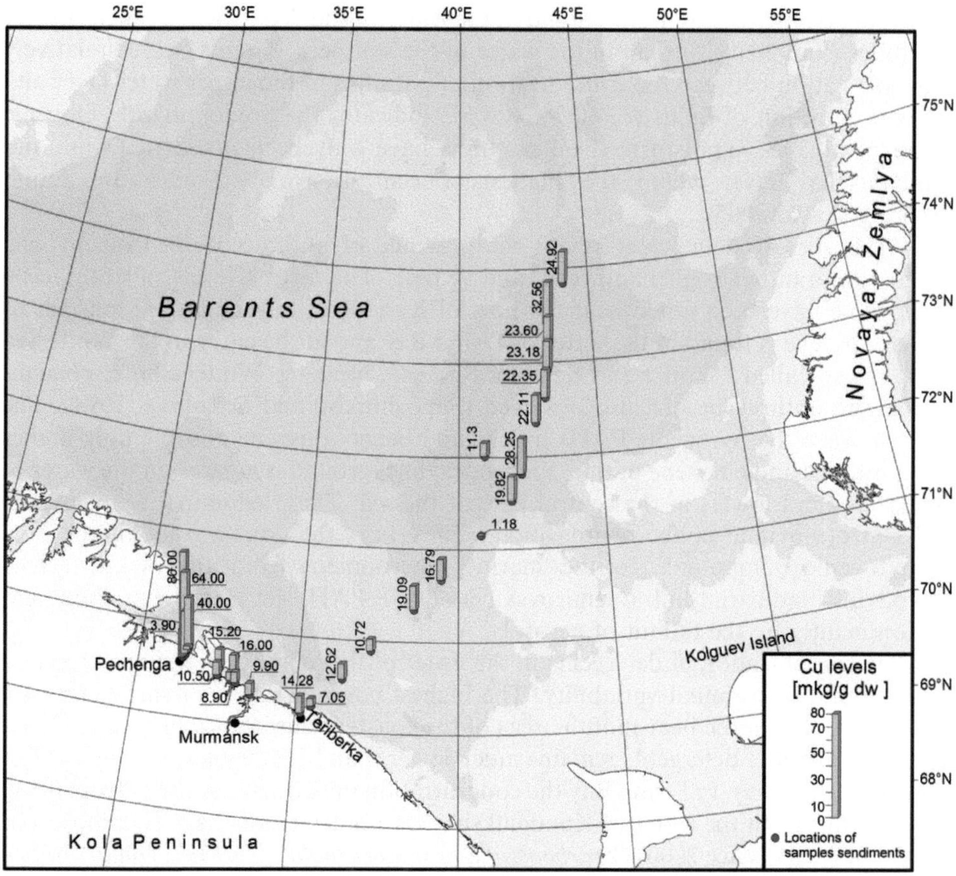

Figure 1.24. Distribution of copper in bottom sediments along a transect across the southern Barents Sea.

European Russia. In addition to the hustle and bustle of ports, the most significant source of pollutants in the White Sea basin is the Severnaya Dvina River. Oil products, phenols, detergents, some pesticides, and heavy metals are brought to a relatively restricted area by river runoff (Figure 1.18).

The water in the rivermouth areas of the Severnaya Dvina and Onega and along the adjoining coasts contains on average about 0.04 mg/L of oil hydrocarbon pollution. A lesser amount (about 0.01 mg/L) of such pollution is typical of Mezen′ Bay. From time to time the maximum concentration of OP in the bays, especially in Kandalaksha and Dvina Bays, can increase to 2 MPC (up to 0.11 mg/L).

In the central part of the sea area the concentration of oil hydrocarbons in water usually varies from 0 mg/L to 0.08 mg/L (i.e., up to 1.5 MPC). However, elevated concentrations are typical of the southern periphery of the sea and are observed episodically (*Yearbook of Sea Water Quality* ..., 1997; *Quality of Sea Water* ..., 2005).

Aliphatic hydrocarbons are found in the water of the White Sea in much smaller quantities (5 to 30 μg/L) than in the water of the southern Barents Sea. A relatively high correlation between the concentration of paraffins in the upper water layer and the concentration of lipids ($r \sim 0.74$, $n = 51$) indicates the predominantly biogenic origin of alkanes. Transformed oil paraffins have only been noted in Dvina and Kandalaksha Bays, where the highest concentrations of alkanes are found (Nemirovskaya, 2005).

PAHs dissolved in water reveal the presence of anthracene, benzo(*a*)pyrene, benzo(*b*)fluoranthene, phenanthrene, and pyrene. The largest levels of anthracene and pyrene have been noted near the port of Kandalaksha. The largest amount of benzo(*a*)pyrene is found in the water of Dvina Bay and at the entry to the White Sea along the so-called "Winter and Summer Shores". Near the Winter Shore, phenanthrene and anthracene are also observed (Sapozhnikov and Sokolova, 1994). The ratio between the pyrogenic PAHs pyrene and benzo(*a*)pyrene and the poly-arenes phenanthrene and chrysene of the geochemical background is greatest in the water of Dvina Bay and lowest in the central part of the sea. The predomination of pyrene and benzo(*a*)pyrene in the composition of PAHs is the consequence of the flow of pyrogenic compounds to the marine environment from the Arkhangelsk–Novodvinsk industrial hub (Nemirovskaya, 2005). PAHs get into the environment predominantly by the fallout of aerosols.

The concentration of detergents in the water of the White Sea is subject to large seasonal and interannual variability. The highest concentrations of surfactants are usually detected in the near-mouth areas of the coasts in surface water layers. Nonetheless, the level of detergents remains much lower than MPC values (0.1 mg/L) (*List of Fishery* ..., 1999). In Dvina Bay the concentration of surfactants varies from 0 mg/L to 0.06 mg/L. In the water of Kandalaksha Bay no detergents have been observed for several years (since 2000) (*Yearbook of Sea Water Quality* ..., 1997; *Quality of Sea Water* ..., 2005).

The spread of phenols in the White Sea is mainly determined by the runoff of large rivers. Therefore, patterns in the distribution of phenols and oil products are very similar. Their concentration in the sea comprises about 0.001 mg/L, which is about the MPC. In the uppermost part of Dvina Bay the concentration of phenols has been known to increase to 0.015 mg/L and correspondingly exceeds the MPC several times. The increased level of phenols in the seawater of the bay is connected with the runoff of Severnaya Dvina. High concentrations of phenols similar to oil hydrocarbons are traced in the brackish water flow from the river runoff into Dvina Bay (Cape Kerets). Under strong northerly winds, pollution spreads along the Summer Shore (Sapozhnikov and Sokolova, 1994). The increased discharge of phenols by the water of the Severnaya Dvina is probably down to discharges from the pulp and paper industry and the timber industry of Arkhangelsk.

The levels of phenols in the water of Kandalaksha Bay vary from 0.001 mg/L to 0.003 mg/L (*Quality of Sea Water* ..., 2005). The composition of phenols reaching the sea has scarcely been investigated and is probably determined by natural phenols produced as a by-product of organic matter transformation in the peat soil of the watershed area. No detailed information on the composition of

Table 1.15. Concentration (μg/L) of heavy metals in the water of Kandalaksha Bay.

Concentration	*Copper*			*Manganese*			*Iron*			*Molybdenum*		
	2000	*2001*	*2002*	*2000*	*2001*	*2002*	*2000*	*2001*	*2002*	*2000*	*2001*	*2002*
Average	7.2	3.6	5.1	0.09	26.6	6.7	0.34	123.0	54.0	—	—	2.1
Maximum	43.2	8.0	9.0	2.0	31.0	11.6	2.2	217.0	112.0	—	—	3.2
MPC		5.0			50			50			1.0	

natural and anthropogenic phenols in the runoff of rivers into the White Sea basin is available.

The spread of pesticides in the water of the White Sea basin has been little studied, but according to current data it is restricted to the coasts and large bays. In Dvina Bay, pesticides of the HCH group have been detected. The average and maximum concentration of the α-HCH isomer is 0.2 ng/L and 0.4 ng/L, respectively, and of the γ-HCH isomer 0.4 ng/L and 0.8 ng/L. Pesticides of the DDT group have not been observed in the bay. In Kandalaksha Bay, HCH and DDT compounds are either not detected or recorded episodically (*Quality of Sea Water* ..., 2005). In the White Sea, even episodically observed maximum concentrations of pesticides are much lower than the MPC level (10 ng/L), and their presence in the sea is determined by the runoff of the largest river, the Severnaya Dvina.

The concentrations of heavy metals in the water is known from studies of Kandalaksha Bay. The dynamics of the concentrations of some metals is given in Table 1.15. Kandalaksha Bay accumulates the runoff of rivers from the Kola Peninsula. Therefore, the levels of metals in the water are determined by the same factors as in the gulfs of the Barents Sea. The average levels of copper, molybdenum, and iron in the water of fishery areas typically exceeds the MPC (*List of Fishery* ..., 1999; *Quality of Sea Water* ..., 2005).

Bottom sediment

The distribution of oil hydrocarbons and PAHs in bottom sediments is similar to their distribution in water. Maximum OP concentrations are observed in the sediment of Dvina Bay (0.5–2.5 mg/kg wet weight) and Kandalaksha Bay (0.5–2.0 mg/kg wet weight). Smaller concentrations are noted in the sediment of Onega Bay (0.5–1.0 mg/kg wet weight) (Sapozhnikov and Sokolova, 1994). According to other data (Dahle *et al.*, 1998), the total concentration of oil hydrocarbons varies from 4.2 mg/kg to 33.2 mg/kg dry weight.

The concentration of alkanes $\sum(C_{15}–C_{40})$ in bottom sediments ranges between 0.8 μg/g and 12.3 μg/g, which corresponds to their levels in other Arctic regions (e.g., the Beaufort and Laptev Seas). The composition of alkanes in Dvina and Kandalaksha Bays is similar to that of oil. In the sediment of the bay, a low ratio of odd and even homologues (CPI = 0.8) and a low ratio of the isoprenoids pristine and phytane (Pr/Pf = 0.1) are noted. Phytane is prevalent in the composition of oils. The chromatographic analysis of hydrocarbons shows a large "unsolved peak" of compounds in the high molecular region of the spectrum.

In the mouths of the bays the composition of alkanes becomes predominantly terrigenous, whereas the composition is predominantly biogenic in the central part of the sea (Nemirovskaya, 2005).

The total concentration of PAHs varies between 13 ng/g and 208 ng/g dry weight throughout the sea area. The highest concentrations of PAHs and benzo(*a*)pyrene (10 ng/g dry weight) were detected in the clayey sediment of Dvina and Kandalaksha Bays.

The bottom sediment of the White Sea is mostly characterized by PAHs of pyrogenic origin such as benzofluoranthene, benzopyrene, and benzoperylene, which is typical of atmospheric fallout near industrial centers. However, in the sediment of the bays the fraction of PAHs of typically terrigenous origin (i.e., perylene and chrysene) is relatively elevated. Terrigenous PAHs provide a sharp increase in the total concentrations of PAHs in areas of river runoff discharge at the heads of the bays. Multi-year observations show continuing accumulation of polycyclic compounds of pyrogenic origin in sediments (Dahle *et al.*, 1998; Nemirovskaya, 2005).

In the sediment of the central part of the sea, both in the basin and at the entry, the concentration of PAHs decreases. In this area there is a regular decrease in concentrations from the deep-water part of the basin towards Gorlo, from 80 ng/g to 47 ng/g, respectively, and even down to 16 ng/g dry weight. Even in the coastal area near the south coast the accumulation of PAHs is less than in the sediment of the basin (about 60 ng/g dry weight). The arenes of oil, which are of petrogenic (natural) and pyrogenic genesis, are detected in the composition of PAHs in the sediment of the open sea, but the poly-arenes, which are of pyrogenic and petrogenic (geochemical background) genesis, predominate in practically the same proportion (Dahle *et al.*, 1998; Savinov *et al.*, 2000).

In general, the accumulation of PAHs even in the most polluted parts of the sea is much less than permissible levels for non-polluted sediments: <300 μg/g dry weight (Molvaer *et al.*, 1997). The concentration of benzo(*a*)pyrene in sediments is between 1 ng/g and 15 ng/g in the main part of the sea and only increases in the middle areas of Kandalaksha and Dvina Bays by between 2.5 ng/g and 4 ng/g dry weight. Observed concentrations are also much lower than the level of unpolluted sediments (<10 μg/g) and less than the MPC of 20 ng/g established for soils. In the sediment at the entry to the sea, the accumulation of benzo(*a*)pyrene decreases even more, down to 0.5 ng/g dry weight (Savinov *et al.*, 2000).

As far as organochlorine pesticides are concerned, bottom sediments, unlike water, have a greater accumulation of compounds of the DDT group, which has to do with their greater persistency (*Geoecology of the Shelf* ..., 2001).

The concentration of heavy metals in the bottom sediment in different regions of the White Sea varies over a relatively wide range. However, in general, their accumulation level is low: 0.5–15.9 μg/L for Cu, 1.9–63 μg/L for Pb, 0.7–19.9 μg/L for Co, 1.2–17.1 μg/L for Cr, 1.6–34.1 μg/L for Ni, and 0.04–0.1 μg/L for Hg. Only nickel shows a tendency for increased concentrations up to borderline for polluted/

unpolluted sediments, <30 μg/g (Molvaer *et al.*, 1997). The elevated concentrations of metals are typical of deposits in Kandalaksha and Dvina Bays. In Onega Bay a lower level of accumulation of heavy metals has been recorded (Sapozhnikov and Sokolova, 1994; *Geoecology of the Shelf ...*, 2001).

1.3.2.3 Kara Sea

In the Kara Sea the distribution and formation of the background accumulation of pollutants is influenced by the advection of Atlantic water and Ob′–Yenisei runoff. Atlantic water carried by warm currents determines the advection of pollutants, predominantly oil hydrocarbons and PAHs, in the northwestern sector of the Kara Sea. In the southern area of the sea the runoff from the Ob′ and Yenisei determines the advection of several pollutants: oil hydrocarbons, PAHs, persistent OCs, phenols, and heavy metals. The total runoff volume of these rivers is enormous and determines the hydrochemical properties of seawater in the entire shelf zone of the water body in summer (around 40% of the sea area). However, the largest effect of the export of pollutants is witnessed in the specific geochemical area where water mixes in the shallow marginal areas, including Ob′ Bay and Yenisei Gulf. The volume of pollutant discharge with river water is shown in Figure 1.18.

The discharge of pollutants, especially oil products and phenols, is determined in many respects by the abundance of developed oil fields and gas fields in the basins of West Siberian Rivers and the trans-shipment of oil in these rivers. Oil from the Kharampurskaya and Urengoyskaya fields in the West Siberian oil–gas province (OGP) is trans-shipped across the Kara and the Barents Seas to the ports of Ob′ Bay: Dudinka and Dikson (Azarova, 2001). For example, oil has been trans-shipped at the roadstead near Cape Kamenny (Ob′ Bay) since 1999.

Oil from the Sredne′Khulymskoye and Sandibinskoye fields is delivered to the trans-shipment complex by river/sea-type tankers (deadweight of 3,000 t). After trans-shipment to ice-class marine tankers of the Astrakhan type, oil is transported to the offshore trans-shipment terminal Belokamenka in Kola Bay (Barents Sea). At the roadstead in Ob′ Bay, 240,000 t of oil is trans-shipped per year (Bambulyak and Frantsen, 2005).

Water exchange with the Barents Sea is only a source of pollutants in localities where this water penetrates (Ilyin, 2007b).

Aquatic environment

The main flow of oil products carried with river runoff is observed in the southern part of the sea where the Ob′ and Yenisei empty into the sea. The concentration varies from 0.00 mg/L to 0.04 mg/L, with a mean value of 0.02 mg/L (i.e., below the MPC threshold).

A wide spectrum of PAHs has been recorded in the water (Table 1.16), with an average concentration of about 115 ng/L. Naphthalenes (typically oil compounds) prevail (27% of $\sum$PAH) in the composition of PAHs. Pyrene, fluoranthene, and alkyl-substituted PAHs of pyrogenic genesis comprise only about 2%. The fraction of PAHs of the natural geochemical background (phenanthrene) comprises on average

Table 1.16. Concentration of polyaromatic hydrocarbons in the water of the Kara Sea (ng/L).

Ingredient	*Naphthalene*	*Biphenyl*	*2-methylnaphthalene*	*2,6-di-methylnaphthalene*	*Phenanthrene*
Average concentration	13.4	21.2	18.8	4.3	17.5

about 15% of the total concentration of PAHs (*Quality of Sea Water ...*, 2005). The main part of PAHs is comprised of technogenous poly-arenes. The fraction of benzo(*a*)pyrene is below 1% of the total quantity of PAHs in water. The total level of PAHs in the water of the Kara Sea by quantitative indicators is slightly less, and by qualitative indicators it is practically the same as in the coastal zone of the Barents Sea, although this sea is remote from industrial centers. The main ingress of PAHs to the basin is probably provided by surface runoff from the watershed territory.

Organochlorine pesticides of the DDT and HCH groups are only observed in very small quantities in the upper water layer. The average concentration of $o'p$-DDT and $p'p$-DDT metabolites is 0.34 ng/L, the average total concentration of $p'p$-DDD and $o'p$-DDD metabolites is 0.18 ng/L, and that of DDE metabolites is 0.27 ng/L. These concentrations are lower than in some regions of the Barents Sea and much less than the MPC threshold (10 ng/L).

In the HCH group the prevailing isomer is the α-isomer whose value of 0.39 ng/L compares with 0.28 ng/L for the γ-HCH isomer. The proportion of DDT/DDE metabolites is about 1, and of α-HCH and γ-HCH isomers the proportion is less than 1. This means that organochlorine pesticides come in a significantly metabolized form relative to initial technical preparations. Chlorobenzene is contained in the water at an average concentration of 0.41 ng/L.

In addition to OCs there are PCBs in the water of the Kara Sea. The average concentration of seven main (Dutch) congeners is 1.0 ng/L (*Quality of Sea Water ...*, 2005). A decrease in PCB concentrations from the western regions of the Barents Sea to its eastern regions and further to the Kara Sea is observed.

The concentration of heavy metals in the upper water layer of the Kara Sea is below the MPC for fisheries (Table 1.17) and much lower than in the water of the Barents Sea (*Quality of Sea Water ...*, 2005). Thus, the significance of the geochemical component and the advection of metals by currents to the Barents Sea is stressed.

Table 1.17. Concentration of heavy metals in the surface water layer of the Kara Sea (μg/L).

Metal	*Mn*	*Zn*	*Cu*	*Ni*	*Pb*	*Co*	*Cd*	*Sn*
Average concentration	1.92	1.74	0.82	0.14	0.52	0.11	0.13	0.14
MPC	50	50	5	10	10	10	10	112

Fluorene	*Anthracene*	*Fluoranthene*	*Pyrene*	*Benzo(a)-anthracene*	*Benzo(b)-fluoranthene*	*Benzo(k)-fluoranthene*
30.1	1.4	3.1	2.9	1.1	0.62	0.87

Bottom sediments

The sedimentary mantle of the Kara Sea shelf is characterized by some particularities contributing to oil–gas accumulation and hence to elimination of hydrocarbons to the upper sediment layer and to the near-bottom water layer. The Jurassic–Cretaceous and Triassic–Upper Paleozoic oil–gas complexes, which comprise the Rusanovskaya, Leningradskaya, Obruchevskaya, and Sharapovskaya oil–gas-bearing structures, serve as a condition for this (Dunayev, 1990; Ilyin, 2005).

In the bottom sediment of the southern periphery of the Kara Sea, the accumulation level of oil products varies between 6.7 μg/g and 7.8 μg/g dry weight. Their average concentration is 7.4 μg/g dry weight (*Quality of Sea Water* ..., 2005). Compared with the Barents Sea, this is a very low accumulation level of oil hydrocarbons, one that matches the cleanest parts of the Barents Sea.

Different groups of oil hydrocarbon compounds form their own distinct patterns in the sea which depending on environmental conditions and sources. In the distribution of paraffins, three areas are distinguished (Figure 1.26): coastal shallows in the southern part of Novozemelskaya trough, the northwestern periphery of the sea, and the central sea area (Ilyin, 2007b).

The sediment of coastal shallows is characterized by the lowest concentration of paraffins, from 150 μg/g to 400 μg/g dry weight. The hydrodynamics of this area is determined by the spread of coastal and mixed water of the Ob′–Yenisei and the East Novozemelsky Currents. The spread of Barents Sea water through Kara Gate Strait is an example of an area with minimum concentrations.

The northwestern periphery of the sea, whose hydrological regime is determined by the exchange of Atlantic water with the Barents Sea and the Arctic Basin by means of deep-water troughs, is characterized by increased levels of paraffins (between 500 μg/kg and 800 μg/kg dry weight).

The sediment of the central part of the sea, where the oil-bearing and gas-bearing Rusanovskaya, Leningradskaya, and Obruchevskaya fields are located, contains the highest levels of paraffins: 1,000 μg/kg to 7,000 μg/kg (Ilyin, 2007b).

Except for the sediment in the central area, the structure of alkanes mainly has compounds of predominantly biogenic origin which are contained in the lipids of planktonic organisms (C_{15}–C_{19}) and waxes of terrestrial plants (C_{23}–C_{29}). In the sediment in the central area of the sea, *n*-paraffins C_{20}–C_{26} and components of metamorphosed organic matter prevail. In the zone where river water spreads, C_{20} is prevalent. Increased concentrations of short-chain paraffins C_{12}–C_{14} are found in the frontal area of Atlantic water coming through St. Anna Strait where there are

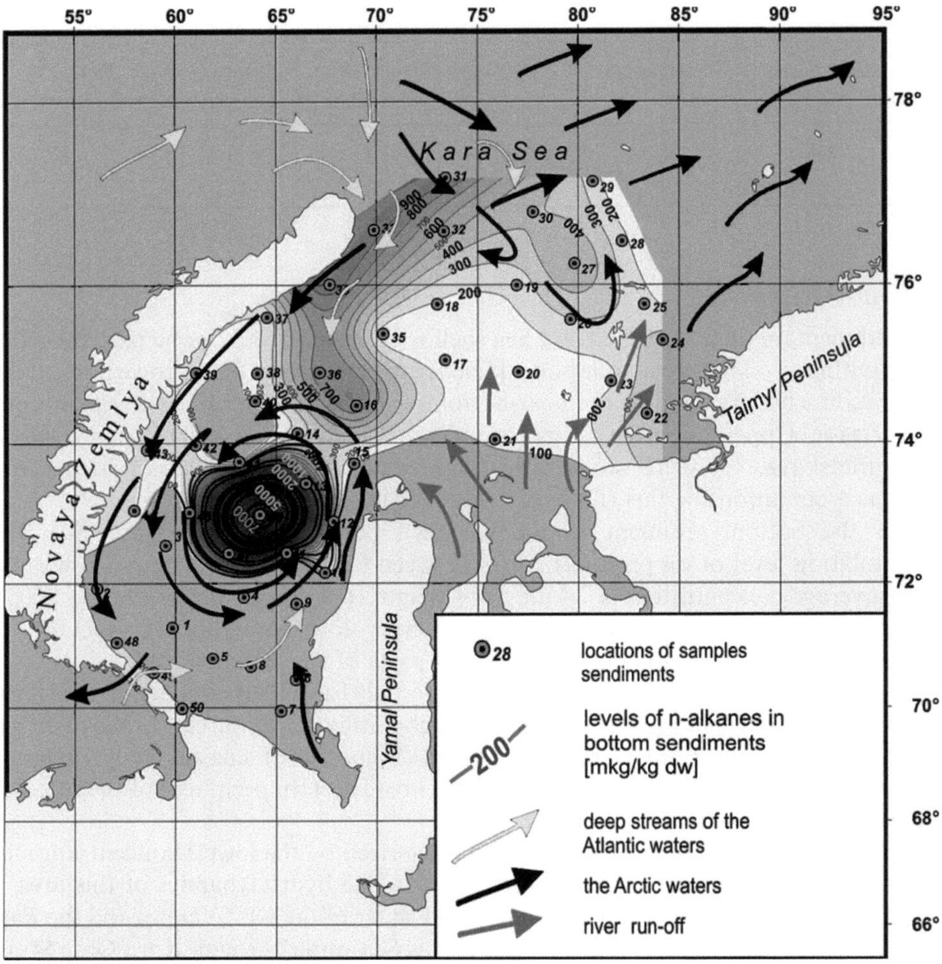

Figure 1.26. Distribution of the accumulation of aliphatic hydrocarbons in the upper layer of the sediment in the Kara Sea.

traces of anthropogenic pollution of Atlantic water by oil by-products, but to some extent this may reflect the invasion of light oil hydrocarbons from the sedimentary mantle (Ilyin, 2007b).

The accumulation of paraffins in the sediment of the Kara Sea is much higher than that in different zones of the Barents Sea. The distribution of alkanes is a peculiarity of the Kara Sea that was not noted in the other northern seas. It is believed that the alkane compound C_{20} can serve as a marker of the spread of brackish water ecosystems and the influence of Siberian rivers.

The spatial fluctuation of the concentration of paraffins depends little on the lithological type of sediments and is determined by the flow of Atlantic and surface waters. Sharp deviations in background geochemical parameters in the area of the oil

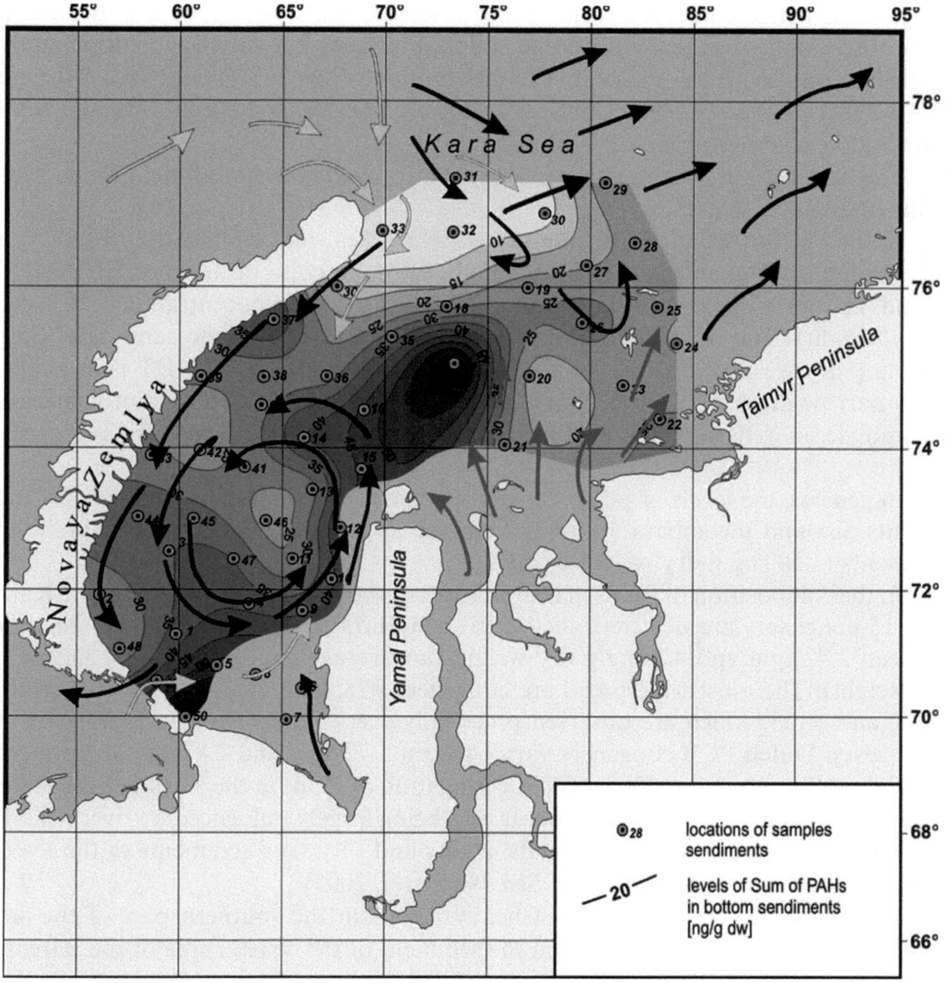

Figure 1.27. Distribution of the accumulation of PAHs in the upper layer of the sediment in the Kara Sea.

fields are supposedly brought about by natural epigenetic processes such as the superposition of the migration flow of light hydrocarbons on the natural geochemical background.

As far as the spread of PAHs (6.3–88.8 ng/g dry weight) is concerned, two areas of increased concentrations are distinguished, both of which are in the southern part of the sea. These are the areas adjoining the Ob′ mouth and in the coastal zone of Vaigach Island (Figure 1.27).

Typically, the oil bi-nuclear PAH naphthalene and its methylated homologues are detected over the entire sea area. The total levels of naphthalenes vary from 10% to 48% of the total levels of PAHs. The high accumulation of naphthalenes is typical of the areas where Ob′–Yenisei runoff mixes with seawater (i.e., frontal zone areas). Over much of the sea area, anthracene and fluoranthene poly-arenes, predominantly of anthropogenic genesis, prevail. The distribution of benzo(*a*)pyrene (0.0–2.4 ng/g dry weight) in the sediment of the Kara Sea coincides in general with the total distribution of PAHs (Ilyin, 2007b).

The structure and character of the spread of PAHs is no different from non-polluted areas of the sea.

In the bottom sediment of the southern Kara Sea, the accumulation of OCs in very insignificant quantities has been noted. The total concentration of HCH compounds is from 0.24 ng/g to 1.02 ng/g dry weight (an average concentration of 0.54 ng/g dry weight). The average total concentration of DDT is at the same level (i.e., 0.41 ng/g dry weight). Spatial variability in the concentrations of DDT is in a relatively narrow range between 0.30 ng/g and 0.6 ng/g dry weight. Hexachlorobenzenes accumulate in sediments in concentrations between 0.09 ng/g and 0.42 ng/g dry weight.

In general, the levels of persistent pesticides is lower than in the sediment of the Barents Sea and incomparably lower than the approximate level for non-polluted sediments (<500 ng/g dry weight by SFT).

In the composition of PCBs in the sediment of the southern periphery of the Kara Sea, 15 congeners are detected, their total concentration changing in the sea area between 2.72 ng/g and 4.24 ng/g dry weight (an average concentration of 3.49 ng/g dry weight). The most widespread are congeners #28, #31, #52, #101, #105, #118, #138, and #153, which are observed practically everywhere. The concentrations of these seven Dutch PCB congeners vary between 1.72 ng/g and 2.83 ng/g dry weight (*Quality of Sea Water ...*, 2005). The accumulation of PCBs in the Kara Sea is similar to that in the coastal area of the Barents Sea, being largely influenced by river runoff.

The concentration of heavy metals in silty and silty sand sediments of the sea is presented in Table 1.18 (*Quality of Sea Water ...*, 2005).

The accumulation level of most heavy metals in the southern part of the sea corresponds to the levels of metals in the sediment of the coastal area of the Barents Sea; as in the Barents Sea, the level is probably determined by the influence of surface runoff. However, in the Kara Sea the increased accumulation of copper, nickel, cadmium, tin, and mercury is more pronounced. Based on the available data, it is impossible to trace the precise localities where there is such a high level of metals and mercury in the sediment. The increase of concentrations may be connected not only

Table 1.18. Concentration of heavy metals and microelements in the surface layer of the bottom sediment of the Kara Sea (μg/g dry weight).

Metal	*Mn*	*Zn*	*Cu*	*Ni*	*Pb*	*Co*	*Cd*	*Sn*	*Cr*	*Hg*	*As*
Average concentration	214	28.3	36.2	34.7	14.2	5.37	0.18	33.5	12.6	0.001	18.7
APC_0 [a]	—	<150	<35	<30	<30	—	<0.25	—	<70	<0.15	<20

[a] APC_0 = approximately permissible concentrations for non-polluted sediments by SFT.

with the local particularities of the geochemical background in the watershed area, but also with the impact of the Norilsk mining–metallurgic complex. Figure 1.18 shows a high discharge of copper and nickel with Yenisei water. Further understanding of the patterns of the distribution of metals in marine sediments will have to wait for additional studies.

The spatial variability of the accumulation of paraffins depends little on the lithological type of sediment, rather it is determined by Atlantic water and surface water flows. Sharp deviations in background geochemical parameters in the areas of oil fields are apparently determined by natural epigenetic processes (i.e., superposition of the migration flow of light hydrocarbons on the natural geochemical background). In the spread of PAHs, there is a tendency for increased concentrations in the zone where surface water and seawater mixes. The main factor of the increased accumulation of PAHs, including the aromatic compounds of oil genesis (naphthalenes), is the Ob′ and Yenisei runoff (Ilyin, 2007b).

1.3.2.4 Laptev Sea

The sources of pollution of the Laptev Sea have been little studied. However, it can be generalized that anthropogenic pollutants accumulate primarily on the southern shelf over which Siberian river runoff spreads. Lena River runoff is considered the main source of pollutants. The export of oil products by individual rivers varies, but, in general, it may comprise between 110,000 t and 390,000 t. More than 98% of this volume of oil products is supplied by Lena River runoff (Figure 1.18) (*Yearbook of Surface Water Quality* ..., 1997; *Quality of Surface Water* ..., 2004). The rivermouth areas of the Lena and Anabar have the highest concentrations of heavy metals, oil products, and DDT and HCH metabolites.

No discharge of organochlorine pesticides with river water has been noted. The existing concentrations of these compounds in the Laptev Sea environment are connected with trans-boundary transfer by sea currents and atmospheric circulation. For example, the role of Atlantic water entering from the Barents Sea and spreading in the intermediate water layer is well expressed in the distribution of PAHs in the Laptev Sea environment (Ilyin *et al.*, 1998a). Some discharge of OCs and PCBs with surface water probably takes place, but it is seasonal and connected with the period of snowmelt and release of aerosol fallout of these pollutants. Ports represent local areas of coastal water pollution, predominantly by oil products (*Yearbook of Sea Water Quality* ..., 1997).

A major polluting center in the region is the port of Tiksi which is involved in the trans-shipment of oil and oil products. Oil trans-shipment in Tiksi was organized in 2001 by the Joint Stock National Oil and Gas Company (JS NOGC) SAKHANEFTEGAS and by the Murmansk Shipping Company. Oil from the Talakansky field, stored at the terminal in the riverport of Vitim, comes for trans-shipment and export to the West. Transportation along the river is carried out by Lenaneft-type tankers. At the terminal in Tiksi, oil is transferred to marine tankers. The transportation of oil products by the Lena River from Vitim is intended mainly to meet the domestic requirements of the Sakha Republic. The volumes of raw

Table 1.19. Concentrations of heavy metals and microelements in the surface water layer of the Laptev Sea (μg/L).

Metal	*Mn*	*Zn*	*Cu*	*Ni*	*Pb*	*Co*	*Cd*	*Sn*
Range of concentrations	0.05–3.4	0.2–11.3	0.1–2.3	0.05–1.0	0.05–0.50	0.03–0.09	0.05–0.38	0.05–0.07
MPC	50	50	5	10	10	10	10	112

materials transported for export are small and comprised 58,000 tons in 2002. In the future, as a result of the increasing production of raw materials it is planned to export oil towards the East (Ilyin and Shavykin, 2007).

Aquatic environment

The average concentration of oil products in the water of the Laptev Sea is no greater than 0.04 mg/L. The maximum concentrations of oil products are observed in the water of Buor-Khaya Bay in the proximity of Tiksi (up to 0.08 mg/L, about 1.5 MPC). The increased concentration of oil products is typical of the water areas of ports.

The levels of OCs range from 0 ng/L to 1.3 ng/L for DDT compounds and from 0.03 ng/L to 2.05 ng/L for HCH compounds. Increased concentrations of OCs are detected in coastal waters impacted by river runoff. An increased concentration of DDT is often recorded in the vicinity of Bolshoy Begichev Island, while HCH concentrations are high in the mouth of the Anabar River, although these concentrations of pesticides still remain well below the MPC. The levels of concentrations of OCs in the water of southern parts of the Laptev Sea and the Barents Sea are in good agreement, although the ratio of DDT/HCH chemicals differs. In the Kara Sea the concentration of OCs is less than in the Laptev and Barents Seas.

The concentration of heavy metals in the upper water layer of the Laptev Sea is low (Table 1.19). In the summer–autumn period, there is spatial variability, with increased concentrations of metals recorded in the mouth areas of the coastal zone (*Yearbook of Sea Water Quality* ..., 2004).

Bottom sediments

Oil products reaching the marine environment with river runoff mostly accumulate in the bottom sediment of the shelf zone. A number of rivermouths form an extensive zone along the coast. In the bottom sediment in this part of the sea the accumulation level of oil products is between 10 μg/g and 180 μg/g dry weight. The highest concentrations of oil products are recorded in the sediment of Buor-Khaya Bay (in the proximity of Tiksi port) where there is up to 180 μg/g dry weight and in Khatanga Bay where there is up to 86 μg/g dry weight (*Yearbook of Sea Water Quality* ..., 1997; *Yearbook of Sea Water Quality* ..., 2004). These concentrations are much higher than in other Arctic Seas (e.g., the White and Kara Seas) and are comparable with the economically developed coastal areas of the Barents Sea.

The deepwater regions of the Laptev Sea have not been studied. The patterns of accumulation of polyaromatic hydrocarbons in sea sediments have been better investigated. In the southern shallow part of the sea, the concentration of PAHs varies between 13 ng/g and 40 ng/g dry weight. In the composition of PAHs the following compounds are detected: fluorene (0.2–3.2 ng/g dry weight), phenanthrene (0.0–10.1 ng/g dry weight), fluoranthene (0.2–6.2 ng/g dry weight), benzo(*b*)fluoranthene (3.2–18.3 ng/g dry weight), benzo(*g*, *h*, *i*)perylene (0.1–12.4 ng/g dry weight), and in the mouth areas benzo(*a*)pyrene (0–0.24 ng/g dry weight) (*Yearbook of Sea Water Quality* ..., 1997, 2004). Although the total accumulation of PAHs in the coastal zone of the sea is close to the level of PAHs in the coastal region of the Barents Sea, naphthalene accompanying low-weathered oil products has not been detected. The composition of PAHs is primarily made up of pyrogenic poly-arenes and petrogenic arenes brought by river runoff.

The concentration of aromatic hydrocarbons in the bottom sediment of the deepwater part of the Laptev Sea is quite low. The total concentration of PAHs in this area varies from 1 ng/g to 66 ng/g dry weight, but only a few sampling points have concentrations greater than 10 ng/g (Figure 1.28). These concentrations are much

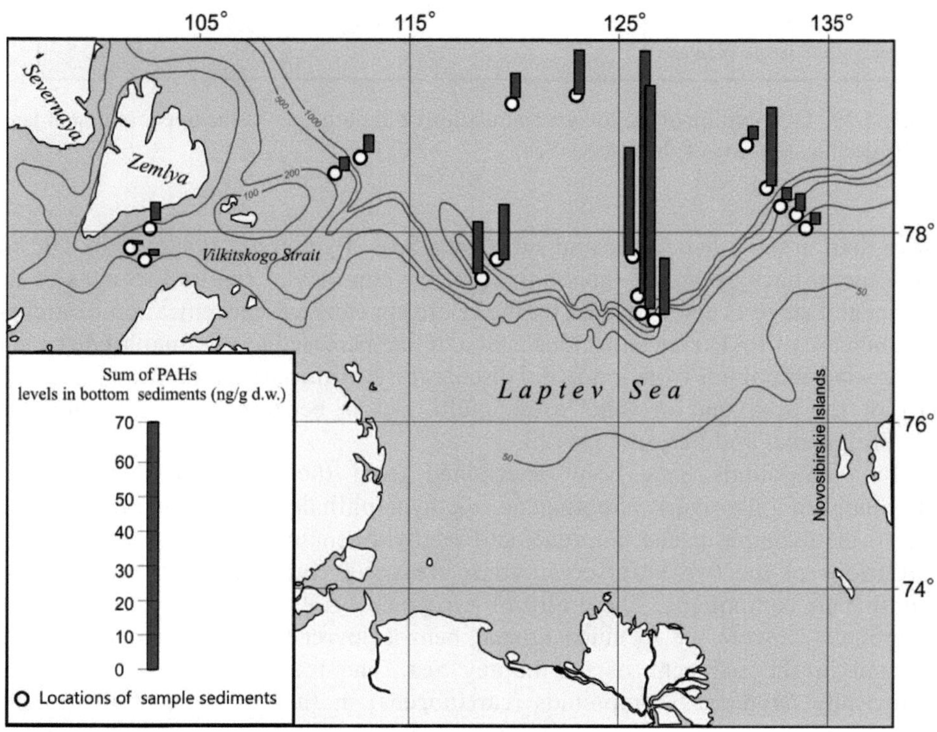

Figure 1.28. Distribution of the accumulation of PAHs in the upper sediment layer of the deepwater area of the Laptev Sea.

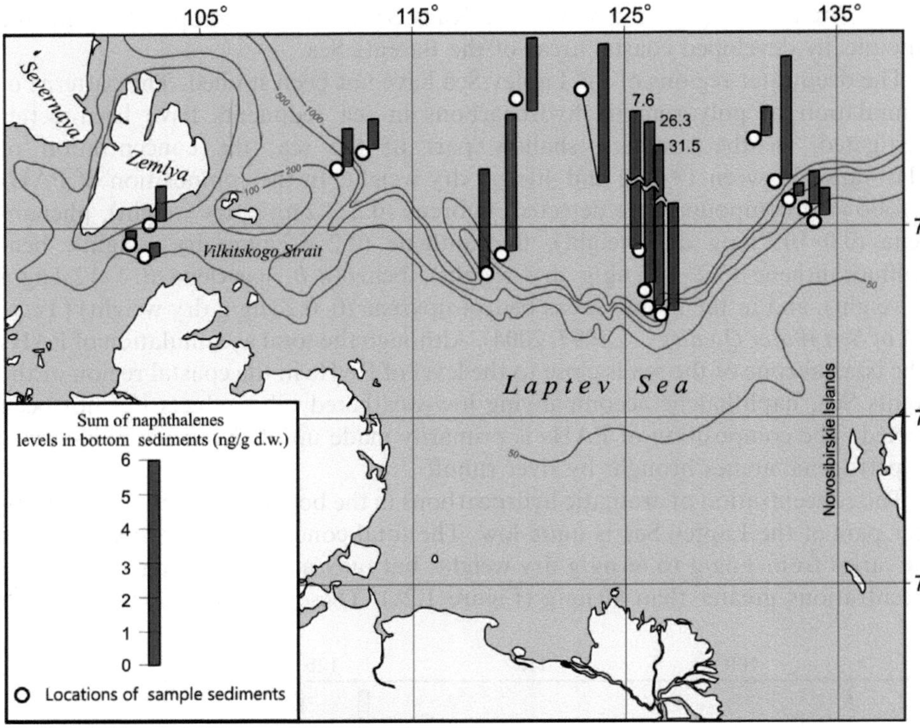

Figure 1.29. Distribution of the total accumulation of naphthalene in the upper sediment layer of the deep-water area of the Laptev Sea.

lower than in the other Arctic and sub-Arctic Seas (Ilyin *et al.*, 1998b; Ilyin, 2007c). The general background of the sediment in the continental rise and troughs of the continental slope is distinguished by the relatively elevated concentration of arenes. The increase of PAH concentrations is due to the increased level of naphthalene and pyrogenic compounds from industrial discharges. Anthracene dominates over a large part of the area, but in some areas multi-nuclear benzo($b+j+k$)fluoranthene, benzo(e)pyrene, and perylene prevail.

Six compounds have been determined from the spectrum of arenes (all potentially of oil origin): naphthalene, methylnaphthalene, di-methylnaphthalene, and tri-methylnaphthalene, fluorene, and methyl-phenanthrene (Figure 1.29). Concentrations of the two latter compounds are small compared with the group of naphthalene compounds. The group of pyrogenic PAHs includes the carcinogenic compounds benzo($b+j+k$)fluoranthene, benzo(a)pyrene, and benzo(a)anthracene, detected in the sediment of the Laptev Sea. The total concentration of these ecologically dangerous compounds (carcinogens) in the sediment of the Laptev Sea is from 0.1 ng/g to 24 ng/g. The predominant accumulation of carcinogens in the study area of the Laptev Sea occurs in the aforementioned area of the continental rise and the slope, including the sediments in slope troughs (Figure 1.30). Continental

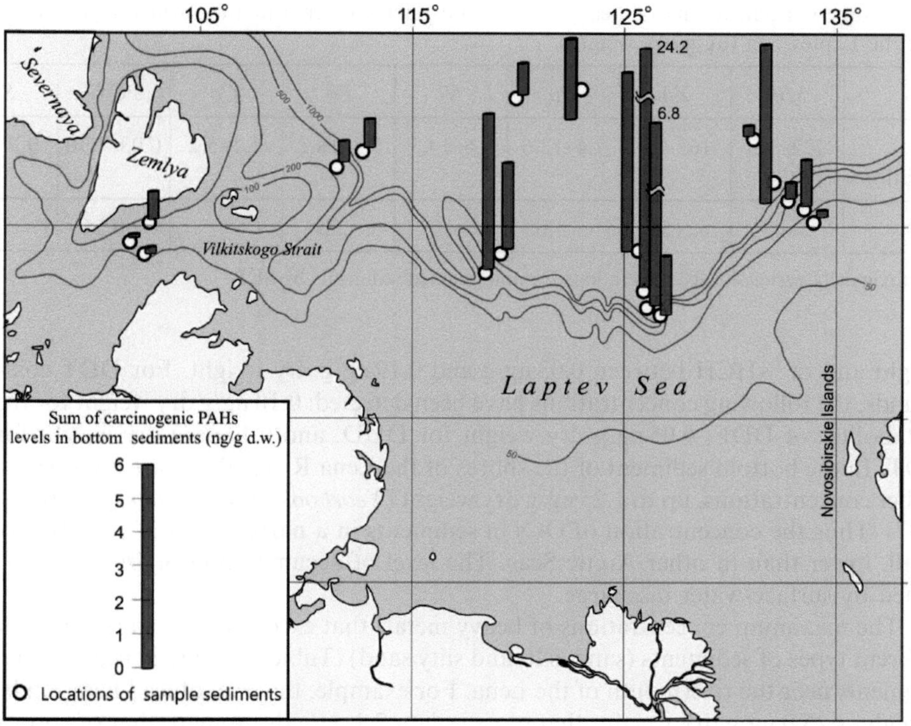

Figure 1.30. Distribution of the total accumulation of carcinogenic PAHs in the upper sediment layer of the deep-water area of the Laptev Sea.

river runoff is a strong source of PAHs reaching the Laptev Sea ecosystem, where there are both anthropogenic and natural PAHs such as chrysene and phenanthrene. Over much of the area, there is an equal proportion of anthropogenic and natural PAHs. Only in a few areas is the dominance of pyrogenic PAHs noted. These areas are probably the consequence of Kara Sea water and river runoff spreading along the continental slope (Ilyin *et al.*, 1998b).

The morphological area in which there are sudden increases in depth determines the formation of a frontal zone where freshwater and seawater mix, while in the deep-water layers, Atlantic water approaches the continental slope (Ilyin *et al.*, 1998a). Sediment flows in the region where there are sudden increases in water depth along the continental troughs determine the areas of increased concentration of PAHs near the base and in the troughs themselves.

At the western margin of the sea, the areas of spread of Kara Sea water through Vilkitsky Strait are probably associated with the increased accumulation of pyrolytic PAHs in sediments.

A zone in which there is a relatively elevated accumulation of organochlorine compounds covers a significant part of the coast from Olenek Bay to Cape Buor-Khaya. The concentrations of α-HCH range between 0.04 ng/g and 0.15 ng/g dry

Table 1.20. Average concentrations of heavy metals and microelements in the surface layer of bottom sediments of the Laptev Sea (μg/g dry weight).

Metal	*Mn*	*Zn*	*Cu*	*Ni*	*Pb*	*Co*	*Cd*	*Sn*
Range of concentrations	222–12	163–21	7.4–13.6	9.6–14.3	7.3–18.2	1.1–5.2	0.03–1.70	0.7–1.8
APC_0 [a]	—	<150	<35	<30	<30	—	<0.25	—

[a] APC_0 = approximately permissible concentrations for non-polluted sediments by SFT.

weight and of γ-HCH between 0.03 ng/g and 0.19 ng/g dry weight. For DDT compounds, the following concentrations have been detected: 0.10 ng/g dry weight for the metabolites of DDE, 0.05 ng/g dry weight for DDD, and 0.12 ng/g dry weight for DDT. In the bottom sediment of the shores of the Lena River, PCBs are recorded in higher concentrations, up to 1.25 ng/g dry weight (*Yearbook of Sea Water Quality* ..., 1997). Thus the concentration of OCs in sediments in a narrow coastal zone is very small, lower than in other Arctic Seas. The level of accumulation of OCs is determined by surface-water discharge.

The maximum concentrations of heavy metals that exceed the average levels for different types of sediments (sand, silt, and silty sand) (Table 1.20) are in near-coastal sediments near the rivermouth of the Lena. For example, the level of lead here can be as high as 25.0 μg/g dry weight, that of cadmium 2.1 μg/g dry weight, that of copper 20.0 μg/g dry weight, and that of zinc 220 μg/g dry weight (*Yearbook of Sea Water Quality* ..., 1997).

The concentration of all metals is below the approximately permissible concentrations (APC_0) for non-polluted sediments (Molvaer *et al.*, 1997). The accumulation of zinc and cadmium is characterized by a relatively increased level compared with the other seas. The significant discharge of zinc occuring with water of the Lena River (Figure 1.18) probably reflects the particular geochemical background of the watershed basin.

2

Study region and environmental datasets

This chapter describes the geography of the study region and the observational environmental data used in these analyses. The geographical and oceanographic features of the study region are described in Section 2.1, which is organized in three sub-sections: the Ob′ and Yenisei River systems, the Kara Sea region, and the Nordic Seas and adjacent seas. Section 2.2 presents an overview of the environmental data that are used in this study, organized in three sub-sections: databases and the information system, natural environmental data (e.g., hydrological, oceanographic, and geophysical), and pollution data.

2.1 GEOGRAPHICAL DESCRIPTION OF THE STUDY REGION

The study area covers a substantial part of the Arctic, a region where there are many sources of radioactive and other industrial pollution in both its marine and terrestrial areas. The marine areas include the deep-water Arctic Basin, the relatively shallow shelf seas of the eastern Arctic, and the seas of the north European sector of the Arctic and North Atlantic—for example, the Nordic Seas (Greenland, Iceland, and Norwegian) and the Barents Sea. The terrestrial areas include the vast basins of the great Siberian rivers the Ob′ and Yenisei. The southern boundary of the region is determined by the location of radioactive contamination sources in Siberia, namely the main enterprises of the Russian Nuclear Complex, and in Europe, Sellafield on the Irish Sea and La Hague on the English Channel.

The total area of the marine region under consideration comprises about 20 million km^2. The watershed area of the rivers carrying freshwater to the Kara Sea alone adds more than 6 million km^2 (*Atlas of the Oceans: Terminology, Definitions, Reference Tables*, 1980; Volkov *et al.*, 2002). A substantial part of the study area is located north of the Arctic Circle. The seas and rivers here are largely ice-covered for much of the year or under the influence of processes associated with the ice cover.

It is worth noting that the fraction of the total mean annual freshwater runoff of the Ob′ and Yenisei predominates (84.9%) the total freshwater runoff volume of rivers flowing into the Kara Sea; this comprises almost half (43.6%) of the volume of river runoff to the Arctic Ocean from the Eurasian continent or more than one-fourth of the entire volume of runoff to the Arctic Ocean. The ratio of mean annual freshwater inflow to the Kara Sea relative to the total sea area is 152 cm, the largest of any sea in the World Ocean. This is primarily due to the vast area of the watershed sea basin in relation to the area of the sea. For the Kara Sea, this ratio is 6.82, whereas it is 2.8 for all the Russian Arctic Seas added together, 1.53 for the Arctic Ocean basin, and 0.33 for the World Ocean (*Atlas of the Arctic*, 1985; Volkov *et al.*, 2002). Therefore, the Ob′ and the Yenisei strongly influence the Kara Sea's oceanographic structure and ice regime, and are the main transfer pathways of radionuclides to the Arctic seas. Within the basins of these rivers, radioactive contaminating components undergo significant transformation.

Three major geographical regions can be delineated as being threatened by anthropogenic activity or hazards in the sense of potential contamination. They are therefore of greatest interest for understanding the real and potential distribution and evolution of radioactive contaminants. These are (1) the Ob′–Yenisei region, (2) the Kara Sea, and (3) the Nordic Seas and the adjacent North Atlantic, Barents Sea, and part of the Arctic Basin.

There are many descriptions of these regions in the scientific literature; among them we mention the three well-known books written in Russian devoted to the oceanographic conditions of the Arctic Ocean that have already become classics (Timofeyev, 1961; *The Soviet Arctic: The Seas and the Islands of the Arctic Ocean*, 1970; Treshnikov *et al.*, 1980). More recently, there is a monograph on the Arctic river regime (Nikanorov *et al.*, 2007), the first ever monograph on the oceanographic regime of the Kara Sea (Volkov *et al.*, 2002), a monograph on the Arctic environment and climate (Bobylev *et al.*, 2003), and *The Nordic Seas: An Integrated Perspective* (Drange *et al.*, 2005a), which contain geographical descriptions of these areas in much detail.

Here we present the characteristics of these three regions that should be of interest to those concerned with the spread and transformation of radionuclides (and other contaminants), and with modeling the processes involved in the transport of contamination.

2.1.1 The Ob′ and Yenisei River systems

The most significant sources of radioactive pollution are situated in Siberia in the catchment basins of the great Siberian Rivers the Ob′ and Yenisei, which are two of the three largest Eurasian rivers and the two main rivers flowing into the Kara Sea (Figure 2.1, see color section). As noted above, together they provide about 84.9% of total freshwater inflow to the sea. The watersheds of the Ob′ and Yenisei Rivers extend from the Kara Sea in the north, to China and Mongolia in the south, and from the Ural Mountains in the west, to the middle of the Central Siberian Plateau in the east. The watersheds of these rivers share a common border south of 62°N and then the rivers run separately to the Kara Sea.

2.1.1.1 The Ob′ River

General description of the river system. The Ob′ River watershed covers the territory of three countries—the Russian Federation, Kazakhstan, and China. Most of the watershed (about 75%) is located within the Russian Federation and comprises two parts—the plains region represented by the West Siberian lowland, and the Altai mountainous area. The area of interest for this study is located completely in the plains region.

The West Siberian plain extends 2,500 km from north to south and 1,500 km from west to east. It has a flat and uniform relief (with a slight elevation northward) and is a combination of lowlands and elevated areas, reaching heights of 250 m to 285 m. The lowest parts are in the middle Ob′ (near the junction of the Ob′ and Irtysh), in the regions of the Ob′ delta, and in the southern part of Ob′ delta. The entire West Siberian lowland is extremely boggy, containing about half of all bogs and marshlands of Russia (*Russia* ..., 1999).

Of all the Russian rivers the Ob′ has the largest watershed area (about 2,975,000 km^2) and is the third largest river by amount of discharge (after the Lena and Yenisei). Average discharge is estimated at 402 km^3/year (Mikhailov, 1997; *Russia*, 1999), or according to other data (Volkov *et al.*, 2002) 406 km^3/year (into Ob′ Bay) and 532 km^3/year (into the sea), with a maximum volume in 1979 of 582 km^3/year and 752 km^3/year, respectively. The length of the Ob′ is 3,680 km from the confluence of the Biya and Katun′ in the Altay region to Ob′ Bay in the Kara Sea. Its hydrographical network is characterized by sharp asymmetry—similar to the Yenisei system, but with the opposite configuration—most of the watershed area (67% of the total area) is located on the left bank. In the region of the lower Irtysh and lower Ob′ there are about 70,000 streams, 89% of them being less than 10 km long (Figure 2.2, see color section).

Another typical feature is the presence of areas of "inner" discharge (i.e., not flowing into the Ob′ River system) which cover 15% of the watershed area. The region also has abundant lakes (over 450,000), mainly small ones with surface areas less than 1 km^2 and depths 2 m to 5 m. The Ob′ is usually divided into three major parts based on hydrographic conditions and the river regime: (1) the upper Ob′ (from the confluence of Biya and Katun′ to the confluence of the Ob′ and the Tom′ near the city of Tomsk); (2) the lower Ob′ (from the Irtysh mouth to Ob′ Bay), and (3) the middle Ob′ (from the Tom' mouth to the Irtysh mouth in the Khanty Mansiysk region). The upper Ob′ is not considered in this study, as it is located upstream from the pollution sources (the Tomsk Chemical Combine and the Mayak PA, Chelyabinsk). The most significant tributary of the Ob′ is the Irtysh (4,248 km in length, with a watershed area of 1,643,000 km^2) which has a well-developed and complicated river network of its own.

Other sources of freshwater inflow to the Kara Sea are the relatively small rivers flowing directly into Ob′ Bay. Though they are not part of the Ob′ River system, information on the three most significant rivers (the Nadym, Pur, and Taz) is included in this description.

Within the area under consideration, the Ob′ is a typical floodplain-type river. The width of the valley is 20 km and the floodplain ranges from 1 km to 5 km. River depths during periods of drought are 2 m to 6 m, and current speeds are 0.3 m/s to 0.5 m/s, with maximum currents up to 2 m/s (*Russia* ..., 1999). After its confluence with the Tom′ the Ob′ flows into the taiga belt.

Downstream from its confluence with the Tom′ the Ob′ flows into a boggy taiga floodplain. The wide and flat interfluves (region of higher land between two rivers in the same drainage system) of the Irtysh and Ob′ are covered by conifer forests and bogs. The valley width increases—from 30 km to 50 km—and the floodplain widens from 20 km to 30 km. Within the floodplain there are numerous lakes and oxbow lakes; the river channel frequently divides into many branches and small channels. Depths during periods of drought are 6 m to 8 m, with current speeds from 0.2 m/s to 0.5 m/s (maximum 1.8 m/s). In this stretch, several large rivers—e.g., the Ket′, Chulym, Tym, Vasyugan, Agan, Vakh, and Irtysh—flow into the Ob′. The Irtysh within the area of interest flows into the taiga zone. Its main tributaries are the Ishim and Tobol. Valley width varies from 5 km to 19 km. After its junction with the Tobol, the Irtysh expands, its valley width increasing up to 35 km at the junction with the Konda, just before the confluence with the Ob′. After its confluence with the Irtysh, the Ob′ becomes a powerful river. During the spring flood, the width of flooded areas sometimes has been known to reach 40 km to 50 km. Maximal depths are 15 m to 20 m, current speeds vary from 0.2 m/s to 0.5 m/s even reaching 1.6 m/s during the spring flood. The most important tributary in this stretch is the Severnaya Sos′va.

The Ob′ delta proper starts after Bol'shiye Yary Island. Here the river consists of several channels, the most significant being the Khamanel′skaya Ob′ and the Nadymskaya Ob′ (both 20 km wide). Ob′ Bay is a typical estuary formed as a result of flooding of part of the river valley. Ob′ Bay includes Taz Bay, into which the Taz and Pur Rivers flow. The length of Ob′ Bay to its exit into the Kara Sea is 760 km, its surface area is 40,800 km^2 (together with Taz Bay 48,550 km^2), its width varies from 35 km to 80 km and its depth from 10 m to 12 m (Mikhailov, 1997).

The northern part of Ob′ Bay is where the frontal zone between fresh river and saline seawater is located. This frontal zone is characterized by sharp gradients in the physical and chemical properties of the water; the frontal divide (halocline) is inclined in the vertical plane toward the bay and its location changes significantly according to the seasonal variability of river inflow. In March/April, when river inflow is minimal, the intrusion of seawater into Ob′ Bay is maximal, and *vice versa* in July/August. In the summer, seawater with salinity over 30‰ enters Ob′ Bay near the bottom and penetrates up to 10 km, in the autumn up to 210 km, and in the winter 340 km or more (Mikhailov, 1997). Seasonal changes in halocline location are about 300 km. Under tidal influence its location changes up to 10 km, and under storm surges it can change up to 100 km. The limit of saline water in Ob′ Bay occurs at the confluence of Ob′ and Taz Bays.

River discharge and water level. The large latitudinal range and complicated structure of the Ob′ River system leads to specific features in the seasonal distribution of river discharge. According to the mean monthly values for the time period considered, at most stations the lowest level of river discharge is observed just before the spring flood (January/March), then comes the height of the spring flood (typically the two months between April and July). At the end of summer and autumn, during the autumnal drought period, discharge gradually decreases (Figure 2.3). The locations of observation points chosen for the study within the Ob′ catchment basin are presented in Figures 2.4 and 2.5, see color section for both).

The Techa (Pershinskoye) and Iset′ (Isetskoye) have maximum discharge in May; in July the discharge decreases and the drought period ensues. For Tobol at Lipovskoye, the spring flood is observed later (no increase in discharge until April) and has a longer duration (May/June). The flood at Irtysh has its source in the upper mountainous part of the watershed; it usually starts at the beginning of April and can be observed for 120–130 days. After the spring flood, the summer/fall drought period starts (50–70 days). The duration of the winter drought period is usually 140–160 days.

The spring flood for the Tom′ at Tomsk occurs between April and June, with the highest values observed in May. For the Ob′ at Kolpashevo and Prokhorkino, the spring flood starts later (in May); in June and July the values decrease but are still high. Due to its northernmost location, the lower Ob′ (Belogorye and Salekhard) has the latest spring flood, with discharge starting to increase only in May, reaching a maximum in June, a slightly lower amount in July, and then starting to decrease. For the Pur, Taz, and Nadym rivers, which flow into Ob′ Bay, the spring flood takes place from late April to early July, peaking in June.

The spring flood raises water levels much more dramatically than during the autumn rain floods. For Irtysh the level goes up 5 m to 11 m at Tobol′sk and 6 m to 10.5 m at Khanty-Mansiysk. During autumnal rain floods the water level at these stations may increase by 1 m to 1.5 m (maximal values 2 m to 2.5 m). During the spring flood, the water level for the Ob′ at Salekhard rises on average some 6.5 m, although only 1 m at the seaside boundary of the Ob′ delta.

Spring flooding of the Pur at Samburg increases the water level by 5 m; closer to Taz Bay (51 km away) this diminishes. In August/September, autumn floods bring an increase of 0.2 m to 0.7 m. Northerly winds result in storm surges with a 1.5 m level increase (0.5 m at Samburg).

Storm surges significantly influence the water level in the Ob′ delta and Ob′ Bay during periods of low river runoff (mid-July to mid-October). Northerly and northeasterly winds result in an increase of water level, while under southerly and southwesterly winds, the level decreases. At the seaside boundary of the delta, storm surges may reach 2.5 m to 3 m and penetrate the river 351 km from the delta boundary. Tidal influences change the water level by 0.3 m to 0.4 m although these changes are not noticeable upstream of the Ob′ delta (Mikhailov, 1997).

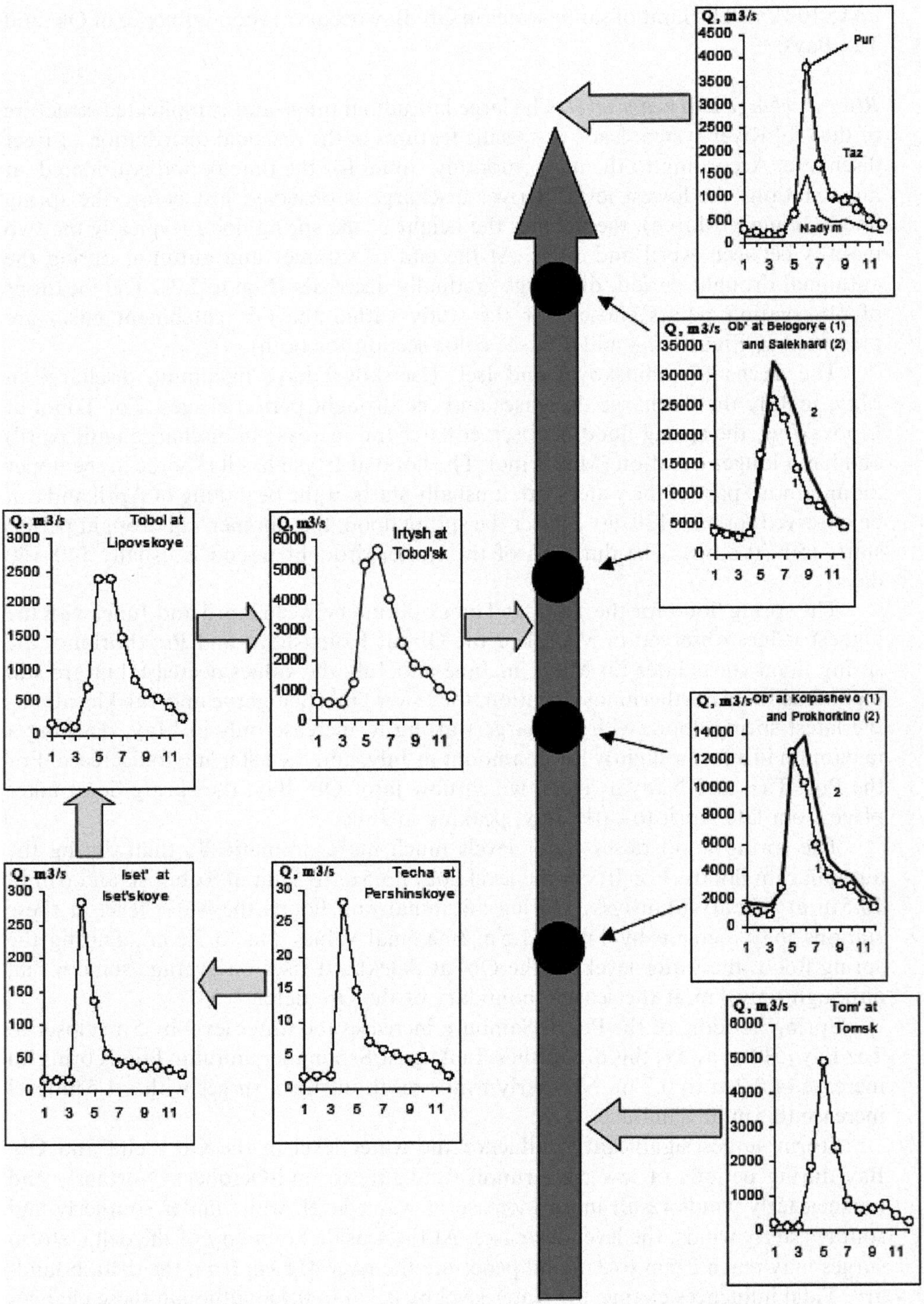

Figure 2.3. Seasonal variations of monthly discharge along the Ob′ River system.

Limited water resources in the steppe and forest–steppe zone and industrial demands led to the creation of reservoirs, which significantly affect the seasonal hydrological regime. The largest reservoir in the Ob′ system (within Russian territory) is the Novosibirsk reservoir. It was created in 1956 for the purposes of managing the seasonal distribution of river runoff and for producing power at Novosibirsk Hydroelectric Power Plant (HPP).

Within the Irtysh River system, the greatest water demand comes from the Ekaterinburg industrial region, including the cities/towns of Ekaterinburg, Verhnyaya Pyshma, and Sredneural′sk. The water supply for industrial and communal purposes comes from water reservoirs constructed along the Iset′ River (e.g., Isetskoye, Verkh-Isetskoye, Gorodskoye, and Nizhneisetskoye).

The Irtysh water regime is significantly affected by discharge regulation at the Bukhtarmisnkaya and Ust′-Kamenogorskaya HPPs and their reservoirs. The Bukhtarminskoye reservoir was created in 1960 and has a surface area of 53 km^2. Water from this reservoir then reaches the smaller Ust′-Kamenogorskoye reservoir with a surface area of 0.8 km^2. The construction of reservoirs has resulted in a decrease in maximum water levels by 1 m.

Hydrological regime: suspended sediments and water turbidity. Rivers transport a large volume of sediments, especially during spring floods (and other significant floods caused by rainfall), which can have a destructive effect on river beds. Floods can alter the river configuration, location of streams, shallows, islands, and the axis of maximum depth.

Seasonal variations in suspended sediment load are very similar to variability in the water discharge regime (State Water Cadastre, 1974). Maximum sediment load and water turbidity occur during the spring flood, when washout from the floodplain as well as bottom and lateral erosion are highest. Lowest sediment load and water turbidity are typical of the winter low-water period. Data on suspended sediment load and water turbidity for the Ob′ River system are shown in Figures 2.6 and 2.7 (see color section for both).

The Irtysh at Tobol′sk has an average annual amount of suspended sediment of about 180,000 t/yr. The average water turbidity in the stretch from Tobol′sk to Khanty-Mansiysk ranges from 25 g/m^3 to 50 g/m^3. Sediment runoff at Tobol′sk is highest during the spring flood (70% to 97% of the annual amount, 89% on average), while during the summer/autumn it is typically 9% to 17%, and in the winter only 2% to 6%. However, in years with significant rain the summer/autumn component can be as large as 20% to 30%.

In the lower Ob′ average turbidity is less than 25 g/m^3. The annual sediment load at Salekhard is 13,000,000 t/yr. Water turbidity is low (33 g/m^3), while during floods it may increase up to 70 g/m^3 or 100 g/m^3 (in rare cases up to 150 g/m^3). Suspended sediment load for the Pur at Samburg is 620,000 t/yr and for the Taz at Sidorovsk 580,000 t/yr (at Taz delta 910,000 t/yr). Water turbidity is very low: 25 g/m^3 for the Pur and 20 g/m^3 for the Taz (Mikhailov, 1997).

Table 2.1. Suspended particle size distribution (% by mass) and median size values.

Particle size (mm) / *Date*	0.5–0.2	0.2–0.1	0.1–0.05	0.05–0.01	0.01–0.005	0.005–0.001	<0.001	*Median value*
Ob′–Novosibirsk								
20.05.1970			2.4	97.6				
24.08.1970	4.0	12.1	19.3	64.6				
14.10.1970			10.7	89.3				
Ob′–Salekhard								
16.05.1978	1.9	3.2	26.1	23.1	45.7			0.02
19.05.1978	6.6	18.2	11.3	23.9	40.0			0.03
03.07.1978	4.6	9.1	8.1	49.3	28.9			0.03
05.08.1978	0.3	33.6	48.2	5.4	6.2	6.3		0.08
09.09.1978	3.6	6.4	19.8	35.1	35.1			0.03
08.1979	0.7	2.4	96.9					
Tobol–Kurgan								
10.04.1974	1.0	5.6	10.6	31.5	51.3			
24.04.1974	0.6	3.4	13.8	32.6	49.6			
05.05.1974	3.8	19.3	35.2	41.7				

Data on particle size distribution are relatively sparse, although an example of such information for 1978 and 1979 (State Water Cadastre, 1974) shows that there are variations between different phases of the river regime (Table 2.1). During high-water periods, particles between 0.05 mm and 0.005 mm are the most common. Smaller size particles are rare and particles less than 0.001 mm were not observed at all during the entire period considered. The share of large particles is low (about 1% to 6%). It is lowest during the summer and winter drought periods and increases during floods due to higher current speeds.

Ice conditions. At the Ob′ estuary, ice appears in mid- or late October, and ice break-up starts on June 1 on average. At the Pur and Taz deltas, ice appears earlier (on average October 11) and remains longer, with break-up starting on June 10. At Samburg, a stable ice cover appears on 15 October. In mid-April the ice thickness is 120 cm at Samburg and 96 cm at Sidorovsk. The ice period for both rivers is 240 days (Mikhailov, 1997).

The ice regime in the Ob′ Bay is very complicated. The natural period of navigation is only 70–90 days and can be prolonged only by the use of icebreakers. Ice appears at the coasts of Ob′ Bay at the end of September/beginning of October. In the first 10-day period of October, the mouths of the Ob′, Nadym, Pur, and Taz Rivers become covered by floating ice. Fast ice appears later, in the second 10-day period of October.

Ice break-up typically starts in the deltas during the first 10 days of June. At the middle and end of June, ice starts to break up in the southern parts of Ob′ and Taz Bays. At the beginning of July these regions become ice-free and ice starts to break up in the rest of Ob′ Bay. The middle of Ob′ Bay becomes ice-free in the middle of July, and by the end of July the whole of Ob′ Bay is ice-free.

Parameters of cross-sections along the Ob′. Geometrical parameters (section surface area, width, depth, level inclination) and current speed at observation points along the Tom′, Ob′, Techa, Tobol, and Irtysh show significant variability related to the different phases of the river regime. Table 2.2 presents statistical estimates of these variations (State Water Cadastre, 1970, 1974).

Measurements show that the water cross-sectional area during the spring flood at large rivers such as the Ob′, Tobol, and Irtysh is about twice as large as the minimum observed during the low-water period; for smaller rivers such as the Techa and Tom′ this ratio can be much higher.

The mean annual amplitude of river width along the Ob′ varies from 200 m at Kolpashevo to over 1 km at Mogochin and Salekhard, reflecting the features of local relief and river regime. The mean depth amplitude of the Ob′ decreases downward from 6 m at Mogochin to 1.6 m at Salekhard. As mentioned earlier, at Novosibirsk, the river regime is under the strong influence of the Novosibirsk reservoir. Maximum values of mean current speed range from 0.6 m/s (Techa) to 1.79 m/s (Tom′), while minimum values are 0.1 m/s to 0.5 m/s. Extreme values are about 2 m/s for the Ob′ at Kolpashevo and Belogorye, and 2.36 m/s for the Tom′. River level inclination on the Tom′ was measured only at Tomsk, with a maximum value of 16 cm/km and a minimum value of 1.4 cm/km.

A more detailed characterization of the Ob′ River basin and its ice conditions, discharge regime, etc., can be found in the Kara Sea monograph (Volkov *et al.*, 2002) and other publications (Antonov, 1970; *Atlas of the North Polar Region Water Balance*, 1996; Nikanorov *et al.*, 2007). Here we focus on some general features of the Ob′ River regime that are significant for modeling radioactive and other man-made contamination.

The Ob′ is a meandering river that runs slowly through marshland. The Ob′ River catchment basin is the lowest part of the catchment area of the Kara Sea; the mean elevation of the Kara Sea catchment area is 415 m, whereas the value for the Ob′ River catchment area is only 274 m (Nikanorov *et al.*, 2007). The natural conditions and the low inclination of the river bed impede the transfer of atmospheric precipitation to river discharge. A considerable proportion of atmospheric precipitation—between 150 mm and 300 mm of the 457 mm annual average—evaporates and much of the rest supplements discharge into marshes and soil saturation. The high percentage of forest land (58%) and marshiness (56% on average, and up to 70% in some places) are additional factors that create unique conditions for the spread and transformation of radionuclides in the Ob′ River basin on the way to the Kara Sea. River water reaching the Ob′ delta and southern part of Ob′ Bay carries a large sediment load—as much as 12×10^6 t/year. River flow in combination with wind-driven and tidal circulation in Ob′ Bay redistributes suspended materials and sedi-

Table 2.2. Annual variations of parameters of water cross-sections in 1970 and 1974.

	Water level[a]	*Water discharge* (m^3/s)	*Water cross-section area* (m^2)	*Width* (m)	*Mean depth* (m)	*Max depth* (m)	*Mean current speed* (m/s)	*Max current speed* (m/s)	*River level inclination* (‰)
Tom′–Tomsk 1970									
Max	654	6,690	3,880	582	6.70	8.20	1.79	2.36	0.16
Min	−14	59	245	226	0.95		0.54		0.014
Amplitude	668	6,631	3,635	356	5.75		1.25		
Ob′–Novosibirsk 1970									
Max	440	5,130	5,260	866	6.1	9.80	1.01	1.34	
Min	130	1,410	2,620	720	3.4		0.46		
Amplitude	310	3,720	2,640	146	2.70		0.55		
Ob′–Mogochin 1970									
Max	980	12,100	10 200	1,270	8.30	16.40	1.19	1.85	
Min	278	108	444	143	2.22		0.24		
Amplitude	702	11,992	9 756	1,127	6.08		0.95		
Ob′–Kolpashevo 1970									
Max	822	11,800	8,570	890	9.6	14.40	1.42	2.14	
Min	218	1,040	3,780	681	5.6		0.26		
Amplitude	604	10,760	4,790	209	4.00		1.16		
Ob′–Prokhorkino 1970									
Max	900	16,300	13,980	1,120	14.2	28.2	1.21	1.6	
Min	236	6,130	7,440	737	10.1		0.22		
Amplitude	664	10,170	6,540	383	4.10		0.99		
Ob′–Belogorye 1970									
Max	957	31,000	24,200	1,680	14.4	26.3	1.32	1.97	
Min	28	3,230	11,800	970	11.3		0.3		
Amplitude	929	27,770	12,400	710	3.10		1.02		
Ob′–Salekhard 1970									
Max	547	39,800	29,400	2,920	12.7	25.5	1.16	1.64	
Min	73	2,920	22,500	1,890	11.1		0.14		
Amplitude	474	36,880	6,900	1,030	1.60		1.02		

ments, and produces very specific conditions for the redistribution of contamination in shallow-water areas.

2.1.1.2 Yenisei River

General description of the river system. The Yenisei is the longest and highest volume river in Russia. It carries water from a vast watershed area with a surface area of

	Water level[a]	*Water discharge* (m^3/s)	*Water cross-section area* (m^2)	*Width* (m)	*Mean depth* (m)	*Max depth* (m)	*Mean current speed* (m/s)	*Max current speed* (m/s)	*River level inclination* (‰)
Techa–Muslyumovo 1974									
Max	187	36.9	39.7	48.1	0.83	1.08	0.93	1.24	
Min	89	0.14	3.78	13.5	0.18		0.06		
Amplitude	98	37	36	35	0.65		0.87		
Techa–Pershinskoye 1974									
Max	222	50.4	133	82.7	1.61	3.15	0.6	1.02	
Min	−26	0.38	4.69	9.8	0.38		0.13		
Amplitude	248	50	128	73	1.23		0.47		
Tobol–Ievlevo 1974									
Max	847	1,780	2,390	520	6.4	10.7	0.78	1.33	
Min	72	57.1	437	167	2.02				
Amplitude	775	1,723	1,953	353	4.38	0.22		0.56	
Tobol–Lipovskoye 1974									
Max	656	3,010	3,280	410	7.9	12.1	0.92	1.22	
Min	−28	145	792	296	2.62		0.17		
Amplitude	684	2,865	2,488	114	5.28		0.75		
Irtysh–Tobol'sk 1974									
Max	707	6,730	7,330	894	9.2	12.5	0.93	1.35	
Min	−10	643	2,340	598	3.82		0.33		
Amplitude	717	6,087	4,990	296	5.38		0.60		
Irtysh–Khanty-Mansiysk 1974									
Max	854	7,400	7,700	990	9.8	16.9	1.03	1.56	
Min	36	935	2,520	517	4.87		0.4		
Amplitude	818	6,465	5,180	473	4.93		0.63		

[a] Level over "zero" mark is specific for each observation point.

2,580,000 km^2, carrying annually 606 km^3—with maximum of 716 km^3 in 1974 according to Volkov *et al.* (2002) or 586 km to 591 km^3 (according to various sources) of freshwater to the Kara Sea, amounting to about half of total inflow to the sea.

The Yenisei valley is almost meridionally aligned and divides the West Siberian plain and Middle Siberian highland. The hydrological network is highly asymmetric (Figure 2.8, see color section). The right bank, located almost entirely in the Middle Siberian highland, is well-developed and its surface is five to six times larger than the left bank, which is a relatively narrow (100–250 km) strip, located on the eastern border of the West Siberian plain. Asymmetry is also typical for many of the Yenisei tributaries, such as the Kureika, Podkamennaya, Nizhnyaya Tunguska, Kan, Mana, and others.

The Yenisei is usually divided into three parts based on the natural features and shape of the river valley and bed: (1) the upper Yenisei (from the river source to the Tuba River); (2) the middle Yenisei (from the confluence of the Tuba to the junction with the Angara); and (3) the lower Yenisei (from the Angara junction to the Kara Sea). The upper Yenisei is not considered in this study, as it is located far upstream from the pollution sources discussed here. The middle Yenisei up to the town of Divnogorsk (upstream of Krasnoyarsk) has lost its natural character due to the immense Krasnoyarsk reservoir and Krasnoyarsk Hydroelectrical Power Plant (HPP) and its high dam that was constructed in 1967. Downstream from Krasnoyarsk, the valley is between 10 km and 11 km wide and asymmetrical. The low-lying left banks are covered by small hills and are partly boggy. The right banks are high and steep, as here the Yeniseiskiy Kryazh Mountains dominate up to the junction with the Angara (near the town of Strelka). From Krasnoyarsk to Strelka Yenisei, there are 28 sills with small depths and high current speed (*Russia: River Basins*, 1999).

The lower Yenisei until its confluence with Podkamennaya Tunguska flows through a valley with an asymmetrical shape similar to the middle Yenisei, with Yaniseiskiy Kryazh bordering the high and steep right banks. Valley width varies from 10 km to 20 km, in some places up to 40 km. Further downstream, the right banks remain mountainous, but not as steep and high as before, and the valley becomes wider varying normally from 20 km to 30 km but reaching up to 100 km. After the confluence with Nizhnyaya Tunguska, the river width varies from 2 km to 5 km. The river starts to divide and form channels, enlarging the Yenisei width up to 7 km to 12 km. The river depth increases from 4–5 m up to 10–15 m, and downstream from Igarka it is 20 m to 40 m (*Russia: River Basins*, 1999). The Yenisei crosses the North Siberian lowland and flows over a flat plain before reaching the Yenisei Gulf, where the Yenisei joins the Kara Sea.

As seen in Figure 2.8 (see color section), there are three main rivers flowing into the Yenisei: the Angara, Podkamennaya Tunguska, and Nizhnyaya Tunguska. They represent half of the water budget of the Yenisei River (Figure 2.9), with the Angara and Nizhnyaya Tunguska being the most important (24% and 18%, respectively), while the Podkamennaya Tunguska brings 8%. The watershed area relative to water abundance is the same for both the Podkamennaya and Nizhnyaya Tunguska (Figure 2.10). However, the ratio of the Angara watershed is significantly larger (42%), which means that the watershed of the Yenisei proper and other rivers represents only 31% of the total area. Correct representation of these three main tributaries is therefore very important for assessing spatial variations in the river regime and for simulating the transport of pollutants through the Yenisei system. The main tributaries and location of observation points chosen for this study are shown in Figure 2.11 (see color section).

River discharge and water level. According to the hydrological classification, the Yenisei is a river characterized by spring/summer flooding fed by snow and rain.

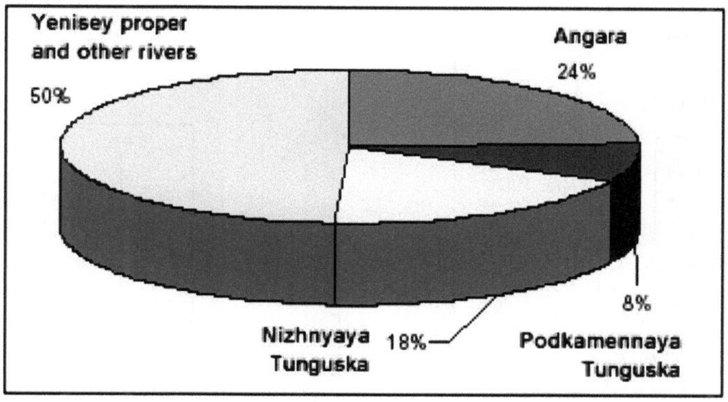

Figure 2.9. Share of main tributaries in the overall Yenisei water budget.

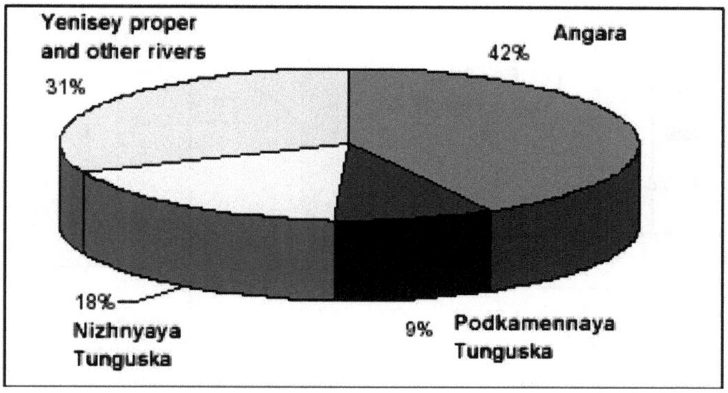

Figure 2.10. Share of main tributaries in the overall Yenisei watershed.

Most of the water reaching the Yenisei is from melting snow. Rainfall has a lesser importance and groundwater even less. However, in wintertime, when the influxes of water from snowmelt and rainfall decrease, the share of the groundwater rises. These features lead to significant seasonal variations in Yenisei discharge (Figure 2.12). The winter low-level period starts in November and lasts 5 months. Increases in discharge occur unevenly due to water inflowing from tributaries, mostly the Angara.

In the middle Yenisei, the spring flood starts in April, while in the lower Yenisei, especially after the junction with Nizhnyaya Tunguska, maximum values can already be observed in May. During the spring flood, water levels rise dramatically. In the middle Yenisei, there are several increases in water level due to temporally and geographically uneven inputs of meltwater from various elevations of the watershed. In the lower Yenisei, a uniform wave of water-level increase forms as a result of the input of water from a vast territory with a similar elevation and, consequently, similar dates of snowmelt onset. During large spring flooding, the water level amplitude may reach 10 m to 15 m in the middle Yenisei and 15 m to 23 m in the lower Yenisei (*Russia: River Basins*, 1999). According to measurements in 1985, the amplitude of monthly level values increased form 3.8 m at Bazaikha to 5 m and 6.9 m at Yeniseisk and Podkamennaya Tunguska, and 14.8 m at Igarka (Long-term observations on regime and resources of surface waters, 1986).

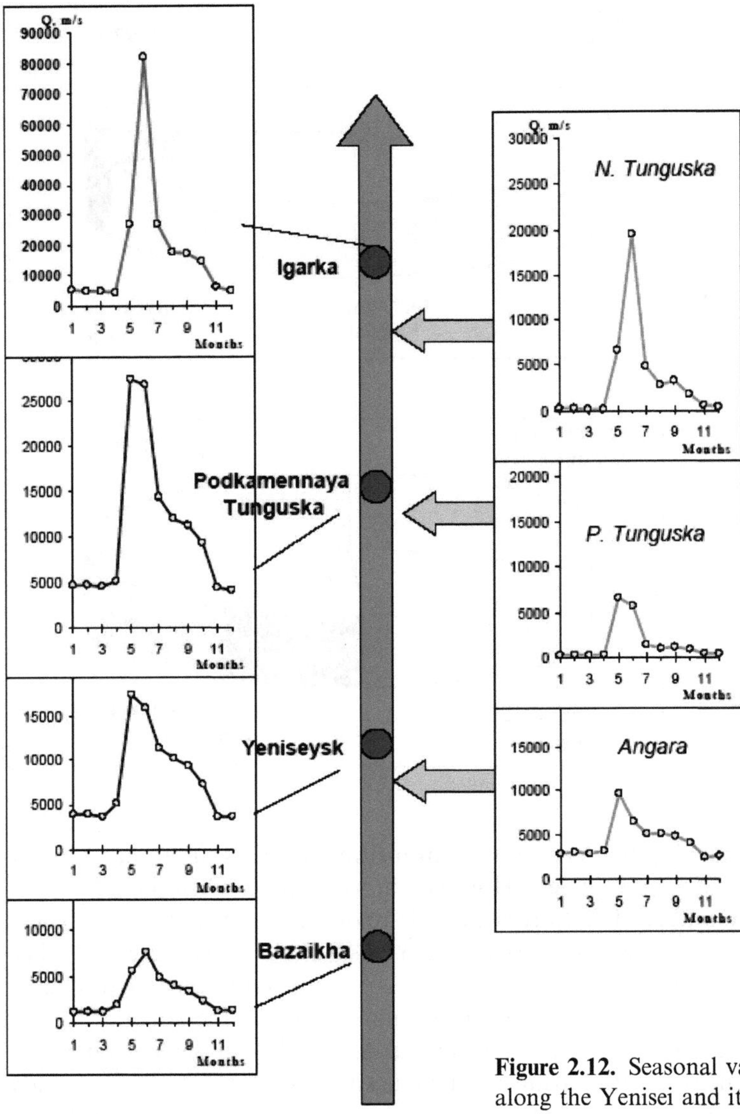

Figure 2.12. Seasonal variations of discharge along the Yenisei and its main tributaries.

The duration of the spring flood is usually 2.5–3.5 months, during which discharge, after initially reaching peak values, first rapidly then gradually decreases and then the summer/autumn low-water period occurs, interrupted by autumn floods caused by rainfall. Toward the end of summer and in autumn, discharge remains high and is twice as large as in winter.

The Yenisei regime at Bazaikha is determined by the water regulatory influence of Krasnoyarsk reservoir. While previously about 60% of annual discharge passed through during the spring flood, now it is more evenly distributed. The spring flood starts in June, resulting in a smooth wave, and discharge decreases gradually almost up to the winter low-water period. High discharge during winter and ice jams cause elevated water levels in winter months.

A similar situation with sharp changes in the seasonal distribution of discharge and water level is observed at Angara downstream from Bratskoye reservoir. During the summer period water levels become lower (0.5–0.8 m) than before, while winter levels have increased (0.14–0.6 m) when compared with the time before the construction of the reservoir. Comparison of the ratios of discharge in June and March before and after construction of the two reservoirs indicates that discharge at Krasnoyarsk became six times and at Yeniseisk two times more "regulated" (*Russia: River Basins*, 1999).

There are periodically high water level increases causing floods on the Yenisei and its tributaries with catastrophic results. Floods may be caused by rainfall, snowmelt, snowmelt and rainfall mixed, or snowmelt combined with ice jams. These are discussed below.

Floods caused by rainfall in the region under consideration are rare and have local importance. Snowmelt causes floods when meltwater comes from vast mountainous areas, such as the Middle Siberian Highland. These floods are especially typical of Yeniseisk, where various tributaries with different hydrological properties create a complicated structure of the river regime.

The most common floods result from a combination of spring snowmelt and ice jams. Ice jams usually occur where the river configuration is complicated by various factors (sharp river bends, islands, narrowing of the river, etc.). At Krasnoyarsk, a significant rise in the water level during ice jams usually take places once every 3 years. The Yenisei, from the Angara junction to its confluence with the Podkamennaya Tunguska, is also frequently affected by floods during ice jams. Due to the low inclination of the river and significant size of ice jams, the water level increases over long stretches of the river, sometimes as long as 200 km to 250 km. At Igarka, Dudinka, and Ust′-Port, the water level rises during ice jams—sometimes from several simultaneous and overlapping ice jams.

Seasonal variations in the water level rapidly decrease toward the sea boundary of the Yenisei estuary. For example, at Selivanikha (853 km from the sea) the level amplitude is 18 m, 16 m at Igarka, 8 m at Ust′-Port (196 km from sea), 5 m at Baikalovo (40 km from the sea), and 2.5 m at the sea boundary.

Storm surges at the sea boundary create water level variations of 2 m to 3 m and during the low-water period may penetrate the Yenisei up to its junction with the Nizhnyaya Tunguska (870 km from the edge of the delta). Tidal oscillations range from 1 m at the delta to 0.08 m to 0.2 m at Ust′-Port. Regulation of Yenisei discharge by reservoirs has led to tidal penetration into the river reducing. According to Mikhailov (1997), tides during the summer/autumn low-water period previously reached the Kureika junction (743 km from the sea), whereas presently it is no more than 445 km. However, according to another source (*Russia: River Basins*, 1999) it is stated that during significant tides the water level increase may be 1 m at Ust′-Port, and in such cases tides reach Igarka (800 km from the delta). Another factor involved in decreasing the propagation of tides is ice cover.

Suspended and bottom sediment. The flow of a river has a significant destructive effect on river beds. During spring floods and significant floods caused by rainfall, rivers transport a large volume of sediment. Rivers in mountainous regions are characterized by strong bottom erosion, whereas lateral erosion is more significant on the plains. River configuration, location of streams, shallows, islands, and line of maximal depth are very often altered after spring floods in particular but also floods in general.

Seasonal variations of suspended sediment load are very similar to changes in the water discharge regime (Figure 2.13, see color section, and Table 2.3). Maximum sediment loads occur during the spring flood, when washout from the floodplain as well as bottom and lateral erosion are greatest. The lowest sediment load is typically during the winter low-water period.

Spatial variations in suspended sediment load along the Yenisei are strongly related to the sediment load regime of its main tributaries. Regulation of the Krasnoyarsk reservoir results in very low (less than 58 kg/s) values and insignificant seasonal variations of sediment load at Bazaikha.

After the junction with the Angara, the sediment load significantly increases and continues to grow until the junction with the Podkamennaya Tunguska, where there are very high values (up to 670 kg/s in the first 10-day period of June in 1985), but this does not apply to the spring flood period (June). In June the highest sediment load is observed at Igarka (up to 1,600 kg/s in the first 10-day period of June), which is related to very high discharge values during the spring flood.

Statistical estimates of suspended sediment transport for two stations are presented in Table 2.4. The variability in sediment transport is substantial, although this is due not only to real differences but also the low accuracy typical of this kind of measurement. The distributions are reasonably symmetrical, such that the Kritsky–Menkel gamma distribution should be used to model multi-year variability.

In the middle Yenisei, relatively low values of suspended sediment and bottom-bed load lead to the formation of gravel–pebble river beds and banks along most of the river. Sand–silt sediments are typical of the junction regions with the rivers Kan, Angara, Kas, Bolshoy Pit, Sym, and Dubches, which carry large amounts of suspended sediment (Nosov and Martynova, 1996). Moreover, high accumulation rates are typical of the natural or artificial creeks (e.g., Kononovo, Pavlovshina, Strelka, and Podtesovo), former river beds, and tailings of islands, where the sediment consists mostly of sand.

Seasonal variations in discharge and suspended sediment load in the Yenisei tributaries lead to bottom sediment having various origins after its junction with the Angara. Thus, according to studies of radioactive pollution of Yenisei sediment, polluted sediment originating from the Yenisei proper are annually covered by "clean" sediment from the Yenisei tributaries, thus creating a complicated "layered" structure of bottom sediment (Nosov and Martynova, 1996; Nosov, 1997).

Data on particle size distribution in summer 1980 (Table 2.5) show that there are large variations between different phases of the river regime. During the spring flood in June particles of 0.5 mm to 0.01 mm dominate (median value 0.2–0.3 mm) and no particles larger than 0.5 mm are observed, which reflects the washout of small

Table 2.3. Discharge of suspended sediment along the Yenisei for 10-day periods in 1985 (*Long-term Observations* . . ., 1986).

10-day period	*JAN*	*FEB*	*MAR*	*APR*	*MAY*	*JUN*	*JUL*	*AUG*	*SEP*	*OCT*	*NOV*	*DEC*	*ANN*
Bazaikha F = 300,000 m^2; P_s = 280,000 t; and M_s = 0.93 t/m^2 (per year)[a]													
1	1.4	1.6	9.7	2.7	6.3	35	0.22	5.5	10	1.1	0.95	0.69	
2	12	3	6.5	3	5.6	34	9.3	8.3	21	2.1	0.6	1.9	
3	7.5	4.8	5.2	4.4	2.5	58	16	13	11	1.4	0.22	4.5	
Mean	11	3.1	7.1	3.4	4.8	42	8.5	8.9	14	1.5	0.59	2.4	8.9
Max													
Min													
Yeniseisk F =1,400,000 m^2; P_s = 2,100,000 t; and M_s = 1.5 t/m^2 (per year)[a]													
1	1.6	9.2	5.6	19	65	120	240	44	170	17	16	5.4	
2	2.7	7.6	1.9	15	70	93	48	27	260	9.3	8.9	5.4	
3	8.8	9	29	31	360	140	200	49	210	9.8	9.6	5	
Mean	4.4	8.7	12	22	170	120	160	40	210	12	12	5.3	65
Max						220	480	81	620	35			620
Min						49	15	10	43	2.1			
Podkamennaya Tunguska F =1,760,000 m^2; P_s = 4,400,000 t; and M_s = 2.5 t/m^2 (per year)[a]													
1				33	99	670	210	200	290	120			
2				39	350	390	180	220	220	65			
3				170	810	300	170	190	160	35			
Mean				81	420	450	190	200	220	73			140
Max				310	1,200	1,100	300	320	400	150			1,200
Min				3.8	15	210	100	83	7.4				
Igarka F = 2,440,000 m^2; P_s = 4,100,000 t; and M_s = 1.7 t/m^2 (per year)[a]													
1	15	32	14	11	34	1,600	28	35	44	61	6.9	12	
2	9.3	22	24	23	64	1,300	61	51	75	47	10	11	
3	50	22	21	9.3	170	490	130	130	57	20	25	9.5	
Mean	25	25	20	14	89	1,100	73	72	59	43	14	11	130
Max													4,600
Min													

[a] F = watershed surface; P_s = sediment runoff; M_s = module of sediment load.

Table 2.4. Statistical parameters of suspended sediment for Bazaikha (1940–1966) and Yeniseisk (1948–1962) (*Long-term Observations* . . ., 1985).

	Maximal daily discharge			*Ordinates of probability curves*									
	Mean	C_v[a]	C_s[a]	0.1	1	5	10	25	50	75	95	99	99.9
Bazaikha	5,100	0.58	1.3	20,000	15,000	11,000	8,900	6,500	4,600	3,200	1,700	1,100	600
Yeniseisk	7,600	0.35	1.82	20,000	16,000	12,000	11,000	8,900	7,100	5,800	4,100	3,200	2,500

[a] C_v = coefficient of variation; C_s = coefficient of skewness.

Table 2.5. Particle size distribution (% by mass) and median size values for the Yenisei at Igarka (*Hydrological Yearbook*, 1980).

Particle size (mm) / *Date*	>100	100–50	50–20	20–10	10–5	5–2	2–1	1–0.5
14.06.1980	—	—	—	—	—	—	—	—
26.06.1980	—	—	—	—	—	—	—	—
08.08.1980	4.0	42.0	22.0	20.0	14.4	12.4	5.2	10.6
Particle size (mm) / *Date*	0.5–0.2	0.2–0.1	0.1–0.05	0.05–0.01	0.01–0.005	0.005–0.001	<0.001	*Median value*
14.06.1980	24.0	5.9	3.0	43.7	10.3	6.8	5.8	0.03
26.06.1980	5.5	16.0	28.3	15.2	5.7	19.4	9.4	0.05
08.08.1980	7.4	1.3	0.7	—	—	—	—	9.5

particles from the floodplain. In August, the range of particles size is greater—up to >100 mm, with a median value of 9.5 mm.

Ice conditions. This description is based on information from *Russia: River Basins* (1999), if not stated otherwise. The middle Yenisei freezes in the second half of November and opens at the end of April, such that the ice cover holds for 150–170 days. The lower Yenisei up to Igarka freezes during the first 10-day period of November (downstream from Igarka in the third 10-day period of October) and opens in the second half of May (downstream from Igarka in the first 10-day period of June); ice is present for 180–200 days and 200–230 days per year, respectively.

Regulatory measures at Krasnoyarsk reservoir have significantly influenced the thermal regime of the middle Yenisei, and hence the ice conditions. The accumulation of water in the reservoir during the spring flood leads to a decrease in temperature downstream of the dam during summer: temperatures become 3.9°C lower than long-term means at Krasnoyarsk and 1.7°C at Yeniseisk. However, in autumn the temperature is higher than long-term means, when summer-heated water is being gradually released from the reservoir. From the Krasnoyarsk reservoir dam to Yeniseisk this increase reaches 5.4°C to 2.3°C.

As a result, the warmer water prevents ice formation between the dam and the junction with the Angara. Directly after the dam there is an ice-free zone, whose length varies from 50 km to 200 km depending on the meteorological conditions and the amount of water released through the dam. Ice appears later than usual. At Krasnoyarsk, ice forms only in cold winters, with the mean date of ice appearance (January 20) offset 2–2.5 months later than before construction of the dam. Between the village of Atamanovo and the village of Pavlovshina, ice appears about December 10–13, and by November 20–25 it reaches the village of Kazachinskoye, 20–40 days later than used to be the case. At Yeniseisk, ice coming from the Angara melts even earlier than at Nazimovo. Ice formation at the Angara junction at Yartsevo takes place 3–5 days later than before. Icemelt now starts 25–40 days earlier, March 16 at Zaliv and April 1–2 at Kazachinskoye. The start of spring ice drift is also delayed—at Yeniseisk it is 15 days later and at Nazimovo 2 days later.

The river-opening regime has also changed. Prior to the construction of Krasnoyarsk reservoir, ice was mechanically broken resulting in a high flood wave (up to 4–7 m in height). Presently, river opening between Zaliv and Yartsevo takes place as a result of a natural opening (150–300 m wide) in the middle of the river that gradually expands downstream. Surface level increases or ice jams, which were previously typical of the Yenisei up to the Podkamennaya Tunguska junction, are observed more rarely. The ice-drifting period now lasts twice as long (40 days at Yeniseisk and 26 days at Nazimovo) which has to do with processes involved in the destruction of residual shore ice. The river remains ice-covered at Zaliv for 82 days, 108 days at Kazachinskoye, and 140 days at Yeniseisk, which are respectively 74, 47, and 23 days fewer than previously. For more details see Volkov *et al.* (2002).

Cross-sectional parameters along the Yenisei. Geometric parameters (section surface, width, depth, level inclination) and current speeds at four observation points along the Yenisei (Bazaikha, Yeniseisk, Podkamennaya Tunguska, and Igarka) show substantial variations that relate to the phases of the river regime. Table 2.6 presents statistical estimates of these variations for 1985.

Data show that for all four stations the maximum water cross-section area during the spring flood is about twice that of the low-water minimum. Annual river width amplitude varies from 29 m at Bazaikha to 130 m to 210 m at downstream stations. Mean depth amplitude increases from 3.6 m at Bazaikha and Yeniseisk to 6.8 m at Podkamennaya Tunguska and 9.8 m at Igarka. Maximum values of mean current speed are normally between 1.44 m/s and 1.8 m/s. However, these values have been known to reach 2 m/s and 2.6 m/s. The amplitude of mean currents is about 1 m for all stations except Igarka, where it is 1.5 m/s for the chosen period.

River level inclination was measured only at Yeniseisk and Podkamennaya Tunguska. According to observations it may reach 9 cm/km, while minimum measured values were 6 cm/km (Yeniseisk) and 4 cm/km (Podkamennaya Tunguska).

Thus, in contrast to the Ob′, the Yenisei is relatively rapid, deep, and straight. There are large floodlands only in the middle and lower parts of the Yenisei. Conditions for the spread and transformation of radionuclides in the Yenisei basin are thus essentially different from the Ob′.

2.1.2 Kara Sea region

The description in this sub-section is based substantially on information from Volkov *et al.* (2002), to which some of the authors of the present book contributed. Here, emphasis is given to those characteristics that are important for studying and modeling the spread of contamination.

Table 2.6. Annual variations in the parameters of water cross-sections along the Yenisei in 1985 ("Long-term observations ...", 1986).

	Water level[a]	*Water cross-section area* (m^2)	*Width* (m)	*Mean depth* (m)	*Max depth* (m)	*Mean current speed* (m/s)	*Max current speed* (m/s)	*River level inclination* (‰)
Bazaikha								
Max	634	5,270	725	7.3	8.8	1.8	2.61	
Min	256	2,570	696	3.69	5	0.82		
Amplitude	378	2,700	29	3.61	3.8	0.98		
Yeniseisk								
Max	762	12,400	1,510	8.2	14.3	1.55	2.08	0.091
Min	262	6,000	1,300	4.62	10.6	0.58		0.062
Amplitude	500	6,400	210	3.58	3.7	0.97		0.029
Podkamennaya Tunguska								
Max	1,305	20,800	1,740	12	19.9	1.44	1.94	0.083
Min	617	8,900	1,610	5.2	7.8	0.44		0.038
Amplitude	688	11,900	130	6.8	12.1	1		0.045
Igarka								
Max	1,462	56,200	1,670	30.1	52.9	1.67	1.98	
Min	–14	29,800	1,460	20.3	3.2	0.15		
Amplitude	1,476	26,400	210	9.8	49.7	1.52		

[a] Level over "zero" mark is specific for each observation point.

2.1.2.1 General description

The main features of the oceanographic regime of the Kara Sea are greatly influenced by its high-latitude location, as the sea is situated entirely north of the Arctic Circle. Hence, solar radiation flux is largely absent in winter and remains small for much of the year. As a result, the Kara Sea is almost completely covered by sea ice all year round.

The Kara Sea is characterized by its strongly irregular coastline, many islands, relative isolation from adjoining sea areas, comparative shallowness, and complicated seafloor relief. All these factors affect its water structure and dynamics. However, it is important to note that they manifest themselves differently in those regions where water exchange with the adjacent seas regions dominate, *vis-à-vis* the central part of the sea.

The geographical location of the sea also influences the intensity of tidal forces, while its morphometry (configuration) determines the tidal characteristics and associated phenomena.

Freshwater runoff and water exchanges with the Kara Sea govern the most pronounced features of its hydrographic structure. According to the ratio "amount of annual river runoff to sea volume" the Kara Sea is subjected to the greatest influence by river runoff of any sea on Earth. Freshwater inflow to the Kara Sea comprises 55% of all freshwater inflow into the seas of the Siberian Shelf.

Another important factor influencing the oceanographic regime of the sea is water exchange with the Barents Sea and the Arctic Basin that takes place at the western and northern boundaries of the Kara Sea, respectively. Our estimates show that the annual inflow of intrusion waters from all sources is comparable with the volume of the sea.

Morphometry, bathymetry, and water exchange conditions. Knowledge of the bottom relief and basin configuration is essential for understanding oceanographic processes in the sea (Figure 2.14, see color section). The sea level depends on bottom relief irregularity and correspondingly influences the currents. The combined influence of relief and shore configuration governs wave and tide transformation in the coastal zone; to reiterate, for the purposes of hydrodynamic modeling of water circulation and pollutant transfer it is vital to have proper bathymetric data.

The Kara Sea, within officially adopted boundaries (*Boundaries of the Oceans and the Sea*, 1960; *Atlas of the Oceans. Appendix: Terminology, Definitions, Reference Tables*, 1980), is 883,000 km^2 in surface area, 98,000 km^3 in volume, and 111 m in average depth, with the greatest depth 600 m in the northern part of the sea (80°26′N, 71°18′E). Its largest range is from northeast to southwest between 68°N and 81°N spanning approximately 1,500 km, with the maximum width of 800 km in the northern region. Its large latitudinal range naturally means that incoming solar radiation and atmospheric conditions are significantly different between the southern and the northern parts of the sea.

The coastline is strongly irregular with large embayments (Baidaratskaya, Gydanskaya, and Ob′ Bays and the Yenisei Gulf) deeply entrenched to the mainland shore. There are many islands covering a total area of about 3,000 km^2 in the Kara Sea, predominantly in its northeastern region (Vorobyev, 1959; *The Soviet Arctic*, 1970). The coastline is 9,045 km along the mainland and 5,653 km around the islands (Vorobyev, 1959; *The Soviet Arctic*, 1970; Nikiforov *et al.*, 1974).

Water exchange is most pronounced in the northwest, whereas in the southwest and at its eastern boundary the sea is connected to adjoining regions by comparatively narrow straits. Such features govern some differences in oceanography in different parts of the sea; indeed, these features explain the traditional sub-division of the Kara Sea by oceanographers into two regions—namely, the southwestern and the northeastern with their boundary passing along the line Cape Zhelaniya–Dikson.

In the southwest, the Kara Sea joins the Barents Sea through the narrow and shallow straits: Yugorsky Shar Strait, Kara Gate Strait, and Matochkin Shar Strait. The northeastern Kara Sea joins the Laptev Sea through the comparatively narrow Krasnaya Armiya and Shokalsky Straits and the wider Vilkitsky Strait separating Yuzhny Island of Severnaya Zemlya from the Taimyr Peninsula.

The wide and very deep water exchange zones at the western and northern boundaries provide free water exchange between this Kara Sea region and the Arctic Basin and the northern Barents Sea. The configuration of the straits governs not only the volume of water exchange, but also the character of water exchange and seawater properties.

The Kara Sea is located almost entirely on the continental shelf and, in accordance with the classification of Zubov (1938) based on the morphology of bottom relief, is a typical continental marginal sea. The bottom relief within the sea itself is complicated (Figure 2.14; *Atlas of the Oceans*, 1980). The bottom relief affects the water circulation structure, which further influences other processes—such as heat and salt advection, transfer of nutrients and gases, and intensity of biological processes in the region (Doronin, 1986). The southeastern part of the sea is the shallowest; however, even the shallow-water regions are distinguished by a significantly irregular bottom relief.

Deep-water regions (deeper than 500 m) in the Kara Sea are estimated to occupy less than 1% of the seabed area according to Baskakov and Shpaikher (1978) and 2.4% from other data (*Atlas of the Oceans. Appendix: Terminology, Definitions, Reference Tables*, 1980). Eastward of Novaya Zemlya, the Novozemelsky trough with depths of 200 m to 400 m extends along the entire coast. An important feature of the bottom relief of the Kara Sea is the presence of two deep-water troughs in its northeastern region. These are the Voronin trough with a depth slightly exceeding 400 m and the St. Anna (*Svyataya Anna*) trough where the maximum depth of the Kara Sea has been recorded. The St. Anna trough is three-branched in shape. One of the most pronounced trough branches is directed to the west into the Barents Sea through the water exchange zone between Franz Josef Land and Novaya Zemlya; the maximum depth here reaches 375 m. Another branch is directed to Cape Zhelaniya, and the third one to the southeast. Between these deep-water troughs there is the central Kara Upland (depths less than 50 m) that connects with the Ob′–Yenisei shallow-water region to the south. Many islands are located on this upland.

Inflow of Atlantic and Barents water. Water of Atlantic origin and Barents Sea water flowing to the Kara Sea (in addition to river runoff) govern the most pronounced features of its hydrological structure. It is known that Barents Sea water flows to the Kara Sea in the surface layer through the shallow straits in the southwestern part of the Kara Sea. The resulting transport of Barents water through Kara Gate is about 20,000 km^3/year (Turanov, 1963; Volkov *et al.*, 2002). The transport of Barents water through Yugorsky Shar according to various authors (Maksimov, 1939; Marsov, 1939; Turanov, 1963) is between 350 km^3/year to 400 km^3/year, which is approximately 50 times smaller than the flow through Kara Gate. A smaller amount (about 42 km^3/year) is transported through Matochkin Shar Strait (Turanov, 1963).

Barents water with slightly different characteristics also flows to the northeastern Kara Sea through its western boundary between Novaya Zemlya and Franz Josef Land. The structure of currents in this strait zone is complicated, with currents running both from and to the Barents Sea. Thus, in deep layers, warm water of Atlantic origin moves westward along the trough, whereas Barents water flows eastward. There is also a transfer of cold water from the Kara Sea. The water exchange estimates obtained by different authors differ significantly. According to some data, the westward and eastward transport balance each other out, whereas the resulting annual transport from the Barents Sea is comparatively small, ranging from 2,000 km^3 to slightly greater than 9,000 km^3 (Turanov, 1963). General publications on regional oceanography often contain references to the estimates of Timofeyev (1963) who estimated the resulting transport to the Kara Sea as 17,100 km^3/year.

The most reliable estimates at the present time are the transport estimates in the water exchange zone between Franz Josef Land and Novaya Zemlya obtained by Norwegian and Russian oceanographers based on the analysis of the yearly current measurement series in 1991–1992 at four moorings (Loeng *et al.*, 1993, 1995; Ozhigin, 1997). These unique data along with the results of accompanying oceanographic surveys allowed a detailed description of the flows and made the results of earlier indirect estimates questionable. It was found that the average annual water transport to the Kara Sea was 1.9 Sv (1 Sv $= 10^6$ m^3/s). Recalculated in terms of annual volume it is almost 60,000 km^3, while net annual inflow (taking into account reverse transport of 0.3 Sv) is about 50,000 km^3.

Direct measurements show that the maximum water flow from the Barents Sea to the Kara Sea is observed in the winter and exceeds the minimum transport (observed in the summer) by nearly a factor of 3. If the calculations are based on summer data only and considering them to be an average value for a year, then one obtains transport values close to Timofeyev's estimates.

Water of Atlantic origin flows to the Kara Sea from the Arctic Basin along the deep-water troughs from the north. Estimates of the Atlantic water inflow volume are 9,400 km^3/year (Nikiforov *et al.*, 1974). Some of this water flows through the western boundary to the Barents Sea. In general, there is water outflow to the Arctic Basin through the northern boundary. Outflow through the northern boundary has been estimated at 12,600 km^3/year (Turanov, 1963). Estimates of the net outflow through the northern boundary are difficult due to the absence of instrumental observation data on currents and limited hydrological information. The outflow of Kara water occurs mainly through the eastern straits of the Kara Sea.

According to the estimates of Turanov (1963), the resulting eastward transport is approximately 10,500 km^3/year through Vilkitsky Strait and 8,400 km^3/year through Shokalsky Strait. The transport through Krasnaya Armiya Strait is unknown. The aforementioned values should be considered as approximations, especially those obtained by indirect calculation methods such as values based on water temperature and salinity distribution data. The water balance of the Kara Sea remains an open question.

Table 2.7 combines data on the water balance of the Kara Sea including freshwater inflow (river runoff and atmospheric precipitation), the most probable transport through the Kara Sea Straits discussed above, and ice export (recalculated for water), which is insignificant according to Borodachev's estimates (Volkov *et al.*, 2002). It can be seen that if the volume of water entering from the western sea boundary was based on the estimates of H. Loeng and co-workers (Institute of Marine Research, Bergen, Norway), inflow would exceed outflow by more than a factor of 2. However, according to Timofeyev (1963), the inflow and outflow water balance components are closer. A cause of such a large discrepancy is underestimation and inadequate knowledge of transport variations in the straits within a year and

Table 2.7. Approximate water balance of the Kara Sea.

Inflow (1,000 km^3/year)		*Outflow* (1,000 km^3/year)	
Kara Gate Strait	20	Shokalsky Strait	8.4
Yugorsky Shar Strait	0.4	Vilkitsky Strait	10.5
Matochkin Shar Strait	0.04	Krasnaya Armiya Strait	?
Franz Josef Land–Novaya Zemlya	50 (17[a])	Franz Josef Land–Severnaya Zemlya	12.6
Rivers	1.3	Ice	0.3
Total	*71.4 (36.74[b])*	*Total*	*31.8*

[a] From Timofeyev's data.
[b] Calculated using Timofeyev's data for Franz Josef Land–Novaya Zemlya.

the water balance of the Kara Sea, which needs further investigation. However, we can state that Barents and Atlantic waters play a predominant role in the formation of the oceanographic regime of the Kara Sea. The total amount of Barents water flowing to the Kara Sea during the year is about 70% of the sea volume, while Atlantic water is about 10% of the sea volume.

Inflow and spread of river water. In addition to the major river water inflow from the Ob′ and Yenisei Rivers, discharges from the Pyasina and Nizhnyaya Taimyra Rivers are significant. On the one hand, the ratio "mean annual freshwater inflow (about 1,340 km^3) to total volume of the sea (883,000 km^3)" appears small. On the other hand, the ratio "mean annual freshwater inflow to total Kara Sea area" is 152 cm, which is greater than any other sea in the world. The Kara Sea is thus greatly impacted by freshwater inflow.

Freshwater from the continental landmass is naturally different from saline seawater in its physico-chemical properties, and these differences affect water mass motion through mixing with saline water. Freshwater forms lighter desalinated water in the surface sea layer, thus preventing the whole water mass of the supercooled surface layer from sinking to the depths (Ivanov, 1976; Ivanov and Nikiforov, 1976).

The river waters typically form one area of freshened surface water, such that a considerable portion of the Kara Sea area is under the direct influence of river water, although the pattern of freshwater spread varies from year to year. However, the boundary of river water usually passes seaward of the generally adopted boundaries of the Ob′, Yenisei, Pyasina, and Nizhnyaya Taimyra estuaries. This suggests that the Ob′–Yenisei–Kara Sea region between the estuary boundaries of these rivers (and the river water isolines with a 95% cumulative probability—Figure 2.15) forms a single Kara Sea mouth region, within which a single estuary ecosystem exists. The ecosystem is strongly influenced by the inflow of river water and pollutants from the rivers (Ivanov, 1995, 1997; Volkov *et al.*, 2002).

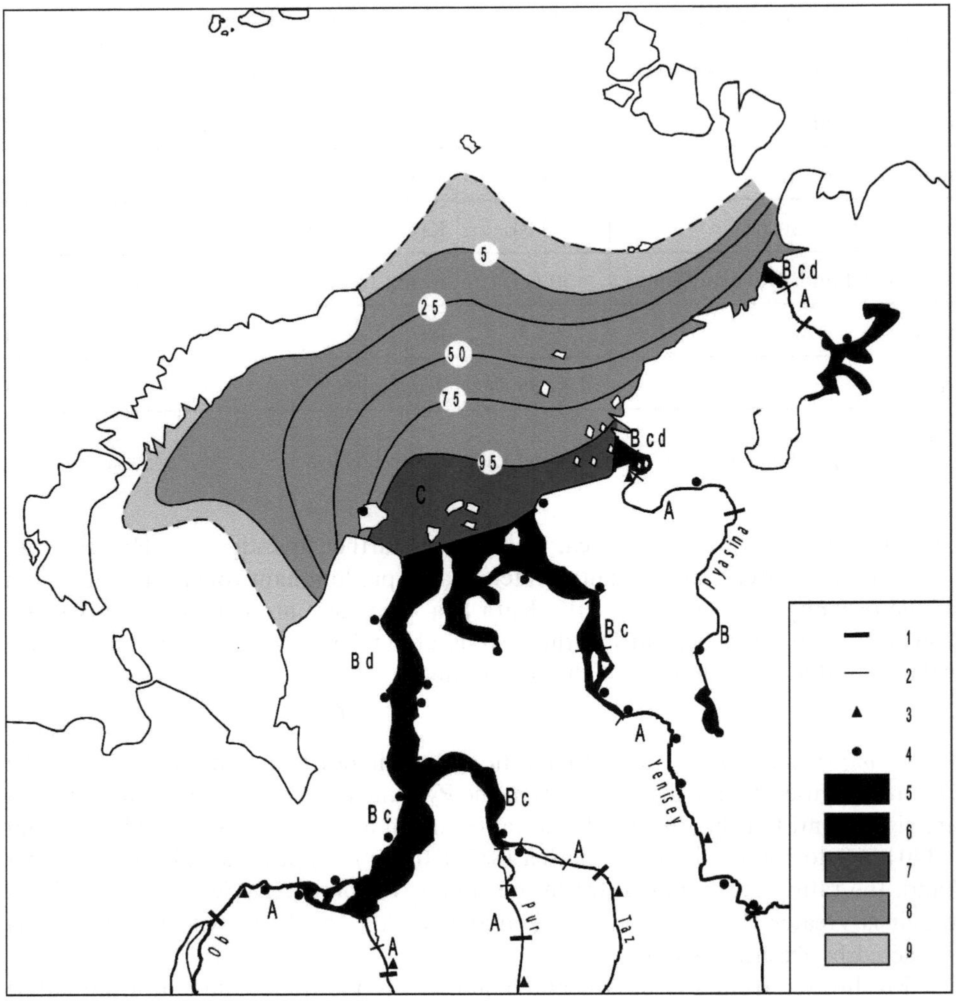

Figure 2.15. Probability of river-water presence (%) in the surface layer of the Kara Sea in the summer. 1—boundaries of the river mouth areas; 2—boundaries of the regions with river mouth areas; 3—hydrological gauges at rivers; 4—hydrometeorological stations; 5—freshwater zone in estuaries (Bc) and in mouth zones (A); 6—zone of transient water in estuaries, in bays, and in gulfs (Bd); 7—zone of freshened water dominance up to 95%; 8—zone of freshened water dominance between 5% and 95%; 9—zone of complete river-water diffusion.

A half-century ago, Antonov (1957) used thermo-chemical characteristics to define the areas of spread of continental runoff water in the Kara Sea and also established the main types of spread. Analysis of data on silicon solubility in surface waters for the summer period 1965–1990 revealed three different types of river water spread: western, northern, and eastern (Figure 2.16a, b, c). The river-water area reaches its maximum during the period August to September, depending on the

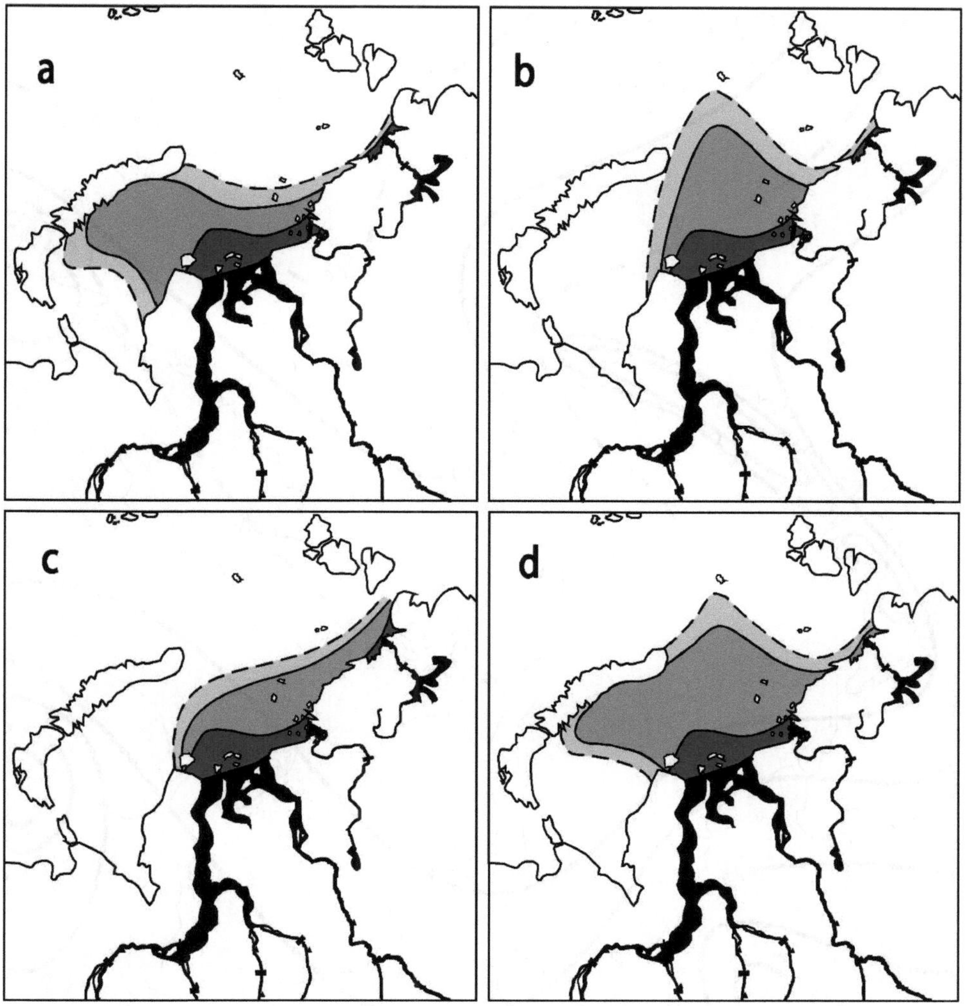

Figure 2.16. Types of river-water areas in the surface layer of the Kara Sea in the summertime: (a) western; (b) northern; (c) eastern; (d) northeastern. For legend see Figure 2.15.

volume of river water that year. The shape of the freshened water area and its transport pathways are predominantly governed by the hydro-meteorological conditions in the sea.

Analysis of annual charts of river-water spread in the Kara Sea indicates a relationship between the type of river-water spread and atmospheric circulation above the sea. The presence of transient types related to the variable meteorological situation over the sea in the latter part of the navigation period (September/October) was also identified. The most pronounced transient type is characterized by the presence of two ridges—to the north and west (Figure 2.16d). However, no reliable

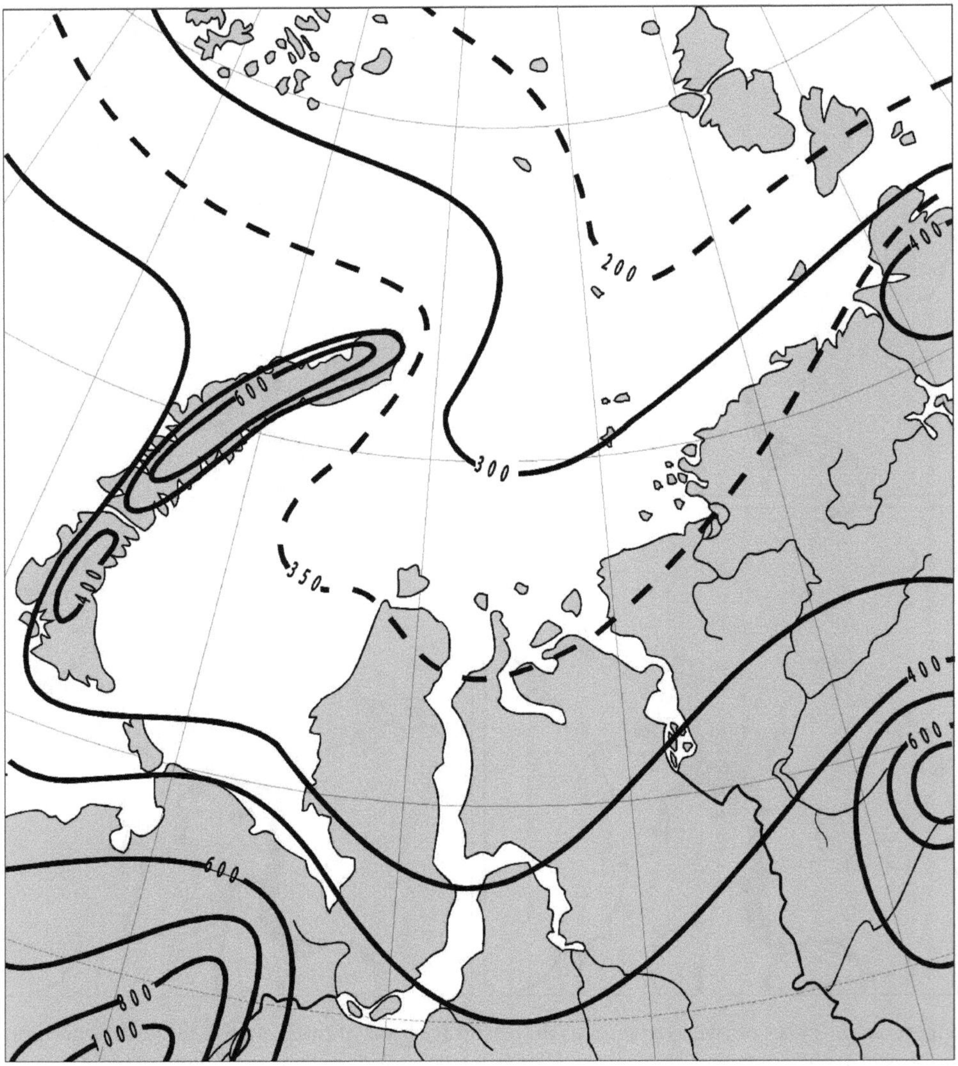

Figure 2.17. Atmospheric precipitation distribution over the Kara Sea (mm/year).

relationships between the volume of total river-water inflow to the sea and type and extent of spread of river water were found.

It should be noted that, in addition to river water, freshwater is input to the sea through atmospheric precipitation. The mean annual precipitation for the entire sea is about 280 mm, varying for different regions from 250 mm to 400 mm a year (Density of the total ..., 1996) (Figure 2.17). The mean annual atmospheric precipitation volume equals 283 km^3, which comprises 17.6% of the total freshwater input to the sea (1,620 km^3). In the warm period of the year (May/October) the ratio "river discharge to total freshwater inflow" is 86.1%, while in the cold period

(November/April) this fraction is only 66.1% to 69.4%. The change in the fraction of river-water inflow and atmospheric precipitation is due to a stronger change in river-water inflow to the sea, rather than to any increase in atmospheric precipitation. It should also be taken into account that river water constantly mixes with seawater year-round, whereas atmospheric precipitation falls to a large extent onto sea ice and much of it stays on the ice cover for a long time.

As shown in an analysis of the interannual variability of freshwater input to the Kara Sea (Volkov *et al.*, 2002), the variability in total mean annual freshwater flow to the sea is governed mainly by variability in the river water; atmospheric precipitation not only has a relatively small value, but also a smaller interannual variability. This predominance of river water remains valid on an intra-annual basis, although the fractions of river runoff and atmospheric precipitation to the sea during a year vary only slightly, due primarily to river-water inflow variability. During the highest runoff month (June) the river water fraction is 93.4%.

Thus, freshwater in the Kara Sea strongly influences the upper layer of the sea bringing about a strong pronounced frontal zone between marine and river origin waters. Dissolved and suspended pollutants are transported with river water to the sea and spread there. Hence, to estimate the levels and the extent of pollution spread in the sea, not only the composition and concentration of pollutants, but the volume of river water inflow to the sea and areas of its spread should be known.

Synoptic atmospheric conditions. Atmospheric pressure and the wind regime are very important in contamination distribution studies because atmospheric forcing can fundamentally influence both the water dynamics of the Kara Sea and vertical turbulent mixing of the water column especially on the shallow shelf (recall that the Kara Sea is located almost entirely on the shelf).

To analyze atmospheric condition variability, the large-scale circulation method of diagnosis and forecasting developed in the Department of Long-range Weather Forecasts of the Arctic and Antarctic Research Institute (AARI), St. Petersburg (Dmitriyev *et al.*, 1989; Dmitriyev, 1994) has been used. It is based on the selection of stages of development of the unidirectional processes of various timescales—from several days (Elementary Synoptic Process or ESP and Homogeneous Circulation Period or HCP) to tens of years (Circulation Epoch). Characteristics of the wind regime depend primarily on ESP and HCP types.

Analyses of many years of data using the ESP observed over the Kara Sea in the warm season (June/October) have shown that, for the Kara Sea region, all ESP variants can be ideally classified into two groups, A and B, based on how they influence the wind circulation of the sea waters (Volkov *et al.*, 2002). The first group (A) is characterized, as a rule, by a south-to-north or east-to-west direction of the pressure gradient over the Kara Sea, and as a result southwesterly, southerly, or southeasterly winds prevail over the sea. This favors wind transport of the Kara Sea surface layer in the coastal continental band toward a generally northeastward direction—from the region of Baidaratskaya Bay to Vilkitsky Strait. The second group (B) is usually characterized by the north-to-south or west-to-east direction of the pressure gradient. In this case northeasterly and northerly winds prevail over the sea, favoring the wind-driven transport of seawater in the coastal continental band, toward a general south-to-west direction—from Vilkitsky Strait to Baidaratskaya Bay. Type A is typical of the early part of the summer and type B prevails during autumn. According to data from polar stations the mean value of the wind speed in January does not exceed 10 m/s, and in July it is slightly less. Maximum values can exceed 30 m/s (*Atlas of the Arctic*, 1985). The greatest number of storms occur in autumn and winter. This is an important consideration because gale force winds usually result in a vertically homogeneous thermohaline water structure.

Mean air temperatures for January are from −16°C along Novaya Zemlya Archipelago to −30°C in the Severnaya Zemlya region. In July, these values range from +8°C in the southern part of the sea to 0°C near the northern boundary of the sea (*Atlas of the Arctic*, 1985).

2.1.2.2 Oceanographic regime

The oceanographic fields of the Kara Sea are influenced by the main factors that were described in the preceding sub-section. The large latitudinal range governs the zonal character of surface-water temperature distribution in the summer according to the amount of incoming solar heat. However, the configuration of the sea and its relative isolation from adjoining areas determine the specific role of water advection from other oceanic regions in forming the particular structure of hydrographic fields. This is most pronounced in the regions of the Kara Sea adjoining the straits.

Water circulation and currents. The total water circulation observed in the Kara Sea represents the sum of three components of different origin: (1) a system of constant (semi-permanent) currents, (2) wind-driven water circulation, and (3) tidal periodic currents.

Tidal periods provide most of the energy within the temporal spectrum of total current variability on timescales from several hours to several days. The total energy contribution of currents at tidal frequencies can comprise 45% to 50%—even greater in some regions. Semi-diurnal oscillations make the most significant contribution to total current speed dispersion. The contribution of daily oscillations does not typically exceed 1% to 2% and that of shallow-water constituents (4-hour and 6-hour periods) does not exceed 3% to 4%. The maximum possible tidal current speed during a 19-year cycle reaches 60 cm/s in the open sea and up to 136 cm/s in the straits (e.g., Yugorsky Shar Strait).

Wind-driven current speeds are typically an order of magnitude greater than those of constant currents; they are similar only in some straits. The significant role of wind-driven water circulation in the coastal continental zone determines an important distinguishing feature of the oceanographic regime in the Kara Sea, as well as in other Arctic Seas—namely, the pronounced non-periodic stochastic oscillations in sea level that in the greater part of the Kara Sea prevail over tidal oscillations. These are attributed mainly to the wind regime, distribution of atmospheric pressure, and to the configuration of the sea (in particular, to a significant development of the continental shoal). The duration of surges ranges between several hours to 10–20 days. Non-periodic sea level oscillations in the Kara Sea increase in the general direction from north to south, considerably increasing in Ob′ Bay and Yenisei Gulf. The amplitude of extreme sea level oscillations near the mainland coast reaches 3 m, but can exceed 4 m at individual points.

Such significant sea level changes can be induced not just by direct wind action, but also by longwaves in the sea caused by rapidly moving atmospheric low-pressure systems. Wind-induced forces create such oceanographic phenomena as seiches, free and forced gravitational waves, and gradient–vorticity waves.

The semi-permanent large-scale currents in the Kara Sea, in particular, and in the Siberian Shelf seas, in general, form a relatively stable circulation system. It is connected to water circulation in the Arctic Basin, to water exchange between the Arctic Ocean and the Pacific and Atlantic Oceans, and to the runoff of the largest rivers of Eurasia. Variability of the general water circulation system in the Arctic Ocean and the marginal shelf seas is related to relatively long-term variability in the large-scale atmospheric pressure field. In the Kara Sea, however, the spatial structure of large-scale water circulation varies little with changes in the large-scale pressure field.

One of the first reliable descriptions of Kara Sea current patterns, based on the generalized observation data of many expeditions and the analysis of buoy drift, was presented by Berezkin and Ratmanov in the *General Current Pattern of the Arctic Ocean and the Adjacent Seas* (Berezkin and Ratmanov, 1940; Berezkin, 1945). An extract is presented in Figure 2.18. Later studies, based on direct current measure-

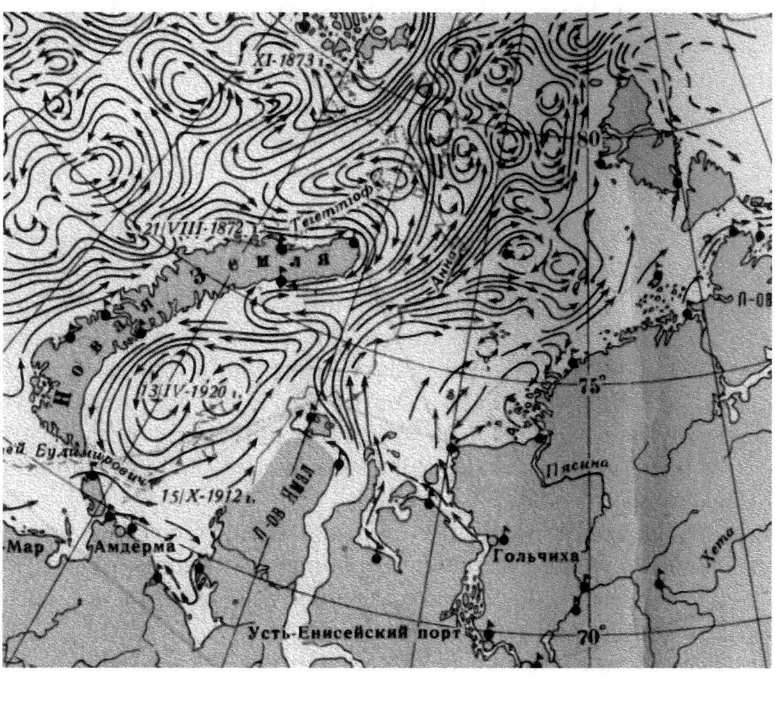

Figure 2.18. Classic schematic depiction of large-scale currents in the Kara Sea (Berezkin and Ratmanov, 1945). Roman numerals refer to the month (e.g., VIII is August).

ments and calculated circulation patterns, simply confirmed this description and added more detail.

Water circulation in the Kara Sea is predominantly cyclonic. Water of Barents Sea origin flows to the Kara Sea via Kara Gate and Yugorsky Shar Straits, reaching the shores of the Yamal Peninsula and moving farther north as the Yamal Current. Near the northern tip of the peninsula the flow is strengthened by the Ob′–Yenisei runoff current, one branch of which goes westward via Malygin Strait while the main flow moves northward. North of Beliy Island, the Yamal Current separates into several branches. One branch turns to the northwest to Novaya Zemlya where it merges with water of the West Novozemelsky Current of the Barents Sea flowing around Cape Zhelaniya, and forms the East Novozemelsky Current going southward and closing the cyclonic gyre in the southwestern part of the sea. Near the southern tip of Novaya Zemlya, part of the East Novozemelsky Current flows through Kara Gate to the Barents Sea (the Litke Current). Another branch of the Yamal Current joins the Ob′–Yenisei flow and continues to move to the northeast forming the resulting outflow to the West Taimyr Current along the shores of the Taimyr Peninsula in the direction of Vilkitsky Strait. A branch of the West Taimyr Current spreads along the western shores of Severnaya Zemlya northward.

In the northeastern part of sea, surface currents flow mostly toward the Arctic Basin. In the deep layers in the northern part of the sea, warm water of Atlantic origin flows from the Arctic Basin along the troughs, mainly along the St. Anna trough.

According to the unique yearly current observation data series from the strait between Franz Josef Land and Novaya Zemlya (Loeng *et al.*, 1993, 1995; Ozhigin, 1997), current direction throughout the year is quite stable and water inflow to the Kara Sea is many times greater than water outflow there. There was a pronounced seasonal variability in current speeds that increases from November to February and decreases in the summer.

Water masses. Typically, six water masses can be identified in the Kara Sea:

- *Surface Arctic water* with a temperature close to freezing point with salinity greater than 32.5 ppt flowing to the Kara Sea from the Kara Basin and partially through the strait between Franz Josef Land and Novaya Zemlya.
- *Surface Kara water* (according to the classification of Nikiforov and Shpaikher, 1980: surface water of the Arctic Seas) with a temperature between −1.5°C to −1.80°C and with a lower salinity of about 32.5 ppt compared with surface Arctic water. It forms at the Kara Sea surface and has to do with the processes of mixing of river runoff water, icemelt and formation, and horizontal mixing with other water masses.
- *River water* forming from continental water outflow with a low salinity. The sea boundary of the river-water area at the sea surface is frequently defined to be the 25 ppt isohaline. River water is characterized by significant seasonal variations of temperature and salinity.
- *Barents water* flowing to the sea through the southern Novozemelsky straits

with a high salinity and temperature that also experience significant seasonal oscillations.

- *Atlantic water* with a temperature greater than 0°C and salinity of 34 ppt to 34.8 ppt and greater, flowing via the deep-water St. Anna and Voronin troughs.
- *Bottom water* with a low temperature between −1.7°C and −1.5°C and high salinity of around 35 ppt forming as a result of winter cooling and salinization of water at ice formation.

Water masses spread throughout the Kara Sea and occupy depths corresponding to their density, creating different types of water mass vertical structure that are quite stable in time.

Similar dynamic transformation processes develop within a region with a uniform water structure. Therefore, defining sea regions based on the types of structures is useful, as it allows us to better understand the role of different physical processes in different regions. According to studies by Nikiforov and Shpaikher (1980) four structural zones can be delineated in the Kara Sea:

- in the southwest (approximately following a line from Cape Zhelaniya to Beliy Island), the water structure is determined by surface Kara water and Barents water, as well as bottom-water masses that form as a result of cooling and salinization during ice formation;
- east of the first region approximately up to the latitude of Cape Zhelaniya, the vertical structure is determined by the predominant influence of continental runoff water, which flows into the Kara water that spreads here in the near-bottom layer;
- the northern sea region is characterized by the intense influence of deep warm Atlantic water, joined from above by surface Arctic water and from below by bottom waters; and
- a small region near Cape Zhelaniya with a structure similar to the northern region, but where Barents water is added in the sub-surface layer.

More complicated transient types of structures can also exist at the boundaries of the regions delineated.

Temperature and salinity fields. The general three-dimensional structure of the water temperature and salinity fields in the Kara Sea is governed by the distribution of water masses, most of which are intrusions; frontal zones form at the boundaries of these water masses. Similarly to water characteristics, the spatial structure of thermohaline fields is also subject to seasonal variations.

Summer hydrophysical processes begin with the end of ice growth and the onset of surface water heating in the regions of discontinuities in the ice cover. Then, young ice starts to melt thus increasing the number of openings, and the water temperature becomes higher and the salinity of the surface layer decreases due to snowmelt and icemelt.

Winter processes begin when the total heat exchange with the atmosphere becomes negative and surface water reaches freezing point at a given salinity and the formation of new ice starts.

The thickness of any layer that is strongly affected by seasonal changes depends not only on geographical latitude and annual variations of air temperature, but also on the vertical distribution of thermohaline characteristics in the surface sea layer. This distribution is in its turn determined not only by water exchange and river runoff, but also by the mixing processes in the upper layer due to wind-driven waves and currents. The presence of the ice cover and its characteristics are also important factors influencing the thickness of this layer.

The northern region of the Kara Sea does not become ice-free every year and seasonal changes in temperature and salinity there are not so noticeable. The large latitudinal range, the non-uniform distribution of solar radiation, and the fact that the main heat inflow with intrusion water is observed in the southern part of the sea produce contrasts in ice and hydrological conditions in summer in the southwestern and northeastern regions of the Kara sea.

Zonal features can be clearly seen in the spatial distribution of water temperature in the surface layer during the summer. Zonality is more pronounced when ice conditions are more favorable. Zonality can be disturbed by the melting of ice massifs and as a result of the heat inflow after intrusion of Barents and river water, which can be clearly seen in Figure 2.19. It should be noted that the zone where 0°C temperatures predominate corresponds to the position of the compacted drifting ice edge that had the general direction at that time from southeast to northwest; isotherms in this region are oriented along the ice edge, thus deviating from zonal distribution.

In ice-free regions, water typically warms up to 6°C to 8°C; the thickness of the heated layer is typically 20 m to 25 m in the comparatively deep-water southwestern region and decreases up to several meters in the shallow eastern regions. In the drifting ice zone, the water temperature only slightly exceeds freezing point.

The salinity of surface water during the summer is predominantly characterized by a significant increase from several parts per thousand to 33 ppt and even greater in the direction from the Yamal shores to Novaya Zemlya and from the Ob′–Yenisei shallow region northward (Figure 2.20). This distribution is due to the influence of river runoff. The frontal zone delineates the region of river-water spread and has a complicated spatial structure, including front meandering, mesoscale eddy currents, and formation of fine hydrological structures. Some of these phenomena have been observed using satellite remote sensing.

In addition to this frontal zone, other less pronounced zones with slightly enhanced horizontal gradients of water temperature and salinity can form, such as off Cape Zhelaniya and near Kara Gate Strait, in connection with the influence of Barents water.

In years when icemelt commences later, the surface water temperature and salinity distribution at this time can be quite non-uniform and background values can be lower. Zones with lower temperature and salinity (resulting from the melting of ice massifs) can also be observed in the areas with temperature above zero and may exist for quite a long time after the ice massif itself has disappeared. Zones of

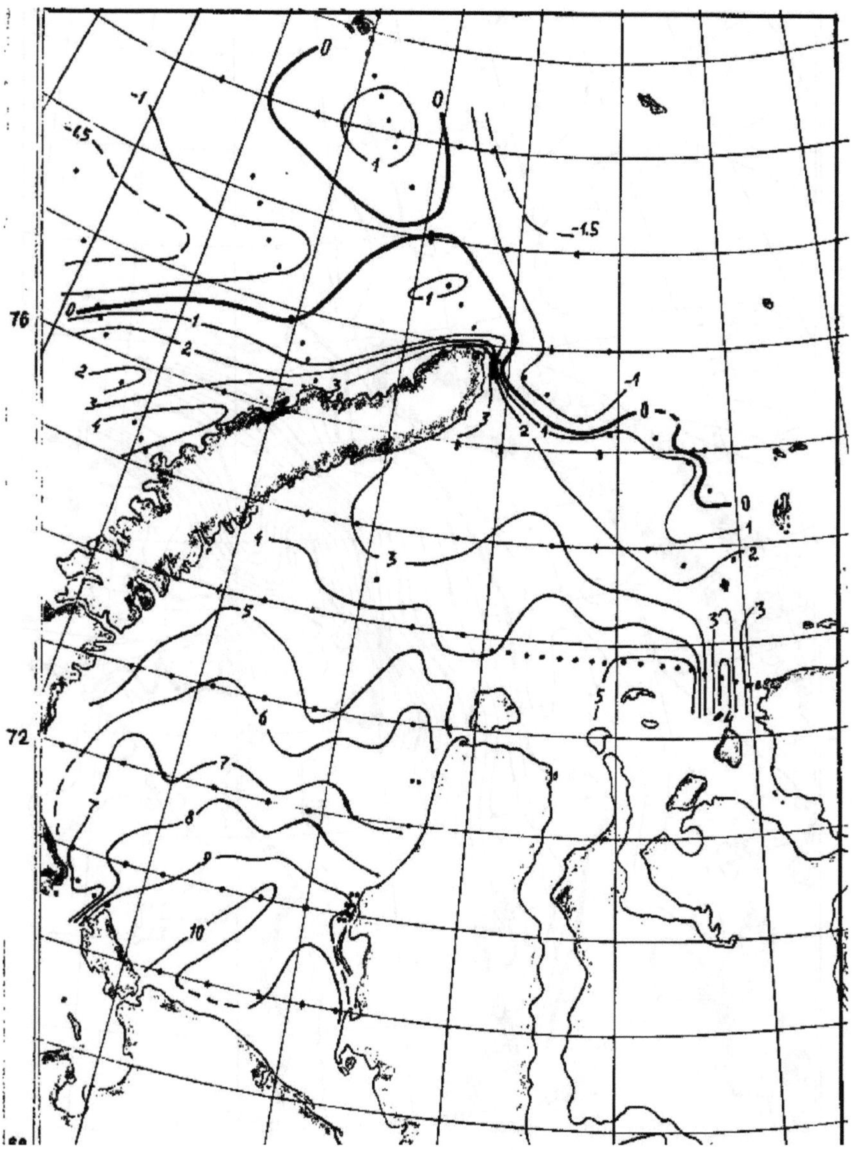

Figure 2.19. Water temperature at the surface (based on measurements of the R/V *Akademik Shuleikin* and the research icebreaker *Otto Schmidt* in August/September 1989, according to V. A. Volkov's data.

freshening can be observed among drifting ice, having salinity several parts per thousand lower than in the ice-free areas.

In the winter, a decrease in river runoff and a concurrent increase in salinization processes due to sea ice formation results in increased salinity in the upper layer that is entrained in vertical convection. Salinity is no less than 25 ppt to 30 ppt in the

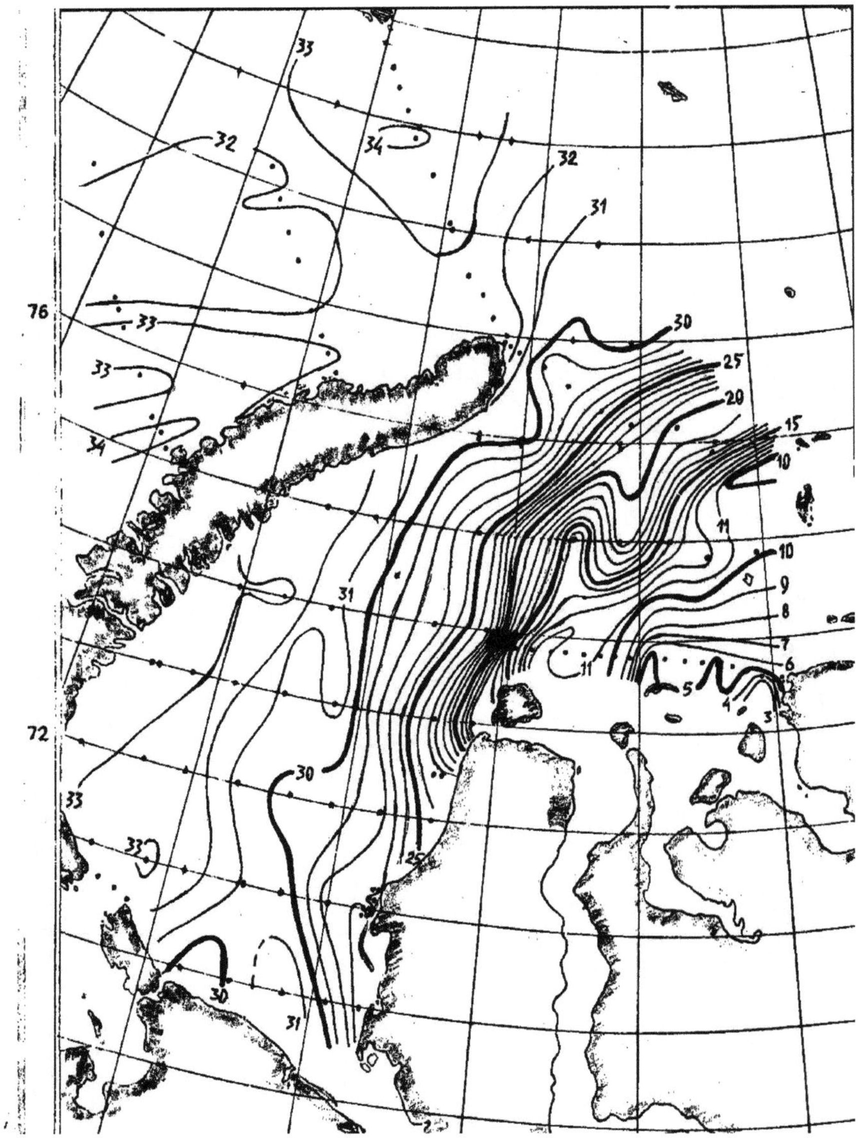

Figure 2.20. Salinity at the sea surface (based on measurements of the R/V *Akademik Shuleikin* and the research icebreaker *Otto Schmidt* in August/September 1989, according to V. A. Volkov's data.

southwest (except for the regions adjoining river mouths) and 33 ppt to 34 ppt in the northern part of the sea and off Novaya Zemlya. Temperature and salinity in the surface layer in some regions have very large seasonal amplitudes reaching 10°C to 12°C and 20 ppt, respectively.

The structure of thermohaline fields in the deep layers is largely governed by the distribution of intruding water masses. Deep frontal divides can form at the boundaries of these water masses. For example, a frontal zone forms when Atlantic water comes into contact with Arctic water in the northern sea region. Its typical width is 8 km to 17 km (5–10 miles), while its horizontal gradients can reach 0.4°C per nautical mile. The contact zone follows the slope of deep-water troughs, because during the autumn/winter season water becomes denser as a result of convective mixing (which rapidly reaches the bottom in shallow regions), and sinks along the trough slopes, displacing less dense water.

Sharply pronounced water stratification is an important feature of the oceanographic structure of the Kara Sea. This is connected with the inflow of buoyant river water and the formation of a quasi-uniform surface layer as a result of wind-induced mixing (especially in the summer) and autumn/winter convection. Density stratification restricts the vertical exchange of heat and matter. Water stratification governed by river water restricts the penetration depth of autumn/winter convection, thus strongly affecting the ice formation process in regions influenced by river runoff. It can also restrict upward heat transfer during cooling of the surface layer and thus create the conditions for preserving heat cores in the deep or near-bottom layers in shallow regions of the Kara Sea. However, strong wind-driven waves, in particular in the coastal zone, can destroy not only a strong pycnocline that forms as a result of river water at the sea surface, but even a relatively strong frontal zone.

It is also important to note that in the summer months, the wave influence is typically insignificant, as strong winds are relatively rare at this time. In years with a smaller ice extent, when the fetch is greatest, the frequency of occurrence of waves with heights 3 m or more is 8% to 10% in July/August and 12% to 15% in September/October. In the southwestern region, strong waves are more frequently observed; maximum wave heights there reach about 8 m (Zatonsky, 1970).

At the same time, it should be recognized that different conditions of water stratification in the central and eastern sea regions create different opportunities for mixing than with the western and northern regions. For example, winter convection reaches the greatest depths in the southwest and can penetrate depths greater than 100 m by the end of the winter. The conditions for water stratification in the northern sea region where warm Atlantic water spreads in the deep layers are favorable for convection development. It should also be noted that the influence of water stratification on hydrophysical conditions in the Kara Sea has been insufficiently studied and requires further studies.

Ice regime. The ice cover in the Kara Sea has a great spatial and temporal variability. In addition to a zonal pattern, there are patterns that arise from the morphometry of the sea (i.e., coastline configuration, bottom relief, islands, etc.), water dynamics in general, and water exchange. Ice conditions in the southwest and the northeast are just as diverse as oceanographic regimes as a whole.

The way in which the ice formation process develops in the Kara Sea at the end of summer is governed by the heat content accumulated by the sea during the summer ice-free period, sea–air heat exchange conditions during cooling of the sea, convective mixing processes, and advection of Barents and river waters. The residual ice plays a significant role, because water temperature in this region in autumn is close to freezing point and ice itself acts as crystallization nuclei, thus accelerating the spread of ice over the sea area.

The development of ice formation in the Arctic Basin reaches the northern boundary of the Kara Sea around mid-September, where at this time the southern limit of close residual ice is located according to mean data (Figure 2.21). The front of stable ice formation moves southward with rates increasing from north to south and decreasing from east to west. Whereas at the western sea boundary the mean rate of ice formation front movement is 0.245 knots or 12.6 cm/s, in the eastern part of the sea it is 3.2 times as high: 0.785 knots or 40.38 cm/s.

The difference in the rate of spread of the stable ice formation front is governed primarily by the different heat content in the active sea layer and heat release conditions. In the western part of the sea, this layer is formed of Barents water with high salinity and enhanced heat content, whereas in the east it is formed of freshened water with a comparatively low heat content. The rate of spread of the ice formation front is significantly influenced by the dates of ice disappearance at the beginning of the summer season and by the presence of residual ice.

In the southwestern Kara Sea, the character of frontal movement changes. The increased heat content of water supported by heat advection from the Barents Sea through the southern Novozemelsky Straits, makes stable ice formation isochrones bend northward. In this part of the Kara Sea the advance rate of the stable ice formation front decreases from 15 miles to 9 miles a day and the way it moves changes from zonal to meridional.

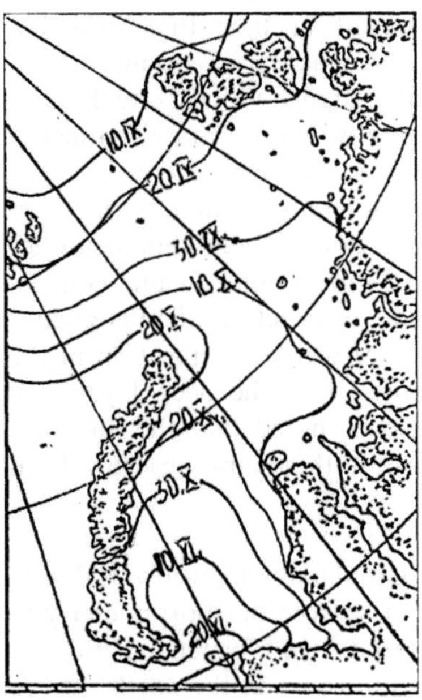

Figure 2.21. Isochrones of average dates of stable ice formation. Roman numbers refer to the month (e.g., IX is September).

Because the meteorological, ice, and hydrological conditions of the summer and autumn have significant interannual variability, the dates of stable ice formation also vary significantly. The largest differences between dates of early and late ice formation are typical of the Zhelaniya Cape region (80 to 125 days) and the regions of the extreme southwestern Kara Sea (70–95 days). The smallest differences are observed in the northern regions, as a result of the constant presence of residual ice and water temperature close to freezing point during the entire summer. In these regions, differences in the extreme dates of ice formation can be 20–30 days. In regions influenced by river water in Ob′ Bay and in Gydan and Yenisei Gulfs, the dates of ice formation vary between 30 and 35 days. Analysis of isochrones of stable ice formation over the Kara Sea indicates that the onset of stable ice formation is primarily zonal, though it is slightly distorted in regions that are influenced by the advection of warm and saline Barents water and water of river origin and in the presence of residual ice as well in some regions.

Ice thickness also differs from region to region. For example, the predominant thickness of the level of first-year ice at the end of winter may be 120 cm in the southwest and 180 cm in the northeast.

In January/March, ice predominately drifts out of the Kara Sea. Ice moves northward/northwestward in the northeastern sea region, and northward/northeastward in the southwestern part of the Kara Sea. The frequency of occurrence of ice drift from the sea is about 50%.

The location of land-fast ice in the Kara Sea is predominantly meridional, governed by the position of the coastline of the islands, archipelagos and peninsulas. Land-fast ice is fully formed by late December. The outer fast ice boundary at this time is established, on average, along a 10 m to 15 m isobath in the southwestern sea region and 20 m to 25 m in the northeastern sea region. The widest fast ice is observed in the northeastern part of the sea where it occupies almost the entire Severozemelsky region. Depending on the combination of hydro-meteorological and geographical factors in specific years, the position of outer boundaries of fast ice and their areas are known to change significantly.

The maximum area of close (ten-tenths) ice in the Kara Sea is observed in December and January to be 570,000 km^2. After this time ice divergence begins. As early as February, the rate of decrease in the close ice area reaches 4,000 km^2/day increasing up to 6,700 km^2/day in March. In May, the divergence of continuous and very close ice results in zones with ice concentration of seven-tenths to eight-tenths. In June, the divergence of continuous ice is irreversible and the ice area decreases.

According to mean multi-year data, stable melting of the snow–ice surface of the Kara Sea begins during the third 10-day period of May off the Amderma coast. The development of ice break-up in the Kara Sea corresponds in general to the zonal distribution of incoming solar radiation.

Ice massifs appear to be the largest ensembles of ice features in the summer. They present vast ice areas more than seven-tenths in concentration. Ice massifs are confined to particular areas of the Arctic Seas, as a result of ocean currents, atmospheric circulation, bottom topography, and location of islands and archipelagos. Three ice massifs are formed and preserved for a long time in the Kara Sea: Novozemelsky, Severozemelsky, and North Kara. The first is situated within the southwestern Kara Sea (i.e., south of the line Dikson Island–Cape Zhelaniya). The other two ice massifs are situated in the northeastern sea over the area north of the indicated line.

2.1.3 The Nordic Seas and adjacent seas

The Nordic Seas comprise the region north of the Greenland–Scotland Ridge and south of the Fram Strait–Spitsbergen–northern Norway transect, covering about 2.5×10^6 km^2. The Nordic Seas region is very dynamic and diverse. The topography of the seafloor is complex with shallow shelves, deep basins, mid-oceanic ridge systems, and steep slopes (Johannessen, 1986). Atmosphere–ocean exchanges of momentum, heat, freshwater, and gases are strong, especially during the cold winter months from November to April. Water masses originating at low and high latitudes meet and interact by means of frontal mixing, deep convective mixing, subduction, and entrainment. Sea ice is formed in the northern and western parts of the Nordic Seas in winter, whereas the region is essentially ice-free during summer except near East Greenland (Johannessen *et al.*, 1994; Drange *et al.*, 2005a, b).

The Nordic Seas are bounded in the south by the shallow North Sea and deep North Atlantic, in the east by the Barents Sea, and in the north by the Arctic Ocean. The bathymetry of the Nordic Seas is very complicated, as are its water dynamics which depending of seafloor relief (Theide and Johannessen, 2008). By means of the approximately 2,600 m deep opening through Fram Strait, the Nordic Seas are the main link between the North Atlantic and the Arctic Oceans. The main bathymetric features of the Nordic Seas are depicted in Figure 2.22 from Blindheim and Østerhus (2005). To the south, the Greenland–Scotland Ridge, which has its deepest sills located in the Faroe Bank Channel (~850 m) and in the Denmark Strait (−620 m), forms the border toward the North Atlantic. Topographic features within the Nordic Seas divide the area into the Greenland, Iceland, and Norwegian Seas.

The most prominent of these features is the mid-ocean ridge. The three main parts of this ridge system are the Kolbeinsey Ridge, the Mohn Ridge, and the Knipovich Ridge. The Kolbeinsey Ridge extends northward from the North Icelandic shelf to the latitude of Jan Mayen, where the Jan Mayen Fracture Zone cuts through it. The Mohn Ridge, between Jan Mayen and approximately 73.5°N, 8°E, has depths ranging between 1,000 m and 2000 m. Its rather complex topography is characterized by many isolated elevations. The Knipovich Ridge stretches from the Mohn Ridge to Fram Strait, its shallower crests being about 1,000 m deep.

Bathymetry divides the Nordic Seas into four major basins. To the west of the mid-ocean ridge are two deep basins in the Greenland Sea. The Greenland Basin is the larger and deeper one, its floor ranging from 3,400 m to 3,600 m (Johannessen *et al.*, 1994). To the north, the Greenland Fracture Zone separates the Greenland Basin from the smaller and shallower Boreas Basin, which has depths around 3,200 m (Johannessen *et al.*, 1991).

From Fram Strait in the north to the Yermak Plateau in the central part, the Greenland Sea borders the deep Nansen Trough extending along the Barents Sea continental slope. The Nansen Trough joins the Barents Sea by means of two meridional troughs. The complicated and diverse bottom relief in different parts of this sub-region is the reason for the heterogeneity of the water dynamics and thermodynamic regime (Doronin, 1986).

In the Iceland Sea, to the south of the Jan Mayen Fracture Zone, the Iceland Plateau is the area between Iceland and Jan Mayen to the east of the Kolbeinsey Ridge. Along its eastern margin, which slopes into the Norwegian Basin, the Jan Mayen Ridge extends southward from Jan Mayen. West of this ridge is a small basin more than 2,200 m deep.

The Norwegian Sea also has two deep basins, the Norwegian and Lofoten

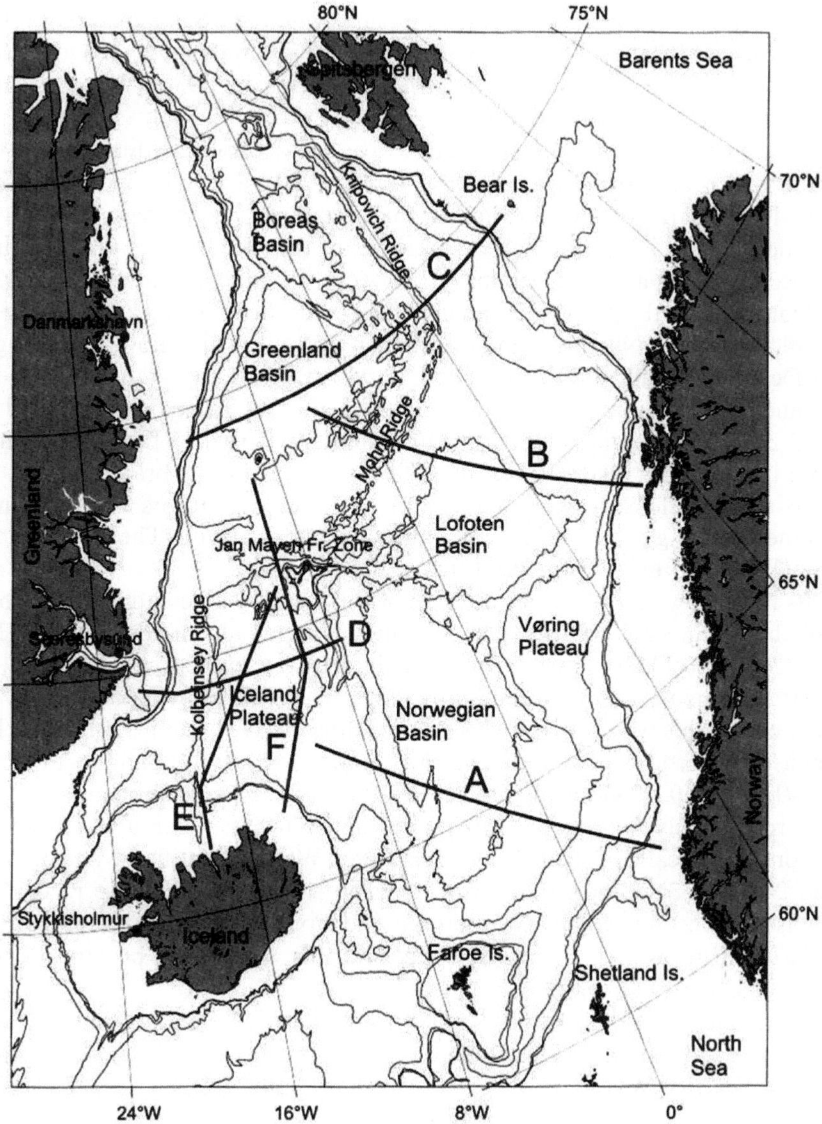

Figure 2.22. Main bathymetric features in the Nordic Seas, showing positions of standard oceanographic sections. Source: Blindheim and Østerhus (2005).

Basins. The former is the largest and deepest basin in the Nordic Seas, extending northward from the Iceland–Faroe Ridge and eastward from the Iceland Plateau to the Vøring Plateau and the continental slope off the Norwegian coast. Its floor lies mainly between 3,200 m and 3,600 m, but depths exceed 3,800 m in a narrow trough around 65°N, 5°W. The shallower and somewhat smaller Lofoten Basin, to the north of the Norwegian Basin and the Vøring Plateau and to the southeast of the Mohn Ridge, is about 3,200 m deep (Blindheim and Østerhus, 2005).

The North Sea is to the south of this region and is where the Sellafield and La Hague nuclear plants are located. The North Sea is relatively shallow with an average depth of 87 m and maximum depth of 725 m (*Atlas of the Oceans*, 1980). Water circulation in the North Sea is generally cyclonic.

The Barents Sea is also shallow, but the average depth is greater (222 m), with a maximum depth of 600 m (*Atlas of the Oceans*, 1980). The bottom relief is rather smooth with a characteristic elevation in the central part of the sea.

The description of fluxes and flow patterns of the Nordic Seas provided here is mostly according to a recent summary overview (Blindheim and Østerhus, 2005). Inflow from the North Atlantic Ocean into the Nordic Seas occurs at three locations. These inflows are the North Icelandic Irminger Current, which flows in through the eastern Denmark Strait; the Faeroe Current, which enters across the Iceland–Faroe Ridge; and the Atlantic inflow entering through the Faeroe–Shetland Channel (Figure 2.23). The potential temperature and salinity properties of these water masses are given in Table 2.8. Estimates of mass fluxes in the various branches of Atlantic inflow to the Nordic Seas have been reported by many authors based mainly on indirect methods (e.g., geostrophic computations, budgets, etc.). They vary from 2 Sv to 12 Sv ($1\ \mathrm{Sv} = 10^6\ \mathrm{m}^3\ \mathrm{s}^{-1}$) for total flux.

In its main flow along the western side of Reykjanes Ridge and the west coast of Iceland, the Irminger Current carries Atlantic Water that derives from the North Atlantic Current (NAC). Downstream of its bifurcation to the west of Iceland, some of its water forms the North Icelandic Irminger Current. Flowing along the North Icelandic coast, this water mixes with Arctic water derived from the East Icelandic Current as well as coastal water, and halfway along the coast (off Siglunes at 18.8°W), the May/June 1950–2000 average of temperature and salinity at 50 m is 3.7°C and 34.7 ppt, respectively. Convective mixing during winter results in a water mass that Stefansson (1962) called North Icelandic Winter Water, identified by temperatures between 2°C and 3°C and salinities of 34.85 ppt to 34.90 ppt. More recently, this water mass has become fresher. Farther east, the mixing with Arctic water continues, particularly during winter; northeast of Iceland this water mass becomes Modified East Icelandic Water (Table 2.8).

The average volume of Atlantic inflow to the North Icelandic Irminger Current was assessed at about 1 Sv. This transport represents a heat flux of approximately 25 TW ($1\ \mathrm{TW} = 10^{12}\ \mathrm{W}$).

The Faeroe Current carries Modified North Atlantic Water into the Nordic Seas. After crossing the Iceland–Faeroe Ridge, it forms the Iceland–Faeroe Front against the Arctic water of the East Icelandic Current and flows eastward over the northern slope of the ridge (Johannessen, 1986). To the east of the Faeroe Plateau, some of this water recirculates into the Faeroe–Shetland Channel and merges with water of the slope current. Most of the water continues, however, into the Norwegian Basin. When it approaches the Voring Plateau, it turns northwest toward Jan Mayen. Near Jan Mayen, a fraction flows into the Iceland Sea, some recirculates along the western slope of the Norwegian Basin, and probably the largest portion flows northeast. Along this path the flow is parallel with the Arctic water of the Greenland Sea, forming the Arctic Front, which to the north of Jan Mayen seems to be fixed to the Mohn Ridge (Johannessen, 1986).

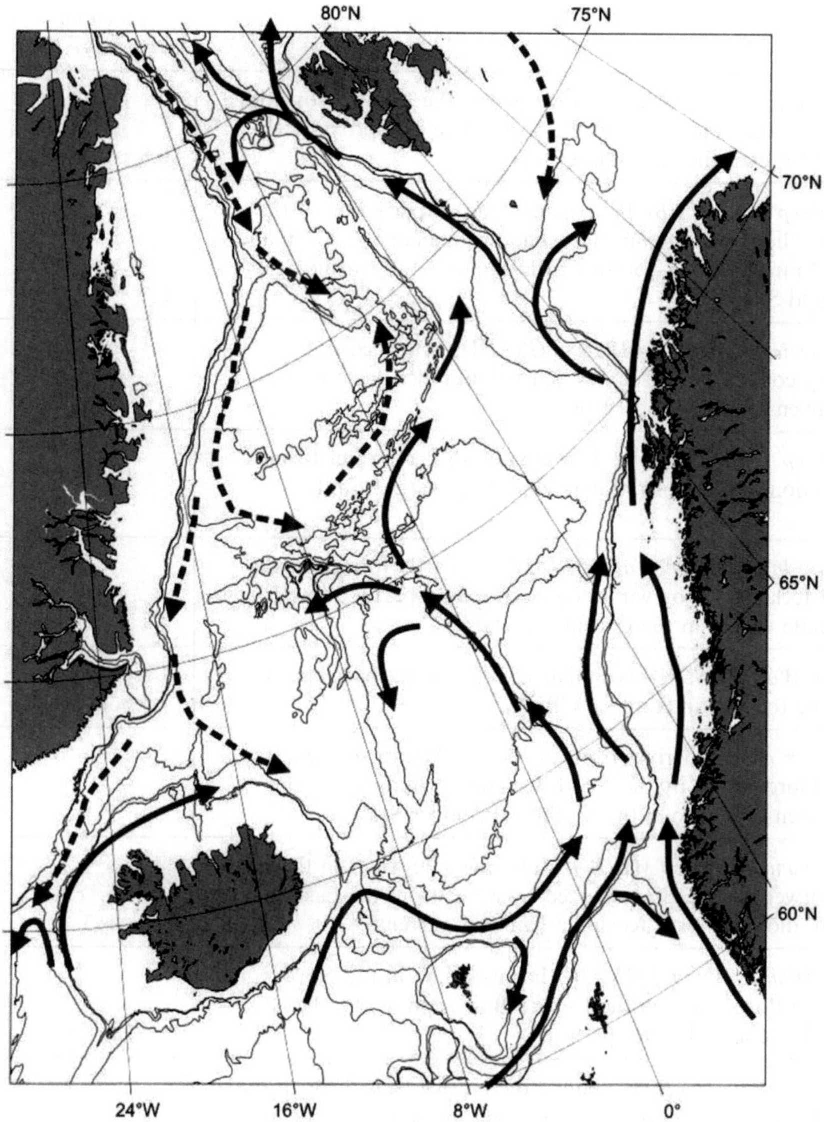

Figure 2.23. Schematic of upper-layer circulation in the Nordic Seas. Solid lines represent Atlantic Water. Broken lines represent Arctic and Polar Water. Source: Blindheim and Østerhus (2005).

Based on ocean current (2009) measurements, the average volume transport carried by the Faeroe Current across 6°W, to the north of the Faeroes, has been assessed to be about 3.5 Sv, which represents a heat flux of 124 TW.

The warmest and saltiest inflow to the Nordic Seas is Northeast Atlantic Water that arrives through the Faeroe–Shetland Channel (Table 2.8). Most of this inflow derives from the slope current to the west of the British Isles and enters mainly

Table 2.8. Major water masses in the Nordic Seas defined by potential temperature and salinity (Blindheim and Østerhus, 2005).

Water mass	*Potential temperature* (°C)	*Salinity* (*S*‰, ppt)
Canadian Basin Deep Water (CBDW) flows across the Lomonosov Ridge north of Greenland and enters the Nordic Seas between approximately 1,500 m and 2,000 m over the Greenland Slope	−0.8 < 0 < −0.5	>34.92
Eurasian Basin Deep Water (EBDW) occupies the deeper strata of the Eurasian Basin in the Arctic Ocean. Between approximately 2,000 m and the sill depth in Fram Strait (~2,500 m) EBDW enters the Nordic Seas over the Greenland Slope	−0.7 < 0 < −0.9	34.92–34.93
Greenland Sea Arctic Intermediate Water (GSAIW) is formed in the Greenland Sea by convection but is less dense than older deep water. The properties given represent the 1990s	−0.9 < 9 < −0.5	34.86–34.89
Greenland Sea Deep Water (GSDW) is a mixture of deep water formed locally by convection during cold winters, and deep water from the Arctic Ocean	<0	34.88–34.90
Iceland Sea Arctic Intermediate Water (ISAIW) is partly formed in the northern/central Iceland Sea by winter convection and partly derived from similar intermediate waters in the Greenland Sea	<1	34.7–34.9
Iceland Sea Deep Water (ISDW) is similar to NSDW but has slightly higher salinity due to the admixture of CBDW	About −1	34.91–34.92
Irminger Sea Water (ISW) derives from the North Atlantic current and is carried into the Nordic Seas by the North Icelandic Irminger current. The properties given are those in the northern Irminger Sea	5–7	35.05–35.10
Modified East Icelandic Water (MEIW) is intermediate water in the southwestern Norwegian Sea characterized by water from the East Icelandic current and the North Icelandic Irminger current	1–3	34.6–34.9
Modified North Atlantic Water (MNAW) has its origin in the North Atlantic current. It flows through the Iceland Basin and across the Iceland–Faeroe Ridge into the Nordic Seas	7.0–8.5	35.1–35.3
North East Atlantic Water (NEAW) enters the Nordic Seas through the Faeroe–Shetland Channel. It derives mainly from the slope current to the west of the British Islands	9.0–10.5	35.3–35.45
Norwegian Sea Arctic Intermediate Water (NSAIW) advects from the Iceland and Greenland Seas into the Norwegian Sea, where it occurs in a layer identified by a salinity minimum between the upper layers, composed mainly of Atlantic Water and deep water	−0.5 < 9 < 0.5	34.7–34.9
Norwegian Sea Deep Water (NSDW) fills the Norwegian Sea beneath the NSAIW	<0.5	34.91
Polar Intermediate Water (PIW) forms a temperature minimum underneath the upper waters of the East Greenland current	<0	34.4–34.7
Recirculating Atlantic Water (RAW) is water of Atlantic origin that has circulated into the East Greenland current from the West Spitsbergen current and the Atlantic Water layer in the Arctic Ocean	0–2	34.9–35.0

through the Rockall Trough. This branch represents an average volume flux of 3.2 Sv and a heat flux of 127 TW.

In total, these recent assessments arrive at an Atlantic volume flux into the Nordic Seas of 7.7 Sv, the largest being the Faeroe Current. Due to the difference in temperature, heat flux through the Faeroe–Shetland Channel is, however, slightly larger than that carried by the Faeroe Current. This results in a total heat flux from the Atlantic into the Nordic Seas of about 275 TW.

The flow that enters through the Faeroe–Shetland Channel continues in the Nordic Seas as a slope current with its core off the shelf break. In Bear Island Trough, a substantial part of its water enters the Barents Sea, while the rest continues northward in the West Spitsbergen Current, partly into the Arctic Ocean, and forms a circumpolar boundary current that flows along the shelf break around the Arctic Basin (Blindheim and Østerhus, 2005).

Some water of Atlantic origin from the deep Arctic Basin penetrates the Arctic shelf seas through deep-water troughs, influencing their thermal regime and ice processes (Matishov *et al.*, 1998a). Although the inflow of Atlantic Water to the Barents Sea is largely topographically steered, the local wind field is also of great importance for the circulation structure. Mean volume flux ranged between 1.3 Sv during summer and 1.7 Sv during winter (Blindheim and Østerhus, 2005).

Water circulation in the Barents Sea is generally cyclonic. Atlantic Water flows along Kola Peninsula and then along Novaya Zemlya Archipelago; some of this water penetrates the Kara Sea as Barents Sea water (*Atlas of the Oceans: Arctic Ocean*, 1980; Matishov *et al.*, 1998a).

It is known that the Barents Sea is important for the formation of dense bottom water by cooling and salt rejection during ice formation (Johannessen *et al.*, 1994). Some dense bottom water that is formed in the eastern and central Barents Sea flows into the Norwegian Sea along the deeper northern slope of Bear Island Trough. Measurements of the current during September/October 1978 indicated a volume flux of 0.8 Sv in this flow. By mixing with less dense waters above, bottom water from the central and eastern Barents Sea becomes less dense along its path within the Barents Sea. In most cases it will therefore mix at intermediate depths in the Norwegian Sea. Dense bottom water formed in Storfjorden (Svalbard) has a shorter route into the Norwegian Sea and is dense enough to mix with bottom water in eastern Fram Strait, hence contributing to deep water in the Arctic Ocean.

Volume exchanges through Fram Strait at 79°N are on the order of 10 Sv in both directions, with a yearly mean net southward flow of approximately 2 Sv to 4 Sv. About half the transport takes place in deep water below 700 m, which is approximately the depth of the 0°C isotherm. The main northward transport of mass and heat takes place in the West Spitsbergen Current. Different authors report a northward transport from 3 Sv up to 7.1 Sv and high variability on monthly and annual timescales. At 79°N, the East Greenland Current (EGC) transports 3 Sv to 4 Sv southward, including 1 Sv of Polar Water. Exchanges through Fram Strait represent a northward net heat transport on the order of 30 TW.

Some Atlantic Water may deflect westward from the West Spitsbergen Current (WSC), into the northern Greenland Basin and the Boreas Basin, but most of the recirculation occurs in Fram Strait (Bobylev *et al.*, 2003). Only a small amount of Atlantic Water eventually enters the Arctic Ocean, where it flows eastward along the Eurasian continental slope. The boundary current continues around the Arctic Ocean, although some of its water is advected into interior basins to form the Atlantic Intermediate Layer. This water exits the Arctic Ocean through western Fram Strait, where it meets Atlantic Water that has deflected westward from the West Spitsbergen Current.

This water of Atlantic origin continues southward in the EGC, where it is known as Recirculating Atlantic Water, and flows parallel with and partly underneath cold and fresh polar surface water, including sea ice. In addition, this current carries intermediate and deep water from the Arctic Ocean (Bobylev *et al.*, 2003). Flowing along the coast of East Greenland, the water of the current gradually mixes with water of the Greenland Basin, and as a result of topographic constraints some water diverts into the basin. The first topographic feature with such effects is the Greenland Fracture Zone, which leads some EGC water into the Boreas Basin. Farther south, in the Jan Mayen Fracture Zone, a large portion of its upper water and all of its water at depths greater than 1,600 m flows into the Jan Mayen Current, forming the southern limb of the cyclonic gyre in the Greenland Basin (Figure 2.23). Farther south, a portion of its upper water masses, and probably most of the bottom water, enters the East Icelandic Current and flows eastward in the Iceland Sea. The remainder flows out of the Nordic Seas through the Denmark Strait.

Besides the inflows of warm Atlantic Water, there is an inflow of fresher and more Arctic character to the Arctic Basin through the Bering Strait. This inflow is to be close to 0.8 Sv.

These inflows have to be balanced by outflows, which take place mainly as deep overflows through the deeper sills across the Greenland–Scotland Ridge. The overflows near the bottom across the sill in Denmark Strait and across the Iceland–Scotland Ridge are both assessed at approximately 3 Sv. There are different views on the origin of the Denmark Strait overflow. Furthermore, a rather uncertain volume of surface outflow from the Arctic Basin has been assessed at approximately 3 Sv, but this number is used to balance the budget rather than being based on measurements. About 1.7 Sv is assumed to pass through the Canadian Archipelago, while the remaining 1.3 Sv are supposed to be carried by the East Greenland Current (Blindheim and Østerhus, 2005).

The northern and western parts of the Nordic Seas are to varying extents covered by sea ice during the winter, usually reaching its greatest extent in March/April (Johannessen *et al.*, 2004, 2008). Some ice forms locally, but to a larger extent it is efflux from the Arctic Ocean. Fram Strait is the main gate for Arctic ice export, and the sea ice carried by the East Greenland Current is the largest and most concentrated meridional ice flow in the World Ocean. The export of ice and liquid freshwater from the Arctic Ocean through Fram Strait represents the major source of freshwater in the Nordic Seas (Bobylev *et al.*, 2003). The amount and distribution of ice and freshwater play a crucial role in deep convection in the Greenland and Iceland Seas. The flux of sea ice through Fram Strait is strongly correlated with local wind. Sea ice distribution in the Nordic Seas depends on the regional pressure field and the under-

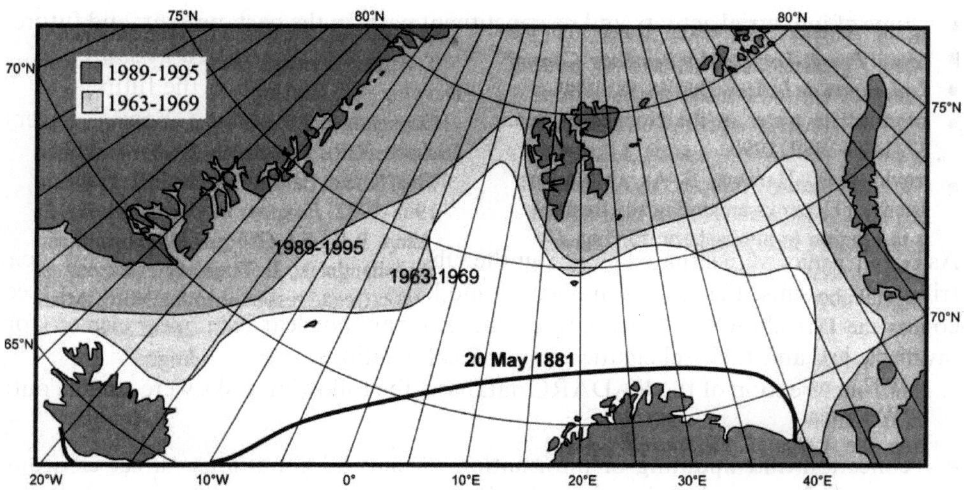

Figure 2.24. Sea ice cover at the end of April for the periods 1963–1969 and 1989–1995 with a minimum and maximum circulation index, respectively, of the North Atlantic Oscillation. Also shown is the extreme sea ice extent in 1881. Source: Drange *et al.* (2005).

lying current system. Figure 2.24 shows the median ice border at the end of April for the periods 1963–1969 and 1989–1995, which were characterized by both a minimum and a maximum, respectively, of the NAO winter index.

2.2 DESCRIPTION OF ENVIRONMENTAL AND POLLUTION DATA

2.2.1 Databases and information system

The overall objective of dataset development was to provide input information necessary to model regions affected by three large-scale enterprises of the Russian Nuclear Complex: the Mayak Production Association (PA), the Siberian Chemical Combine (SCC Tomsk-7), and the Krasnoyarsk Mining and Chemical Combine (MCC Krasnoyarsk-26), as well as the Sellafield plant and other point sources such as submarines that have sunk taking nuclear reactors with them. The dedicated database was used for (1) validation of three-dimensional numerical ice–ocean circulation models for the Arctic Seas; (2) simulation of radionuclide transport through river systems from external sources to the Kara Sea and within the Arctic Ocean using a generic model system (GMS); (3) analysis and selection of the most probable scenarios of accidental radionuclide releases; and (4) risk assessment of radioactive contamination.

The databases included historical environmental (atmosphere–ice–ocean) *in situ* and remote-sensing data, environmental data for radionuclide sources, and radiation pollution data for the Russian Arctic coastal zone and other adjacent regions relevant to the study.

Selection of the sources of radioactivity for the study was made by assessing the following parameters:

- type of industrial activity and its constituent parts in the past, present, and future;
- amount and isotopic composition of radionuclides accumulated at sources;
- accidents that took place in the past or potential accidents in the future;
- presence of man-made and natural barriers on the Ob′ and Yenisei Rivers hindering the release of radionuclides into the environment; and
- geographical location of facilities.

As the facilities were situated either on the Ob′ or Yenisei Rivers proper or on their tributary systems, the task of numerical modeling was simplified. Figure 2.25 (see color section) shows the facilities of the Russian Nuclear Complex chosen for investigation and their schematic geographical locations.

During creation of the RADARC database, the following tasks were carried out:

- collection and updating of information on natural conditions, direct environmental effects, and environmental risk at the most potentially dangerous sources of radioactive contamination in the Ob′ and Yenisei watersheds;
- analysis of the data obtained, developing and designing the contents and structures of the databases and graphical layers;
- informational filling (content creation) of the GIS, including processing of the databases concerned with thematic graphical layers;
- optimization of the data for application to numerical modeling of radionuclide transport to the Kara Sea;
- preparation of the RADARC information system for subsequent efficient imagery of the results of numerical modeling;
- adjustment of the developed RADARC information system for automated data-processing technology of thematic information; and
- production of digital maps to display the state of the region under consideration.

The RADARC information system is a customized GIS application based both on ArcView GIS 3.x and MapInfo Professional 6.0 software for use on PCs running Microsoft Windows. The ArcView GIS 3.x software developed by ESRI (Environmental Systems Research Institute, Redlands, CA) and MapInfo Professional 6.0 (MapInfo Corporation One Global View Troy, NY) are widely employed by GIS users and provide sufficient functions and tools to store, analyze, and visualize spatial data geographically. The information system collects, stores, generalizes, analyzes, and prepares spatial data for use in numerical modeling.

This information system has mostly been used as a display tool to highlight important features about the data and as input for numerical modeling. Figure 2.26 reflects the main tools applied in this work. Much data were obtained during project realization, although data were not available for all years or for all radionuclides.

The first basic step was to create a base map, as at the time the RADARC project started there was no available digital base map for the particular area of interest (i.e., the Ob′ and Yenisei River watersheds and the Russian Arctic coastline).

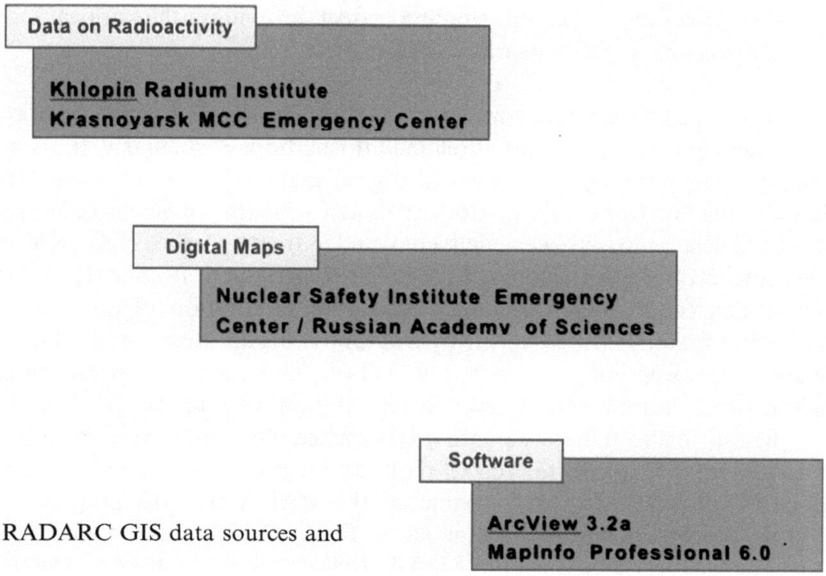

Figure 2.26. RADARC GIS data sources and software.

Therefore, this base map was manually assembled in an ArcView environment with major features such as coastlines, rivers, lakes, urban areas, land cover, etc.

ArcView GIS works with views, tables, charts, layouts, and scripts, conveniently stored in one file called a project (.apr). A project file stores information about each project component, including the storage location of the datasets used in the project. Cartographic data are provided in shapefile format and can be directly read from a CD-ROM. All shapefiles have an associated ArcView GIS legend file (.avl) and projection file (.prj) for use with ArcView GIS 3.x running under MS Windows.

Sources of the basic cartographic data for use with ArcView GIS are:

- *Digital Chart of the World* (DCW, ESRI Inc.). This is a comprehensive 1:1,000,000-scale vector base map of the world. It consists of geographic, attribute, and textual data that can be accessed, queried, displayed, and modified with ARC/INFO software. The database was originally developed by ESRI in Vector Product Format (VPF) for the U.S. Defense Mapping Agency (DMA). The primary source for the database is the DMA Operational Navigation Chart (ONC) series that is produced by Australia, Canada, the U.K., and the U.S. The DCW is considered a good topographic base for creation of GIS in the event that the map of 1:1,000,000 scale is convenient as a general chart.
- ESRI data and maps. The basic mapping datasets of different scales delivered by ESRI with ArcView 3.2 software.
- *World Topography and Bathymetry* (MrSID image by ESRI) that represents a color hill-shaded digital elevation model (DEM)–based image of the region under

consideration. The hill-shading effect provides the appearance of three dimensions (also known as 2.5-D).

The specific map sets compatible with MapInfo GIS software were chosen for more detailed representation of related information. Particularly, the sources of these cartographic data were commercial digital maps of 1:200,000 scale from the map bank of the Nuclear Safety Institute, Russian Academy of Sciences, Moscow (IBRAE RAS). These map sets were delivered under treaty to the V. G. Khlopin Radium Institute, Ministry of Atomic Energy, St. Petersburg, Russia (KRI MINATOM). These charts are used for more detailed representation of data and include the following major features: relief, hydrography, urban areas, industrial objects (pipelines and power lines), main roads, railway line network, landscape constituents, aerodromes, administrative boundaries, captioning data, etc.

MapInfo Pro 6.0 works with a workspace file, which contains a list of all used files and also MapInfo references required for recovery of previous session appearance of all windows. An example of the RADARC GIS graphical interface in MapInfo environment is represented on Figure 2.27 (see color section).

The datasets used during RADARC project implementation have been taken up by participants, namely NIERSC (responsible partner), and the others: KRI, IMMSP, AMAP, Risø National Laboratory, and NERSC, from different open available sources (including the Internet) were uniformly classified into the corresponding database. A schematic of the RADARC information system structure is shown in Figure 2.28.

The environmental and radioactivity data describing the environmental state of the area chosen for investigation into the sources of radioactive contamination of the Kara Sea were available from RADARC project participants in different file formats. It was necessary to convert these data to an identical file format, in our case, that of the ArcView GIS Database (.dbf) or MapInfo internal file format (.tab). The database contains both environmental and radioactivity datasets that are not presented in the AMAP database. In addition, the environmental database contains data on hydrological and meteorological parameters of the watershed areas under study as well as environmental data for radionuclide sources obtained during the INTAS 97-31278 Project *Study of Influence of Land-based Sources of Radionuclides on Radioactive Contamination of the Kara Sea through Ob′ and Yenisei River Systems* and some other projects mentioned in the Preface. The radioactivity part of the database contains historical and *in situ* data and information about radionuclide presence in river–marine systems, and data on nuclear facilities' operational history (total amount and discharges into environment, accidental events, etc.). A schematic representation of the required/used data and database organization is shown in Figure 2.29.

For several reasons, the most important data on radioactivity were available for regions affected by the Krasnoyarsk Mining and Chemical Combine (MCC Krasnoyarsk-26). For the other two facilities under consideration, the Mayak PA and Tomsk-7, the data on radioactivity are much scarcer. Figure 2.30 represents an example of an extract of data on Krasnoyarsk-26 routine releases.

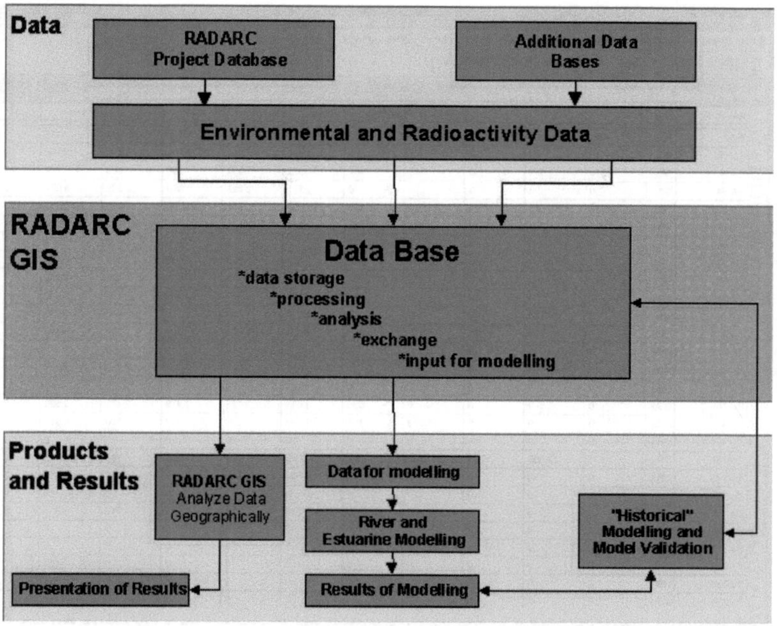

Figure 2.28. The RADARC GIS concept and inputs and outputs.

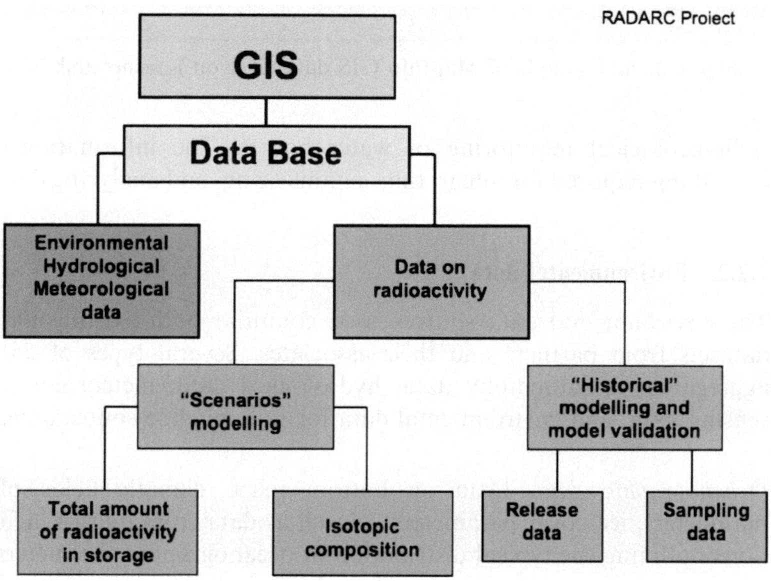

Figure 2.29. Schematic representation of the required/used data and database organization.

Data on the results of numerical modeling and model validation obtained during implementation of the final part of the project are also integrated into the project database.

In summary, it may be stated that the RADARC information system is an indispensable tool for a comprehensive approach to solution of the global tasks of

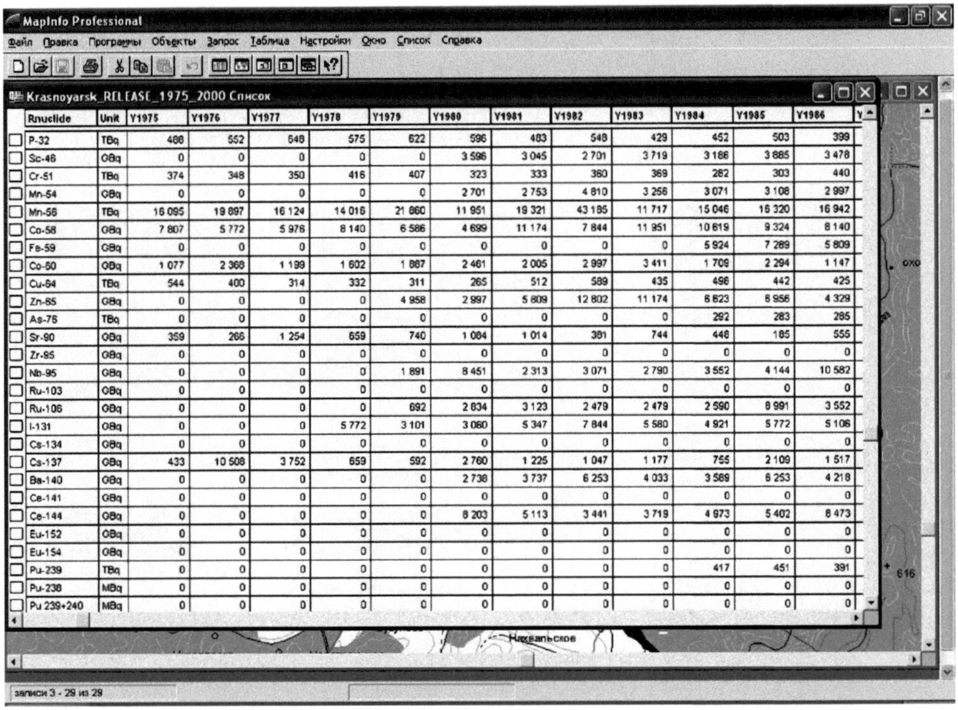

Rnuclide	Unit	Y1975	Y1976	Y1977	Y1978	Y1979	Y1980	Y1981	Y1982	Y1983	Y1984	Y1985	Y1986
P-32	TBq	488	552	648	575	622	596	483	548	429	452	503	399
Sc-46	GBq	0	0	0	0	0	3 596	3 045	2 701	3 719	3 186	3 885	3 478
Cr-51	TBq	374	348	350	416	407	323	333	360	369	282	303	440
Mn-54	GBq	0	0	0	0	0	2 701	2 753	4 810	3 256	3 071	3 108	2 997
Mn-56	TBq	16 095	19 897	16 124	14 016	21 860	11 951	19 321	43 185	11 717	15 046	16 320	16 942
Co-58	GBq	7 807	5 772	5 976	8 140	6 586	4 699	11 174	7 844	11 951	10 619	9 324	8 140
Fe-59	GBq	0	0	0	0	0	0	0	0	0	5 924	7 289	5 809
Co-60	GBq	1 077	2 368	1 199	1 602	1 887	2 461	2 005	2 997	3 411	1 709	2 294	1 147
Cu-64	TBq	544	400	314	332	311	265	512	589	435	496	442	425
Zn-65	GBq	0	0	0	0	4 958	2 997	5 809	12 802	11 174	6 623	6 956	4 329
As-76	TBq	0	0	0	0	0	0	0	0	0	292	283	285
Sr-90	GBq	359	266	1 254	659	740	1 084	1 014	381	744	446	185	555
Zr-95	GBq	0	0	0	0	0	0	0	0	0	0	0	0
Nb-95	GBq	0	0	0	0	1 891	8 451	2 313	3 071	2 790	3 552	4 144	10 582
Ru-103	GBq	0	0	0	0	0	0	0	0	0	0	0	0
Ru-106	GBq	0	0	0	0	692	2 834	3 123	2 479	2 479	2 590	8 991	3 552
I-131	GBq	0	0	0	5 772	3 101	3 060	5 347	7 844	5 580	4 921	5 772	5 106
Cs-134	GBq	0	0	0	0	0	0	0	0	0	0	0	0
Cs-137	GBq	433	10 508	3 752	659	592	2 760	1 225	1 047	1 177	755	2 109	1 517
Ba-140	GBq	0	0	0	0	0	2 738	3 737	6 253	4 033	3 589	6 253	4 218
Ce-141	GBq	0	0	0	0	0	0	0	0	0	0	0	0
Ce-144	GBq	0	0	0	0	0	8 203	5 113	3 441	3 719	4 973	5 402	8 473
Eu-152	GBq	0	0	0	0	0	0	0	0	0	0	0	0
Eu-154	GBq	0	0	0	0	0	0	0	0	0	0	0	0
Pu-239	TBq	0	0	0	0	0	0	0	0	0	417	451	391
Pu-238	MBq	0	0	0	0	0	0	0	0	0	0	0	0
Pu 239+240	MBq	0	0	0	0	0	0	0	0	0	0	0	0

Figure 2.30. Example of MapInfo GIS data table on Krasnoyarsk-26 routine releases.

radio-ecological monitoring of water bodies. The information system provides everything required for integrating, summarizing, and analyzing data geographically.

2.2.2 Environmental data

The environmental data sources used comprise both existing databases and new datasets from partners and their associates. Several types of datasets have been aggregated: oceanography data, hydrological data, meteorological data, remote-sensing data, and environmental data for radionuclide source areas.

Oceanographic data. Data on bottom relief, climatic fields of oceanographic parameters, ice cover parameters, as well as data from quasi-synchronous observations outlining the typical distribution of oceanographic parameters, and water and ice circulation. Long-time series of sea level, and currents and ice cover distribution data in SIGRID format were also used. Assessment of interannual variability and the main types of circulation of water and ice in the Arctic Ocean was made.

NIERSC's regular digital bottom relief model of the Kara Sea was supplemented using a new collection of bottom topography maps made by the Russian Hydrographical Service. The climatic fields of oceanographic parameters, ice cover parameters, as well as data from quasi-synchronous observations outlining the typical distribution of oceanographic parameters and water and ice circulation for the Kara Sea and Arctic Basin were collected from a comprehensive range of sources, such as previous reports of partners and published sources.

The database includes oceanographic surveys of the Kara Sea, water temperature, and salinity (gridded, 5 km resolution) from the Kara Sea Marine Information System created under the INTAS 93-814 project with the participation of AARI (Johannessen *et al.*, 1996, 1997). The Kara Sea Marine Information System contains information about the summer and winter oceanographic regime, and reflects the main typical variability in distribution of oceanographic parameters in the Kara Sea. The oceanographic data include oceanographic surveys carried out in 1977 (summer), 1978 (summer), 1979 (winter), 1980 (summer), as well as original data derived under the Soviet/Russian Norwegian Oceanographic Program (SNOP/RUSNOP, 1988–1995): 1989 (summer), 1993 (summer), 1994 (summer), and 1995 (summer).

An important criterion of the state of oceanographic conditions in the Kara Sea is the type of river-water distribution. Having analyzed Kara Sea hydrological and ice conditions for the whole period under study, it was established that the type of river-water distribution over the sea is a reliable indicator of the dynamic condition of the sea; close correlations between ice and hydrological conditions were found by dividing into types of freshwater distribution. According to this classification, the data used for the period from 1977 to 1980 describe each type of river-water distribution: western (1977 and 1979), eastern (1978), and central (1980).

A gridded dataset for temperature and salinity was produced using data from the 1977–1980 AARI oceanographic surveys, supplemented by climate data in order to close gaps in the gridpoints in the northern sea covered with ice. Table 2.9 provides insight into the details of the initial information.

The original oceanographic survey data obtained under SNOP/RUSNOP for 1989 and 1993–1995 were extracted from accessible sources and used for validation of hydrodynamic models. Each survey includes about 100–150 points. An example of oceanographic station distribution is presented in Figure 2.31.

Table 2.9. Information about expeditions in the Kara Sea.

Vessel, expedition	*Latitude* *N* (min–max)	*Longitude* *E* (min–max)	*Dates* (from–to)	*Number of stations*	*Number of observations* T/S
R/V *Storm*	68°26′–78°53′	55°38′–86°30′	05.08–30.09.77	128	940/926
R/V *Storm*	68°51′–79°57′	54°52′–104°06′	15.08–09.10.78	136	1,078/1,073
HAE Sever	66°53′–84°57′	56°25′–104°10′	11.03–17.05.79	71	541/538
R/V *Otto Schmidt*	75°10′–81°04′	58°00′–90°00′	10.08–28.08.80	74	762/762
R/V *Storm*	69°59′–74°00′	56°00′–83°22′	25.08–27.09.80	73	496/495

Note: R/V = research vessel; HAE = high-latitude aerial expedition; T = temperature; S = salinity.

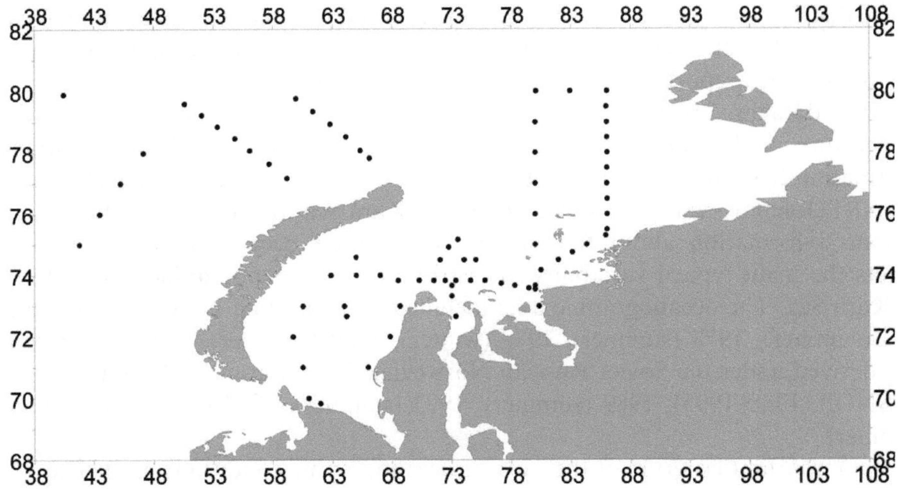

Figure 2.31. Location of oceanographic stations during the KAREX-95 expedition onboard the *Ivan Petrov* in the summer of 1995.

A long time series of sea level (10 polar stations, from 8–50 years' duration) and currents (10 moorings and 16 levels for the summer and winter period of 1977–1980, Table 2.10) for key points along the Kara Sea coast was collected and used for model validation.

Sea ice data in SIGRID format, which were derived from initial data on total ice concentration, were obtained for the Kara Sea ice conditions between 1977 and 1980 in the form of integrated 10–day period ice charts prepared under the INTAS 93-814 project at the AARI Center for Ice Hydrometeorological Information. These integrated ice charts contain information both on the main ice cover parameters (total and partial concentration, stage of ice development, and form of sea ice) and on parameters such as surface characteristics (stages of melting, snow concentration), the navigational situation (cracks, fractures, polynyas), and manifestations of dynamical processes (hummock and ridge concentration). These data were incorporated in the environmental database and also used for model validation.

Hydrological data. These include hydrological parameters, such as river discharge, water level, river and sea ice, topography, and bottom sediment for key points along the river, as well as information on the climatic and meteorological parameters for the watershed area of the rivers under study. In addition, data from various field observations of the distribution of hydrochemical parameters and pollutants along the river systems were also needed to understand the patterns of sediment transfer with river discharge.

A new regular Digital Bottom Relief Model for the Ob′ and Yenisei estuaries was created based on detailed bottom topography navigation maps made by the Russian Hydrographical Service. A long time series of river discharge parameters for the Ob′ and Yenisei watershed area, as well as temperature, salinity, and level data, was

Table 2.10. List of the current observation stations in the summer (a) and in the winter (b) periods, 1977–80.

Cadastral number of station	*Position data*		*Depth*	*Observation period*		*Duration of observation*
	Latitude N	*Longitude E*	(m)	*Start*	*End*	(days)
(a) Summer						
527	71°10′	65°56′	18	15.08.77	09.09.77	26
529	73°16′	69°01′	10	26.08.79	27.09.79	33
530	74°00′	79°43′	25	05.09.79	26.09.79	22
138	70°20′	57°28′	10	21.10.79	28.10.79	7
			25	21.10.79	28.10.79	7
			Near-bottom	21.10.79	28.10.79	7
139	70°40′	57°52′	10	14.10.80	29.10.80	15
			25	14.10.80	29.10.80	15
140	70°32′	58°32′	10	16.10.80	30.10.80	15
			25	16.10.80	30.10.80	15
(b) Winter						
131	76°19′	95°04′	10	02.04.78	22.04.78	20
132	76°21′	95°11′	10	02.04.78	22.04.78	20
359	78°04′	103°37′	10	10.04.79	01.05.79	20
			Near-bottom	10.04.79	25.04.79	15
360	77°47′	104°10′	10	10.04.79	25.04.79	15
			Near-bottom	10.04.79	01.05.70	20

formed and processed; as a result a unique new river discharge dataset for 19 points for up to 50–60 years was made (Tables 2.11 and 2.12).

Meteorological data. NCAR/NCEP reanalysis data for the Arctic have been acquired. These data consist of wind velocity, air temperature, cloudiness, humidity, and precipitation. Systematization of atmospheric pressure and other meteorological parameter data for the Kara Sea region was carried out according to AARI's classification of atmospheric circulation developed at NIERSC in the framework of earlier projects (INTAS 93-814 and 97-1277) and provided as input into the corresponding dataset. Data sources included sets of synoptic maps of the Arctic, datasets collected during earlier projects (e.g., the Global Historical Climatological Network or GHCN dataset, and others). NCAR/NCEP reanalysis data were processed for the Ob′ and Yenisei estuaries to create files in the necessary format for computations.

In addition, a new dataset for surface air temperature (SAT) was created. The new gridded SAT dataset (Kuzmina *et al.*, 2008) for the region north of 40°N was created for the period 1900–2000 using objective analysis (OA) methods. The dataset uses all available data from land meteorological stations, ARGOS buoys, drifting

Table 2.11. Description of hydrological stations used for the study, Yenisei.

Distance from Bazaikha (km)	*River*	*Station*	*Bank*	*Distance from mouth* (km)	*Discharge* (m^3/s)	*Watershed surface area* (km^2)	*Observation period*
0	Yenisei	Bazaikha	Transit	2,468	2,000	299,000	1936–1995
113	Kan	Podporog	Right	17	296	36,800	1938–1991
343	Angara	Tatarka	Right		4,510	1,040,000	1953–1995
414	Yenisei	Yeniseisk	Transit	2,054	7,450	1,420,000	1936–1995
898	Podkamennaya Tunguska	Kuz'movka	Right	209	1,620	218,000	1938–1995
900	Yenisei	Podkamennaya Tunguska	Transit	1,568	10,600	1,520,000	1936–1995
1,504	Nizhnyaya Tunguska	Bol'shoy Porog	Right	125	3,560	438,000	1938–1995
1,531	Turukhan	Yanov Stan	Left	309	110	10,100	1941–1995
1,771	Yenisei	Igarka	Transit	697	18,200	2,380,000	1936–1995

Table 2.12. Description of hydrological stations used for the study, Ob′ River.

No.	*River*	*Station*	*Bank*	*Distance from mouth* (km)	*Discharge* (km^3/s)	*Watershed surface area* (km^2)	*Observation period*
1	Iset	Koliutino	Transit	516	(16)	3,470	1961–1962, 1964–1990
2	Iset	Kataisk	Transit	386	(28)	11,200	1958–1990
3	Miass	Kargapoliye	Right	24	18	18,800	1936, 1949–1990
4	Iset	Mekhonskoe	Transit	204	62	477,000	1941–1990
5	Tobol	Yalutorovsk	Transit	426	107	185,000	1936–1995
6	Tobol	Kurgan	Right	708	32	107,000	1936–1990
7	Tavda	Nizhnyaya Tavda	Left	123	(424)	86,100	1967–1969, 1971–1993, 1995
8	Tobol	Lipovka	Transit	104	(869)	367,000	1936–1984
9	Irtysh	Tobolsk	Left	654	2,050	969,000	1936–1995
10	Ob′	Belogoriye	Transit	1,152	9,934	2,180,000	1936–1994
11	Severnaya Sos′va	Igrim	Left	147	(774)	87,800	1958–1995
12	Synya	Ovgort	Left	88	(95)	9,880	1963–1988, 1990–1995
13	Ob′	Salekhard	Transit	287	12,813	2,450,000	1936–1995

stations, and patrol ships. The results obtained by OA were compared with other gridded SAT data already in use. The advantage of the new dataset is its enhanced spatial coverage in high northern latitudes, due to the ability of OA to optimize the data when information is scarce.

Remote-sensing data. For model validation, various types of remote-sensing data were used. These data provide information on the mixing processes of river and sea water in the Ob′ and Yenisei mouths, the dynamics of frontal zones, current velocities in the mouths, ice cover and drift, sea surface topography, etc. The satellite imagery used includes observations in the visible, infra-red, and microwave, with a focus on synthetic aperture radar (SAR) data. These types of data were used for qualitative comparison and analysis of numerical calculation results.

Environmental data for radionuclide source areas. Data on the natural characteristics of the environment (geological, hydrological, meteorological, ecological, and other variables) are contained in the data archives of RADARC Project participants. These data were used in modelling scenarios which consider the influence of environmental parameters.

All data were distributed to the partners of the project so that they could be included in the GIS Project and the AMAP Database.

2.2.3 Radioactive and non-radioactive pollution data

Radioactive pollution. During the RADARC Project, several types of datasets not present in the AMAP (Arctic Monitoring and Assessment Program) Database were developed. Information on radiation pollution characteristics served as input data for numerical modeling (Chapters 3 and 4). Special attention was placed on identifying the possibility, types, and scale of potential natural and technological accidental releases. The most possible pathways of radioactivity releases from nuclear facilities (under normal functioning and in the case of accidents) were described and evaluated providing quantitative and qualitative characteristics of these pathways. The choice of the most potentially hazardous land-based sources of Kara Sea radioactive contamination was made using the following parameters: type of industrial activity and its constituent parts in the past, present, and future; amount and types of radionuclides at the sources; accidents that took place in the past or potential accidents in the future; and description of measures and man-made and natural barriers hindering the release of radioactivity into the environment.

Taking into account the above-stated points, the following three large-scale enterprises of the Russian Nuclear Complex, namely the Mayak PA, the SCC Tomsk-7, and the MCC Krasnoyarsk-26, were chosen as the main sources for this research. Detailed descriptions are presented in Chapter 1.

Collection and systematization of environmental radioactivity data for the Arctic region that are not presented in the AMAP Database was coordinated by the Khlopin Radioactivity Institute using open information sources. The studies under this topic have as their objectives: (1) to obtain existing environmental levels, amount, and isotopic composition of radioactive contamination that are not presented in the AMAP Database, and (2) to describe the main radioactive sources of radioactivity not only in the Russian Arctic coastal zone (i.e., including sources located at the Kola Peninsula, within the watersheds of the Ob′ and Yenisei Rivers, and in the Barents and Kara Seas, where significant amounts of radionuclides are being stored, used, processed, and/or transported) but also those in Europe. All available historical information from published sources was used as input data.

Data descriptions include type of industrial activity and its constituent parts in the past, present, and future, amount and isotopic composition of radionuclides accumulated at the sources, accidental events that took place in the past or potential accidents in the future, presence of man-made and natural barriers on the Ob′ and Yenisei Rivers hindering the release of radionuclides into the environment, etc. Specific attention was paid to identifying secondary sources of radionuclides (reservoirs of liquid radioactive waste storage, polluted territories, river plain soils, natural and artificial water bodies, bottom sediment, etc.), and evaluation of radionuclide behavior in the regions of these sources.

The most detailed validation data on radioactivity were available for the regions affected by the MCC Krasnoyarsk-26. For the other two facilities under consideration—Mayak PA and Tomsk-7—the data on radioactivity are much scarcer.

During the implementation of this work, we were able to obtain some new and important data for numerical modeling and model validation. For example, new data on Krasnoyarsk-26 routine releases for the period 1975–2000 were successfully used in numerical modeling and model validation for the Yenisei River. These data from the early years had been classified but recently have become publicly open within the framework of global nuclear disarmament. Figure 2.29 represents an example of the tabulated data on Krasnoyarsk-26 routine releases.

Non-radioactive pollution data. Data on non-radioactive pollution were used to demonstrate the feasibility of the GMS to investigate the propagation and dispersion of certain anthropogenic pollutants in aquatic environments (see Chapter 5). This portion of the database was not incorporated in the RADARC Database, as RADARC focused on radioactive pollution. Some datasets used in the present work were collected from the archives of the Ukrainian Institute of Mathematical Machines and System Problems; to a lesser degree, the data of the Marine Biology Institute in Murmansk were also used. A description of the pollution data is given in the latter part of Chapter 1 of this monograph.

3

Generic model system (GMS) for simulation of radioactive spread in the aquatic environment

This chapter presents a set of numerical modeling techniques for simulating the spread of radioactivity in the aquatic environment, in both marine and inland waters. Section 3.1 describes the concept and structure of the modeling system. Section 3.2 presents a model for the Atlantic and Arctic Oceans. Section 3.3 presents a shelf sea model for the Kara Sea. Section 3.4 describes in detail the river and estuary models for the Ob′ and Yenisei Rivers. These comprise a one-dimensional model for simulation of the transport of radionuclides in a river system (RIVTOX), and a numerical model for three-dimensional dispersion simulation of radionuclides in stratified water bodies (THREETOX). For each model, we present results from validation of the models against comparable measurement data and knowledge based on observations.

3.1 RATIONALE, CONCEPT, AND STRUCTURE OF THE GMS

As discussed in Chapter 1, the main sources of radioactivity in Arctic seas are atmospheric fallout, releases from reprocessing plants in Western Europe and Western Siberia, fallout from Chernobyl, and local fallout from nuclear testing at Novaya Zemlya. The potential sources of radioactivity can be releases due to accidents on reprocessing plants and vessels with nuclear engines, ship transportation of nuclear waste or decommissioned reactors, leakage from dumped fuel reactors and containers with radioactive wastes. The processes of radionuclide dispersion in the Arctic are complicated because of multiple pathways through catchments, rivers, estuaries, shelf areas, and deep waters. In rivers, radionuclides are transported in solute and attached to suspended sediments, whereas in the estuaries they are deposited and re-entrained by tides. In the sea, radioactivity is transported mainly as solute but drifting ice can transport contamination for long distances in the Arctic Ocean. The processes of radionuclide dispersion in the Arctic strongly differ in their spatial and temporal scales. The travel time of radioactivity through large Siberian river systems is on the order of months, whereas travel time through coastal seas is years, and the characteristic flushing time of the Arctic Ocean may have a century timescale.

A number of models have been developed during the past 20 years to describe the transport and fate of radioactivity in the Arctic Basin. Models of radionuclide transport by the Ob′ and Yenisei Rivers were developed by Paluszkiewicz *et al.* (1997, 2001) and Maderich *et al.* (2005). Modeling of radionuclide transport in the Kara Sea due to the potential release of radioactivity from dumped waste and scuttled reactors in the bays of Novaya Zemlya has been carried out by Harms (1997) and Koziy *et al.* (1998, 2000), while transport of radionuclides in the North Atlantic–Arctic has been simulated by Gao *et al.* (2004). However, each of these models describes only a narrow part of the possible spectrum of processes of dispersion in the marine, riverine, and river–marine environments. Therefore, there is a need for a comprehensive modeling system that can describe significantly different temporal and spatial dispersion processes.

The approach is to combine different existing models in a generic model system (GMS) that can simulate the transport and fate of radionuclides through different pathways and at different temporal and spatial scales. In addition to advancing scientific understanding, this system can be used to support decision making for: risk prevention, forecasting and preparedness for probable crisis, alert and detection, relief and mitigation, and damage assessment. The GMS includes a geographic information system (GIS) and an interactive, hydrodynamic model chain for simulating radionuclide transport from particular land-based or sea-based sources. The system can describe transport through rivers and estuary systems and in the shelf seas, as well as in the Arctic and North Atlantic Oceans. The GMS is organized as a hierarchical system that consists of four levels of models, ranging from a large-scale ocean circulation model to a one-dimensional river model. The models are dynamically coupled through one-way nesting (i.e., data flow from the model at a higher level to the model at a lower level). An exception is river inflow, which is used at all levels. The radionuclide data flow has the opposite direction: from the lower to higher level. The system allows efficiently organized simulations at different timescales and spatial scales.

3.1.1 GMS structure and data streams

The GMS is organized as a hierarchical system that consists of models at four levels. The configuration of developed GMS consists of four adapted and validated hydrodynamic models: a medium-resolution (50 km) three-dimensional numerical ice–ocean model for the Arctic and North Atlantic Oceans (Section 3.2); nested with a high-resolution (5 km) model for the Kara Sea including a sediment module (Section 3.3); nested with the river model system including a one-dimensional model of river dynamics and radionuclide transport (Section 3.4.1); and a three-dimensional model of Ob′ and Yenisei estuarine and coastal area hydrodynamics and radionuclide transport (Section 3.4.2). The GMS structure and data streams are shown in Figure 3.1.

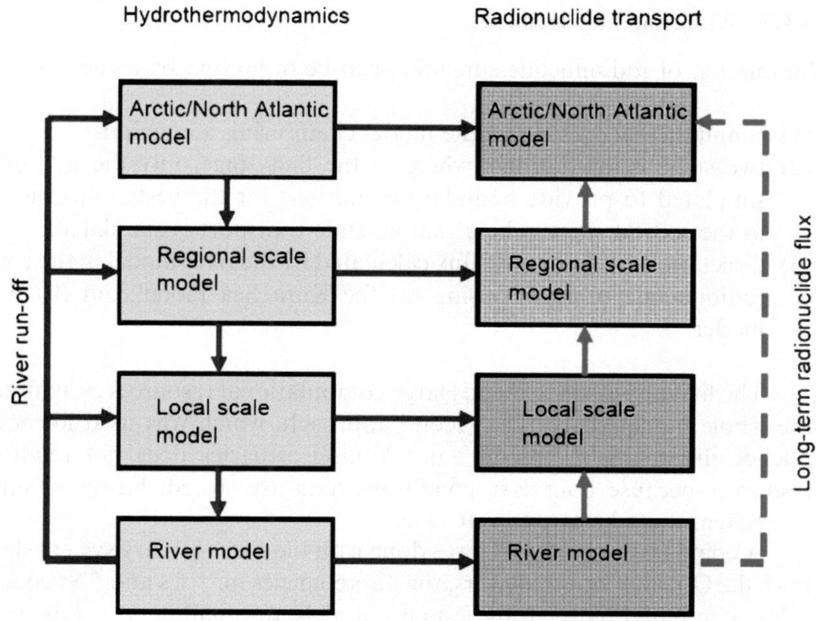

Figure 3.1. Generic model system (GMS) structure and data streams.

The models have a similar structure, consisting of a hydrodynamics module and a radionuclide transport module. The hydrodynamics modules of the first three models are similar. Free surface primitive equation models are employed with different parameterizations of turbulent mixing. The ice dynamics–thermodynamics model is based on the Hibler (1979) approach. Therefore, models are compatible in the nesting procedure. The main differences in the models are related to the numerical implementation and use of different coordinate systems. River and local models include a module of sediment transport that is important in the transfer of radionuclides absorbed/adsorbed on sediments in shallow waters, whereas large-scale and regional models describe only the transport of dissolved radionuclides.

The data streams are organized as follows (Figure 3.1). The large-scale model provides hydrodynamics boundary conditions for the nested Kara Sea model which in turn provides outer-boundary conditions for the estuary model. The estuary model provides sea level boundary conditions for the river model. River discharge is prescribed in the estuary models. As an option, river discharge can be prescribed in the Kara Sea model or in the large-scale model (blue dashed line in Figure 3.1). The data stream of sediment concentration and radioactivity is opposite to the hydrodynamics data stream. In river mouths, river sediment concentration and radionuclide concentration in the solute and suspended sediments calculated by the river model provide input for the estuary model, which in turn provides data for the Kara Sea model, which provides input data for the large-scale model on radionuclide fluxes from Siberian rivers.

3.1.2 Modeling management

Simulation of radionuclide spreading can be organized by means of

(1) simultaneous running of the model chain using a coupler;
(2) two-stage computations, where in the first stage only the hydrodynamics are simulated to provide boundary conditions for the nested models that are used in the second stage, where radionuclide transport is simulated;
(3) direct use of radionuclide flux calculated by the river model in the computation of radionuclide transport using (a) the Kara Sea model and (b) the large-scale model.

The first approach demands large computational resources to simultaneously run the whole model chain. The second approach, which was used in modeling radionuclide dispersion in the Ob′ and Yenisei estuaries does not require such large resources because boundary conditions data are stored; however, simulations for the system must be carried out twice.

A comparative study that was done with the river–estuary system showed that for both the Ob′ and Yenisei Rivers, the discrepancies in ^{137}Cs and ^{90}Sr concentrations in solute computed by river and estuary models are small (e.g., a difference of around 10% in the concentration of ^{137}Cs in suspended sediments was found in the Yenisei estuary—see Section 3.4). Therefore, to improve the computational efficiency of the GMS, output from the river model was input in the estuary mouth for most simulations. This was used in Kara Sea modeling and in the long-term runs of the large-scale model (dashed red line in Figure 3.1).

3.2 ATLANTIC AND ARCTIC OCEAN MODEL

Ocean models covering a large geographical region are required to model atmospheric fallout and long-range transport of radionuclides from sources located outside the Arctic Ocean. Accordingly, an ocean circulation and tracer model has been set up, validated, and used for the Atlantic Ocean, the Nordic Seas and the Arctic Ocean. This ocean model has been used to simulate the impact of nuclear releases from sources in Europe (e.g., the U.K. Sellafield and the French Le Hague reprocessing plants), as well as the contribution of atmospheric fallout. The model has also been used to simulate the impact of various climate change scenarios, including a doubling of atmospheric CO_2 content, on the distribution of radioactive contaminants in the Arctic Ocean.

3.2.1 General model description

Over the past decades, several primitive equation ocean general circulation models (OGCMs) have been developed and the governing equations are commonly solved in a Cartesian framework in which fixed depths serve as the vertical coordinate, and these models are often referenced as "level" models. Level models have a common deficiency in that either meridional heat transport is too weak or the thermocline is too diffuse (Bryan, 1987). Model defects are believed to be associated with too strong mixing in the horizontal plane, whereas mixing and advection occur predominantly along isopycnals (planes of constant density with respect to a given reference pressure), or along neutral surfaces (i.e., locally defined isopycnal surfaces) in the interior ocean (McDougall, 1987). Therefore, in regions of sloping isopycnals, spurious diapycnal buoyancy fluxes occur in level models due to horizontal mixing (Hu, 1997). In practice, iso-neutral mixing and advection are approximated in OGCMs by either use of an isopycnal diffusivity tensor (Redi, 1982) in level models, or use of "isopycnic" ocean models. In level models, use of rotated local isopycnal diffusivity tends to mix along neutral surfaces; however, advection along local isopycnal surfaces does not occur naturally (Hu, 1997). Based on vertical density stratification of the ocean, isopycnic OGCMs are developed in which potential density serves as the vertical coordinate. In isopycnic models, both mixing and advection naturally occur along isopycnal surfaces. The isopycnic concept is appealing for several reasons (Drange, 1994): the preferred plane of advective flow in the ocean is along, rather than across, planes of constant density (Garcon *et al.*, 1993); the magnitude of the mixing is several orders of magnitude weaker in the diapycnal than in the isopycnal direction (Sarmiento *et al.*, 1976; Ledwell *et al.*, 1993; Toole *et al.*, 1994), and this is embedded in the isopycnic concept *per se*; only two-dimensional prognostic equations have to be solved, so there is no numerical smoothing of dynamic fields in the diapycnal direction, and the inclusion of a thermodynamic mixed layer represents a natural extension to the isopycnic concept.

In our experiments, the Miami Isopycnal Coordinate Ocean Model (MICOM, Bleck *et al.*, 1992), fully coupled to a sea ice module consisting of the Hibler (1979) rheology in the implementation of Harder (1996), and the thermodynamics of Drange and Simonsen (1996) is used for simulating the transport and spread of radionuclides in the North Atlantic–Nordic–Arctic Seas.

MICOM is a primitive equation model. It uses Eulerian coordinates in the horizontal and potential density as a Lagrangian coordinate in the vertical. The model equations are spatially discrete on a C-grid in the horizontal plane and are stepped forward in time with a split explicit numerical scheme. The closure of prognostic variables is matched by means of a prognostic equation for the hydrostatic pressure balance in the vertical, a state equation that links temperature, salinity, and potential density, and a turbulent kinetic closure equation for the upper mixed layer (ML).

The model has an uppermost ML with temporally and spatially varying density, and layers with fixed potential densities below the ML. The specified potential densities of the sub-surface layers were chosen to ensure a realistic representation of the major water masses in the North Atlantic–Nordic Seas region. In the horizontal, the model is configured with a local orthogonal grid mesh with one pole over North America and one pole over the western part of Asia (Bentsen *et al.*, 1999), yielding an improved horizontal grid spacing in the North Atlantic–Nordic Seas region.

The vertically homogeneous ML utilizes the Gaspar (1988) bulk parameterization for the dissipation of turbulent kinetic energy, and has temperature, salinity, and layer thickness as the prognostic variables. In the isopycnic layers below the ML, temperature and layer thickness are the prognostic variables, whereas salinity is diagnostically determined by means of the simplified equation of state of Friedrich and Levitus (1972). A flux-corrected transport scheme (Zalesak 1979; Smolarkiewicz and Clark, 1986) is used to advect the model layer thickness and tracer quantities. The mixed layer is allowed to exchange mass with other isopycnic layers by detrainment and entrainment. The water mass transfer between isopycnals is achieved by diapycnal mixing. The diapycnal mixing coefficient K_d ($m^2\,s^{-1}$) is parameterized according to the Gargett (1984) expression with K_d inversely proportional to the Brunt–Väisälä frequency N (s^{-1}). Numerical implementation of diapycnal mixing follows the scheme of McDougall and Dewar (1998).

Bathymetry is computed as the arithmetic mean value based on the ETOPO-5 data base (Data Announcement 88-MGG-02, *Digital Relief of the Surface of the Earth*, NOAA, National Geophysical Data Center, Boulder, Colorado, 1988).

The model was initialized using the January Levitus and Boyer (1994) and Levitus *et al.* (1994) climatological temperature and salinity fields, respectively, a 2 m thick sea ice cover based on climatological sea ice extent, and an ocean at rest. The model was first spun up for 180 years with monthly mean atmospheric forcing fields derived from National Center for Environmental Prediction/National Center for Atmospheric Research (NCEP/NCAR) reanalysis (Kalnay *et al.*, 1996). In this spin-up, both sea surface salinity (SSS) and sea surface temperature (SST) were relaxed towards the monthly mean surface climatology. Relaxation was carried out by applying fluxes of heat and salt proportional to the SSS and SST differences between model and climatology, respectively, with an e-folding timescale of 30 days for an ML of 50 m or less, decreasing linearly with thicker ML depths. The spin-up was then continued with daily NCEP/NCAR forcing, repeating the period 1974–1978 twice. Relaxation for salinity only was applied in these two integrations, and the deduced salinity adjustment flux was stored from the latter integration to produce seasonal-averaged restoring fluxes for salt. The period 1974–1978 was chosen because of the relatively neutral North Atlantic Oscillation (NAO) (Hurrell, 1995) conditions of these years. Finally, the model was integrated with daily forcing for the period 1948 up to the present with no relaxation but with diagnosed restoring fluxes for salt.

Two global versions of MICOM, with horizontal resolution of 40 km and 80 km in the North Atlantic–Nordic Seas, have been used to simulate the spread and transport of anthropogenic radionuclides (Gao *et al.*, 2004, 2005). In the vertical, the 40 (80) km model has respectively 25 (23) isopycnal layers below the ML. Specifically, the diffusive velocities (diffusivities divided by the size of the grid cell) for layer interface diffusion, momentum dissipation, and tracer dispersion in the 40 km version are $0.015\,m^2\,s^{-1}$, $0.01\,m^2\,s^{-1}$, and $0.0025\,m^2\,s^{-1}$ and the diapycnal mixing coefficient K_d ($m^2\,s^{-1}$) is parameterized as $K_d = (5 \times 10^{-8})/N$, where N (s^{-1}) is the Brunt–Väisälä frequency. Consequently, the values of tracer dispersion and diapycnal mixing are a factor of 2 and 6, respectively, below those used in the 80 km version.

In addition, a nested, medium-resolution regional version of MICOM with a horizontal resolution of 20 km in the North Atlantic–Nordic Seas was also used for radionuclide dispersion (Orre *et al.*, 2007). The regional model domain covers the northern North Atlantic Ocean and the Nordic Seas (from about 30°N to about 80°N). At the lateral boundaries, weekly resolved velocity, temperature, and salinity fields from a global 40 km version of the same model (e.g., Nilsen *et al.*, 2003; Drange *et al.*, 2005) are forced onto the regional model. The OGCM is driven by daily atmospheric forcing fields provided by NCEP/NCAR for the period 1948 to the present (Kalnay, 1996).

Primitive equations

The primitive equations involved in MICOM are the horizontal momentum equation, the mass continuity equation, or the layer thickness tendency equation and conservation equations for heat and salt. All of the equations are written in terms of a generalized vertical coordinate (Bleck, 1978) to make them applicable to both interior isopycnic layers and to the top mixed layer.

Momentum equation

The momentum equation reads:

$$\frac{\partial \vec{V}}{\partial t} + \nabla_s \frac{\vec{V}^2}{2} + (\zeta + f) K \times \vec{V} = \left(\dot{s} \frac{\partial p}{\partial s} \right) \frac{\partial \vec{V}}{\partial p} + \nabla_\alpha M = -g \frac{\partial \tau}{\partial p} + \left(\frac{\partial p}{\partial s} \right)^{-1} \nabla_s \left(v \frac{\partial p}{\partial s} \nabla_s \vec{V} \right) \tag{3.1}$$

where $\vec{V} = (u, v)$ is the horizontal velocity vector; p is the pressure; $\zeta = \left(\frac{\partial v}{\partial x}\right)_s - \left(\frac{\partial u}{\partial y}\right)_s$ is the relative vorticity; f is the Coriolis parameter; K is the vertical unit vector; α is the specific volume; $M = gz + p\alpha$ is the Montgomery potential; gz is the geopotential; v is the variable eddy viscosity coefficient; and finally τ is the wind-induced and bottom drag–induced shear stress. Subscripts indicate which variable is held constant during partial differentiation.

The expansion of the above vector equation has the following nonlinear form in x, y, s coordinates under adiabatic flow conditions (Bleck and Smith, 1990):

$$\frac{\partial u}{\partial t} + \frac{\partial}{\partial x} \frac{V^2}{2} - (\zeta + f) v = -\frac{\partial M}{\partial x} + \frac{1}{\Delta p} [g \Delta \tau_x + \nabla \cdot (\nu \Delta p \nabla u)] \tag{3.2}$$

$$\frac{\partial v}{\partial t} + \frac{\partial}{\partial y} \frac{V^2}{2} - (\zeta + f) u = -\frac{\partial M}{\partial y} + \frac{1}{\Delta p} [g \Delta \tau_y + \nabla \cdot (\nu \Delta p \nabla v)] \tag{3.3}$$

where the second term on the right-hand side includes wind stress and bottom-related stress, and momentum diffusion.

Continuity equation

The continuity equation with s serving as the vertical coordinate reads:

$$\frac{\partial}{\partial t_s}\left(\frac{\partial p}{\partial s}\right) + \nabla_s \cdot \left(\vec{V}\frac{\partial p}{\partial s}\right) + \frac{\partial}{\partial s}\left(\dot{s}\frac{\partial p}{\partial s}\right) = 0 \tag{3.4}$$

After integration over a coordinate layer bounded with two s surfaces s_{top} and s_{bot}, the continuity equation becomes a prognostic equation for the layer weight per unit area, $\Delta p = p_{\text{bot}} - p_{\text{top}}$ (Bleck *et al.*, 1992):

$$\frac{\partial}{\partial t_s}\Delta p + \nabla_s \cdot (\vec{V}\Delta p) + \left(\dot{s}\frac{\partial p}{\partial s}\right)_{\text{bot}} - \left(\dot{s}\frac{\partial p}{\partial s}\right)_{\text{top}} = 0 \tag{3.5}$$

where $s\frac{\partial p}{\partial s}$ represents the vertical mass flux taken to be positive if it is in the $+p$ (or downward) direction across an s surface.

Hydrostatic equation

Expressed in terms of the Montgomery potential M, the hydrostatic equation becomes:

$$\frac{\partial M}{\partial \alpha} = p \tag{3.6}$$

Heat and salt conservation equations

The conservation equation for temperature, salinity, and buoyancy can be put in the following form:

$$\frac{\partial}{\partial t_s}\left(\frac{\partial p}{\partial s}T\right) + \nabla_s \cdot \left(\vec{V}\frac{\partial p}{\partial s}T\right) + \frac{\partial}{\partial s}\left(\dot{s}\frac{\partial p}{\partial s}T\right) = \nabla_s \cdot \left(\upsilon\frac{\partial p}{\partial s}\nabla_s T\right) + \kappa_T \tag{3.7}$$

where T denotes any one of the thermodynamic variables (temperature, salinity, and buoyancy); and κ_T represents the diabatic source of T.

3.2.2 Radionuclide tracer module

The spatial and temporal evolution of the concentration C and the age A (s) of a conservative and non-decaying tracer are governed by the linear Eulerian transport equations (England, 1995; Delhez *et al.*, 1999; Delhez and Deleersnijder, 2002):

$$\frac{\partial C}{\partial t} + \nabla \cdot (\mathbf{v}C) = \nabla \cdot (\mathbf{K} \cdot \nabla C) + F_c \tag{3.8}$$

$$\frac{\partial AC}{\partial t} + \nabla \cdot (\mathbf{v}AC) = \nabla \cdot (\mathbf{K} \cdot \nabla AC) + C \tag{3.9}$$

where t (s) is time; $\mathbf{v}$ (m s^{-1}) is the velocity field; $\mathbf{K}$ (m^2 s^{-1}) is the diffusivity tensor; ∇ (m^{-1}) is the grad operator; and F_C is the source rate for the tracer C. For a radioactive tracer (e.g. ^{137}Cs) the concentration is commonly expressed in terms of $Bq\,m^{-3}$, and the above equations are modified by appropriate decay terms on the right-hand side of the equations. The age A of the tracer can be viewed as an inherent property, or an attribute, of the tracer. The age is set to zero for source waters, and increases linearly with time since the release. For any model grid cell, the Eulerian concentration C from (3.143) represents the sum of sub-tracers that have been released at different times and have followed different pathways. By incorporating the age tracer, each of these sub-tracers can be given their own age. The age A derived from (3.9) is therefore the concentration-weighted mean age of each sub-tracer in any model grid cell, hereafter the "apparent age". It should be noted that the age tracer is included and implemented only in the 40 km and 20 km versions of MICOM. Furthermore, to identify the relative importance of the tracer source function on the time evolution of the tracer distribution, idealized tracers with a constant release rate have been included in the regional model.

3.2.3 Model validation results

A detailed description of the simulated circulation and hydrography in the region of interest (e.g. the northern North Atlantic and the Nordic Seas) will not be given here as these quantities have been addressed in detail in studies by Furevik *et al.* (2002), Gao *et al.* (2003, 2004, 2005), Nilsen *et al.* (2003), Eldevik *et al.* (2005), Hatun *et al.* (2005a, b), Mauritzen *et al.* (2006), and Orre *et al.* (2007) for ocean model systems. In general, the model system has demonstrated skill in simulating the ocean circulation and thermodynamics in the Atlantic and Arctic region of interest.

Briefly, the main ocean circulation in the region of interest is complex—see the characteristics based on observations described in Chapter 2—and is characterized by warm and saline surface Atlantic Water (AW) flowing through the Denmark Strait, passing the ridge between Iceland and the Faroe Islands, and entering east of the Faroes (Hansen and Østerhus, 2000; Østerhus *et al.*, 2005). The AW continues northward along the Norwegian coast as the Norwegian Atlantic Current (NAC). One branch of the AW enters the Barents Sea, whereas the remaining AW continues to the Fram Strait as the West Spitsbergen Current (WSC). Here a fraction of the AW subducts and enters the Arctic Ocean, whereas the rest re-circulates and flows southward.

3.2.4 Extension and validation of the Arctic/North Atlantic model

Description of the new version of MICOM

The numerical methods and thermodynamics of MICOM as summarized in Section 3.2.1 are documented in detail in Bleck and Smith (1990) and Bleck *et al.* (1992). An improved version of MICOM has been developed, and several important aspects differ from the original model.

The original MICOM uses potential density with reference pressure at 0 db as vertical coordinates (σ_2-coordinates). This ensures that the very different flow and mixing characteristic in neutral and dia-neutral directions is well represented near the surface, because isopycnals and neutral surfaces are similar near the reference pressure. However, this does not hold for pressures that differ substantially from the reference pressure. In this study, we choose a reference pressure of 2,000 db (σ_2-coordinates). The non-neutrality of isopycnals in the world ocean is then reduced compared with having the reference pressure at the surface (McDougall and Jackett, 2005).

For the advection of tracers (potential temperature, salinity, and passive tracers) and layer thickness we use incremental remapping (Dukowicz and Baumgardner, 2000) adapted to the grid staggering of MICOM. The algorithm is computationally rather expensive compared with other second-order methods with limiters for a single tracer, but the cost of adding additional tracers is modest. In contrast to the original transport methods of MICOM, incremental remapping ensures the monotonicity of tracers.

Traditionally, MICOM expresses the pressure gradient force (PGF) as a gradient of a potential on an isopycnic surface. This is only accurate if the density can be considered as a function of potential density and pressure alone, which is not the case (de Szoeke, 2000). Inspired by the recent work of Rainer Bleck (pers. commun.), we have based our formulation on Janjic (1977) where the PGF is expressed as a gradient of the geopotential on a pressure surface. This allows us to use a more accurate representation of density in the PGF formulation.

For diapycnal mixing we follow the MICOM approach in which there is background diffusivity that is dependent on local stability implemented using the scheme of McDougall and Dewar (1998). To incorporate shear instability and gravity–current mixing, we add a Richardson number–dependent diffusivity to background diffusivity. This greatly improves the water mass characteristics downstream of overflow regions. Lateral turbulent mixing of momentum and tracers is parameterized by Laplacian diffusion, and the layer interface is smoothed with biharmonic diffusion.

A sea ice module is incorporated, with dynamics using the viscous–plastic rheology (Hibler, 1979) as implemented by Harder (1996), and the thermodynamics of Drange and Simonsen (1996).

The model grid covers the global domain and by using a conformal mapping (Bentsen *et al.*, 1999) the poles have been placed in such a way as to enhance resolution in the North Atlantic and Arctic region. Grid size in the Nordic Seas is about 40 km and increases smoothly to about 500 km on the opposite side of the Earth. In the vertical, the model has 35 layers the uppermost being a mixed layer with temporal and spatial varying density. Potential densities of the isopycnal layers range between 1,030.12 kg m^{-3} and 1,037.80 kg m^{-3}.

The model is forced with daily NCEP/NCAR reanalysis fields (Kalnay *et al.*, 1996), and the spin-up consists of two full NCEP/NCAR reanalysis cycles covering the period 1948–2005, giving in total 116 years of spin-up.

Validating the new model using simulated and observed ^{129}I from Sellafield and La Hague

Iodine-129 (^{129}I) is a highly soluble and long-lived ($T_{1/2} = 15.7$ Myr) radioactive isotope with a relatively large anthropogenic component. The pre-anthropogenic amount of ^{129}I in the oceans was roughly 50 kg, atmospheric fallout from nuclear weapons testing added 50 kg to 150 kg, while direct releases from the nuclear fuel–reprocessing plants at Sellafield in the Irish Sea and at La Hague on the French coast in the English Channel added more than 5,000 kg (Aldahan *et al.*, 2007). There are some indications that contaminated Russian rivers from Siberia can be a source of ^{129}I, but this contribution is nevertheless of minor importance for the large-scale distribution in the North Atlantic and Arctic Oceans (Cooper *et al.*, 1999). By far the main source of ^{129}I entering the oceans over the last half century, and also the main source for the oceanic inventory, has been direct discharges from Sellafield and La Hague.

Observations of ^{129}I have shed light on many of the characteristic features of ocean circulation along pathways from the source regions. Assuming that the bulk of tracers from both Sellafield and La Hague are being transported into the North Sea and then entrained in the Norwegian Coastal Current (NCC) (Raisbeck and Yiou, 1999), mixing between the Norwegian Atlantic Current (NwAC) and the NCC along the Norwegian coast should cause a net transport of tracers from the NCC to the NwAC. Gascard *et al.* (2004) estimated that roughly half of the ^{129}I load initially carried by the NCC is lost to the NwAC at about 70°N. In the Arctic Ocean, ^{129}I can be seen as Atlantic-derived waters, where a pronounced front is found between waters of Pacific and Atlantic origin. This is because ^{129}I concentrations are more than an order of magnitude higher in Atlantic waters (Smith *et al.*, 1999). Going deeper, the Pacific–Atlantic front is observed farther into the Canada Basin, in line with the observed distribution of intermediate Atlantic Waters (e.g., Rudels *et al.*, 1994). Alfimov (2004b) observed a doubling of ^{129}I concentration between 1996 and 2001 in the Amundsen Basin of the Arctic Ocean down to the 1,000 m depth, while in the Nansen and Makarov Basins concentrations increased in a shallower layer, indicating the slower ventilation of Atlantic Waters in these basins. In the western Nordic Seas, observed ^{129}I and hydrography revealed the complex structure of different water masses and the way in which they are transformed, being transported southwards with the East Greenland Current (EGC) (Alfimov *et al.*, 2004a). Farther south in the Irminger and Labrador Seas, ^{129}I concentrations are highest in the deepest layers, providing a direct signature in the overflow waters from the Nordic Seas (Smith *et al.*, 2005).

The timing of the release rate signal of ^{129}I from Sellafield and La Hague is favorable for studying ocean processes that have occurred since 1990, since discharges increased sixfold during the 1990s (Figure 3.2). This ramping up of increased concentration is presently being observed by ongoing programs in North Atlantic subpolar seas that are tracking deep overflow waters from the Nordic Seas.

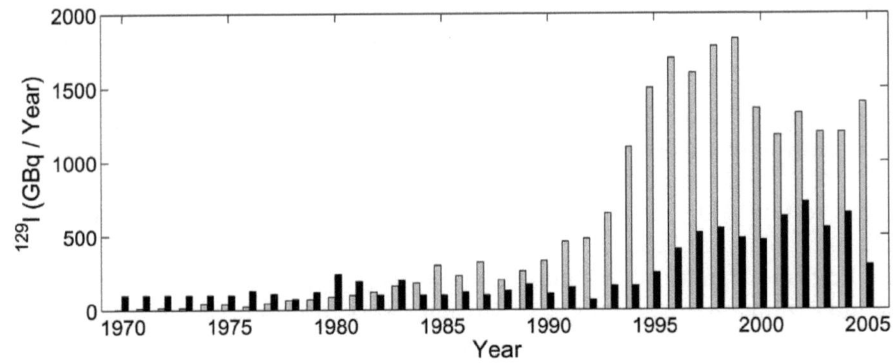

Figure 3.2. Annual releases of ^{129}I (10^9 Bq) from the reprocessing plant at Sellafield (black bars) and La Hague (gray bars).

Simulating the dispersion of ^{129}I from its source regions to the Nordic Seas and farther into intermediate layers in the Arctic Ocean and deep layers in the Irminger and Labrador Seas is a very challenging task, requiring a reasonable representation of numerous ocean processes—some of which are only parameterized in the current generation of OGCMs. Failure to represent the transport realistically in only some areas could potentially preclude a simulated large-scale distribution pattern. Surface values of ^{137}Cs, ^{90}Sr, and ^{99}Tc have been reasonably well reproduced by models representative of the current generation of OGCMs used in ocean climate studies (e.g., Gao *et al.*, 2004; Karcher *et al.*, 2004; Orre *et al.*, 2007), while the vertical dispersion of these tracers has received less attention.

The objective of this analysis is to use both observed and simulated distribution of ^{129}I to extract some characteristic circulation features and timescales of tracers in the northern North Atlantic Ocean. Because the applied OGCM does not explicitly resolve mesoscale processes, we are primarily interested in large-scale transport characteristics, rather than details on the local scale.

Figure 3.3 (see color section) shows the horizontal distribution of ^{129}I at three different depths in 1995 and 2005. At the surface, which is here defined to cover the uppermost 200 m of the water column, concentration of ^{129}I is highest in the shallow North Sea and along the Norwegian coast in the eastern Nordic Seas. In the western Nordic Seas, one sees tracers having much smaller concentrations being transported southwards with the EGC along the east Greenland coast. The surface transport route through the Barents and Kara Seas, and the entrance to the Arctic Ocean through the St. Anna Trough is also clearly depicted—even more so at intermediate depths between 500 m and 1,500 m. At these depths, relatively high concentrations extend from the St. Anna Trough into the Nansen Basin of the Arctic Ocean. Southward transport along the east Greenland coast indicates that the EGC carries tracers both at the surface and at intermediate depths. At deeper layers, here defined as depths below 2,000 m, almost all traces of ^{129}I are found in the deep North Atlantic subpolar seas. This is a very clear signature of the pathways of overflow waters from the Nordic Seas spilling over the Greenland–Scotland Ridge. For the latter, the largest contribution passes through the Denmark Strait, but the contributions from the Iceland–Faroe Ridge and the deep Faroe–Shetland Channel are also clearly seen.

Figure 3.4 (see color section) shows observed (Alfimov *et al.*, 2004b) and simulated surface concentrations of ^{129}I from the NCC to the North Pole sampled in July 2001. Along the Norwegian coast and in the Barents Sea the concentrations from the OGCM match observed values quite closely. The OGCM slightly overestimates the values just north of Spitsbergen in the Nansen Basin, and slightly underestimates the values on the Nansen–Gakkel Ridge, while in the Amundsen Basin close to the North Pole, the OGCM shows no tracers at all. The latter is in contrast to observations showing slightly elevated concentrations in the Amundsen Basin compared with the Nansen Basin.

In the Labrador Sea, vertical profiles show a very different vertical distribution. Figure 3.5 (see color section) displays the depth profiles of ^{129}I concentrations at Station 17 on the ARW7 (World Ocean Circulation Experiment or WOCE) section, where the observed values for 1997, 1999, and 2001 are obtained from Smith *et al.* (2005). The observed values for 2003 and 2005 have not previously been published. Both observed and simulated values show a slight decrease from the surface at the approximately 3,000 m depth, then a pronounced increase down to the deepest layer at about 3,700 m exceeding surface concentrations by more than a factor of 2. At the surface and at intermediate depths down to 3,000 m, the OGCM slightly exaggerates ^{129}I concentrations, although the general vertical structure is maintained. In the observations, a rapid increase is seen at deep layers from 1997 to 2001, then ^{129}I concentrations level off, but with the highest concentrations found in 2005. The OGCM captures both the magnitude and the time evolution of ^{129}I concentrations at deep layers in a fairly realistic way.

We have demonstrated that the OGCM reproduces the vertical structure and the temporal evolution of ^{129}I in the deep Labrador Sea quite reasonably, although it slightly exaggerates values at the surface and intermediate depths. We identify those isopycnals in the OGCM that have the highest concentrations of ^{129}I, as these layers are representative for overflow waters from the Nordic Seas into the deep North Atlantic Ocean. Figure 3.6 (see color section) shows ^{129}I at σ_2-layers 37.07 and 37.12 in the North Atlantic subpolar seas in December 2005. It can be readily seen that the Denmark Strait branch is dominant in terms of the southward ^{129}I transport, but also the Iceland–Faroe Ridge and the Faroe–Shetland Channel branch are clearly identified. Interestingly, from the age distribution in the deep Labrador Sea, the youngest ^{129}I is found along the rim of the basin, and older ^{129}I in the interior, confirming the faster transport route along the shelf break (Getzlaff *et al.*, 2006; Schott and Brandt, 2007). The primary pathways of overflow waters from the Nordic Seas in the subpolar North Atlantic are therefore visible in Figure 3.6 (see color section) as tongues of relatively high concentrations of ^{129}I. The ^{129}I distribution of overflow isopycnals clearly illustrates that the main pathways of dense waters spilling over the Greenland–Scotland Ridge are constrained to the western boundary in the North Atlantic, consistent with the observed Deep Western Boundary Current. In less dense layers, the two eastern overflow branches (the Iceland–Faroe Ridge and the Faroe–Shetland Channel) cross the North Atlantic Ridge. Previous versions of MICOM utilizing σ_0-

coordinates have all displayed a serious deficiency in that the two eastern overflow branches were trapped on the eastern side of the North Atlantic Ridge (Gao *et al.*, 2003).

Summary

Simulated and observed values of ^{129}I agree rather well along the Norwegian coast and in the Barents Sea. In the Arctic Ocean, the comparison is less favorable, which is ascribed to too-vigorous vertical mixing southwest of Spitsbergen and an unrealistic Atlantic Inflow into the Arctic Ocean through Fram Strait. In the North Atlantic subpolar seas, the OGCM reproduces the observed vertical profile of ^{129}I quite reasonably, with highest concentrations at the deepest layers.

The isopycnic coordinate ocean model (MICOM) utilizing σ_2-coordinates greatly improves the distribution of overflow waters from the Nordic Seas in the North Atlantic subpolar seas compared with previous versions utilizing σ_0-coordinates. It is therefore recommended for use in future studies of the large-scale three-dimensional spread of radioactivity, at least in the Nordic Seas and adjacent subpolar seas.

3.3 KARA SEA SHELF SEA MODEL

This model is intended to simulate seasonal and interannual variability in hydrologic characteristics and transport of radionuclides in the Kara Sea. A more detailed description of the model is given elsewhere (Neyelov and Chalikov, 1981; Neyelov, 1982; Johannessen *et al.*, 1996; Neelov and Kouraev, 1996). Here we focus on the version of the model used here, which includes hydrodynamics and thermodynamics, including the ice module.

Calculations were made over the time period 1948–2000. The simulations begin with boundary conditions relating to water temperature, salinity, and free water surface. The boundary itself is considered as a liquid one. These boundary conditions were obtained as a result of simulations conducted with the use of a large-scale model of the North Atlantic–Arctic model (Gao *et al.*, 2002).

3.3.1 General model description

Ocean circulation model

The model is based on the equations of movements of a viscous and incompressible liquid in spherical coordinates in the approximations of Boussinesq and hydrostatics:

$$\left.\begin{aligned}\frac{\partial u}{\partial t} &= -\frac{g}{r\sin\theta}\frac{\partial\eta}{\partial\lambda} - \frac{g}{\rho_0 r\sin\theta}\frac{\partial}{\partial\lambda}\int_{-z}^{0}\rho\,dz + fv + \frac{\partial}{\partial z}K_M\frac{\partial u}{\partial z} + A_M\Delta u\\ \frac{\partial v}{\partial t} &= -\frac{g}{r}\frac{\partial\eta}{\partial\theta} - \frac{g}{\rho_0 r}\frac{\partial}{\partial\theta}\int_{-z}^{0}\rho\,dz - fu + \frac{\partial}{\partial z}K_M\frac{\partial v}{\partial z} + A_M\Delta v\end{aligned}\right\}\tag{3.10}$$

The equation of heat and salt transfer is:

$$\frac{\partial(T,S)}{\partial t}+\frac{1}{r\sin\theta}\left[\frac{\partial u(T,S)}{\partial\lambda}+\frac{\partial v(T,S)\sin\theta}{\partial\theta}\right]+\frac{\partial w(T,S)}{\partial z}=\frac{\partial}{\partial z}K_T\frac{\partial(T,S)}{\partial z}+A_T\Delta(T,S) \tag{3.11}$$

The equation of continuity is:

$$\left[\frac{\partial U}{\partial\lambda}+\frac{\partial V\sin\theta}{\partial\theta}\right]\frac{1}{r\sin\theta}+\frac{\partial w}{\partial z}=0;\qquad U=\int_{-z}^{0}u\,dz;\qquad V=\int_{-z}^{0}v\,dz \tag{3.12}$$

The equation of hydrostatics is:

$$\frac{\partial p}{\partial z}=-\rho_w g \tag{3.13}$$

The equation of state in the UNESCO form is:

$$\rho_w=f(T,S,P) \tag{3.14}$$

where λ is longitude; θ is co-latitude; z is a vertical coordinate directed upwards and originating on the surface of the ocean; t is time; u, v, w are components of the velocity vector projected on the axes λ, θ, z, respectively; η is the free surface; r is the Earth's radius; g is gravitational acceleration; T is temperature; S is salinity; $f=2\Omega\cos\theta$ is the Coriolis parameter; Ω is the angular velocity of the Earth's rotation; $\Delta u=\frac{1}{r^2\sin\theta}\left[\frac{\partial}{\partial\theta}\left(\sin\theta\frac{\partial u}{\partial\theta}\right)+\frac{1}{\sin\theta}\frac{\partial^2 u}{\partial\lambda^2}\right]$ is the Laplacian; K_M, K_T and A_M, A_T are coefficients of vertical and horizontal turbulent exchange; p is pressure; ρ_w is density; and ρ_0 is mean density.

At the surface of the ocean, the components of wind stress are:

$$\tau_u=k\rho_0\frac{\partial u}{\partial z};\qquad \tau_v=k\rho_0\frac{\partial v}{\partial z} \tag{3.15}$$

The fluxes of heat and moisture are:

$$K_T\frac{\partial(T,S)}{\partial z}=F_{T,S} \tag{3.16}$$

and the following kinetic condition is required:

$$w_0=\frac{\partial\eta}{\partial t}+\frac{u_0}{r\sin\theta}\frac{\partial\eta}{\partial\lambda}+\frac{1}{r}v_0\frac{\partial\eta}{\partial\theta} \tag{3.17}$$

At the bottom, the condition of thermal isolation is imposed:

$$K_T\frac{\partial(T,S)}{\partial z}=0 \tag{3.18}$$

as well as the condition of streamlining:

$$w_0=\frac{u_H}{r\sin\theta}\frac{\partial H}{\partial\lambda}+\frac{1}{r}v_H\frac{\partial H}{\partial\theta} \tag{3.19}$$

The components of bottom stress are obtained from the formulas:

$$\left.\begin{aligned} \tau_x^H &= -\rho_0 c_H V_H (v_H \sin\alpha + u_H \cos\alpha) \\ \tau_y^H &= -\rho_0 c_H V_H (u_H \cos\alpha - v_H \sin\alpha) \end{aligned}\right\} \tag{3.20}$$

where V_H is the velocity module; $c_H = 0.0026$; and $\alpha = 20^\circ$.

At horizontal boundaries, free slip conditions are imposed:

$$v_n = 0; \qquad \frac{\partial v_\tau}{\partial n} = 0 \tag{3.21}$$

where n is normal with respect to the shoreline; and v_n, v_τ are normal and tangential components of the velocity vector.

Also, at the coastline, a condition of heat and salt isolation is fixed:

$$\frac{\partial(T, S)}{\partial n} = 0 \tag{3.22}$$

The following equation is used for determining the free surface:

$$\frac{\partial\eta}{\partial t} = -\frac{1}{r\sin\theta}\left(\frac{\partial U}{\partial\lambda} + \frac{\partial V \sin\theta}{\partial\theta}\right) \tag{3.23}$$

where

$$U = \int_{-H}^{\eta} u\, dz; \qquad V = \int_{-H}^{\eta} v\, dz$$

This condition is obtained by integrating the equation of continuity (3.12) over the distance extending from the surface to the bottom upward, taking into account boundary conditions (3.15) and (3.17).

At the liquid boundary, depending upon the concrete task at hand, the following conditions are imposed:

Active boundary conditions, when temperature and salinity are imposed at the boundary of the domain:

$$\zeta^* = \zeta(t, \Gamma); \qquad T^* = T(t, \Gamma); \qquad S^* = S(t, \Gamma) \tag{3.24}$$

where t is time; and Γ are coordinates of the liquid boundary. As a rule, such conditions are used for modeling tidal processes.

Passive boundary conditions are used when the required information about liquid boundary parameters is unavailable. In this case, the boundary condition is assigned based upon analyses of the solution behavior inside the domain. The most common procedure to generate such conditions is to use a wave equation in order to obtain radiative conditions at the boundary. The simplest option of such conditions is the Orlanski radiative conditions (Orlanski, 1976). In our model we use the two-dimensional modified option of imposing boundary conditions reported in Marchesiello *et al.* (2001). A radiative condition for a prognostic variable φ can be given as:

$$\frac{\partial \phi}{\partial t} = c_x \frac{\partial \phi}{\partial x} + c_y \frac{\partial \phi}{\partial y} \tag{3.25}$$

where (c_x, c_y) are, respectively, normal and tangential velocity components with respect to the boundary. Phase velocities c_x, c_y can be calculated as follows:

$$c_x = -\frac{\partial \phi}{\partial t} \frac{\partial \phi/\partial x}{\partial^2 \phi/\partial y^2 + \partial^2 \phi/\partial x^2} \tag{3.26}$$

and

$$c_y = -\frac{\partial \phi}{\partial t} \frac{\partial \phi/\partial y}{\partial^2 \phi/\partial y^2 + \partial^2 \phi/\partial x^2} \tag{3.27}$$

For calculating the boundary conditions from Formulas (3.25–3.26), an implicit scheme of first-order time and space coordinates is used.

For closing the equation system, a methodology analogous to that of Galperin *et al.* (1988) is used; that is, a local equation of turbulent energy balance is expressed as:

$$\frac{\partial k}{\partial t} = K_M \left[\left(\frac{\partial u}{\partial z} \right)^2 + \left(\frac{\partial v}{\partial z} \right)^2 + \frac{g}{\rho_0} \alpha_p \frac{\partial \rho}{\partial z} \right] - \varepsilon + \frac{\partial}{\partial z} K_b \frac{\partial k}{\partial z} \tag{3.28}$$

where k is a specific kinetic energy of turbulence; α_ρ is a parameter; ε is the dissipation of turbulent energy as specified by the Kolmogorov–Obukhov hypothesis:

$$\varepsilon = \frac{c_\varepsilon k^2}{K_b}; \qquad K_b = 0.2L\sqrt{k}; \qquad K_m = S_{\text{mom}} L\sqrt{k}; \qquad K_T = S_{\text{heat}} L\sqrt{k} \tag{3.29}$$

where $c_\varepsilon = 0.09$; and L is the scale of turbulence:

$$L = \min(L_d, L_u, L_b) \tag{3.30}$$

where $L_d = (\chi/H) Z_H Z_\eta Z_0$; Z_H is the distance between the bottom and the point for which the calculation is performed; and Z_η is the distance between the water surface and the same point:

$$Z_0 = 1 - \beta_1 H^{-2} Z_H Z_\zeta \tag{3.31}$$

$0 \le \beta_1 \le 4$ is the truncation parameter:

$$L_u = \frac{c_d \sqrt{k}}{\sqrt{\left(\frac{\partial u}{\partial z} \right)^2 + \left(\frac{\partial v}{\partial z} \right)^2}} \tag{3.32}$$

$$L_b = \frac{c_b \sqrt{k}}{\sqrt{-\frac{g}{\rho_0} \frac{\partial \rho_w}{\partial z}}} \tag{3.33}$$

where $c_d = 0.53$; and $c_b = 1.41$:

$$S_{\text{mom}} = \frac{(Sm_1 - Gh \cdot Sm_2)}{(1 - Sm_3 \cdot Gh)(1 - Sm_4 \cdot Gh)} \tag{3.34}$$

$$S_{\text{heat}} = \frac{Sh_1}{1 - Sh_2 \cdot Gh} \tag{3.35}$$

where $(sH_1, sH_2) = (0.493928, 34.6764)$; and $(Sm_1, Sm_2, Sm_3, Sm_4) = (0.42747,$ $5.805, 34.6764$, and $6.1272)$:

$$Gh = -\frac{L^2 N^2}{b} \tag{3.36}$$

where N^2 is the Brunt–Vaisälä frequency; and function Gh is specified within the interval:

$$-0.28 < Gh < 0.0233 \tag{3.37}$$

Ice cover model

The sea ice model and method of calculation of ice drift is given later in Section 3.4.2.

Atmospheric impact

Daily meteorological data from NCEP/NCAR for the time period 1948–2000 are used as model-forcing parameters. These meteorological data are subjected to linear interpolation into the simulated domain.

Initial and boundary conditions

As initial and boundary conditions for simulating seasonal variability for the Kara Sea, data on water temperature and salinity are obtained on a monthly basis from the large-scale North Atlantic–Arctic model (Gao *et al.*, 2002). River discharges from the Ob′ and Yenisei are set as monthly climatology.

3.3.2 Model validation results

The three-dimensional circulation model developed for the Kara Sea was validated using two sources of information: (1) available scientific publications and (2) data from *in situ* current observations. The first validation was done separately for the "validation" periods: 1977, 1978, 1980 (summer), and 1979 (winter) hydrological seasons, because for this period we had detailed *in situ* information about the distribution of temperature and salinity of water, and data from current measurements. The following steps were developed for model validation:

- selection of key regions of the Kara Sea showing a highly expressed character of currents and water circulation, as well as the oceanographic field structure;
- qualitative analysis of the structure of modeled water circulation compared with published sources;

- comparison of calculated current characteristics with the results of direct measurements of the currents.

The following regions were chosen as "key regions": Kara Gate and Vilkitsky Straits, as well as areas near the coast of the Yamal Peninsula (near Cape Harasawey and Beliy Island, in the zone of the Yamal current), regions in the vicinity of Sverdrup Island (north of Dikson, in the zone of the Ob′–Taimyr current and the St. Anna current), in the Matissen Strait (in the zone of the West Taimyr current), in the strait between Franz Josef Land and Novaya Zemlya in the region of Cape Zhelaniya, east of Novaya Zemlya (in the zone of the East Novaya Zemlya current), and in the Malygin Strait. The present understanding of the characteristics of water circulation in these regions is based on decades of observations using various techniques (direct measurements, analysis of data from drifting buoys, floats, ships, etc.), as well as indirect methods and calculations. In the present scientific literature, there is reasonable consensus on the dynamics of waters in these regions. As follows from description of the data used in the analysis, data from current measurements were found for almost all the chosen regions except for the last three.

To analyze the results of calculations for the southwestern part of the sea, T–S analysis was additionally performed in order to confirm the existence of the transport of transformed Barents Sea waters to the Kara Sea. To assess the structure of deep circulation, one can take the regions of the deep-water St. Anna and Voronin troughs in the northern part of the sea as representative.

Results of the calculations made showed satisfactory agreement with observational data and present knowledge about the dynamics of Kara Sea waters. This fact reveals the high quality of initial information (interpolated fields of temperature and salinity and the unique Digital Bathymetry Model of the Kara Sea, with its high spatial resolution on a 5×5 km grid), and the advantages of the model applied (especially its high vertical resolution in the bottom layer).

Historical data were compared with results of calculations obtained using initial boundary conditions on the calculation grid area (the free surface inclination was set to 0 cm). The circulation of sea surface waters during the summer hydrological season in August of 1977, 1978, and 1980 (when calculations were made under August atmospheric conditions, assumed to reflect average conditions for the summer) agree satisfactorily with the known state of the art of the large-scale structure of circulation in the Kara Sea. In the southwestern part of the sea, cyclonic circulation can be clearly seen from the vividly expressed flows of the east Novaya Zemlya and Yamal currents, as well as the Ob′–Yenisei current and the St. Anna current transporting freshened surface waters to the north. In the Malygin Strait, the flow is oriented westward and in the Vilkitsky Strait the eastern component is constantly observed—this is the coastal current of Russia's Arctic seas. The inflow of waters from the Barents Sea in the region of Cape Zhelaniya can be detected, which was most clearly expressed in 1978.

Interannual differences in the structure of water circulation manifest themselves as changes in the intensity of some elements of the large-scale circulation of the Kara Sea. These differences are associated with variability in atmospheric circulation and river runoff. In particular, changes in the intensity of Barents Sea inflow have already been observed in the region of Cape Zhelaniya. In some cases, Kara Sea waters can flow to the Barents Sea (which is observed in some years on our maps of currents), particularly when B-type (according to the P. A. Gordiyenko and D. B. Karelin classification system) atmospheric circulation prevails and when the polar anticyclone shifts closer to Canada, while the Siberian High matures and shifts to eastern Arctic seas and the Icelandic Low shifts northward. Moreover, one can observe interannual variations of the (a) intensity of the east Novaya Zemlya current, (b) flow in the Vilkitsky Strait (minimum velocities were observed in 1977), and (c) changes in the structure and intensity of the branch of the Ob′–Yenisei current skirting around Beliy Island (maximum southward deviations of this current and maximum velocities were also indicated on our maps in 1977, when the distribution of hydrological characteristics indicated a western propagation of river waters).

In September, when atmospheric processes and systems start reforming, the structure of currents changes to some degree, although its major features are preserved. Differences are manifested in the sharp attenuation of the major elements of circulation, primarily in the east Novaya Zemlya current, which can be explained by the predominance of northeastern winds rather than the southwestern ones in September. The velocities of calculated currents for all the experimental years can exceed 20 cm/s, but in most cases velocities range between 5 cm/s and 13 cm/s according to *in situ* observations.

As mentioned above, the August and September fields of atmospheric pressure have been used in calculations with seasonally averaged fields of water density for qualitative comparison of the effect of atmospheric conditions on the formation of the fields of currents. To more accurately estimate the purely wind constituent calculations of the fields of density-driven currents (in the absence of wind) and wind-driven currents as a difference of total and density currents have been made using August 1977 as an example. Results show that winds contribute most to the formation of cyclonic circulation in the southwestern part of the sea and to the outflow in the north.

Deep circulation in the sea has been poorly studied and the literature gives information on circulation in only some of the regions listed above. The southward transport of waters from the Arctic Basin to the mainland coastline over the whole region is a major feature of deep circulation (as shown on our maps of the currents). The zone where currents intensify along the western shore of Yamal to the region of Cape Harasawey is particularly clearly expressed. In the area of this bottom current, sharp changes in bottom depths (from 25 m to 150 m or more) oriented across the flow were observed. Calculated velocities exceed 13 cm/s. The maps also clearly show the flow of warm Atlantic waters to the sea by means of the Voronin and St. Anna deep-water troughs. The fact that it was possible to obtain countercurrents of salt waters flowing from the north and river runoff waters from the south in the bottom layer of the northern parts of the Ob′ and Yenisei Bays is very important evidence of the quality and reliability of model calculations. The maps of bottom circulation obtained using September fields of atmospheric pressure differ only in the attenuation of current velocities along the slope of Ob′–Yenisei shallow waters. The interannual variations of circulation in the bottom layer manifested themselves as decreasing intensities of southern flows from 1977 to 1980.

The structure of currents in the winter hydrological season has been poorly studied, because it is difficult (and expensive) to carry out oceanographic observations at this time of year. In this project we made use of the materials of one of the unique surveys of the Kara Sea carried out aboard aircraft-based laboratories that were able to land on drift ice and pack ice to take measurements. The implementation of such data in modeling calculations is apparently the first for the winter period performed over such a detailed calculation grid.

In the winter of 1979 (March and April), the circulation of surface waters undergoes considerable changes. The intensity of cyclonic circulation in the southwestern part of the sea decreases sharply; in the remaining part of the sea, northward outflows are observed everywhere except for the region west of Severnaya Zemlya, where the summertime southward transport remains.

In the bottom layer in the southwestern part of the sea, cyclonic circulation becomes anticyclonic; in the northeastern part, southward transport is observed everywhere, with its intensity being comparable with that during the summer period.

To finalize qualitative comparison of the historical data and modeling calculation of ocean circulation in the Kara Sea, it can be concluded that major differences were observed in only two regions: Kara Gate Strait and to a lesser degree the Severnaya Zemlya region. For Severnaya Zemlya, differences are on a small scale and they could be caused by the lack of data for these regions and by the difficulties of data interpretation. The principal discrepancies were found in Kara Gate Strait, for which there have been numerous observations. Such discrepancies may be explained by an insufficiently exact set-up of the initial boundary conditions on the borders of calculation grid.

All the current-meter data that were used refer only to shallow coastal regions and do not characterize the entire Kara Sea basin. Nevertheless, stations were located in key regions, enough to demonstrate the major elements of the large-scale structure of circulation of Kara Sea waters. These key areas represent the regions of straits and a zone of the major coastal current of eastern Arctic seas. Thus, even these limited data will be very useful, especially taking into account that the period of observations coincided with the time of oceanographic surveys, whose data have been used in diagnostic calculations.

The method—developed under the aegis of Professor V. A. Rozhkov—was used to analyze current measurement data. The method is based on the most substantiated existing model of temporal variations of the vector estimates of current velocity as a random function with values in two-dimensional Euclidean space (Belyshev *et al.*, 1983). The statistical characteristics of the non-periodic constituent of currents at different stations have been obtained.

For the region of the Kara Strait, we had data from three levels: current measurements made in October 1979 (depths of 10 m, 25 m, and the bottom) and in October 1980 (depths of 10 m and 25 m). The measurements refer to the end of the summer hydrological season. Average currents in the Kara Strait are sufficiently stable at that time of year, such that they can be conditionally taken for analysis.

It should be noted that all stations lay outside the zone of the Litke current, and the *in situ* data show that at every depth constant currents run from the Barents Sea to the Kara Sea. Velocities are much higher in the southern part of the strait and exceed 25 cm/s at a depth of 10 m, as follows from generally accepted knowledge of water circulation. From the north station toward the south, there is a decrease in the relative contribution of the variable constituent to the non-periodic current.

These results contrast with the preliminary results of calculations, which demonstrate the inflow of waters from the Kara Sea to the Barents Sea, a condition that should be observed only in the narrow shallow-water northern part of the strait. Such a disagreement can be explained by difficulties in setting boundary conditions at the western boundary of the area, because there are no comparable data of observations of the sea level at the different shores of the strait.

Summarizing the preliminary analysis of the currents, it is obvious that in many cases qualitative comparison of current structure and velocities has shown good results. However, in some regions current measurement data and model results disagree—even wildly (e.g., Kara Gate). Therefore, the main goal was to carry out additional analysis and model calculations to improve comparison for such regions as the Kara Gate Strait, Cape Harasawey, Beliy Island, and Matissen Strait.

Therefore, to confirm the inflow of Barents Sea waters to the Kara Sea and to assess the variability of this process, an additional study of T–S characteristics was carried out and the areas of propagation of transformed Barents Sea waters to the Kara Sea in 1977–1980 were determined. It is known that water structure in the southwestern part of the Kara Sea is influenced by three main water types: (1) cold freshened surface Arctic water of the Kara Sea, (2) more saline Barents Sea water passing through Kara Gate and Yugorsky Shar Straits, and (3) bottom water formed during winter cooling and brine formation during ice freezing. Barents Sea water occupies an intermediate position between surface Arctic water and bottom water.

The main propagation of Barents Sea water usually goes northward along the coast of Yamal Peninsula, and near Beliy Island this flux merges with continental discharge water. Further, these waters are entrained in the cyclonic circulation that can predominate over the entire western part of the Kara Sea, especially under conditions of a well-developed Siberian High and a relative northern position of the Icelandic Low. However, when high pressure over the Arctic dominates, this cyclonic circulation occurs only in the southwestern part of the Kara Sea, and water current velocities get slower.

It is apparent that this area can vary substantially in different years, but general transport runs from the Barents Sea toward the Kara Sea. The largest area occupied by these waters was observed in summer 1977. Comparison of the variability of areas of transformed Barents Sea water and model simulation of water dynamics are in good agreement (e.g., when the area is smaller, cyclonic circulation in the southwestern part of the sea is less pronounced). Model simulations in 1978 and 1980 show strengthening of the Yamal current (a westward turn and then a southward turn) in the region south of Malygin Strait in comparison with 1977. Thus in these years the area covered by cyclonic circulation was smaller. The close relation between the results of T–S analysis and simulated values gives confidence in the capabilities of the model.

At the next stage of model validation, an attempt was made to correct boundary conditions on the western boundary of the grid (the Kara Gate Strait) for August 1977. This was done to improve the relation between *in situ* measurements and model results in the southwestern region of the Kara Sea and especially in Kara Gate Strait. Proceeding from the results of *in situ* analysis, "distortions" of the free surface level in the Kara Gate Strait were assigned to be equal to 10 cm, 12 cm, 15 cm, and 30 cm.

As a result, the model proved to be very accurate and sensitive. The most realistic results were obtained by inclining the free surface level by 12 cm and 15 cm. The distortion of the level resulted in the inflow of water from the Barents Sea to the Kara Sea, intensification of the Yamal current as well as of the coastal current in the region of Dixon and Matissen Strait, which considerably improved the agreement of data of both calculations and measurements, especially in the region of Harasawey.

Summarizing the analysis of model validation given above, we conclude that even on this restricted grid, strictly bordering the Kara Sea region, modeling results agree well with the results of T–S analysis and *in situ* current measurements. The overall conclusion is that—by correcting boundary conditions—the ocean circulation model of the Kara Sea can be successfully used to model the transport and dilution of radioactive waste and dissolved pollutants in the Kara Sea.

Thus, the overall conclusion from analyzing the model calculation of three-dimensional current fields, *in situ* current measurements, and general knowledge of the water dynamics structure of the Kara Sea is that the Kara Sea water circulation model—as developed and implemented—can be reliably used for water circulation simulation. The high resolution of the model grid, increased resolution near the bottom, and correct approach to parameterization of natural phenomena are significant advantages of the model. Further improvements should concern appropriate assessment of the input values for model boundary conditions and full inclusion of the wind–wave interaction module and ice module into the model. More details about first-step validation can be found in Volkov *et al.* (2002).

General calculations to assess pollution distribution were made on a $10'$ lat. $\times$ $20'$ long. spatial grid. The temporal increment (time step) was set to 45 minutes. Calculations were made for the time period 1948–2000. As mentioned above, as initial and boundary conditions for simulating seasonal variability for the Kara Sea, data on water temperature and salinity were obtained on a monthly basis from the large-scale North Atlantic–Arctic model.

Some examples of the results of calculations for key regions, average monthly ice drift (cm/s), and ice thickness (m) for April are illustrated in Figure 3.7, and surface current distribution in August in the Kara Gate region is represented in Figure 3.8. The cyclonic circulation in the southwest part of the Kara Sea and countercurrent flow in the Kara Gate region that is apparent in Figure 3.8 correspond well with knowledge of current structure for this region based on *in situ* oceanographic data. Bottom-current distribution is represented in Figure 3.9. Cyclonic circulation near the bottom is not as well pronounced as on the surface; only the east Pechora current into the Kara Sea is evident in the Kara Gate and the speed of its current is greater

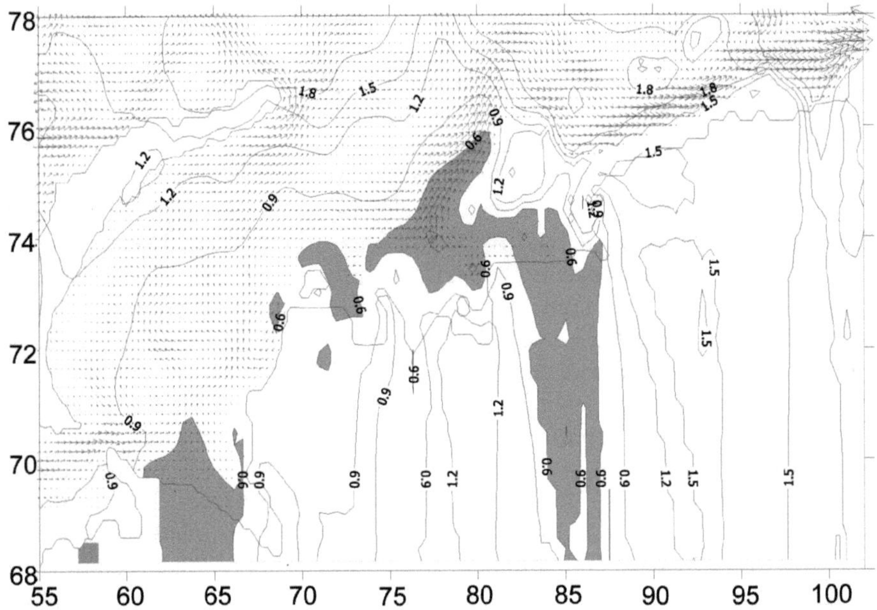

Figure 3.7. Average monthly ice drift (cm/s) and ice thickness (m) in April (averaged over 1948–2000).

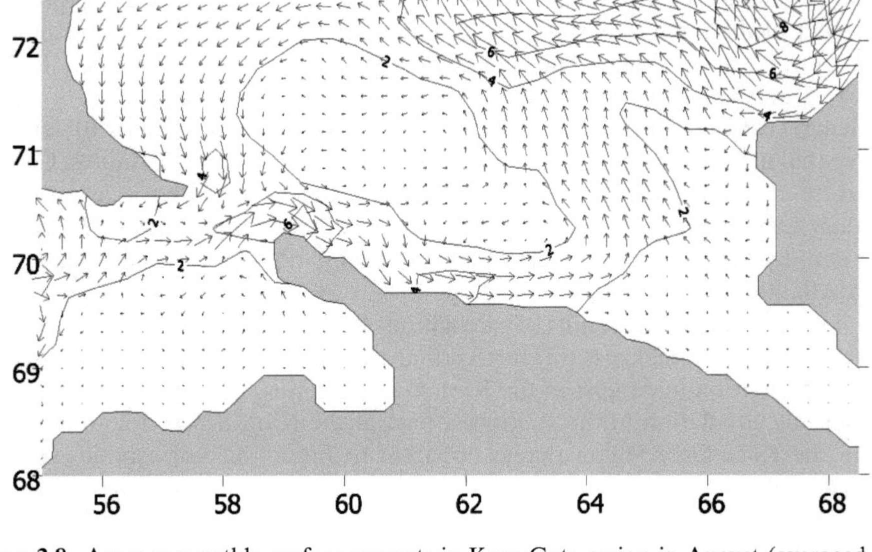

Figure 3.8. Average monthly surface currents in Kara Gate region in August (averaged over 1948–2000).

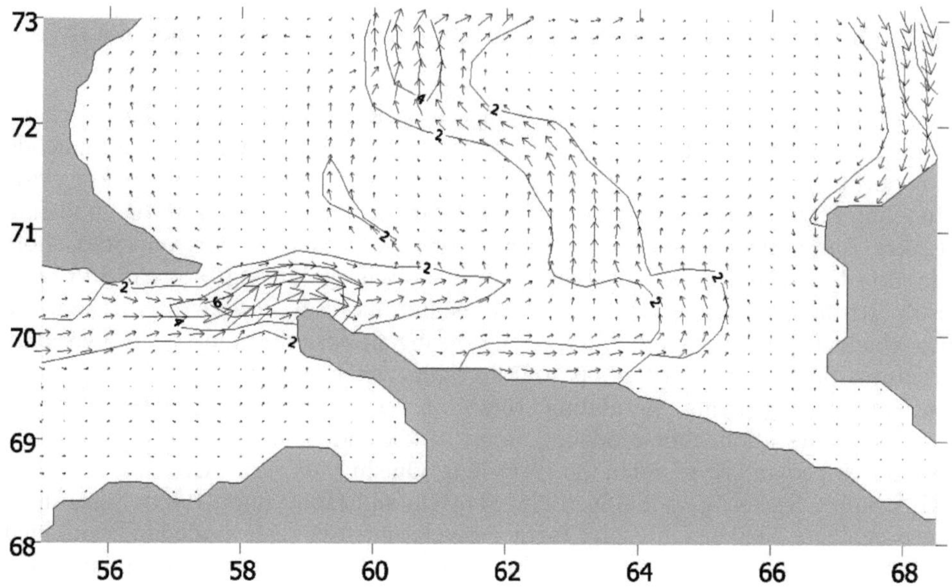

Figure 3.9. Average monthly bottom currents in Kara Gate region in August (averaged over 1948–2000).

than that in the surface of the sea, which also corresponds well with measurement results.

3.4 THE OB′ AND YENISEI RIVER AND ESTUARY MODELS

This section describes two linked models for simulating the transport of radionuclides from external sources through river systems to the sea. These models are components of the GMS with a medium-resolution (50 km) three-dimensional numerical ice–ocean model for the Arctic and North Atlantic Ocean nested with a high-resolution (5 km) model for the shelf seas. The river–estuary model chain includes a one-dimensional model of river dynamics and radionuclide transport and a three-dimensional model of estuarine and coastal area hydrodynamics and radionuclide transport. The hydraulic part of the one-dimensional model is based on Saint-Venant equations. Sediment transport in the one-dimensional and three-dimensional models is simulated on the basis of the advective dispersion equation with semi-empirical parameterization of erosion–sedimentation fluxes. The hydrodynamic module of the three-dimensional model is a free surface primitive equation baroclinic model with an ice model. Both the one-dimensional and three-dimensional models describe the transport of radionuclides in solute and on suspended sediments and the exchange processes involved in bottom deposition.

3.4.1 One-dimensional model to simulate the transport of radionuclides in a river system—RIVTOX

Studies of the environmental impact of nuclear accidents demonstrate that, after the initial atmospheric fallout, large river systems are the main pathways for radionuclide transport from the point of deposition to places some hundreds of kilometers away. Modeling radionuclide dispersion in rivers differs in some details from the modeling of lakes. Radionuclide dispersion in rivers is affected by different flow velocities, short retention times, and large variability in water discharge during the year and, as a result, strong temporal variations in sedimentation–resuspension rates. There are also channel and floodplain interactions during floods, the impacts of hydraulic structures on flow parameters, as well as rapid water level changes as a result of reservoir management. Simulating these processes requires special particular approaches in radionuclide modeling. Some of these modeling methods had already been developed and reviewed in the 1980s (e.g., Onishi *et al.*, 1981; Codell *et al.*, 1982; IAEA Safety Series No. 50SGS6, 1985; Santschi and Honeyman, 1989). Since then, however, there has been intensive further development of river modeling, motivated by further advances in computer technology and the urgent need to increase the predictability of models to provide adequate information for decision making concerning remedial measures in the most contaminated water bodies after the Chernobyl accident.

Mathematical models describing radionuclide transport and dispersion in rivers and reservoirs can be classified according to two different approaches: (1) spatial averaging of variables, and (2) individual treatment of variables describing radionuclides in different physicochemical forms.

Variables averaged over compartments or "boxes" represent the highest level of averaging and, as a result, are used in models with the lowest spatial dimension. These box-type (zero-dimensional) models treat the entire body of water (including the sediment layer, etc.) or a part of the entire body (e.g., one box for water and one for sediments) as a homogeneous compartment. Cross-sectionally averaged variables are often used in channel models and in models for narrow reservoirs. This one-dimensional approach is used to simulate pollutant transport from distances larger than about one-tenth of the river width downstream from the point source term (i.e., after full mixing of contaminant over the cross-section) up to some hundreds of kilometers. The timescale for river one-dimensional models ranges from minutes to decades (e.g., for long-term sedimentation studies). Two-dimensional vertical models operate by means of width-averaged variables. These models are used to describe currents, suspended sediments, and radionuclide transport, in cases where there is significant variability in channel depth. Depth-averaged variables are used in horizontally two-dimensional models that describe the flow pattern and radionuclide dispersion in shallow reservoirs, parts of river channels, and floodplains. The lowest level of averaging takes place in three-dimensional models that solve primitive or basic governing equations. The real spatial averaging scale of these models is based on the width of the computational grid—not on the particular parameterization or averaging procedure.

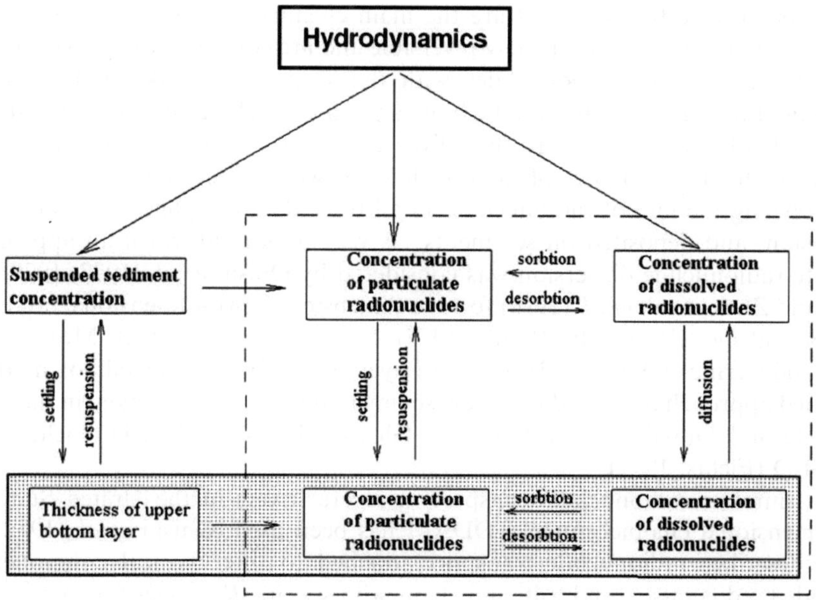

Figure 3.10. Key processes of radionuclide transport in surface waters.

The one-dimensional model used here (RIVTOX) was taken as the basic model to simulate radionuclide transport in the river network in the short-term and middle-term post-accident phase. The main processes governing radionuclide transport in river systems are presented in Figure 3.10. Pollutants in rivers are transported by the water flow (advection processes) under the simultaneous influence of turbulent diffusion processes. Radionuclides can interact with suspended sediments and bottom depositions. Pollutant transfer between river water and suspended sediment is described by adsorption–desorption processes. The transfer between river water and the upper layer of bottom deposition depends on adsorption–desorption and diffusion processes. Sedimentation of contaminated suspended sediments and bottom erosion represent an important pathway of "water column–bottom" radionuclide exchange. Different types of river models describe these processes at different levels of parameterization.

River models, regardless of their spatial resolution, include two main sub-models: hydraulic sub-models that describe water, suspended sediment, and bottom dynamics, and sub-models that simulate the fate of radionuclides in the different phases driven by these hydraulic processes.

The main physical exchange mechanisms are sedimentation of contaminated suspended matter into the river bed and resuspension of sediments into the water. They are controlled by hydraulic factors (e.g., river flow, sediment transport), and strongly depend on sediment-size fractionation (e.g., clay, silt, sand, and gravel). The diffusion of radionuclides through interstitial water is a process that accounts for transfers not related to sediment transport. Adsorption and desorption of a radionuclide by surface bed sediment are the main chemical exchange processes. These processes are not always completely reversible and are controlled by the geochemical reactions of dissolved radionuclides with the sediment. Uptake and subsequent excretion of radionuclides by aquatic biota and, in general, perturbation of sediments as a result of organisms living their lives represent biological processes that are responsible for the exchange of radionuclides between water and the bed sediment.

Modeling the fate of radionuclides in all three different phases—in solution, in suspension, and deposited on sediments—is very important. Such an approach to simulate radionuclide dispersion was considered by Onishi *et al.* (1977, 1979, 1981, 1982) and Zheleznyak *et al.* (1990) for one-dimensional, two-dimensional, and three-dimensional models, and by Booth (1975), Schückler *et al.* (1976), Monte (1992), Benes and Cernic (1990), and Hofer and Bayer (1993) for fully mixed box models. A simplified approach to the radionuclide–sediment interaction was taken in the models of White and Gloyna (1969), Shih and Gloyna (1970), CHNSED (Fields, 1976), HOTSED (Fields, 1977).

To simulate radionuclide transport in several rivers in the United States, the one-dimensional channel model TODAM has been used (Onishi *et al.*, 1982). The model describes radionuclide transport attached to three typical components of suspended sediments—sand, silt, and clay—with specific K_d values for each of them. The radionuclide transport module is supported by the comprehensive suspended sediment transport module that describes the transport of both cohesive and non-cohesive sediments. TODAM does not include a river hydrodynamic module; its use was based on river hydrodynamics calculated by specific codes (DKWAV, HEC-2, or CHARIMA). TODAM was used to simulate ^{239}Pu transport during flash flood events in the Mortandad Canyon, New Mexico (Whelan and Onishi, 1983), to reconstruct bottom contamination of the Clinch River/Tennessee River system due to releases from the Oak Ridge National Laboratory (Onishi, pers. commun.), to simulate ^{90}Sr and ^{137}Cs transport in Dnieper Reservoir after the Chernobyl accident (Zheleznyak, Blaylock *et al.*, 1995).

A one-dimensional model called SPA "TYPHOON"—developed by the State Hydrometeorological Committee of the U.S.S.R. (Borzilov *et al.*, 1989)—used empirical data on sediment transport rate and flow. The model includes more detailed descriptions of the transfer between different forms of radionuclides. Model parameters have been identified from experimental data of Prypyat River spring floods.

The one-dimensional model by Smitz and Everbecq (1986) considers the kinetics of the radionuclide interaction for two size fractions of suspended solids. The model was verified after accurately simulating the migration of radionuclides in the Meuse River, and subsequently was extensively applied elsewhere (Smitz, pers. commun.).

Several one-dimensional numerical models have been developed to simulate non-radioactive pollutant transport in rivers. These models ignore how radionuclide transport and radionuclide interaction with sediments is specified; however, some of them could in principal be modified for such purposes. One of the most well-known European one-dimensional modeling systems of pollutant transport in the river network is MIKE-11, developed by the Danish Hydraulic Institute (Havno *et al.*, 1995). MIKE-11 is a modeling system designed to simulate the flows, sediment transport, and water quality in estuaries, rivers, irrigation systems, and other water bodies. MIKE-11 was designed to have an integrated modular structure with basic computational modules for hydrology, hydrodynamics, advection–dispersion, water quality, and cohesive and non-cohesive sediment transport. It also includes modules for surface runoff. MIKE-11 has a well-developed graphical user interface integrated with pre-processors and post-processors that support system interaction with the GIS (Sorensen *et al.*, 1996a).

The RIVTOX model was developed by the Institute of Mathematical Machine and System Problems (IMMSP) to solve water contamination–use problems of Ukrainian rivers after the Chernobyl accident (Zheleznyak at al., 1992; Tkalich *et al.* 1994). The radionuclide transport part of the model is similar to the TODAM model (Onishi, 1982); however, simplifications have been made to create a practical, applicable model that does not depend on so much input data. On the other hand, RIVTOX presently includes a more detailed description of adsorption–desorption processes (e.g., non-equal rates of desorption and adsorption, different K_d's for bottom sediments and suspended sediments) that have been incorporated in the model, based on validation using Chernobyl data. RIVTOX was validated for the Clinch River tributary of the Tennessee River in the framework of the IAEA VAMP program (Zheleznyak *et al.*, 1995) and for the Dudvah River/Vah River tributary of the Danube (Slavik *et al.*, 1997). During the latter study, the possibility of using two-step kinetics in RIVTOX was analyzed.

To simulate radionuclide transport in rivers, it is necessary to estimate river flow and suspended sediment transport driven by hydrodynamic processes beforehand. There are many models that simulate river hydraulics and hydrodynamics. Overviews of these methods are presented in Orlob (1983), Cunge *et al.* (1986), and others; a more contemporary state of the art is presented by Rutherford (1994) and demonstrated by Jobson (1989) and Zheleznyak and Marinets (1993).

One-dimensional models based on cross-sectional averaged variables appear to be the best at determining river flow. Well-known computer codes include HEC-6 and HEC-2SR (Hydrological Engineering Center, 1977, 1982), REDSED (Chen, 1988), FLUVIAL 11 (Chang, 1988), IALLUVIAL (Karim *et al.*, 1981, 1987), and CHARIMA (Holly *et al.*, 1990), which expands the IALLUVIAL approaches, as well as MIKE-11 (Danish Hydraulics Institute), and TELEMAC (Laboratory of Hydraulics, EDF, France). The latter two are commercially distributed modeling systems that include models of different dimensions. HEC-2SR, FLUVIAL 11, CHARIMA, MIKE-11, and TELMAC contain river hydraulic modules based on numerical solution of the Saint-Venant equation. The possibility of efficiently estimating river hydraulics is based on numerical solution of the "diffusive wave" equation and the simplified version of the Saint-Venant equation, where suspended sediments act as a carrier of radionuclides in river–reservoir flow. The amount of radionuclide transported by sediments depends on suspended sediment concentration in river flow and the K_d value. After the Chernobyl accident, for example, up to one-half of ^{137}Cs transported by the Prypyat River from the vicinity of the Chernobyl Nuclear Power Plant (NPP) was bound on suspended sediments (Voitsekhovitch *et al.*, 1992). Sedimentation and bottom erosion processes play a key role in flow self-purification and in secondary contamination.

Mathematical modeling of sediment and transport is a major branch of hydraulics—overviews can be found in Raudkivi (1967), Ackers and White (1973), Engelund and Fredsoe (1976), Grishanin (1976), Karim *et al.* (1981, 1987), van Rijn (1984), Cunge *et al.* (1986), Mehta *et al.* (1989), Holly *et al.* (1990), Onishi (1993). Under steady-state conditions, sediment discharges are calculated by empirical and semi-empirical formulas that connect the sediment discharge with sediment parameters, flow velocities, and river cross-section characteristics or shear stress acting on the bed. In the case of cohesive sediments (finest silt and clay) and cohesive bonds between particles, this has to be taken into account (Mehta *et al.* 1989). The variability of streams and sediment parameters has led to the present situation where several different formulas are used for practical applications. It was demonstrated within validation studies (Onishi, 1993) that the approaches of Ackers-White, Engelund-Hansen, van Rijn, and Toffaletti show the most acceptable results for non-cohesive sediments under a wide range of flow and sediment conditions. However, for an individual river the best result can also be obtained by empirical formulas specially tuned for that particular case.

Sediment transport models are based on the suspended sediment mass conservation equation (advection–diffusion equation with a sink–source term describing the sedimentation resuspension rate) and the equation of bottom deformation (Exner equation). The most important problem for modeling to overcome is parameterization of the sedimentation and resuspension rates. A physically based approach calculates these rates as a function of the difference between the actual and equilibrium concentration of suspended sediments. This is often termed "suspended sediment capacity" and can be derived on the basis of the above-mentioned formulas.

The suspended sediment model includes different formulas to calculate equilibrium sediment concentration. The most comprehensive models (e.g., CHARIMA) contain modules for river hydraulic computation and methods to simulate bottom erosion dependent on the sediment grain distribution in the upper bottom layer (bottom-armoring calculation) and to calculate bottom friction dependent on the simulated dynamics of bottom forms.

Sub-model of river hydraulics

RIVTOX includes two sub-models to simulate cross-sectional averaged flow velocity and water elevation in a network of river channels. The first is based on the full set of Saint-Venant equations. The second is the "diffusive wave" simplified form of Saint-Venant equations. The latter has been demonstrated (e.g., Jobson, 1989; Marinets and Zheleznyak, 1993) to provide accurate results for flood routing in a river network that does not include structures (dams, gates) that could have significant upstream influence on flow parameters. However, full Saint-Venant equations should be used when there is significant upstream influence of the river structure, including pumping to the water intakes of irrigation channels, for example. The hydraulic parameters of a stream (depth, sectional area, velocity) can be calculated using the full dynamic model or the diffusive wave model. The full dynamic model is used for rivers with dams and other obstacles in the channel and for accelerated flows. The diffusive wave approximation of a full dynamic model can be used in cases where there is insignificant upstream influence of obstacles in a channel.

The following assumptions are valid when using Saint-Venant equations for water flow modeling: (1) the flow is mainly one-dimensional (the velocity is constant in the cross-section), and the fluid is incompressible and homogeneous; (2) the curvature of current flow lines is small, and vertical acceleration is insignificant; (3) the bottom slope is small, and parameters change slightly along a stream; (4) turbulence and friction influence could be taken into account in accordance with resistance laws for constant flows.

The system of Saint-Venant equations include the continuity equation (mass conservation law) and the momentum equation:

$$\frac{\partial A}{\partial t} + \frac{\partial Q}{\partial x} = q_l \tag{3.37}$$

$$\frac{\partial Q}{\partial t} + \frac{\partial}{\partial x}\left(\frac{Q^2}{A}\right) + gA\left(\frac{\partial h}{\partial x} + S_f\right) = 0 \tag{3.38}$$

The components of Equation (3.38) are local acceleration, hydrostatic gradient, gravity, and friction. Total water depth and bed slope are defined as follows:

$$h = y - y_b \tag{3.39}$$

$$S_0 = -\frac{\partial y_b}{\partial x} = \operatorname{tg} \alpha \tag{3.40}$$

The momentum equation (3.38) could be rewritten on the basis of the above notation as follows:

$$\frac{\partial Q}{\partial t} + \frac{\partial}{\partial x}\left(\frac{Q^2}{A}\right) + gA\left(\frac{\partial y}{\partial x} + S_f - S_0\right) = 0 \tag{3.41}$$

The friction slope S_f is calculated using one of the empirical resistance laws, such as Chezy's or Manning's, for example:

$$S_f = \frac{Q|Q|}{K^2} \tag{3.42}$$

where K is stream-metering characteristics. The usual approach in river hydraulics is to use the empirical Chezy's friction coefficient C_{Cz}:

$$K = C_{Cz} A \sqrt{R} \tag{3.43}$$

The hydraulic radius of flow R is defined as A/P, where P is the "wetted perimeter" of the stream. For a wide river channel the P value is close to river width b ($P \approx b$) and then $R \approx h$. On the basis of Mannings's empirical "friction parameter" n (average value is 0.02–0.03 for plain rivers) C_{Cz} is determined as:

$$C_{Cz} = \frac{1}{n} R^{1/6} \tag{3.44}$$

In kinematic wave approximation, only the last two terms are considered in Equation (3.41) (i.e., $S_f = S_0$), which leads to the formula:

$$K = \frac{Q}{\sqrt{S_0}} \tag{3.45}$$

In this approximation, which uses (3.42)–(3.44), Chezy's formula for steady-state flow can be obtained:

$$\frac{Q}{A} = C_{Cz}\sqrt{RS_0} \tag{3.46}$$

Substitution of (3.46) into the mass balance equation (3.37) using assumptions $A = bR$, $R \approx h$, and Manning's formula (3.44) will lead to the kinematic wave equation:

$$\frac{\partial(bh)}{\partial t} + S_0^{1/2} n^{-1} \frac{\partial(bh^{10/6})}{\partial x} = q_l \tag{3.47}$$

Sediment transport sub-model

Suspended sediment transport in river channels is described by the one-dimensional advection–diffusion equation that includes a sink–source term describing sedimentation and resuspension rates and laterally distributed inflow of sediments:

$$\frac{\partial(AS)}{\partial t} + \frac{\partial(QS)}{\partial x} - \frac{\partial}{\partial x}\left(E_S \frac{\partial(AS)}{\partial x}\right) = \Phi_b + \Phi_l \tag{3.48}$$

where Φ_b is the vertical sediment flux at the bottom, describing sedimentation or resuspension processes from their dependence on flow dynamical parameters and size of bottom sediments:

$$\Phi_b = \frac{A}{h} \cdot (q_{\text{res}} - q_{\text{sed}}) \tag{3.49}$$

Fluxes are calculated as thw difference between actual and equilibrium suspended sediment concentration multiplied by the fall velocity of sediment grains:

$$\left.\begin{aligned} q_{\text{res}} &= \beta w_0 \cdot F(S_* - S) \\ q_{\text{sed}} &= w_0 \cdot F(S - S_*) \end{aligned}\right\} \tag{3.50}$$

where $F(x)$ is a function of argument x defined as:

$$F(x) = \frac{x + |x|}{2} \equiv \begin{cases} x, & x > 0 \\ 0, & x < 0 \end{cases} \tag{3.51}$$

The coefficient of erodibility β characterizes bottom protection from erosion as a result of cohesion and natural armoring of the upper layer of the river bed (i.e., vegetation). This empirical coefficient usually has values of magnitude 0.1–0.01.

The equilibrium suspended sediment concentration (flow capacity) S_* can be calculated by means of different approaches. The first empirical formula used in RIVTOX (Zheleznyak *et al.*, 1993) was taken from Bijker (1998).

The equilibrium discharge of suspended sediments $p = QS_*$ is calculated as a function of bed load p_b:

$$p = 1.83 p_b \left(I_1 \ln \frac{h}{z_0} + I_2 \right) \tag{3.52}$$

where I_1 and I_2 are functions of the non-dimensional fall velocity $w_* = w_0/(kU_*)$ and bottom roughness $r_* = r/h$

$$I_1 = 0.216 \frac{r_*^{w_*-1}}{(1-r_*)^{w_*}} \int_{r_*}^{1} \left[\frac{1-z'}{z'} \right]^{w_*} dz' \tag{3.53}$$

$$I_2 = 0.216 \frac{r^{w_*-1}}{(1-r_*)^{w_*}} \int_{r_*}^{1} \left[\frac{1-z'}{z'} \right]^{w_*} \ln z' \, dz' \tag{3.54}$$

where $z' = z/h$; $r = 30r_0$; r_0 is the typical size of bottom inhomogeneity; k is the von Karman parameter ($k = 0.4$); and U_* is bottom shear stress velocity:

$$U_* = \sqrt{\frac{T_c}{\rho_w}} = \sqrt{ghS_f} \tag{3.55}$$

Bottom sediment equilibrium flow is described as follows:

$$\vec{p}_b = 5D \frac{\vec{T}_c}{\sqrt{\vec{T}_{cw}}} \left(\frac{\mu}{\rho} \right)^{0.5} \exp\left(-0.27 \frac{\rho D}{|\vec{T}_{cw}|} \right) \tag{3.56}$$

where $\vec{T}_c$ is current-driven bottom shear stress; $\vec{T}_{cw}$ is bottom shear stress driven by the joint action of currents and waves; D is averaged size of sediments; and μ is the parameter of bottom ripples. Formula (3.56) was proposed for coastal areas. In recent versions of RIVTOX van Rijn's methods of calculating suspended sediment transport is included in the module library.

The lateral inflow of suspended sediments to the river network Φ_l in Equation (3.48) can be presented as follows:

$$\Phi_l = q_l S_l \tag{3.57}$$

Substitution of (3.49) and (3.57) into (3.48) leads to the following equation:

$$\frac{\partial(AS)}{\partial t} + \frac{\partial(QS)}{\partial x} - \frac{\partial}{\partial x}\left(E_S \frac{\partial(AS)}{\partial x} \right) = \frac{A}{h}(q_{\text{res}} - q_{\text{sed}}) + q_l S_l \tag{3.58}$$

which takes the water balance equation (3.37) into account, presented as:

$$\frac{\partial S}{\partial t} + U \frac{\partial S}{\partial x} - \frac{1}{A}\frac{\partial}{\partial x}\left(AE_S \frac{\partial S}{\partial x} \right) = \frac{1}{h}(q_{\text{res}} - q_{\text{sed}}) + \frac{q_l}{A}(S_l - S) \tag{3.59}$$

where $U = Q/A$ is cross-sectionally averaged flow velocity; and E_S is the dispersion coefficient for suspended sediments. For suspended sediments it is assumed that the same dispersion coefficient could be used as for a soluble tracer in water flow. An overview of the methods of calculating the dispersion coefficient for one-dimensional channel flow was presented by Holly (1985) and fairly recently by Won and Sung (1998).

The widely used formulas are as follows (Elder, 1958):

$$E_x = \alpha_E U_* h \tag{3.60}$$

The calibration constant α_E has a constant value of 5.9. According to the Fischer (1973) formula (presented by van Majzik,1992):

$$E_x = \alpha_F \frac{b^2}{h} \frac{U^2}{U_*} \tag{3.61}$$

The calibration constant α_F (e.g., for the Rhine) has a value of 0.011 (Won and Sung, 1998):

$$E_x = \alpha_W \left(\frac{b}{h}\right)^{1.23} \left(\frac{U}{U_*}\right)^{1.25} \tag{3.62}$$

The calibration constant α_W is given in the literature with a value of 0.64. The simplest approach—based on the Elder formula (3.60), shear stress velocity definition (3.55), and Chezy's and Manning's formulas (3.39)–(3.42)—is used in RIVTOX:

$$E_s = E_x = \alpha_E U_* h = \alpha_E \sqrt{g} n U h^{-5/6} \tag{3.63}$$

The values of parameter α_E should be calibrated if trace experiment data are available for the river used for model implementation. Other longitudinal dispersion formulas could be used optionally.

The fluxes of sedimentation and the resuspension (3.50) control the dynamics of the uppermost contaminated layer of bottom sediments. The thickness of this layer Z^* can be calculated from the mass balance equation:

$$\rho_s(1-\varepsilon)\frac{dZ^*}{dt} = q_{\text{sed}} - q_{\text{res}} \tag{3.64}$$

After simulation of stream hydrodynamics, the numerical solution of Equations (3.57), (3.60) is used with the appropriate empirical formulas to model suspended sediment transport in river flow.

Sub-model of radionuclide transport

This sub-model of RIVTOX describes the advection–diffusion transport of cross-sectionally averaged concentrations of radionuclides in solution C, the concentration of radionuclides on suspended sediments C^s, and the concentration C^b in the top layer of bottom deposits. Adsorption–desorption and diffusive contamination transport in the systems "solution–suspended sediments" and "solution–bottom deposition" is treated via the K_d approach to the equilibrium state, additionally taking into account the exchange rates $a_{i,j}$ between the solution and particles for more realistic simulation of kinetic processes.

The present version of RIVTOX uses different values for sorption and desorption rates $a_{1,2}$ and $a_{2,1}$ for the system "water–suspended sediment" and $a_{1,3}$ and $a_{3,1}$ for the system "water–bottom deposits" because this fits better with the real physicochemical behavior of radionuclides in water systems. Furthermore, the use of different exchange rates gives a better fit in simulations:

$$\left.\begin{aligned}
\frac{\partial C}{\partial t}+\frac{Q}{A}\frac{\partial C}{\partial x}-\frac{1}{A}\frac{\partial}{\partial x}\left(AE_C\frac{\partial C}{\partial x}\right) &= f^C(S,C,C^S,C^b,Z^*,\vec{p}^C)+f^{C_l}(C,C_l)\\
\frac{\partial C^S}{\partial t}+\frac{Q}{A}\frac{\partial C^S}{\partial x}-\frac{1}{A}\frac{\partial}{\partial x}\left(AE_C\frac{\partial C^S}{\partial x}\right) &= f^{C^S}(S,C,C^S,C^b,Z^*,\vec{p}^{C^S})+f^{C_l^S}(C,C^S,C_l,C_l^S)\\
\frac{\partial C^b}{\partial t} &= f^{C^b}(S,C,C^S,C^b,Z^*,\vec{p}^{C^b})\\
\frac{\partial Z^*}{\partial t} &= f^{C^b}(S,\vec{p}^{Z^*})
\end{aligned}\right\} \tag{3.65}$$

where

$$\left.\begin{aligned}
f^C &= -\lambda C - a_{1,2}(K_{dS}SC - C^S) - a_{1,3}(K_{db}C - C^b)\frac{\rho_S(1-\varepsilon)Z^*}{h}\\
f^{C_l} &= \frac{Q}{A}(C_l - C)\\
f^{C^S} &= -\lambda C^S + a_{1,2}(K_{ds}C - C^S) + \frac{q_{\text{sed}}(C^b - C^S)}{hS}\\
f^{C_l} &= \frac{QS_l(C_l^S - C^S)}{AS}\\
f^{C^b} &= a_{1,3}(K_{db}C - C^b) - \frac{q_{\text{res}}(C^b - C^S)}{\rho_S(1-\varepsilon)Z^*} - \lambda C^b
\end{aligned}\right\} \tag{3.66}$$

The numerical solver for the system of equations (3.66) is included in the RIVTOX code. The values of the radionuclide transport parameters for river systems are given in Table 3.1.

Boundary and initial conditions in river network

In a complex river network, the above models are applied to the river network graph. Each equation is solved for the respective "simple channel" (branch) with relevant initial conditions on a branch and the appropriate boundary condition in a junction of the river network. Each branch is considered as a river channel; the term "simple channel" means that inside this branch there are no junctions with two or more inflows or outflows, sharp changes in depth, width of stream, or water discharge. These points are named "special" points and define a scheme of the river network for simulation purposes.

For the numerical solution of model equations, each branch is covered by a set of grid nodes (computational mesh) as presented in Figure 3.11. The configuration of the graph and the size of the branches can be taken from the GIS (e.g., MapInfo map). The information about river hydrographical characteristics (e.g., mean depth

Table 3.1. Radionuclide transport parameters.

Nuclide	K_{db} (m^3/kg)	K_{ds} (m^3/kg)	$a_{1,2}$ (L/da)	$a_{2,1}$ (L/da)	$a_{1,3}$ (L/da)	$a_{3,1}$ (L/da)
^{137}Cs	3	15	1	0.02	0.01	0.002778
^{90}Sr	0.25	0.8	1	0.02	0.04	0.002778
^{3}H	0	0	0	0	0	0
^{60}Co	5	20	1	0.02	0.01	0.002778
^{131}I	0.01	0.01	1	0.02	0.04	0.002778
^{239}Pu	200	800	1	0.02	0.01	0.002778
^{106}Ru	4	15	1	0.02	0.01	0.002778

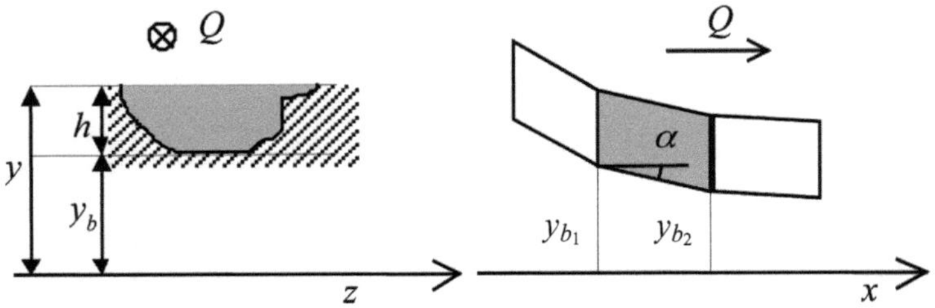

Figure 3.11. River stream parameters.

and width, typical discharge, cross-section data—$A = A(h)$ or $b = b(h)$ tables) is prepared for some river cross-sections so that they can be interpolated in the computational nodes in the modeling system.

Boundary and initial conditions

(a) In all models, for all influx points of the river network, source term conditions should be set:

$$\left.\begin{aligned} Q(t)\Big|_{x=0} &= \tilde{Q}(t) \\ S(t)\Big|_{x=0} &= \tilde{S}(t) \\ C(t)\Big|_{x=0} &= \tilde{C}(t) \\ C^S(t)\Big|_{x=0} &= \tilde{C}^S(t) \end{aligned}\right\} \quad (3.67)$$

where $\tilde{Q}(t), \tilde{S}(t), \tilde{C}(t), \tilde{C}^S(t)$ are the source functions.

(b) At outflow nodes, the conditions of free propagation are:

$$\left.\begin{aligned}
\left(\frac{\partial Q}{\partial t}+\frac{Q}{A}\frac{\partial Q}{\partial x}\right)_{x=l} &= f^Q\Big|_{x=l}\\
\left(\frac{\partial S}{\partial t}+\frac{Q}{A}\frac{\partial S}{\partial x}\right)_{x=l} &= f^C\Big|_{x=l}\\
\left(\frac{\partial C}{\partial t}+\frac{Q}{A}\frac{\partial C}{\partial x}\right)_{x=l} &= f^C\Big|_{x=l}\\
\left(\frac{\partial C^S}{\partial t}+\frac{Q}{A}\frac{\partial C^S}{\partial x}\right)_{x=l} &= f^{C^S}\Big|_{x=l}
\end{aligned}\right\} \qquad (3.68)$$

where l is the branch length.

(c) At the conjunction node (inflow from more than one branch and outflow through one branch) the conditions of free propagation are imposed at the inflow branch. The mass balance conditions are:

$$\left.\begin{aligned}
(Q)_{\text{out}}\Big|_{x=0} &= \sum_i (Q)_i\Big|_{x=l_i}\\
(QC)_{\text{out}}\Big|_{x=0} &= \sum_i (QC)_i\Big|_{x=l_i}\\
(QS)_{\text{out}}\Big|_{x=0} &= \sum_i (QS)_i\Big|_{x=l_i}\\
(QSC)_{\text{out}}\Big|_{x=0} &= \sum_i (QSC)_i\Big|_{x=l_i}
\end{aligned}\right\} \qquad (3.69)$$

(d) At the fork nodes (inflow from one branch and outflow through two or more branches), the free propagation conditions (3.68) are imposed at the inflow branch. At outflow branches, the source term conditions (3.67) are imposed, where source functions are equal to the output from the inflow branch. Initial conditions should also be set for all parameters in all points of the numerical mesh at the start time.

Numerical solution

The implicit finite difference scheme (Fletcher, 1988) was used for numerical solution of the parabolic advection–dispersion equations on the river network grid. The finite difference scheme can be written in general form as follows:

$$M_x\left(\frac{\Phi^{n+1}_{\nu,i}-\Phi^{n}_{\nu,i}}{\Delta t}\right)+(V^{n+1}_{\nu,i}\cdot L_x - E^{n+1}_{\nu,i}\cdot L_{xx})\cdot 0.5\cdot(\Phi^{n}_{\nu,i}+\Phi^{n+1}_{\nu,i}) = f^{n}_{\nu,i} \qquad (3.70)$$

where ν is the number of the branch; i is the number of the point on each branch; n is the time layer number; and the functions Φ, f present the model variables Q, S, C, C^S in the equations and source term functions.

The finite difference operators M_x, L_x, and L_{xx} are applied to three neighbor nodes of the spatial mesh $\{i-1, i, i+1\}$ and can be formally written as:

$$\left.\begin{aligned} M_x &\equiv \{\delta, (1-2\delta), \delta\} \\ L_x &\equiv \frac{1}{2\Delta x}\{-1, 0, 1\} \\ L_{xx} &\equiv \frac{1}{\Delta x^2}\{1, -2, 1\} \end{aligned}\right\} \tag{3.71}$$

where $0 < \delta < 1$; formally, this scheme assumes the second order and is stable when $\delta \leq 0.25$.

Assuming $s = E_{\nu,i}^{n+1}\,\Delta t/\Delta x^2$ and $C = V_{\nu,i}^{n+1}\,\Delta t/\Delta x$ (Courant number), Equation (3.70) can be transferred into a system of linear equations with the following three-diagonal matrix:

$$\begin{aligned} &[\delta - 0.25C - 0.5s] \cdot \Phi_{\nu,i-1}^{n+1} + [(1-2\delta) + s] \cdot \Phi_{\nu,i}^{n+1} + [\delta + 0.25C - 0.5s] \cdot \Phi_{\nu,i+1}^{n+1} \\ &\quad = [\delta + 0.25C + 0.5s] \cdot \Phi_{\nu,i-1}^{n+1} + [(1-2\delta) - s] \cdot \Phi_{\nu,i}^{n+1} + [\delta - 0.25C + 0.5s] \cdot \Phi_{\nu,i+1}^{n} \\ &\quad\quad + \Delta t \cdot f_{\nu,i}^{n} \end{aligned} \tag{3.72}$$

where $\nu = 1, \ldots, N$; and $i = 2, \ldots, I_\nu - 1$.

Solution of the system (3.72) using the finite difference approximation of boundary conditions (3.67)–(3.69) is obtained on the basis of the three-diagonal sweep procedure.

3.4.2 Numerical model for three-dimensional dispersion simulation of radionuclides in stratified water bodies—THREETOX

In this sub-section, we present the modeling system THREETOX to simulate the dispersion of radionuclides and other contaminants on local scales. The system includes sub-models of hydrodynamics, ice dynamics–thermodynamics and sub-models of sediment and radionuclide transport. The prognostic variables of the hydrodynamics sub-model are the three components of the velocity fields: temperature, salinity, and surface level elevation; kinetic energy of turbulence; and dissipation rate. The ice sub-model predicts ice drift, thickness, and ice concentration. The exchanges between water, ice, and atmosphere are parameterized by bulk aerodynamic formulas. The transport, deposition, and resuspension of several fractions of sediments are calculated by the sediment transport sub-model. The radionuclide concentration in solute, in suspended sediments, and at the bottom is predicted by the radionuclide transport sub-model.

The components of the system were developed in several papers (Margvelashvili *et al.*, 1997, 2002; Koziy and Maderich, 1997; Koshebutsky *et al.*, 2004; Maderich *et al.*, 2008). The system was applied to estuaries (Margvelashvili *et al.*, 1998, 1999, 2002; Maderich *et al.*, 2008), coastal seas (Koziy *et al.*, 1998, 2000), and rivers, lakes, and reservoirs (Margvelashvili *et al.*, 2002; Koshebutsky *et al.*, 2004; Maderich *et al.*, 2008).

Hydrodynamics

Hydrodynamics is simulated on the basis of the three-dimensional, time-dependent, free surface, primitive equation model in the Boussinesq approximation. Model equations are written in curvilinear orthogonal coordinates. The governing Reynolds averaged equations of continuity, horizontal momentum, conservation equations for temperature T and salinity S, state equation and hydrostatic relation can be written in Cartesian coordinates (x, y, z) as:

$$\vec{\nabla} \cdot \vec{U} = 0 \tag{3.73}$$

$$\frac{\partial u}{\partial t} + \vec{U} \cdot \vec{\nabla} u - fv = -\frac{1}{\rho_0}\frac{\partial p}{\partial x} + \frac{\partial}{\partial z}\left(\nu_t \frac{\partial u}{\partial z}\right) + \frac{\partial}{\partial x}\left(K_M \frac{\partial u}{\partial x}\right) + \frac{\partial}{\partial y}\left(K_M \frac{\partial u}{\partial y}\right) \tag{3.74}$$

$$\frac{\partial v}{\partial t} + \vec{U} \cdot \vec{\nabla} v + fu = -\frac{1}{\rho_0}\frac{\partial p}{\partial y} + \frac{\partial}{\partial z}\left(\nu_t \frac{\partial v}{\partial z}\right) + \frac{\partial}{\partial x}\left(K_M \frac{\partial v}{\partial x}\right) + \frac{\partial}{\partial y}\left(K_M \frac{\partial v}{\partial y}\right) \tag{3.75}$$

$$\frac{\partial T}{\partial t} + \vec{U} \cdot \vec{\nabla} T = \frac{\partial}{\partial z}\left(\nu_t' \frac{\partial T}{\partial z}\right) + \frac{\partial}{\partial x}\left(K_T \frac{\partial T}{\partial x}\right) + \frac{\partial}{\partial y}\left(K_T \frac{\partial T}{\partial y}\right) + \frac{(1-A)}{\rho_W c_{pW}}\frac{\partial I}{\partial z} \tag{3.76}$$

$$\frac{\partial S}{\partial t} + \vec{U} \cdot \vec{\nabla} S = \frac{\partial}{\partial z}\left(\nu_t' \frac{\partial S}{\partial z}\right) + \frac{\partial}{\partial x}\left(K_T \frac{\partial S}{\partial x}\right) + \frac{\partial}{\partial y}\left(K_T \frac{\partial S}{\partial y}\right) \tag{3.77}$$

$$\rho_W = \rho_W(T, S, p) \tag{3.78}$$

$$p(x, y, z, t) = p_a + g\rho_0\eta + g\int_z^0 \rho_W(x, y, z', t)\, dz' \tag{3.79}$$

where t denotes time; $\vec{U} = (u, v, w)$ is velocity; η is surface elevation; f is the Coriolis parameter; p_a is atmospheric pressure; ρ_W is water density; ρ_0 is undisturbed density; c_{pW} is water heat capacity; and $(1 - A)I$ is solar insolation absorbed in the water. Here ice concentration (compactivity) A is the fractional area covered by ice, whereas $(1 - A)$ is the fractional area of open water (Mellor and Kantha, 1989). The concept of eddy viscosity–diffusivity is used to determine turbulent stresses. The vertical eddy viscosity $\nu_t \geq \nu_f$ and eddy diffusivity $\nu_t' \geq \nu_f$ are modeled as a product of the turbulent velocity scale and turbulent length scale l, whereas ν_f is the constant background viscosity–diffusivity value when turbulence is suppressed by buoyancy forces at stable stratification. The turbulent velocity scale is proportional to the square root of kinetic turbulent energy (KTE) k, therefore $\nu_t = c_\mu\sqrt{k}l$, $\nu_t' = c_\mu'\sqrt{k}l$. Here c_μ, c_μ' are stability functions (Canuto *et al.*, 2001). The UNESCO equation of state in the form proposed by Mellor (1991) was used.

The two-equation $k-\varepsilon$ turbulence model is used to calculate k and the dissipation rate $\varepsilon=(c_\mu^0)^3k^{3/2}/l$. Here c_μ^0 is constant. The equations for KTE and the dissipation rate are:

$$\frac{\partial k}{\partial t}+\vec{U}\cdot\vec{\nabla}k-\frac{\partial}{\partial z}\left(\nu_k\frac{\partial k}{\partial z}\right)-\frac{\partial}{\partial x}\left(K_M\frac{\partial k}{\partial x}\right)-\frac{\partial}{\partial y}\left(K_M\frac{\partial k}{\partial y}\right)=P+B-\varepsilon \tag{3.80}$$

$$\frac{\partial \varepsilon}{\partial t}+\vec{U}\cdot\vec{\nabla}\varepsilon-\frac{\partial}{\partial z}\left(\nu_\varepsilon\frac{\partial \varepsilon}{\partial z}\right)-\frac{\partial}{\partial x}\left(K_M\frac{\partial \varepsilon}{\partial x}\right)-\frac{\partial}{\partial y}\left(K_M\frac{\partial \varepsilon}{\partial y}\right)=\frac{\varepsilon}{k}(c_{\varepsilon_1}P+c_{\varepsilon_3}B-c_{\varepsilon_2}\varepsilon) \tag{3.81}$$

where $\nu_k=\nu_t, \nu_\varepsilon=\nu_t/\sigma_\varepsilon$; $P=\nu M^2$ is shear production; $B=-\nu_T N^2$ is buoyancy production; $N^2=-\dfrac{g}{\rho_0}\dfrac{\partial\rho}{\partial z}$ is the Vääsäla frequency; and $M^2=\left(\dfrac{\partial u}{\partial z}\right)^2+\left(\dfrac{\partial v}{\partial z}\right)^2$. Constants of the turbulence model are $c_{\varepsilon_1}=1.44$, $c_{\varepsilon_2}=1.92$, $c_{\varepsilon_3}(B<0)=1.0$, $\sigma_\varepsilon=1.08$, $c_\mu^0=0.5562$ (Burchard, 2002). Horizontal turbulent viscosity–diffusivity is parameterized by the Smagorinsky (1963) formula:

$$(K_m,K_h)=(C_m,C_h)\frac{1}{2}\Delta x\,\Delta y\sqrt{\left(\frac{\partial u}{\partial x}\right)^2+\frac{1}{2}\left(\frac{\partial v}{\partial x}+\frac{\partial u}{\partial y}\right)^2+\left(\frac{\partial v}{\partial y}\right)^2} \tag{3.82}$$

where constants $C_m=0.1$–0.2, $C_h=0.2C_m$, $\Delta x,\Delta y$ are scales of motion.

At the free surface $z=\eta$ the kinematic boundary condition is:

$$\frac{\partial\eta}{\partial t}+u\frac{\partial\eta}{\partial x}+v\frac{\partial\eta}{\partial y}=w \tag{3.83}$$

The surface fluxes required by the model are the fluxes of momentum, heat, and salt. The turbulent fluxes of momentum at $z=\eta$ are:

$$\nu_t\frac{\partial u}{\partial z}=(1-A)\frac{\tau_{AW}^{(x)}}{\rho_W}+A\frac{\tau_{IW}^{(x)}}{\rho_W},\qquad \nu_t\frac{\partial u}{\partial z}=(1-A)\frac{\tau_{AW}^{(y)}}{\rho_W}+A\frac{\tau_{IW}^{(y)}}{\rho_W} \tag{3.84}$$

where $\tau_{AW}^{(x)}$ and $\tau_{AW}^{(y)}$ are atmospheric stresses directly imposed on water; and $\tau_{IW}^{(x)}$ and $\tau_{IW}^{(y)}$ are stresses between ice and water.

The boundary conditions for KTE are formulated at some distance $\tilde{z}$ from the surface of the sea or ice under the assumption of the law of the wall (Burchard, 2002), as:

$$k=(1-A)\left(\frac{u_*^{AW}}{c_\mu^0}\right)^2+A\left(\frac{u_*^{IW}}{c_\mu^0}\right)^2 \tag{3.85}$$

$$\frac{\nu_t}{\sigma_\varepsilon}\frac{\partial\varepsilon}{\partial\tilde{z}}=(c_\mu^0)^3\frac{k^{3/2}}{\kappa(\tilde{z}+z_0)^2} \tag{3.86}$$

where u_*^{AW} and u_*^{IW} are the friction velocities at the atmosphere–ocean and the ice–ocean interface, respectively; κ is the von Karman constant; and z_0 is the roughness parameter.

Turbulent heat flux through the water surface is:

$$\nu'_t \frac{\partial T}{\partial z} = F_T / \rho_w C_{pW} \tag{3.87}$$

where F_T is heat flux from the water. In the absence of ice it is equal to atmospheric net heat flux Q_{AW}, which includes latent heat Q_{EW}, sensible heat Q_{HW}, and longwave radiation Q_{BW} as

$$Q_{AW} = Q_{EW} + Q_{HW} + Q_{BW} \tag{3.88}$$

Solar insolation Q_{SW} is absorbed in the water column as:

$$I(z) = Q_{SW}(a \exp(z/h_1) + (1-a)\exp(z/h_2)) \tag{3.89}$$

where a, h_1, h_2 are constants. Insolation Q_{SW} is calculated from solar radiation at the top of the atmosphere and then corrected for relative humidity, cloudiness, and inclination (see the review of Simonsen and Haugan, 1995) as:

$$Q_{SW} = \begin{cases} (1-\alpha) S_0 K T_R \left(\dfrac{\bar{d}}{d}\right)^2, & \cos Z \geq 0 \\ 0, & \exp Z \leq 0 \end{cases} \tag{3.90}$$

where $S_0 = 1{,}370\,\mathrm{W\,m^{-2}}$ is the solar constant; $\alpha = 0.06$ is the water albedo; K is cloud cover correction; $d/\bar{d}$ is the normalized Earth–Sun distance; and T_R is the clear sky transmittance formula (Zillman, 1972), modified for polar regions by Shine (1982) as

$$T_R = \frac{\cos Z}{\beta_1 \cos Z + (\beta_2 + \cos Z) e(T_a) \beta_3 + \beta_4} \tag{3.91}$$

where Z is solar elevation. The pressure of saturated vapor (given in Pascals) is related to surface air temperature T_a (given in Kelvins) as:

$$e(T_a) = 611 \cdot 10^{a(T_a - 273.15)/(T_a - b)} \tag{3.92}$$

where the constants are $a = 7.5$, $b = 35.86$ for $T_s > 273.15$, and $a = 9.5$, $b = 7.66$ for $T_s < 273.15$ (Murray, 1967). The empirical constants in (3.55) are $\beta_1 = 1.0$, $\beta_2 = 1.0$, $\beta_3 = 10^{-5}$, $\beta_4 = 0.046$ (Shine, 1982). Cloud cover correction (Reed, 1977) is $K = 1 - 0.62n + 0.0019 Z_{\mathrm{noon}}$, where n is cloudiness and Z_{noon} is solar elevation at noon in degrees. The ratio of annual mean to actual distance is estimated in terms of Julian day β for the present-day orbit:

$$\left(\frac{\bar{d}}{d}\right)^2 = 1.00011 + 0.00128 \sin\beta + 0.034221 \cos\beta + 0.000077 \sin 2\beta + 0.000719 \cos 2\beta$$

where $\beta = 2\pi J_d/365$ is the angle of sunset on orbit starting at $J_d = 0$; and J_d is Julian day. Solar elevation is calculated in terms of declination δ, latitude φ, and time from midday $\tilde{t}$ as:

$$\cos Z = \sin\delta \sin\varphi + \cos\delta \cos\varphi \cos\psi$$

$$\psi_* = \frac{2\pi\tilde{t}}{86{,}400}$$

Declination δ and noon solar elevation Z_{noon} are determined from approximated relations as:

$$\delta = 0.06918 + 0.070267 \sin\beta - 0.399912 \cos\beta + 0.000907 \sin 2\beta$$
$$- 0.006758 \cos 2\beta + 0.00148 \sin 3\beta - 0.002697 \cos 2\beta$$
$$\sin Z_{\text{noon}} = \sin\delta \sin\varphi + \cos\delta \cos\varphi$$

Longwave radiation is parameterized (Zapadka *et al.*, 2001) as

$$Q_{BW} = \sigma_T[\varepsilon_W T_s^4 - 0.732 T_a^4(1 - \exp(-0.47 e(T_a)))(1 - 0.067n + 0.301n^2)] \quad (3.93)$$

where $\sigma_T = 5.7 \cdot 10^{-8}\ \text{W}\,\text{m}^{-2}\,\text{K}^{-4}$ is the Stefan–Boltzmann constant; T_s is surface temperature; and $\varepsilon_W = 0.97$ is water emissivity. Sensible and latent heat fluxes into the water are estimated by the bulk formulas of Blackadar (1979), as:

$$Q_{HW} = C_{pA}\rho_a \kappa u_*^{AW} \frac{T_a - T_s}{\ln\dfrac{z_a}{z_0} - \psi_h} \quad (3.94)$$

$$Q_{EW} = \rho_a L_v \kappa u_*^{AW} \frac{(q_a - q_s)}{\ln\left(\dfrac{\kappa u_*^{AW} z_a}{\chi_a} + \dfrac{z_a}{z_0}\right) - \psi_h} \quad (3.95)$$

where q_a is specific humidity at height z_a over the sea surface; q_s is specific humidity at the sea surface; C_{pA} is the heat capacity of the air; T_a and T_s are temperatures at height z_a over the sea surface and at the sea surface; $\chi_a = 2.4 \cdot 10^{-5}\ \text{m}^2\,\text{s}^{-1}$ is background molecular diffusivity; and L_v is the heat of vaporization. The relations for q_a and q_s are:

$$q_a = \frac{0.622 r e(T_a)}{P_a - 0.378 r e(T_a)}, \qquad q_s = \frac{0.622 e(T_s)}{P_a - 0.378 e(T_s)} \quad (3.96)$$

Friction velocity is calculated from the bulk formulas of Blackadar (1979) as

$$u_* = \max\left(\frac{\kappa V}{\ln\dfrac{z_a}{z_0} - \psi_m}, 0\right) \quad (3.97)$$

Nondimensional parameters ψ_m and ψ_h are functions of the bulk Richardson number:

$$Ri_B = \frac{g z_a}{\theta_a}\frac{T_{va} - T_{vs}}{V^2} \quad \text{and parameter} \quad \ln\frac{z_a}{z_0}$$

(Blackadar, 1979). Here T_{va} and T_{vs} are virtual temperatures; $T_v = T(1 + 0.6078q)$; V is the velocity parameter, which is given by $V = \sqrt{V_a^2 + V_c^2}$, where V_a is horizontal wind speed at height z_a and V_c is convective velocity, which is defined from the relation:

$$V_c = \begin{cases} 0, & T_s < T_a \\ \sqrt{(T_s - T_a) g z_a / T_a}, & T_s \geq T_a \end{cases} \quad (3.98)$$

Turbulent salt flux through the water surface is:

$$\nu'_t \frac{\partial S}{\partial z} = F_S \tag{3.99}$$

where F_S is the salt flux from the water. In the absence of ice, $F_S = -(\dot{P} - \dot{E})S_W$, where S_W is sea surface salinity, $\dot{P} - \dot{E}$ is the difference between precipitation $\dot{P}$ and evaporation $\dot{E}$.

Under the assumption of a layer of constant fluxes, the standard boundary conditions are formulated at some distance $\tilde{z}$ from the bottom $z = -H$ as:

$$-u\frac{\partial H}{\partial x} - v\frac{\partial H}{\partial y} = w \tag{3.100}$$

$$\nu_t \frac{\partial u}{\partial z} = C_D\sqrt{u^2 + v^2}u, \qquad \nu_t \frac{\partial v}{\partial z} = C_D\sqrt{u^2 + v^2}v \tag{3.101}$$

$$\nu'_t \frac{\partial T}{\partial z} = \nu'_t \frac{\partial S}{\partial z} = 0 \tag{3.102}$$

$$k = \left(\frac{u_*^b}{c_\mu^0}\right)^2, \qquad \frac{\nu_t}{\sigma_\varepsilon}\frac{\partial \varepsilon}{\partial \tilde{z}} = (c_\mu^0)^3 \frac{k^{3/2}}{\kappa(\tilde{z} + z_b)^2} \tag{3.103}$$

where u_*^b is friction velocity near the bottom; and z_b is the bottom roughness parameter:

$$C_D = \max\left(0.0025; \frac{\kappa}{\ln \tilde{z}/z_b}\right) \tag{3.104}$$

At land boundaries, normal and tangential velocities and normal fluxes of scalars are set to zero:

$$u_n = 0, \qquad v_n = 0, \qquad \frac{\partial}{\partial n}(T, S, k, \varepsilon) = 0 \tag{3.105}$$

A variety of open lateral boundary conditions for external and internal modes are used, among them are radiation conditions (Flather, 1976; Orlanski, 1976; Blumberg and Kantha, 1985).

Ice thermohydrodynamics

Following Hibler (1979, 1980) and Mellor and Kantha (1989) we consider a simple ice dynamic–thermodynamic sub-model that is applicable to simulate the seasonal cycle of moving ice in coastal seas. An early version of the model was developed by Koziy and Maderich (1997). The ice sub-model describes momentum balance, ice rheology, mass balance, ice concentration, and ice strength. The variable thickness of ice h_I is averaged over ice cover and Ah_I is the averaged ice thickness over the total area (Mellor and Kantha, 1989).

Momentum balance describes the forces that determine the drift and deformation of ice cover:

$$\frac{\partial(mu_I)}{\partial t}+\frac{\partial(mu_Iu_I)}{\partial x}+\frac{\partial(mu_Iv_I)}{\partial y}-mfv_I+\frac{\partial\sigma_{xx}}{\partial x}+\frac{\partial\sigma_{xy}}{\partial y}=A(\tau_{AI}^{(x)}+\tau_{OI}^{(x)})-mg\frac{\partial\eta}{\partial x} \quad (3.106)$$

$$\frac{\partial(mv_I)}{\partial t}+\frac{\partial(mv_Iu_I)}{\partial x}+\frac{\partial(mv_Iv_I)}{\partial y}-mfu_I+\frac{\partial\sigma_{xy}}{\partial x}+\frac{\partial\sigma_{yy}}{\partial y}=A(\tau_{AI}^{(y)}+\tau_{OI}^{(y)})-mg\frac{\partial\eta}{\partial y} \quad (3.107)$$

where $\vec{U}_I = (u_I, v_I)$ is ice drift velocity; m is ice mass per unit area $m = A\rho_I h_I$ related to the thickness h_I and density ρ_I of ice; and the internal ice stress tensor σ_{ij} is:

$$\sigma_{ij} = 2\xi\dot{\varepsilon}_{ij} + \left[(\zeta-\xi)\sum_k \dot{\varepsilon}_{kk} - \frac{P}{2}\right]\delta_{ij} \quad (3.108)$$

where the strain tensor $\dot{\varepsilon}_{ij}$ is defined as

$$\dot{\varepsilon}_{xx} = \frac{\partial u_I}{\partial x}, \qquad \dot{\varepsilon}_{xy} = \frac{1}{2}\left(\frac{\partial u_I}{\partial y}+\frac{\partial v_I}{\partial x}\right), \qquad \dot{\varepsilon}_{xx} = \frac{\partial v_I}{\partial y} \quad (3.109)$$

where ξ and ζ are nonlinear shear and bulk viscosities; and P is ice pressure. Ice pressure as a function of ice concentration and ice thickness (Hibler, 1979) is

$$P = P^* A h_I \exp[-C(1-A)] \quad (3.110)$$

where P^* and C^* are empirical constants. Following Hibler (1979), ξ and ζ are taken as:

$$\zeta = \frac{P}{2\Delta}, \qquad \xi = \zeta/e^2$$

where

$$\Delta = \left[(\dot{\varepsilon}_{11}^2+\dot{\varepsilon}_{22}^2)\left(1+\frac{1}{e^2}\right)+\frac{4}{e^2}\dot{\varepsilon}_{12}^2+2\dot{\varepsilon}_{11}^2\dot{\varepsilon}_{22}^2\left(1-\frac{1}{e^2}\right)\right]^{1/2}$$

where $e = 2$ is the ratio of principal axes of the ellipse of the elliptical yield curve. If $\Delta = 0$, then $\zeta = 2.5 \cdot 10^8 P$.

For a mean per-unit-area thickness h and ice concentration A, the following continuity equations are used (Mellor and Kantha, 1989):

$$\frac{\partial Ah_I}{\partial t}+\frac{\partial(u_IAh_I)}{\partial x}+\frac{\partial(v_IAh_I)}{\partial y}=\frac{\rho_w}{\rho_I}[A(W_{IW}-W_{AI})+(1-A)W_{AW}+W_{FR}]+K_i\vec{\nabla}^2 h \quad (3.111)$$

$$\frac{\partial A}{\partial t}+\frac{\partial(uA)}{\partial x}+\frac{\partial(vA)}{\partial y}=\frac{\rho_w}{\rho_I}[\Psi A(W_{IW}-W_{AI})H(W_{AI}-W_{IW})$$
$$+\Phi(1-A)W_{AW}+(1-A)W_{FR}]+K_i\vec{\nabla}^2 h \quad (3.112)$$

where W_{IW} is the ice production–melting rate at the ice–water interface; W_{AI} is the ice production–melting rate at the atmosphere–ice interface; W_{AW} is the ice production rate at the atmosphere–water interface; and W_{FR} is the frazil ice production rate that triggers ice cover formation at salinity $S > 0$; and $H(x)$ is the Heaviside function

$\Phi = 4, \Psi = 0.7$. The diffusion terms, which are small, have been added for computational stability.

Heat flux Q_{AI} is expressed as:

$$Q_{AI} = Q_{SI} + Q_{LI} + Q_{BI} + Q_{HI} \tag{3.113}$$

Insolation Q_{SW} is calculated from (3.54) with snow albedo $\alpha_s = 0.75$. Longwave radiation is parameterized as (3.93) with ice emissivity $\varepsilon_I = \varepsilon_W$. Sensible and latent heat flux into the ice and friction velocity are estimated by the bulk formulas of Blackadar (1979) (i.e., Equations (3.94)–(3.95)) with the latent heat of sea ice sublimation $L_S = 2.834 \cdot 10^6\ \mathrm{J\,kg^{-1}}$. Conductive heat flux through ice is:

$$Q_c = Q_{IW} = \frac{2k_I}{h_I}(T_f - T_s) \tag{3.114}$$

where k_I is conductivity; the temperature of the lower surface of ice is assumed to be the freezing point of water T_f. It can be expressed as a linear function of salinity S and pressure (depth) (Mellor and Kantha, 1989):

$$T_f = m_f S + n_f z \tag{3.115}$$

where $m_f = -5.43 \times 10^{-2}\,^\circ\mathrm{C\,m^{-2}}$; and $n_f = 7.59 \times 10^{-4}\,^\circ\mathrm{C}$.

The ice production/melting rate at the ice–water interface W_{IW} is determined from the Stefan condition as:

$$\rho_W L_W W_{IW} = Q_{IW} - F_{TI} \tag{3.116}$$

where

$$F_{TI} = -\rho_W c_{pW} C_{Tz}(T_0 - T_s) \tag{3.117}$$

Parameter C_{Tz} is the heat transfer coefficient (Mellor and Kantha, 1989).

Ice surface temperature T_s is determined from the balance between flux at the cost of conductivity Q_c and Q_{AI}. If $T_s > 0$ then Q_{AI} is calculated at $T_s = 0$ to provide the Stefan condition for the upper surface of ice melting as:

$$\rho_W L_W W_{AI} = -Q_{AI} + Q_c \tag{3.118}$$

where $Q_c = 2k_I(T_1 - T_s)/h_I$; and T_s can be found from solution of the nonlinear equation

$$\frac{T_s - T_1}{h_I/2k_I} + Q_{AI}(T_s) = 0$$

An equation for the mean temperature of ice T_i is:

$$\frac{\partial T_1}{\partial t} = \frac{Q_{IW} - Q_{12}}{\rho_I h_I (C_{pI} - mS_I L_F / T_1^2)}$$

Over open water with $A > 0$ the Stefan condition is:

$$\rho_W L_f W_{AW} = Q_{AW} - F_{TW} \tag{3.119}$$

where $F_{TW} = -\rho_W c_{pW} C_{TzW}(T_{0W} - T_s)$.

The volume flux of seawater corresponding to ice growth over open water or melting existed ice is:

$$F_S = AF_{SI} + (1 - A)(F_{SW} - F_{SP}) \tag{3.120}$$

where $F_{SI} = -C_{SzI}(S_{0I} - S_W)$; $F_{SW} = -C_{SzW}(S_{0W} - S_W)$; $F_{SP} = (E - P)S_{0W}$; $T_{0I} = mS_{0I}$; and $T_{0W} = mS_{0W}$, and:

$$S_{0I} = \frac{C_{SzI}S_W - (W_{IW} - W_{AI})S_I}{C_{SzI} - (W_{IW} - W_{AI})}, \qquad S_{0W} = \frac{C_{SzW}S_W}{C_{SzW} + P - E} \tag{3.121}$$

where S_{IW} is salinity at the ice–water interface in areas where $A > 0$; and C_{Sz} is the salt transfer coefficient (Mellor and Kantha, 1989).

The parameter C_{Tz} is given by:

$$C_{Tz} = \frac{u_*}{P_{rt}\ln|\tilde{z}/z_0|\kappa + B_T} \tag{3.122}$$

where $B_T = b(z_0u_*/\nu)^{1/2}P_r$; $P_{rt} = 0.85$ is the turbulent Prandtl number; $\tilde{z}$ is the distance from the lower ice boundary or water surface; ν is molecular viscosity; $P_r = 12.9$ is the molecular Prandtl number; and $b = 2.14$ is the empirical constant.

The parameter C_{Tz} is:

$$C_{Sz} = \frac{u_*}{P_{rt}\ln|\tilde{z}/z_0|\kappa + B_T} \tag{3.123}$$

where $B_T = b(z_0u_*/\nu)^{1/2}S_c$; and $S_c = 2.432$ is the molecular Stanton number.

The production of frazil ice W_{FR} in the water column from the surface to depth H_F, where $T < T_f$, is calculated according to Mellor and Kantha (1989):

$$W_{FR} = \frac{1}{\delta t}\int_0^{H_F} \gamma\, dz$$

where local frazil ice production is

$$\gamma = \frac{C_{pW}[T_f - T(z)]}{L_{FR} - C_{pW}m_f(S(z) - S_I)}$$

where $L_{FR} = L_F + (C_{pW} - C_{pI})T$ is the latent heat of fusion; and δt is the time required to transform the water column from a super-cooled state to a state with frazil ice formation and the remaining water has been returned to the freezing state. Changes of temperature and salinity occurring in the time step δt are

$$\delta T = \gamma L_{FR}/C_{pW}$$

$$\delta S = \gamma(S - S_I)$$

where S_I is the mean salinity of sea ice ($S_I = 4$).

Ice–water stresses are calculated as:

$$\left.\begin{aligned}\tau_{IW}^{(y)} &= \rho_W \frac{\kappa u_*}{\ln|z/z_0|}(v_I - v_W)\\ \tau_{IW}^{(x)} &= \rho_W \frac{\kappa u_*}{\ln|z/z_0|}(u_I - u_W)\end{aligned}\right\} \tag{3.124}$$

The roughness length z_0 at the air–water boundary can be estimated from the Charnock relation as

$$z_0 = 0.032\frac{u_*^2}{g} + 10^{-4} \tag{3.125}$$

while at the water–ice boundary, it is proportional to the ice thickness:

$$z_0 = m_I h_I/3 \tag{3.126}$$

where m_I is the slope ($m_i = (2\text{–}5) \times 10^{-2}$).

Sediment transport

The model simulates the transport of noncohesive sediments, cohesive sediments, and a mixture of fractions of different size cohesive–noncohesive sediments. The water column and bottom are divided into a set of layers: water layer, several active bed layers, and a nonerodible bed layer. If the transport thickness of the upper active bed layer becomes zero (as a result of erosion and/or bedload), then the lower active bed layer interacts with the water layer. When the thickness of the upper active bed layer exceeds some value (as a result of deposition), then a new active bed layer arises. The model accounts for mixed grain–size sediments with an arbitrary number of sand-size fractions. The range of simulated grain sizes is determined by limits in sediment transport, resuspension, and deposition parameterizations. It is assumed that all sediment particles of a given size class inside active bed layers are equally exposed to the water flow and entrainment in the active layer. The sediments are transported in the water layer as suspended sediments. Suspended sediment transport is described by the advection–diffusion equation, taking into account the settling velocity of sediment grains:

$$\begin{aligned}&\frac{\partial S_{di}^w}{\partial t} + U\frac{\partial S_{di}^w}{\partial x} + V\frac{\partial S_{di}^w}{\partial y} + (W - W_{gi})\frac{\partial S_{di}^w}{\partial z}\\ &\qquad = \frac{\partial}{\partial z}\left(\nu_t'\frac{\partial S_{di}^w}{\partial z}\right) + \frac{\partial}{\partial x}\left(K_T\frac{\partial S_{di}^w}{\partial x}\right) + \frac{\partial}{\partial y}\left(K_T\frac{\partial S_{di}^w}{\partial y}\right)\end{aligned} \tag{3.127}$$

where S_{di}^w is the concentration of the ith fraction of sediment in the water column; and W_{gi} is the settling velocity of solid particles. At the free surface $z = \eta$, zero vertical sediment flux is assumed; that is:

$$(W - W_{gi})S_{di}^w = \nu_t'\frac{\partial S_{di}^w}{\partial z} \tag{3.128}$$

The bottom boundary condition describes sediment resuspension or settling–sedimentation, which is dependent on the ratio between equilibrium and actual near-bottom suspended sediment concentration. The vertical flux of suspended sediments at the bottom $z = -H + z_b$ is equal to the difference between resuspension and sedimentation rates:

$$\nu'_t \frac{\partial S^w_{di}}{\partial z} + W_{gi} S^w_{di} = q^w_i - q^b_i \tag{3.129}$$

where z_b (m) is bottom roughness; and q^w_i, q^b_i are the sedimentation and resuspension rates, respectively.

The concentration of sediment in the active bottom layer is described by the following equation:

$$\frac{\partial(m^b_{di})}{\partial t} = q^w_i - q^b_i \tag{3.130}$$

where $m^b_{di} = S^b_{di} Z_*$ (kg/m^2) is the mass of the ith fraction of bottom sediment per unit area; and Z_* is the depth of the active bottom sediment top layer. The thickness of the active bottom layer and bottom porosity are defined from the relationships:

$$\left.\begin{aligned} \tilde{Z} &= \frac{1}{1-\varepsilon} \sum_{i\,\text{solid}} \frac{m^b_{di}}{\rho_{Si}} \\ \tilde{\varepsilon} &= \frac{V_{\text{tot}} - \sum_{i\,\text{solid}} V_i}{V_{\text{tot}}} = \frac{Z_{\min} - \sum_{i\,\text{solid}} \frac{m^b_{di}}{\rho_{Si}}}{Z_{\min}} \\ Z_* &= \begin{cases} \tilde{Z}, \varepsilon = \varepsilon_0 = \text{const}, \tilde{Z} > Z_{\min} \\ Z_{\min}, \varepsilon = \tilde{\varepsilon}, \tilde{Z} \le Z_{\min} \end{cases} \end{aligned}\right\} \tag{3.131}$$

where ε is porosity of the bed layer; V_i is the volume of the ith fraction of sediments; and ρ_{Si} is the density of solid particles. From Equations (3.130) and (3.131) the concentration of the ith fraction of sediment in the bottom layer is defined as:

$$S^b_{di} = \frac{m^b_{di}}{Z_*}$$

Sedimentation and resuspension rates in (3.129) and (3.130) for fine noncohesive sediment can be estimated as follows:

$$\left.\begin{aligned} q^w_i &= \begin{cases} W_{gi}(S^w_{di*} - S^w_{di}), & S^w_{di} > S^w_{di*} \\ 0, & S^w_{di} < S^w_{di*} \end{cases} \\ q^b_i &= \begin{cases} 0, & S^w_{di} > S^w_{di*} \\ E_r W_{gi}(S^w_{di*} - S^w_{di}) F_{Si}, & S^w_{di} < S^w_{di*} \end{cases} \end{aligned}\right\} \tag{3.132}$$

where E_r is the erodibility coefficient; S^w_{di} is the actual sediment concentration at the bottom level $z = -H + a$; S^w_{di*} is the near-bottom equilibrium sediment concentration at reference level a that corresponds to the sediment capacity of a steady and uniform flow with the same local parameters. The parameter $F_{Si} = \dfrac{6S^b_{di}}{\rho_{Si}\pi}$, where $\dfrac{S^b_{di}}{\rho_{Si}} \le \dfrac{\pi}{6}$ is included in Formula (3.95) to take into account of the multi-fractional structure of suspended sediment. It is equal to the ratio of the ith sediment fraction volume to the maximum possible volume that can be covered by the same fraction of sediment (because each sediment fraction is assumed to consist of spherical solid particles with a fixed diameter). The ratio is equal to 1 for a single sediment fraction and varies from 0 to 1 for the multi-fractional sediment.

The equilibrium concentration in (3.132) is calculated by an empirical formula (van Rijn, 1984) as

$$S^w_{di^*} = 0.015\rho_{Si}\frac{d_i T_i^{1.5}}{aD^{0.3}_{*,j}}, \qquad T_i = \frac{u_*^2}{u^2_{*cr,i}} - 1, \qquad D_* = d_i\left[\frac{g(\rho_{Si} - \rho)}{\rho\nu^2}\right]^{1/3} \tag{3.133}$$

where d_i is the diameter of the solid particles; ρ_{Si} is the density of the particles; and $u_{*i,cr}$ is the critical bed shear velocity for initiation of bed motion that is computed using the Shields criterion. The reference level $a = \max\{0.01h, k_s\}$, where k_s is roughness height. The settling velocity W_{gi} is calculated according to van Rijn (1984).

Sedimentation and resuspension rates for fine cohesive sediment are described following Krone (1962) and Partheniades (1962) as:

$$q_i^w = \begin{cases} W_{gi}\left[1 - \left(\dfrac{u_*}{u_{*\text{crit}}}\right)^2\right] \cdot S^w_{di}, & u_* \le u_{*\text{crit}} \\ 0, & u_* > u_{*\text{crit}} \end{cases} \tag{3.134}$$

$$q_i^b = \begin{cases} 0, & u_* \le u_{*\text{crit}} \\ M\left[1 - \left(\dfrac{u_*}{u_{*\text{crit}}}\right)^2\right] \cdot F_{Si}, & u_* > u_{*\text{crit}} \end{cases} \tag{3.135}$$

where

$$(u_{*\text{crit}})^{5/3} = 0.06g\left(\frac{\rho_{Si} - \rho}{\rho}\right)\nu^{1/2} d_i^{1/2}$$

Coefficient M is the dimension of the erosion rate (kg m^{-2} s^{-1}) and varies from one mud to another.

Radionuclide transport

The sub-model of radionuclide transport describes specific water–sediment sorption processes. It includes the advection–diffusion equations for dissolved (C^w_s) and adsorbed by suspended sediment (C^w_p) radioactivity in the water column, and equations for the concentration of dissolved (C^b_s) and adsorbed (C^b_p) radioactivity in bottom deposits:

$$\frac{\partial C_s^w}{\partial t}+\frac{\partial UC_s^w}{\partial x}+\frac{\partial VC_s^w}{\partial y}+\frac{\partial WC_s^w}{\partial z}$$
$$=\frac{\partial}{\partial z}\left(\nu_t'\frac{\partial C_s^w}{\partial z}\right)+\frac{\partial}{\partial x}\left(K_T\frac{\partial C_s^w}{\partial x}\right)+\frac{\partial}{\partial y}\left(K_T\frac{\partial C_s^w}{\partial y}\right)$$
$$-\lambda C_s^w-a_{1,2}^w(S_d^wK_d^wC_s^w-C_p^w), \tag{3.136}$$

$$\frac{\partial C_p^w}{\partial t}+\frac{\partial UC_p^w}{\partial x}+\frac{\partial VC_p^w}{\partial y}+\frac{\partial (W-W_g)C_p^w}{\partial z}$$
$$=\frac{\partial}{\partial z}\left(\nu_T\frac{\partial C_p^w}{\partial z}\right)+\frac{\partial}{\partial x}\left(K_T\frac{\partial C_p^w}{\partial x}\right)+\frac{\partial}{\partial y}\left(K_T\frac{\partial C_p^w}{\partial y}\right)$$
$$-\lambda C_p^w-a_{1,2}^w(S_d^wK_d^wC_s^w-C_p^w), \tag{3.137}$$

$$\frac{\partial m_p^b}{\partial t}=\left(\frac{C_p^w\sum_{i\,\text{solid}}q_i^w}{\sum_{i\,\text{solid}}S_{di}^w}-\frac{C_p^b\sum_{i\,\text{solid}}q_i^b}{\sum_{i\,\text{solid}}S_{di}^b}\right)+a_{1,2}^b\left(\frac{S_d^b}{\varepsilon}K_d^bC_s^b-C_p^b\right)\cdot Z_*-\lambda m_p^b \tag{3.138}$$

where $m_s^b=C_s^bZ_*$; $m_p^b=C_p^bZ_*$; $S_d^w=\sum_{i\,\text{solid}}S_{di}^w$; $C_p^w=\sum_{i\,\text{solid}}C_{pi}^w$; $W_g=\sum_{i\,\text{solid}}W_{gi}S_{di}^w/S_d^w$; $S_d^b=\sum_{i\,\text{solid}}S_{di}^b$; $C_p^b=\sum_{i\,\text{solid}}C_{pi}^b$; λ is the radionuclide decay constant; and F_{dif} is the diffusion coefficient. The governing equations of the model are similar to those used in the Flow, Energy, Salinity, Sediment, and Containment Transport (FLESQOT) model (Onishi *et al.*, 1989).

The exchanges between different phases are described by diffusion, sorption, and sedimentation–resuspension processes. Adsorption and desorption of radionuclides between liquid and solid phases are described by the radionuclide exchange rates $a_{1,2}^w, a_{1,2}^b$, and by the distribution coefficients K_d^w, K_d^b (Onishi *et al.*, 1981; Santschi and Honeyman, 1989; Carrol and Harms 1999; IAEA, 2003), which are defined, under equilibrium conditions, as:

$$S_d^wK_d^w=\lim_{t\to\infty}\left(\frac{C_p^w}{C_s^w}\right),\qquad \frac{S_d^b}{\varepsilon}K_d^b=\lim_{t\to\infty}\left(\frac{C_p^b}{C_s^b}\right) \tag{3.139}$$

The solid–solution partitioning of radionuclides depends on salinity, with distribution coefficients decreasing with increasing salinity. This effect is important in estuaries and river plumes. The simple interpolation formula of Laissaoui *et al.* (1998) is used for the dependence of distribution coefficients on salinity, as:

$$K_d=K_d^{fw}(1-\delta) \tag{3.140}$$

where

$$\delta=\frac{S}{S+S_0}$$

In these equations, K_d^{fw} is the freshwater value of the distribution coefficient; S is salinity; and S_0 is the salinity value at which 50% saturation occurs (Laissaoui *et al.*, 1998).

At the free surface $z = \eta$ the boundary conditions are:

$$\nu_T \frac{\partial C_s^w}{\partial z} = W C_s^w, \qquad (W - W_g) C_p^w - \nu_T \frac{\partial C_p^w}{\partial z} = 0 \tag{3.141}$$

Fluxes into the bottom $z = -h + z_0$ are

$$\nu_T \frac{\partial C_s^w}{\partial z} = \frac{F_{\text{dif}}}{Z_*} \left(C_s^w - \frac{C_s^b}{\varepsilon} \right) \tag{3.142}$$

$$\nu_T \frac{\partial C_p^w}{\partial z} + (W - W_g) C_p^w = \frac{C_p^w \sum_{i\,\text{solid}} q_i^w}{\sum_{i\,\text{solid}} S_{di}^w} - \frac{C_p^b \sum_{i\,\text{solid}} q_i^b}{\sum_{i\,\text{solid}} S_{di}^b} \tag{3.143}$$

Numerical setup

The numerical algorithm was implemented in the horizontal curvilinear–orthogonal coordinate system. Fitting of the coordinate system to bottom topography by means of σ coordinate transformation (Blumberg and Mellor, 1987) allows us to accurately describe shallow-water flow and the boundary layer over realistic bottom topography. However, terrain-following models using the σ coordinate have difficulties in simulating flows over steep topography because of pressure gradient errors, errors in diffusion terms, and errors in representation of the thin thermocline formed by spreading cooling water. Furthermore, the number of σ layers should be minimal in shallow intertidal areas for efficient parameterization of wetting–drying processes. Therefore, the water column was split by a horizontal plane into two vertical domains. In the upper shallow domain, the σ coordinate is used which allows accurate description of the structure of flow and density in the upper layer using a limited number of σ layers. In the lower domain, two coordinates are optionally used: a second σ coordinate (Beckers, 1991) and a z vertical coordinate (Barron *et al.*, 2006).

Governing equations together with boundary conditions are solved by finite difference techniques. The model equations are solved on an Arakawa C-grid with all scalars located at the cell centroid, while velocity components are defined at the center of the faces of cells. Temporal differencing is a second-order leapfrog scheme with an Asselin filter. The vertically integrated equations of continuity and momentum (external mode) are separated from equations for the vertical structure of flow (internal mode). Splitting into external and internal modes was done following Blumberg and Mellor (1987). The equations for external mode were solved explicitly using a short time step Δt_e to satisfy the Courant–Friedrichs–Lewy condition for fast barotropic longwaves. The three-dimensional velocity and scalar fields (temperature, salinity, turbulent quantities) were computed semi-implicitly with a larger step Δt_i. The implicit treatment of vertical viscosity and diffusion terms is used, whereas advective terms, horizontal viscosity, and diffusion are computed in the previous time step. The advection of scalars is approximated by the high-order TVD scheme (van Leer, 1979).

Wetting and drying of coastal areas occurs in low-lying coastal zones. This can be caused by tidal or storm flooding and river floods. The effects of wetting and drying can play an important role in heat dispersion on tidal flats. Therefore, the efficient wetting and drying scheme developed by Oey (2006) for the Princeton Ocean Model (POM) was implemented in THREETOX. This scheme is based on a cell face–blocking methodology. "Dry" cells are defined as regions with depth less than a user-defined "critical depth". Primitive equations are solved in the thin film at a critical depth on the order of centimeters as well as in other regular wet cells. The Oey (2006) scheme only requires flux-blocking conditions across cell interfaces when wet cells become dry. The temperature and salinity of the thin film are the same at the moment of drying. Oey (2006) assumed that there can be no vertical (baroclinic) velocity structures across a dry cell's faces.

Embedding a finer resolution grid within a coarser grid (nesting) allows description of cooling water transport in the near and intermediate zone. The development of a two-way nesting procedure in THREETOX followed the approaches of Zhang *et al.* (1986), Oey and Chen (1993), and Miyazawa and Minato (2000) with some modifications. The nesting ratio between coarse-grid and fine-grid lengths and time steps on coarse and fine grids was chosen as 3 : 1, after Zhang *et al.* (1986). The vertical resolution on coarse and fine grids is the same. The nesting algorithm includes calculations on coarse and fine grids and an exchange of information between them. At each internal and external time step on the coarse grid, large-scale forcing is provided to the embedded fine grid on the coarse-grid boundary using bilinear interpolation on the input interface. Calculated fine-grid variables averaged over the nested grid replace coarse values on the output interface. Here, unlike Oey and Chen (1993) and Miyazawa and Minato (2000), the input interface is separated from the feedback interface, which prevents numerical noise (Zhang *et al.*, 1986).

3.4.3 River model validation results

The one-dimensional model RIVTOX and three-dimensional model THREETOX have been tested in a number of case studies (Zheleznyak *et al.*, 1995; Slavik *et al.*, 1997; Koziy *et al.*, 1998; Margvelashvili *et al.*, 1999, 2002; Koziy *et al.*, 2000; Koshebutsky *et al.*, 2004; Monte *et al.*, 2006). Here we present the results of a comparison with measurements of radionuclide contamination of the Dnieper (Dnipro) River and Dnieper–Bug (Buh) Estuary (DBE) caused by the accident at the Chernobyl NPP in April 1986. The Chernobyl NPP is located near the mouth of one of the Dnieper River's tributaries—the Prypyat River (Figure 3.12). The Dnieper River system downstream of the Prypyat mouth includes six large artificial reservoirs (Kiev, Kanev, Kremenchug, Dneprodzerzynsk, Dniprovske, and Kakhovka) and some tributaries (the largest tributary is the Desna River). The reservoirs are shallow-water bodies except for the Dniprovske reservoir. The length of the Dnieper River from the Prypyat confluence is about 980 km. Mean annual discharge from the

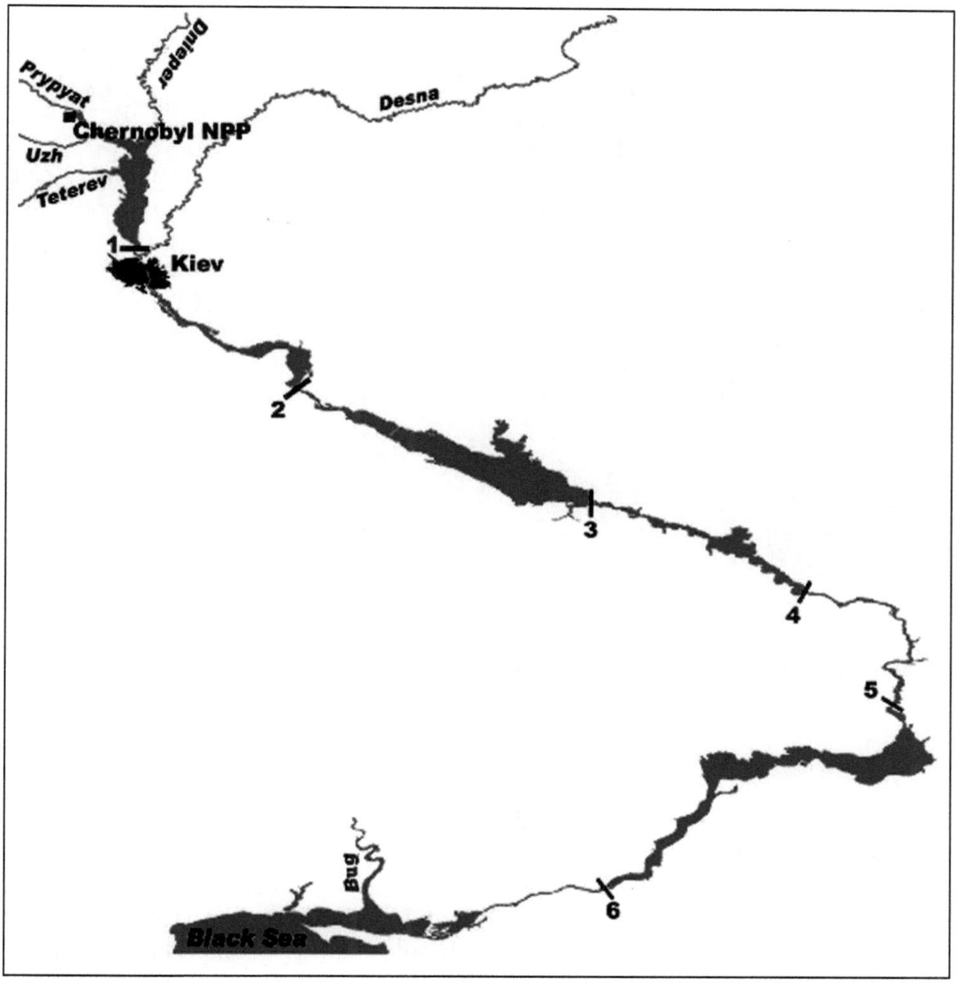

Figure 3.12. Map of the Dnieper River and estuary. The dams (1—Kiev, 2—Kanev, 3—Kremenchug, 4—Dneprodzerzynsk, 5—Dniprovske, and 6—Kakhovka) with gauging stations are shown.

mouths of the Dnieper, Prypyat, and Desna Rivers are 1,700 m^3/s, 450 m^3/s, and 360 m^3/s, respectively. Spring snowmelt flooding of the Prypyat, Desna, and Dnieper rivers is usually observed in early March to mid-May; however, flows through the system of Dnieper reservoirs are regulated. During winter, the entire river system is ice-covered in December–March.

The Dnieper–Bug Estuary (DBE) (Figure 3.13) connects the Dnieper River with the Black Sea. It is the largest of all the Black Sea estuaries, with a surface area of 1,006.3 km^2 and a volume of 4.24 km^3 (Zhykinsky *et al.*, 1986). The DBE water system consists of the Dnieper Estuary and the adjoining Bug River Estuary. The length of the DBE is 63 km with a width of up to 15 km. The DBE is connected with

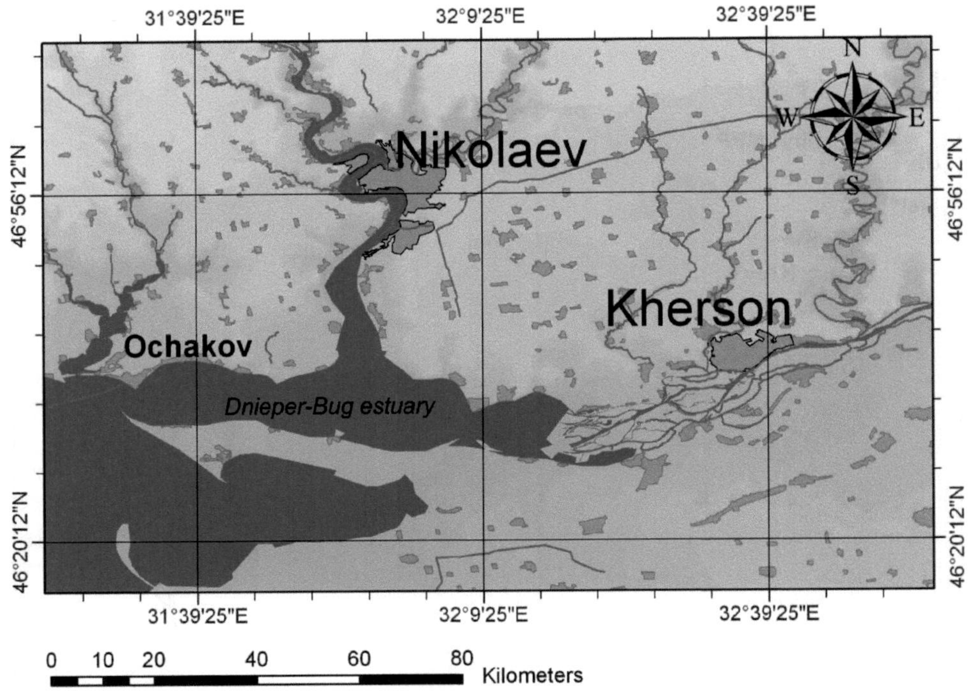

Figure 3.13. Map of the Dnieper–Bug estuary.

the Black Sea through the Strait of Kinbourn. The average depth of the DBE is 4.4 m. There is a narrow, 10 m to 12 m deep channel suitable for shipping along the Dnieper and Bug Estuaries to the Black Sea ports of Mykolayiv and Kherson (Figure 3.14). The bottom is covered mainly by clay (50%) and sand. The main factor affecting the system is the mixing of fresh river waters with saline marine waters. This forms a salt wedge in the estuary, which in the summer months can reach the city of Kherson. Stratification in the estuary ranges from almost none in the eastern part at the Dnieper mouth to a defined two-layer system in the western marine part of the DBE. These processes are highly season-dependent. This drowned river estuary varies from stratified in summer to partially mixed in winter. The monthly average discharge of the Dnieper River ranges from about 400 m^3/s in the summer to about 6,000 m^3/s in the spring, whereas the monthly average discharge of the Southern Bug ranges from 80 m^3/s to 1,000 m^3/s. Dnieper River discharge, unlike the Southern Bug, is not simply seasonal because it is regulated at the Kakhovka reservoir dam situated 70 km from the Dnieper mouth. In summer, the salt wedge penetrates much farther into the estuary than in spring. In addition, the salinity of the upper layer in summer is much higher than in spring. In addition to freshwater input, other key factors governing the transport of contaminants are air temperature, wind, and sea level variability. The estuary is ice-covered in January–February and wind surges cause short-term excursions of the salt wedge into the river mouths (Kostyanitsyn, 1964).

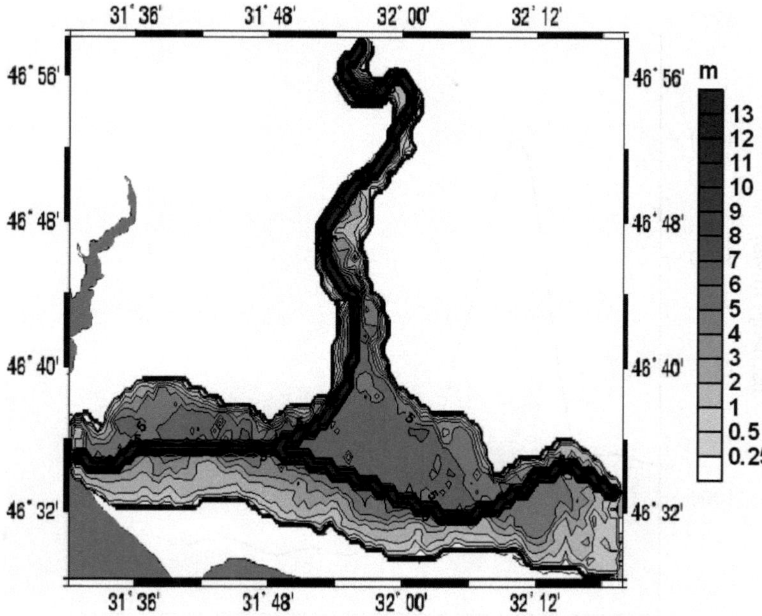

Figure 3.14. Bathymetry of the Dnieper–Bug estuary.

Together these factors produce complicated three-dimensional time-dependent stratified flows in the DBE.

The Chernobyl accident occurred on April 30, 1986 and caused atmospheric release of a large amount of radioactivity during the first 10 days. The highest level of contamination was found in the area near the NPP and particularly on the floodplain of the Prypyat River. This floodplain and contaminated water bodies were the main secondary source of radionuclide contamination of the Prypyat and Dnieper Rivers. Water flows and sedimentation in the Dnieper reservoirs govern the transport of radionuclides into the Black Sea. More than 80% of the suspended matter transported by river was deposited in the reservoirs. The Dnieper reservoirs therefore act as an efficient trap for particulate radionuclides.

The results of two simulations using the RIVTOX and THREETOX models are presented in this section. In the first case study, the transport of ^{90}Sr through the Dnieper River from the Prypyat mouth was simulated by RIVTOX for 1999. This year was chosen because of a strong flood that washed radioactivity from floodplains near the Chernobyl NPP. In the second case study, the transport of ^{137}Cs and ^{90}Sr through the DBE was simulated by THREETOX for 1986–1987.

In the first case study, the river network was divided (for modeling purposes) into eight branches in compliance with the location of dams and the main tributaries, which are considered as internal boundary conditions (Figure 3.12). Observation points are located at the dams. The beginning of the modeled river network is the confluence of three rivers: the Dnieper, Prypyat, and Uzh in the upper part of Kievskoe reservoir. The end of the modeled river network is the Kahovka dam. The length of the simulated river network is 843 km. The space step employed was

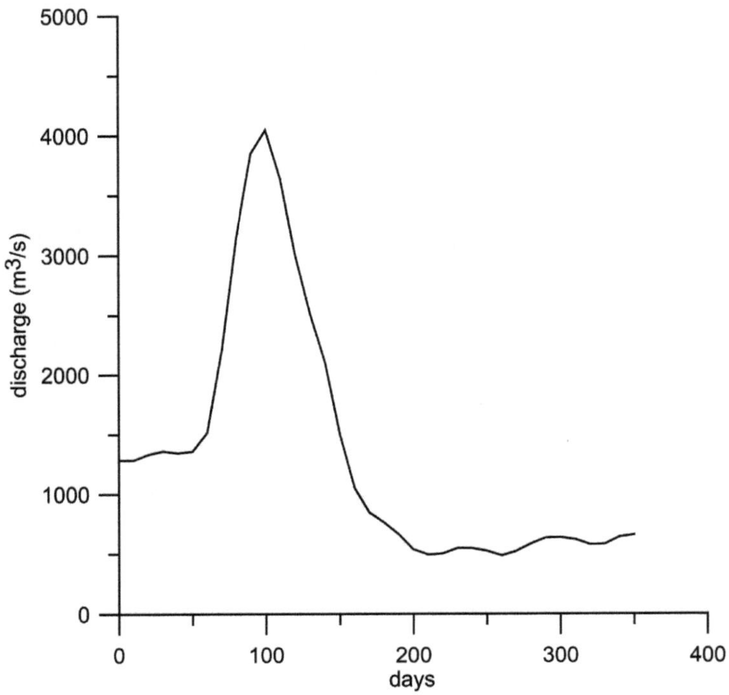

Figure 3.15. Upstream boundary conditions for Dnieper discharge in 1999.

2 km. The upstream boundary condition is the sum of discharges of the Dnieper, Prypyat, and Uzh, which were averaged in 10-day periods (Figure 3.15). The downstream boundary conditions for each branch are the levels at the dams, which were also averaged in 10-day periods. The concentrations of ^{90}Sr in the rivers Prypyat (Chernobyl NPP), Dnieper (Nedanchichi), and Uzh (mouth), provided by the Ukrainian Hydrometeorological Institute (UHMI), were multiplied by the discharges, then these fluxes were summed and divided by the sum of discharges of these rivers to produce the concentration in the upstream section in Figure 3.16. Sediment size was 0.05 mm. The value K_d for suspended sediments was $0.8\,m^3/kg$, whereas the K_d for bottom sediments was $0.25\,m^3/kg$, with $a_{1,2} = a_{2,1} = 0.5$ L/day, $a_{1,3} = 0.02$ L/day, and $a_{3,1} = 0.003$ L/day. Simulation results are compared with UHMI measurements on the dams in the Kiev, Dniprovske, and Kakhovka reservoirs in Figures 3.17–3.19. As seen in the figures, simulations agree well with measurements.

The second case study was a simulation of radionuclide transport in the DBE using the THREETOX model. The estuary was contaminated by radionuclides introduced in the environment following the accident that occurred at the Chernobyl NPP. The contamination was caused by direct deposition into the estuary and by radioactive substances transported by the Dnieper River and, to a much lesser extent, by the Bug River. The simulation was carried out for 1986–1987. The following input data were used: bathymetry of the DBE; daily discharges of the Dnieper and Bug

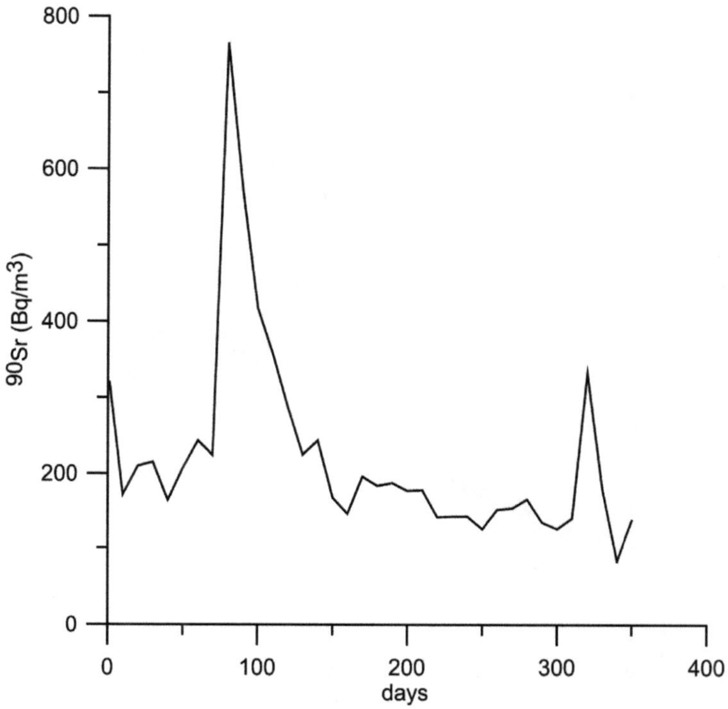

Figure 3.16. Upstream boundary conditions for ^{90}Sr in 1999.

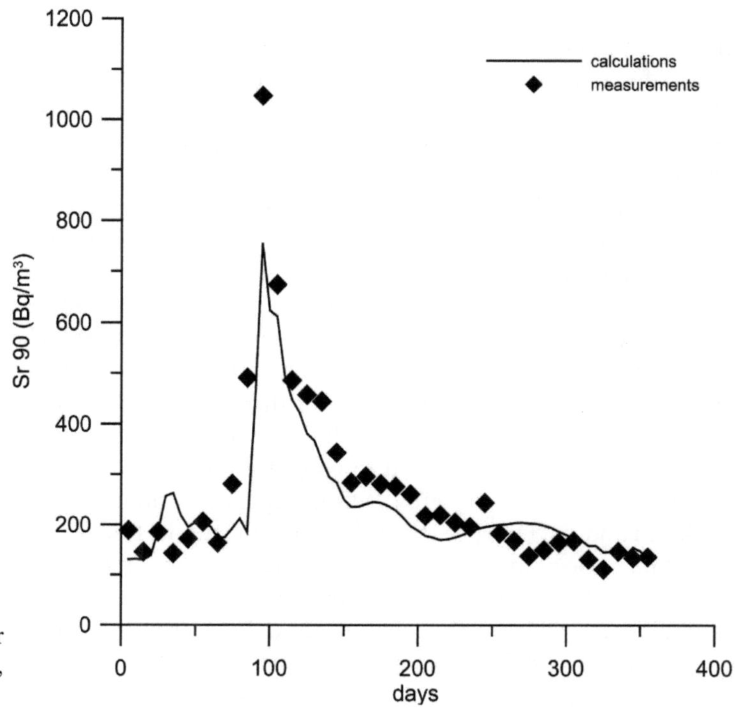

Figure 3.17. Calculated and measured concentration of ^{90}Sr in the Kiev reservoir, 1999.

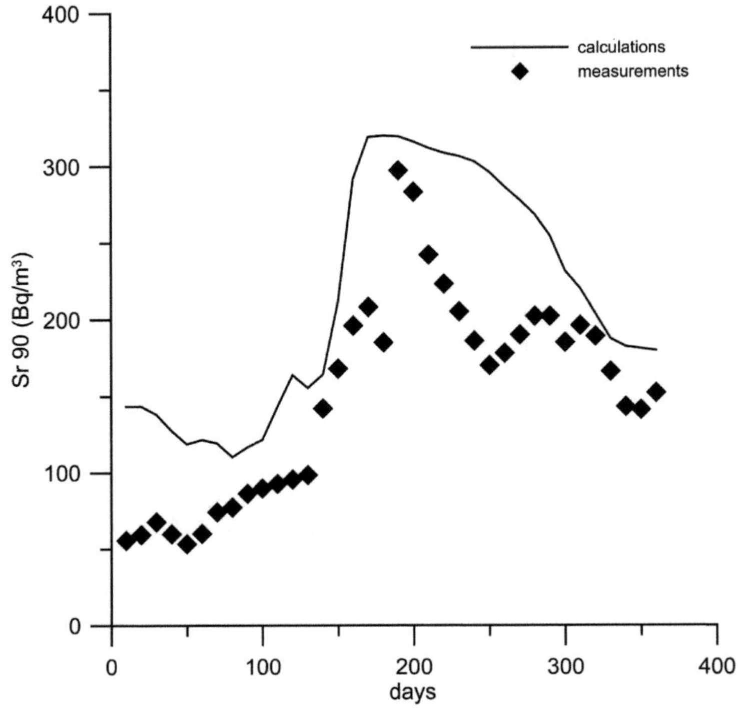

Figure 3.18. Calculated and measured concentration of ^{90}Sr in the Dneprovskoe reservoir, 1999.

Rivers; sea level, daily water temperature, and salinity in the Kinbourn Strait; 3-hour values of wind, air temperature, relative humidity, and cloudiness from the Ochakov meteorological station. Monthly averaged concentrations of suspended sediment at the mouth of the Dnieper River and at the mouth of the Southern Bug River were specified according to data from the State Water Cadastre. The sediment size was 1 μm. The initial thickness of the upper bottom sediment layer was set at 2 cm. Other parameters of DBE hydrology were not available; therefore, temperature in the Dnieper River and monthly averaged concentrations of suspended sediment at the mouth of the Dnieper River and at the mouth of the Southern Bug River were specified according to data from the State Water Cadastre. The temperature of the Bug River, and temperature and salinity profiles in the Kinbourn Strait were also unavailable; therefore, information on their seasonal changes was used. The 1986 survey data on temperature and salinity and the 1987 survey data on temperature, salinity, and velocity were used to compare the results of simulations.

The input data used were the time-dependent radionuclide concentrations in the Black Sea and the Dnieper River. Data for the model tests were selected for ^{137}Cs and ^{90}Sr from measurements. These data include contamination at the mouth of the Dnieper, contamination of the near-surface layer and near-bottom layer extracted from different sources (Kulebakina and Polikarpov, 1991; Katrich *et al.*, 1992; Kanivets *et al.*, 1997). The constant values K_d and $a_{1,2}$ for ^{137}Cs were $K_d = 15\,m^3/kg$ on suspended sediments, $K_d = 3\,m^3/kg$ in the bottom, $a_{1,2} = 1/365\,da^{-1}$, and for

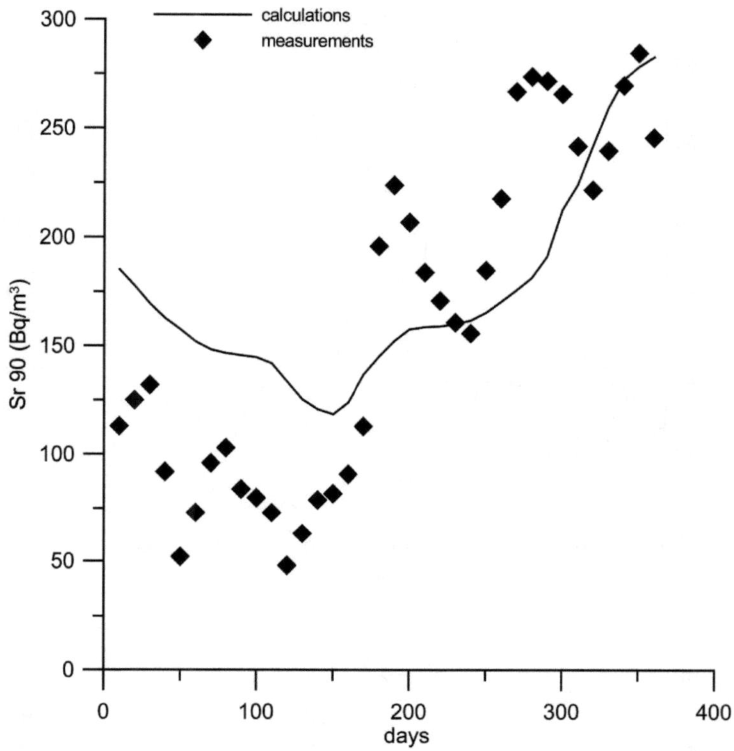

Figure 3.19. Calculated and measured concentration of ^{90}Sr in the Kahovka reservoir, 1999.

^{90}Sr they were $K_d = 0.3\,\mathrm{m^3/kg}$ on suspended sediments, $K_d = 0.25\,\mathrm{m^3/kg}$ in the bottom, $a_{1,2} = 1/100\,\mathrm{da^{-1}}$. Horizontal resolution was $606 \times 926\,\mathrm{m}$ and 20 σ-levels were used. The internal and external time steps were 126 s and 21 s, respectively. The simulations started on January 1, 1986 with background concentrations of 3.5 Bq/m^3 of ^{137}Cs and 23 Bq/m^3 of ^{90}Sr in the water. On May 1, the day after the incident, the concentration of ^{137}Cs was 5,000 Bq/m^2 whereas the deposition of ^{90}Sr was negligible. The results of simulations are given in Figures 3.20–3.30. The computed surface and bottom fields of salinity are given in Figures 3.30 for June 1987. These fields are compared with the UHMI survey data that are represented in the figure by the values with decimals. Saltwater intrusion into the DBE takes place mainly in the summer season, when the water discharge from the Dnieper River is low. Along the navigational channel of the DBE, the density-induced deep undercurrent results in a wedge of salty Black Sea water. The measurements and simulations agree reasonably well considering the incompleteness of the input data.

The spatial and temporal distributions of ^{137}Cs and ^{90}Sr are essentially different. In Figures 3.21–3.26 the surface fields of ^{137}Cs and ^{90}Sr are shown together with fields of bottom contamination by ^{137}Cs and ^{90}Sr. Comparison of simulations and measurements in the Dnieper Estuary is given in Figures 3.17–3.30. The simulated surface fields of ^{137}Cs and ^{90}Sr were averaged in the western and eastern areas of the Dnieper Estuary. These values were compared with values of ^{137}Cs and ^{90}Sr at distances

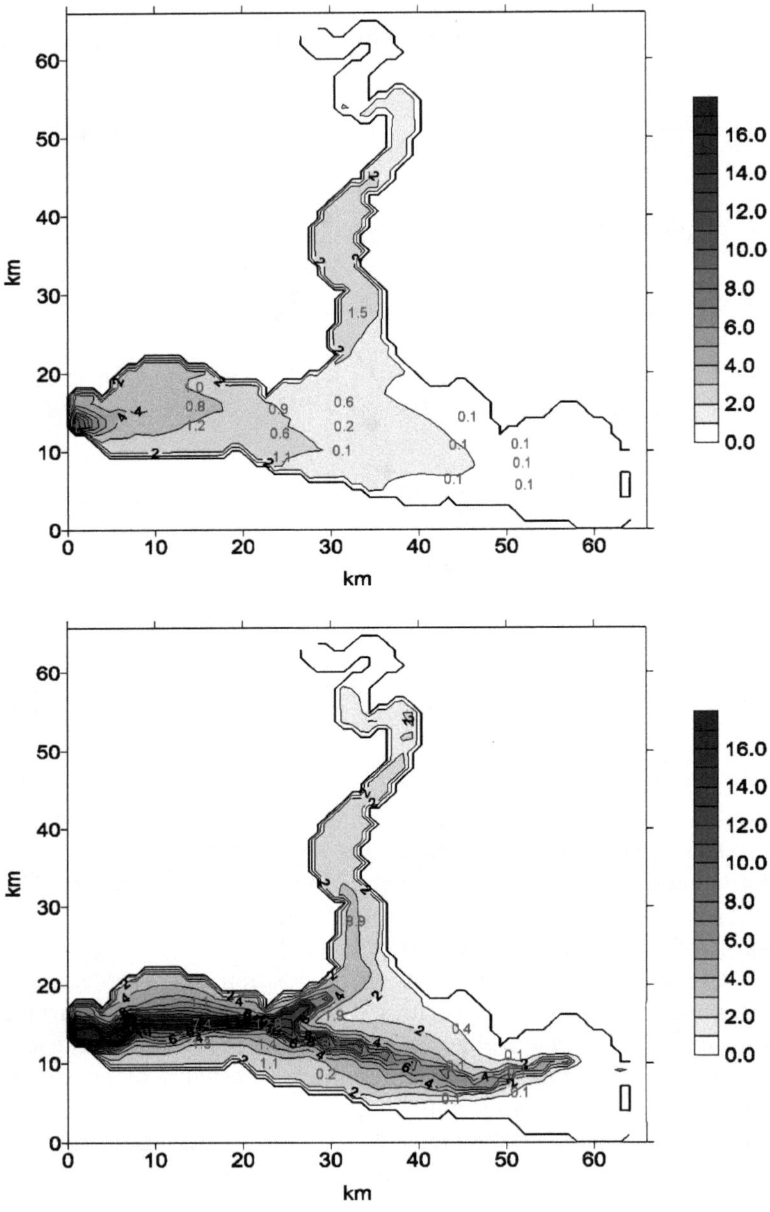

Figure 3.20. Calculated and observed salinity (a) at the surface and (b) on the bottom of the Dnieper–Bug estuary, June 24, 1987.

20 km, 30 km, 40 km, 50 km, and 60 km from Ochakov, as given by Katrich *et al.* (1992). Measurements and simulations agree well. The highest water contamination from ^{90}Sr was inside the DBE, whereas for ^{137}Cs water contamination increased seaward. Hence, the DBE was contaminated by ^{137}Cs and purified from ^{90}Sr as a result of the intrusion of saline water from the Black Sea, and it was contaminated by ^{90}Sr and purified from ^{137}Cs as a result of the influx of the fresh river water. This can

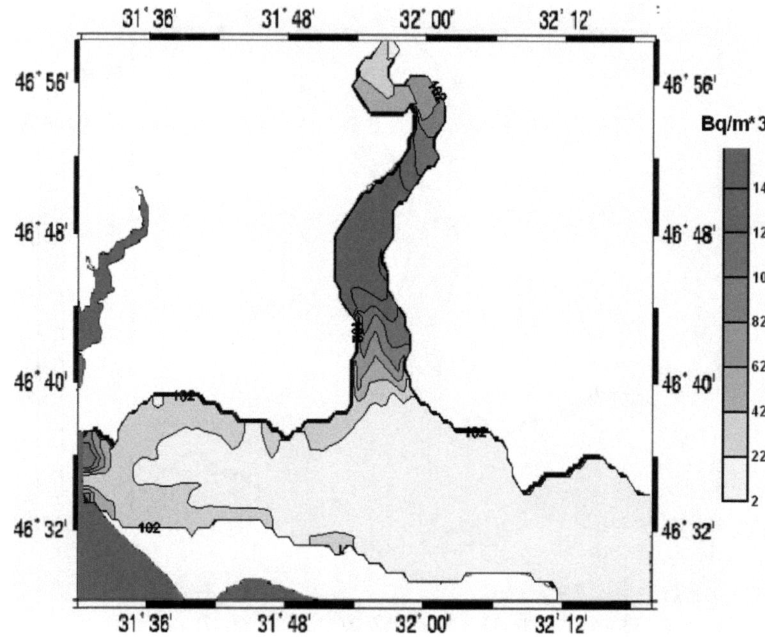

Figure 3.21. Simulated concentration of ^{137}Cs in water at the surface, June 10, 1986.

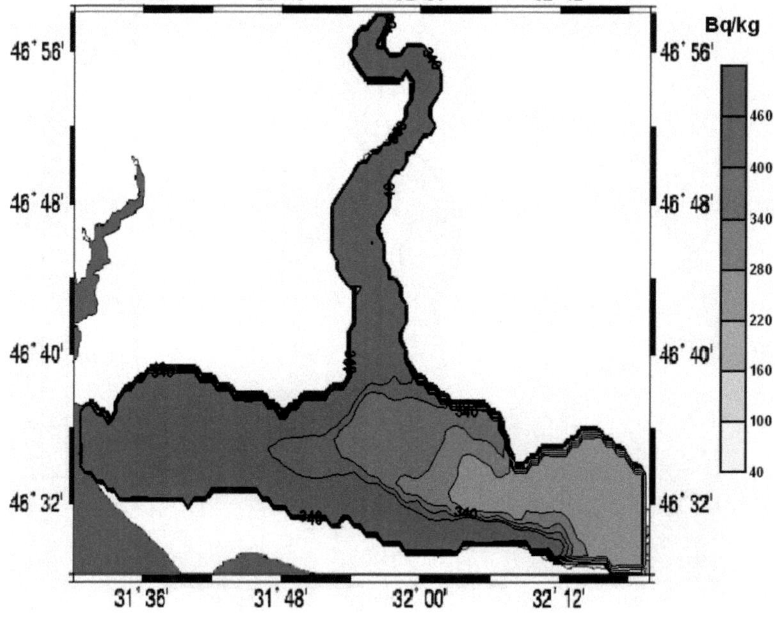

Figure 3.22. Simulated concentration of ^{137}Cs in bottom sediment, June 10, 1986.

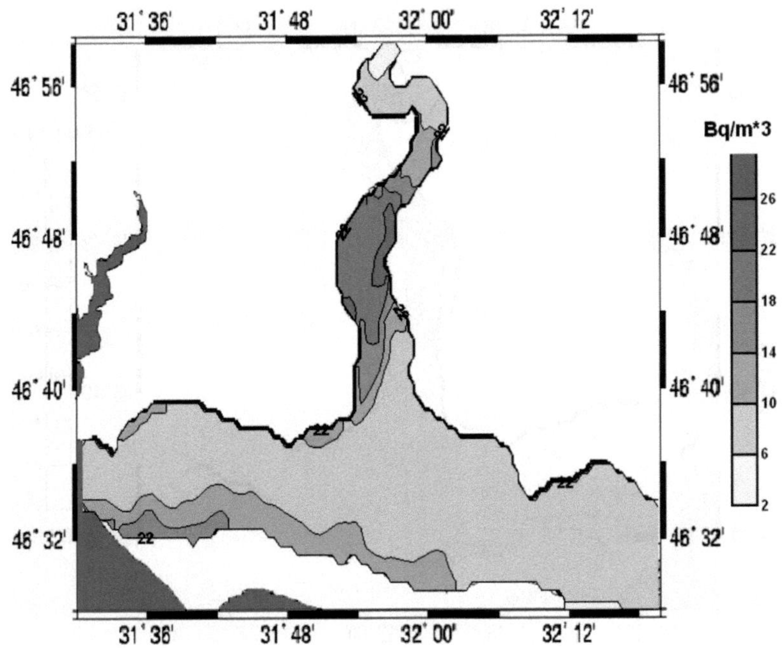

Figure 3.23. Simulated concentration of ^{137}Cs in water at the surface, June 10, 1987.

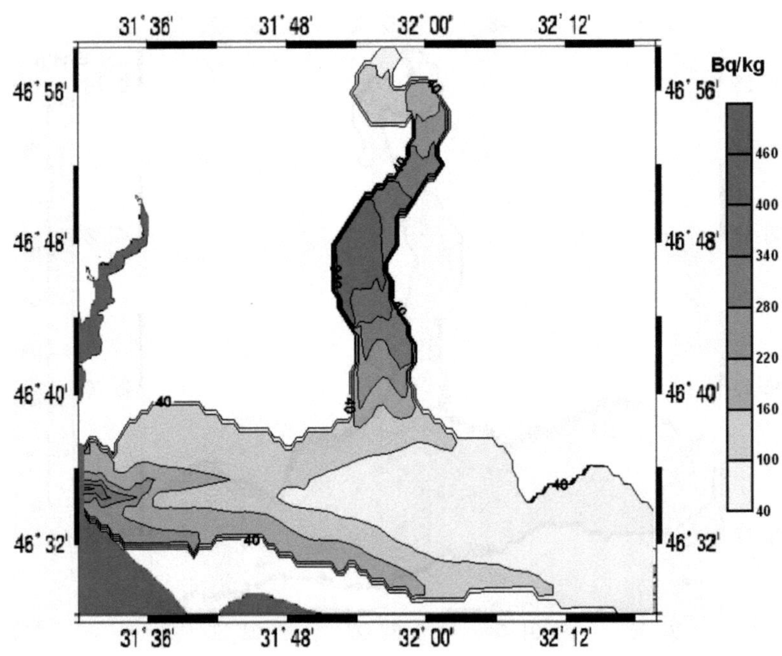

Figure 3.24. Simulated concentration of ^{137}Cs in bottom sediment, June 10, 1987.

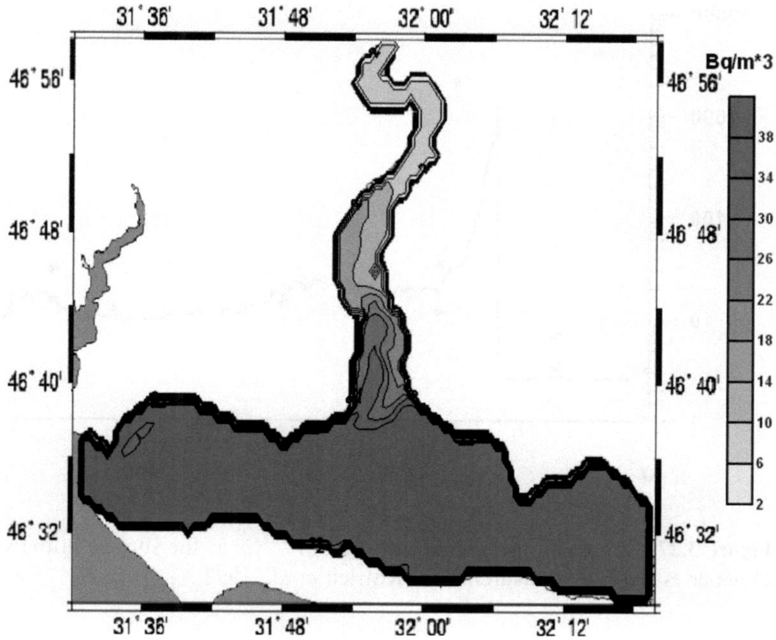

Figure 3.25. Simulated concentration of ^{90}Sr in water at the surface, June 10, 1986.

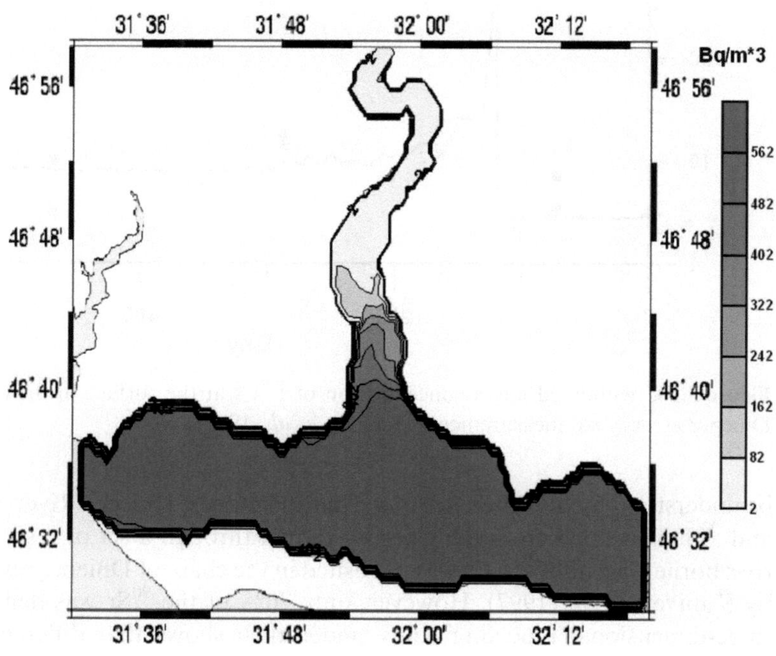

Figure 3.26. Simulated concentration of ^{90}Sr in water at the surface, June 10, 1987.

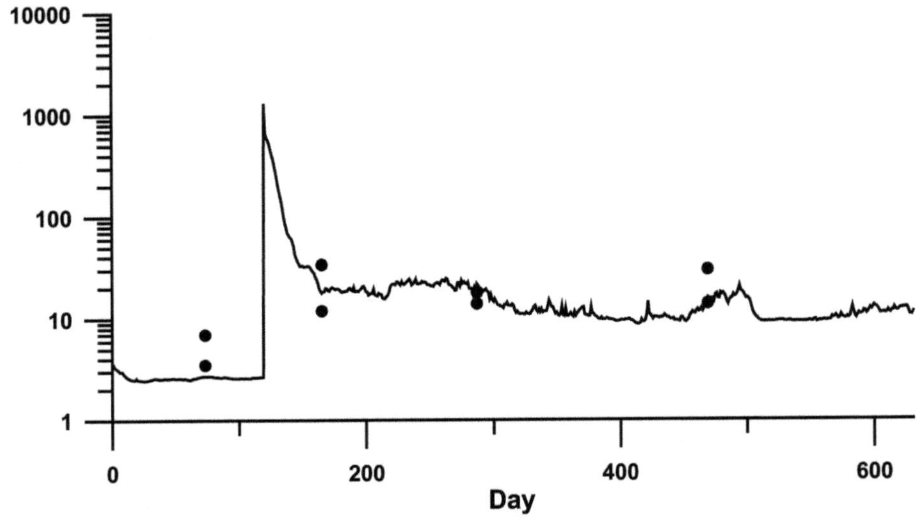

Figure 3.27. Simulated mean concentration of ^{137}Cs at the surface in the western part of the Dnieper estuary vs. measurements (Katrich *et al.*, 1992).

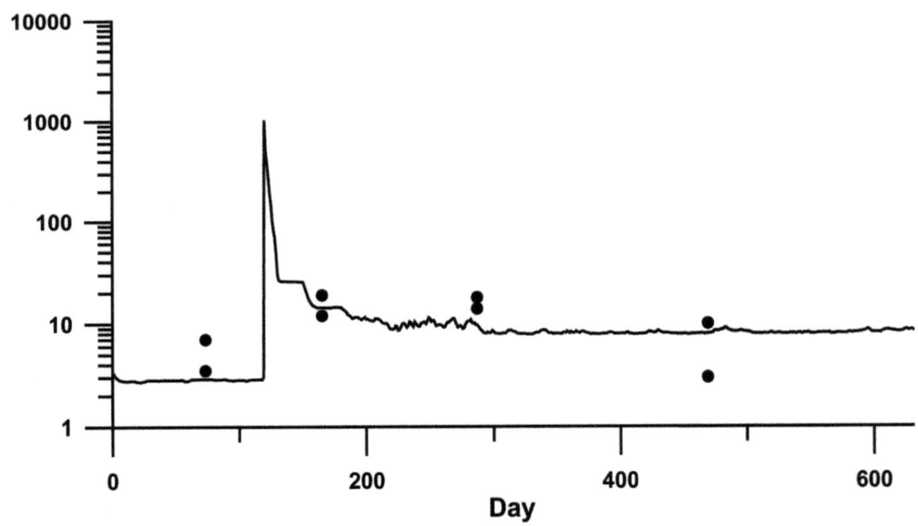

Figure 3.28. Simulated mean concentration of ^{137}Cs at the surface in the eastern part of the Dnieper estuary vs. measurements (Katrich *et al.*, 1992).

be understood by the difference in self-purification of Dnieper River water from ^{137}Cs and ^{90}Sr that flows from the Chernobyl area through a set of reservoirs. Nearly all riverborne Chernobyl ^{137}Cs was deposited in the chain of Dnieper reservoirs, as noted by Kanivets *et al.* (1997). However, only 70% of the ^{90}Sr was deposited here. The three-dimensional modeling study undertaken shows that differences in the total

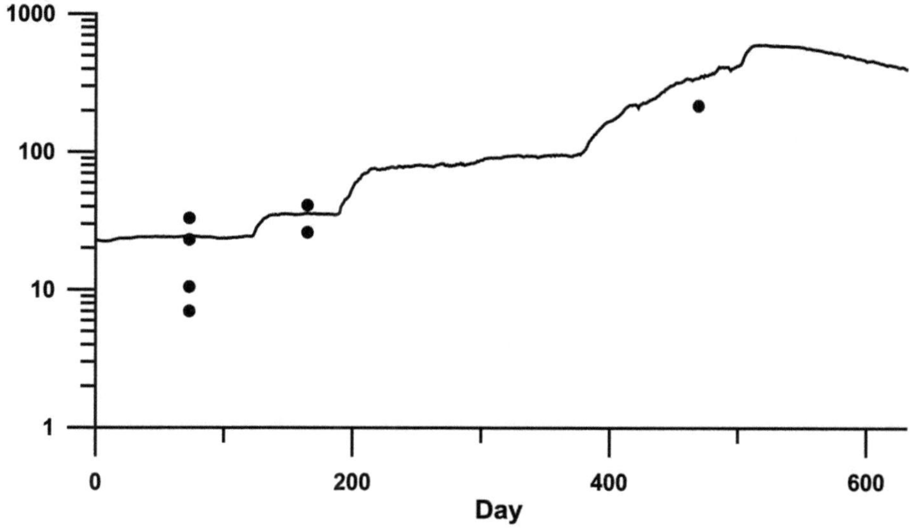

Figure 3.29. Simulated mean concentration of ^{90}Sr at the surface in the western part of the Dnieper estuary vs. measurements (Katrich *et al.*, 1992).

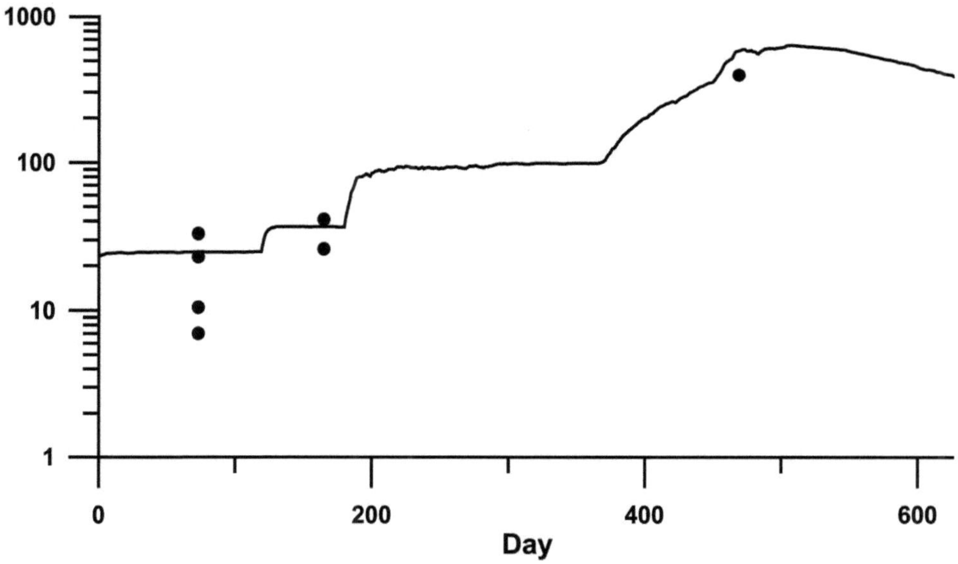

Figure 3.30. Simulated mean concentration of ^{90}Sr at the surface in the eastern part of the Dnieper estuary vs. measurements (Katrich *et al.*, 1992).

fluxes and distribution of radionuclides having mid-magnitude (^{137}Cs) and low-magnitude (^{90}Sr) distribution coefficient values can be quantified on the basis of the chosen schematization of sediment transport and of radionuclide–sediment exchange processes.

4

Studies of potential radioactive spread in the Nordic Seas and Arctic using the generic model system (GMS)

This chapter is dedicated to study of the spread of radioactivity in the Arctic using the generic model system (GMS) described in Chapter 3. Two sets of numerical experiments were carried out: (1) simulations or "hindcasts" of *past* contamination by anthropogenic radionuclides, originating from nuclear bomb testing, atmospheric fallout from Chernobyl, discharges from the Sellafield Reprocessing Plant, and radionuclide transport by river from nuclear plants; (2) simulations of contamination as a result of *potential* accidents in nuclear plants and submarines.

The first series of experiments allowed reconstruction of radionuclide pollution of the Ob′ and Yenisei Rivers and the flux of radioactivity to the Kara Sea from nuclear plants in 1949–1993 and 1958–1994, respectively. The GMS model chain was applied to simulated pathways of radioactivity in 1950–1994 from different sources in the Arctic and North Atlantic. The simulation provides a detailed distribution and evolution of radionuclides over the integration time. The model chain was validated against the data available in this period.

In the second series of experiments, the most probable scenarios of accidental release of radionuclides influencing the Arctic environment and their transport from external sources to the Arctic marine environment are considered, with the objective of choosing and evaluating various realistic scenarios for numerical modeling of hypothetical situations at selected Russian industrial plants and submarines resulting in significant radioactivity release into the environment. The scenarios of hypothetical accidental radionuclide release into the environment from Russian radiochemical plants—the Mayak Production Association (the Mayak PA), Krasnoyarsk Mining and Chemical Combine (MCC), and Tomsk Siberian Chemical Combine (SCC)—are considered. Five different situations resulting in radionuclide pollution to rivers were simulated using GMS to predict or forecast their spread in the Arctic for the present-day scenario and a CO_2-doubling global warming scenario ($2 * CO_2$). In addition, two scenarios based on accidents onboard the submarines *Kursk* and *K-159* were simulated to show the capability of using GMS as an operational tool in decision support systems.

4.1 SIMULATION OF PAST CONTAMINATION OF THE NORDIC SEAS AND ARCTIC FROM ANTHROPOGENIC RELEASES

4.1.1 River and estuary transport and dilution of radioactive pollutants from rivers to the Kara Sea

The long-term transport and fate of radionuclides released by nuclear plants in the basins of the Ob′ and Yenisei Rivers depend on river and estuary hydrology and sediment transport. Because of the large scale of these river systems and sorption on sediments, radionuclides can take several decades to be transported through the entire system (Paluszkiewisz *et al.*, 2001). This section addresses the long-term modeling of the pathways of transport of radionuclides to the Kara Sea from the Mayak Production Association (the Mayak PA) located in the Ob′ basin and from the Krasnoyarsk Mining and Chemical Combine (MCC) situated on the bank of the Yenisei River; these sources of radionuclides are considered in Chapter 1 (see also NATO, 1998). Some aspects of the problem have been considered in papers by Paluszkiewisz *et al.* (1997, 2001), Novitsky *et al.* (1999), and Mokrov (2000). A map of these river systems is given in Figure 4.1. The hydrodynamic regime of the Ob′ and Yenisei estuaries has been simulated by Ivanov and Sviyatsky (1987), Doronin and Ivanov (1997), and Harms *et al.* (2000), primarily to study saltwater intrusions that depend on river discharge variability and seasonal factors. Here a chain of river estuary models RIVTOX–THREETOX was used to reconstruct the flux of ^{90}Sr and ^{137}Cs through the Ob′ River from activities at the Mayak PA for the period 1949–1994 and through the Yenisei River from activities at the MCC for the period 1959–1994. Preliminary results of the simulations have been published in papers by Johannessen *et al.* (2002) and Maderich *et al.* (2005).

Reconstruction of past contamination of the Ob′ River by the Mayak PA nuclear plant

The RIVTOX–THREETOX model chain was customized for the Ob′ basin in order to simulate the transport of radioactivity towards the Arctic Ocean from the Mayak PA located on the Techa River and the SCC on the Tom River (Figure 4.1). Simulations of *potential* radionuclide transport from the SCC and the Mayak PA are presented later (Section 4.3), whereas results of the reconstruction of *past* contamination are considered here.

The Ob′ River and Ob′ Bay are described in detail in Section 2.1.3, and so here we simply confine ourselves to supplementing that description with information pertinent to model customization. The Ob′ hydrology has strong seasonal changes in runoff (Figure 4.2), which has an important effect on contamination transport. The annual average outflow from the Ob′ River to the Kara Sea is about 395 km^3, with a minimum outflow of 268 km^3 and a maximum of 561 km^3 (Ivanov and Osipova,

Figure 4.1. Pathways of radionuclide contamination from PA Mayak, SCC, and MCC to the Kara Sea.

1972). Dilution is therefore strongly dependent on season. The interannual amplitude of mean annual values of Ob′ River runoff is also large (see Figure 4.3) and reaches up to 30% to 40% of the mean (Ivanov and Osipova, 1972), thereby substantially affecting dilution of transported contaminant in the river.

The pathways of radionuclides in the Ob′ River system can be considered in more detail. The SCC is located along the small Romashka River which is a tributary of the Tom River (segment length 75 km, mean annual discharge $Q = 1{,}075\,\mathrm{m^3/s}$) which discharges into the Ob′ River (2,273 km from the delta). The pathway of radionuclides from the Mayak PA site to the Kara Sea include the Techa River (184 km, $Q = 6.6\,\mathrm{m^3/s}$), Iset River (359 km, $Q = 53\,\mathrm{m^3/s}$), Tobol River (426 km, $Q = 816\,\mathrm{m^3/s}$), Irtysh River (636 km, $Q = 2{,}120\,\mathrm{m^3/s}$), Ob′ River (1,000 km) and Ob′ Bay (750 km). The Ob′ delta consists of several channels that connect the river with the immense (more than 760 km long) estuary—Ob′ Bay. The bathymetry of this shallow estuary is given in Figure 4.4.

In the one-dimensional RIVTOX model, the pathway from the Mayak PA to the Irtysh–Ob′ confluence was divided into nine branches (segment length): the Techa River from the Mayak PA to Muslumovo (a village) (22 km), Muslumovo–Pershino (another village) (135 km), Pershino to the confluence with the Iset River (23 km),

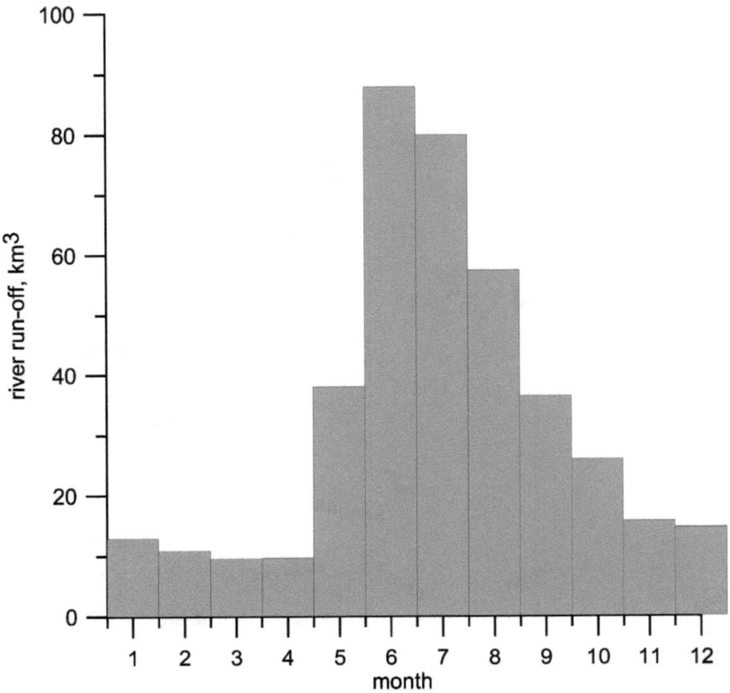

Figure 4.2. Seasonal variability of mean discharge (1934–1995) of Ob′ River.

Techa mouth to Shadrinsk (62 km), Isetskoe–Yalutorovsk (Iset mouth) (105 km), Yalutorovsk–Levlevo (262 km), Irtysh–Tobolsk (Tobol mouth) (3 km), Tobolsk–Khanty–Mansiysk (616 km), and the Khanty–Mansiysk–Irtysh mouth (10 km). The Ob′ River pathway includes part of the Tom River from the SCC to the mouth of the Tom, and the middle and lower parts of the Ob′ from the Tom to the beginning of Ob′ Bay 212 km downstream at Salekhard, with the main tributaries considered as internal boundary conditions.

The Ob′ River from Tom to Ob′ Bay was divided into eight branches to comply with the locations of observation points and the main tributaries such that every observation point is the beginning of the corresponding branch (except the seventh branch, the beginning of which is the mouth of Severnaya Sos′va). The last branch includes Ob′ Bay, which was modeled using both RIVTOX and THREETOX to estimate the accuracy of a one-dimensional model for a shallow and long estuary. Observation points include the Tom at Tomsk, the Ob′ at Mogochin, Kolpashevo, Prokhorkino (in the stretch of Tomsk–Irtysh–Ob′ confluence), and the Ob′ at Belogorye and Salekhard (in the stretch Irtysh–Ob′ confluence–Ob′ Bay) (Akimenko *et al.*, 2001). Because of reduction of the river network in the model realm, the total number of tributaries were reduced to those that are located near the observation points in such a way that their discharges correspond to the sum of their own discharges plus those of all the nearest drop-out tributaries. For numerical simulation, the monthly river discharge values and parameters of water cross-sections for

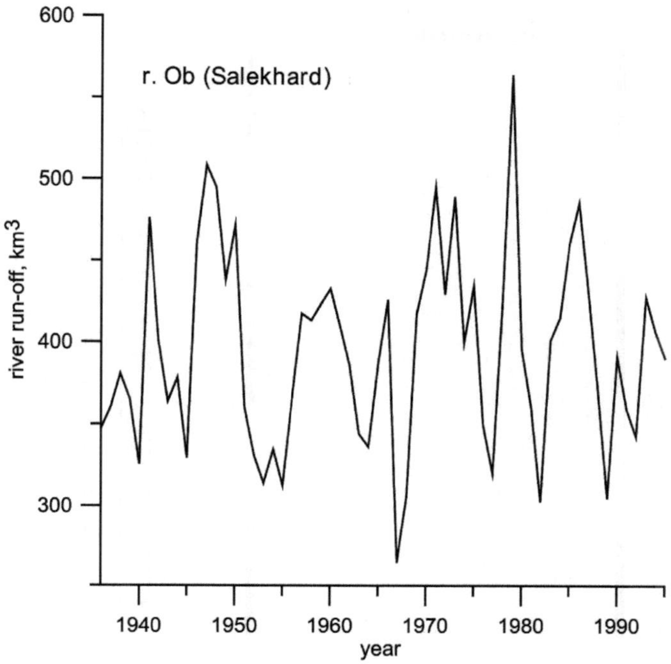

Figure 4.3. Interannual variability of mean year discharge of Ob′ River.

observation points were taken from Akimenko *et al.* (2001a). River cross-sections in the computational points were obtained using linear interpolation between observation points. The length of the river networks Techa–Ob′ and Tom–Ob′ Bay are 1,564 km and 2,348 km, respectively. Model grid resolution is 1 km.

Simulations of the dynamics and radionuclide transport by THREETOX in Ob′ Bay were carried out for two periods: (1) 1993–1994, to compare temperature and salinity fields with data from the KAREX-94 expedition (Vinje and Volkov, 1997), and (2) 1949–1956, to reconstruct contamination as a result of direct discharge of radionuclides from the Mayak PA. The THREETOX model was customized in the bay using the 1-minute bathymetry grid from the General Bathymetric Chart of the Oceans (GEBCO, 2003) database (Figure 4.4). Input meteorological data for THREETOX were air temperature, wind, cloudiness, humidity, and surface pressure, which were interpolated from reanalysis of NCAR/NCEP (Kistler *et al.*, 2002). Ob′ River discharge, sediment concentration, and radionuclide concentration for 1949–1956 were provided by the RIVTOX model, whereas for 1993–1994 observed hydrological parameters were used. Hydrological parameters in the mouth of Ob′ Bay were adopted from the Kara Sea model for simulations 1949–1956 and climatological data and available expedition data for 1993–1994. Temperature and salinity seasonal variations were reconstructed on the base of surveys in 1977–1979 (Johannesen *et al.*, 1998) and KAREX-94 data (Vinje and Volkov, 1997).

Comparison of RIVTOX model results with measurements of hydraulic parameters (discharges and suspended sediment concentration) are given for 1970

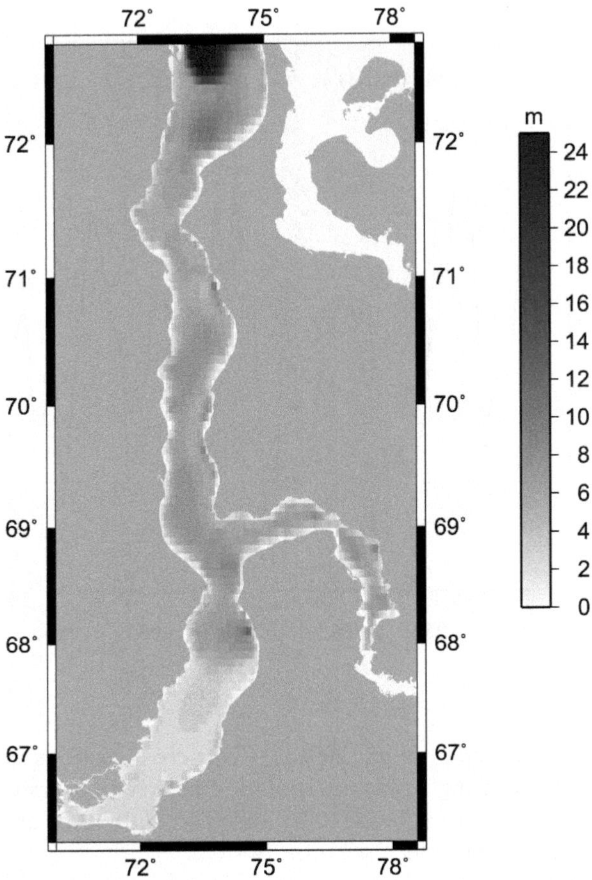

Figure 4.4. Bathymetry of the Ob′ River estuary.

in Figures 4.5–4.8. As seen in these figures the model describes the seasonal dynamics of flow in the Ob′ River system reasonably well. Simulations using THREETOX were carried out for 1993–1994 for comparison with KAREX-94 survey data. The results of comparison between surface temperature and salinity for September 1994 are given in Figure 4.9. Ice thickness and concentration of sediments are given in Figure 4.10. The results from the simulation agree with the KAREX-94 survey data and descriptions of the Ob′ Estuary given by Antonov and Maslyayeva (1965), Zhizhanov (1980), Solovieva and Antonova (1995), and Stanovoy (2006). Estuarine circulation is driven by tidal forcing at the outer boundary, discharge from the Ob′ and Taz-Pur, and wind. Baroclinic effects are small except in the outer part of the estuary. The water in most of this shallow bay is well mixed and nearly as fresh ($S‰ < 4$) in winter as in summer. Suspended sediment concentrations are in the range $0.05\,\mathrm{kg\,m^{-3}}$ to $0.005\,\mathrm{kg\,m^{-3}}$ and decrease steeply from the Ob′ delta to the estuary. Ice thickness changes are in the range 1.2 m to 0.7 m, which is close to the average value of 1 m given by Solovieva and Antonova (1995).

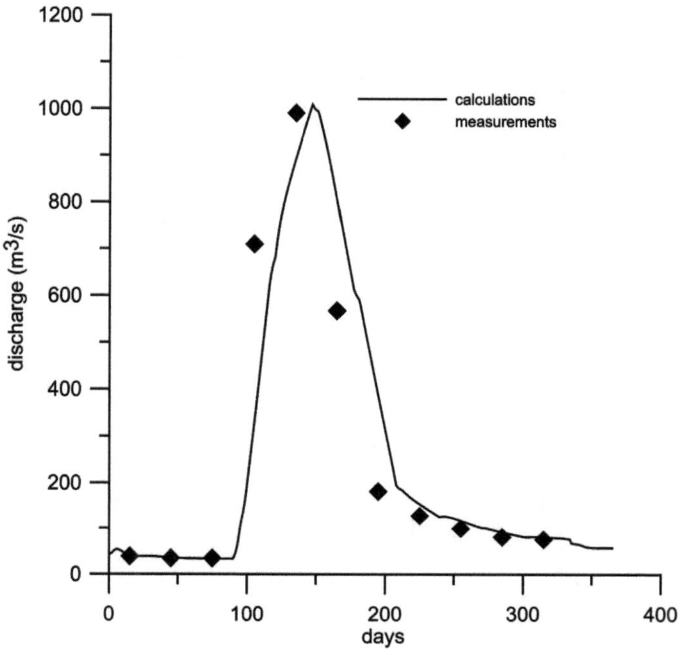

Figure 4.5. Simulated vs. observed Tobol River discharge in Yalutorovsk in 1970.

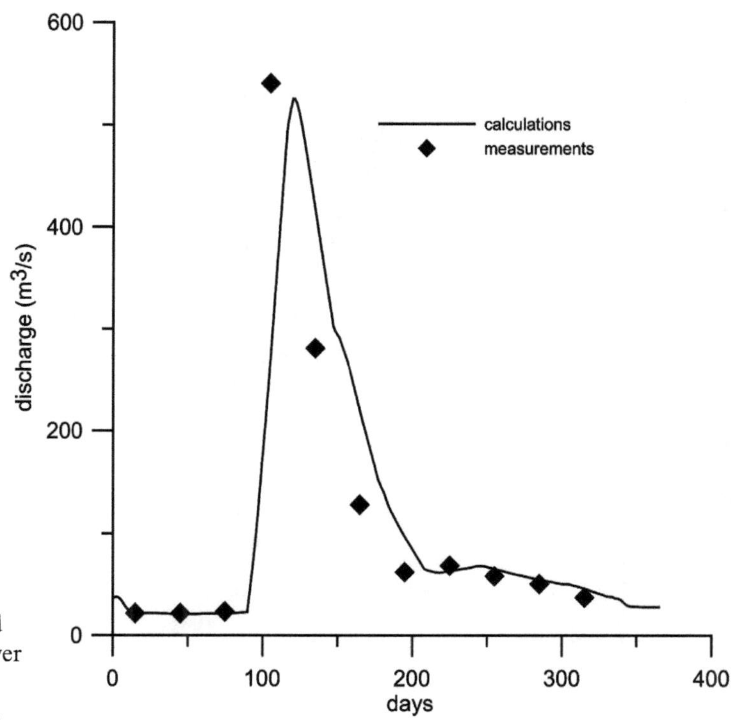

Figure 4.6. Simulated vs. observed Iset′ River discharge in Mekhonskoe in 1970.

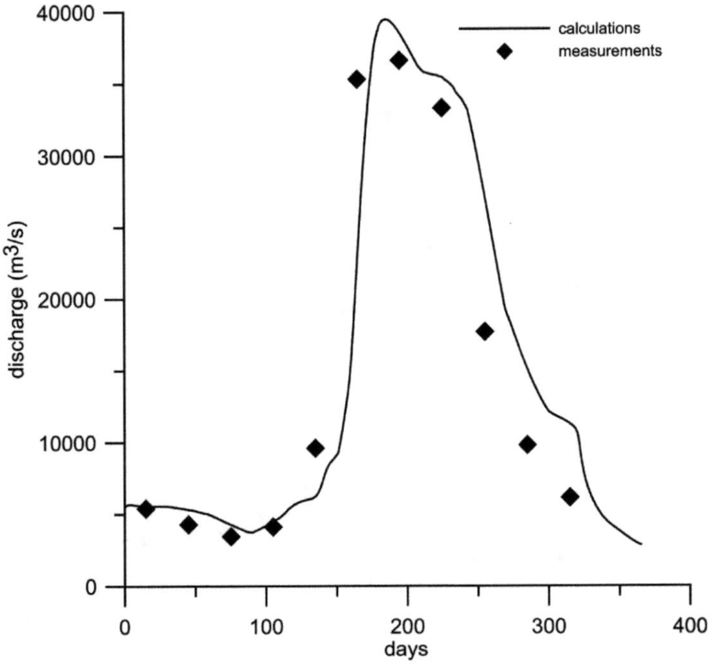

Figure 4.7. Simulated vs. observed Ob′ discharge in the Salekhard in 1970.

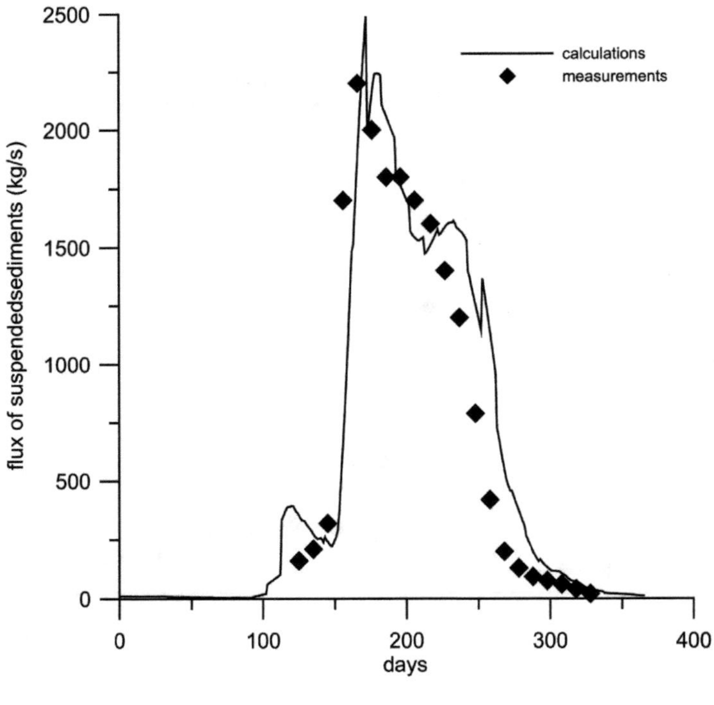

Figure 4.8. Simulated vs. observed suspended sediment concentration in the Salekhard in 1970.

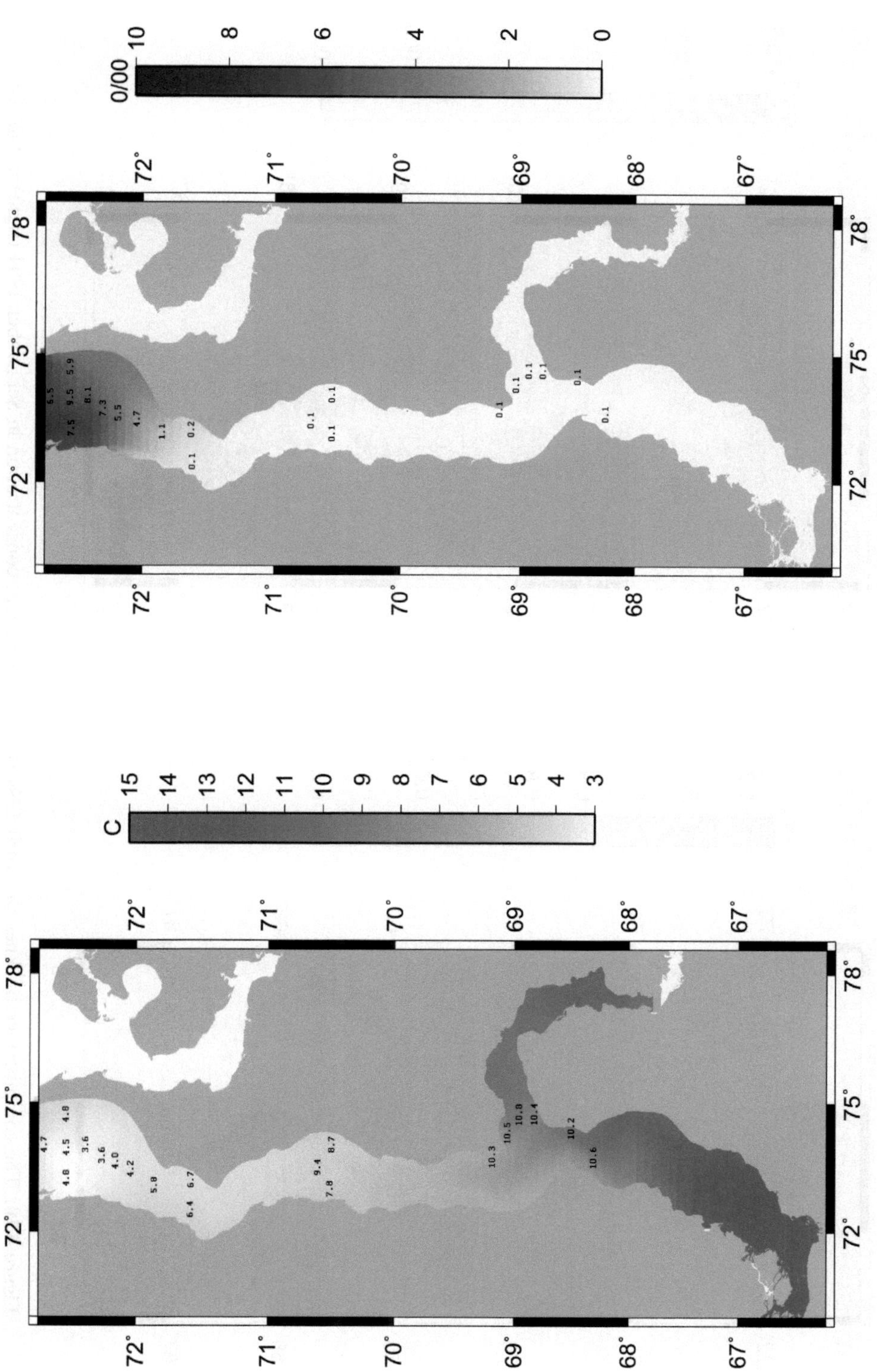

Figure 4.9. Computed surface fields of temperature (left) and salinity (right) vs. KAREX-94 survey data in Ob′ Bay on September 6, 1994.

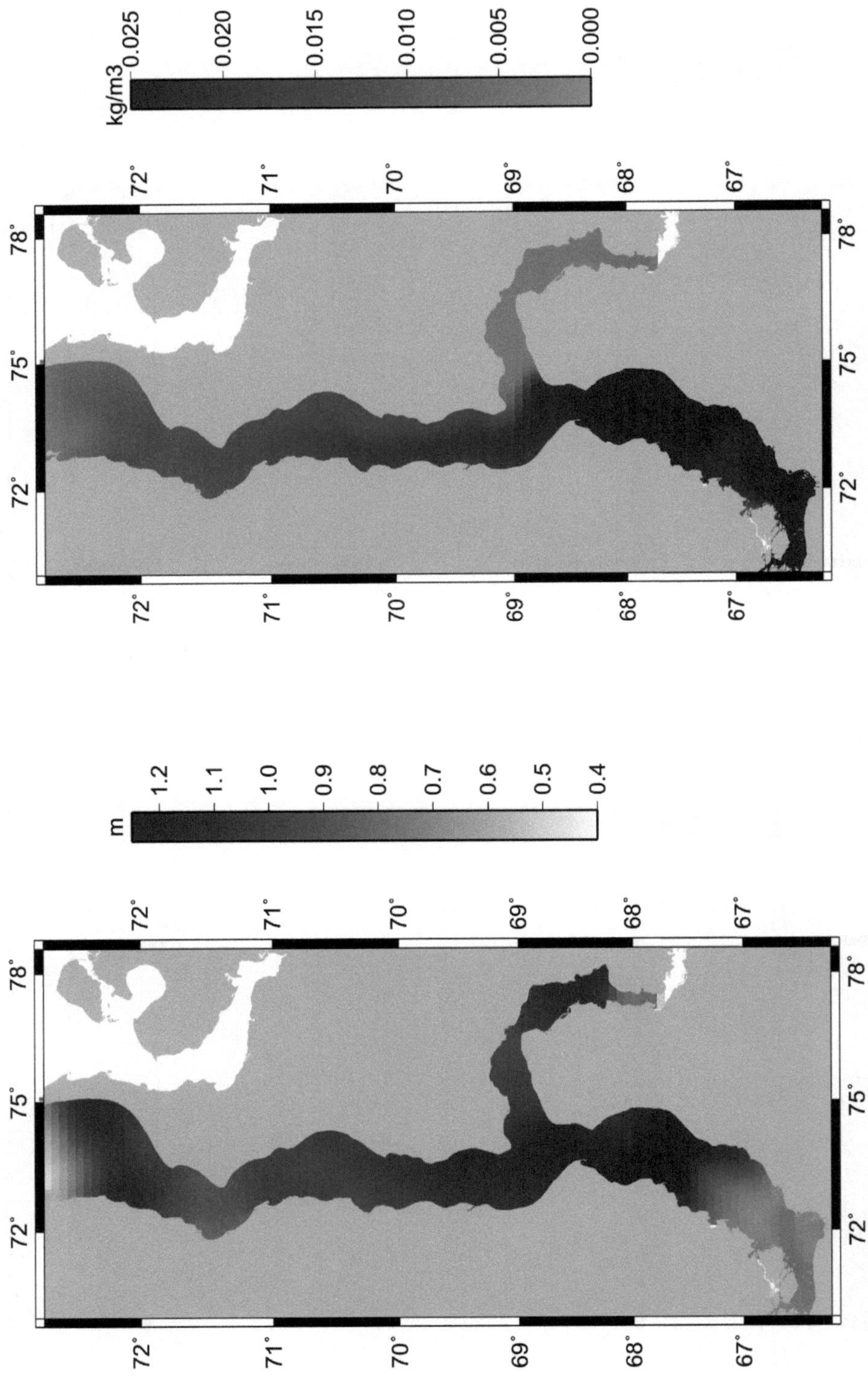

Figure 4.10. The computed ice thickness in April 1994 (left) and sediment concentration in September 1994 (right) in Ob′ Bay.

Table 4.1. Assessment of radionuclide discharges into the Techa River in 1949–1956 (Degteva *et al.*, 1994).

Year	^{90}Sr (GBq/yr)	^{137}Cs (GBq/yr)
1949	1.9×10^5	3.0×10^5
1950	6.0×10^6	6.4×10^6
1951	6.2×10^6	6.5×10^6
1952	9.5×10^4	3.5×10^4
1953	2.0×10^4	7.4×10^3
1954	8.1×10^3	3.0×10^3
1955	5.0×10^3	1.8×10^3
1956	1.3×10^4	4.8×10^3

Simulations of the flux of ^{90}Sr and ^{137}Cs through the Ob′ River and estuary from activities at the Mayak PA for the period 1949–1994 were carried out using available data on radionuclide sources. Radionuclide fluxes from the Mayak PA into the Techa River were prescribed according to Tables 4.1 and 4.2. Table 4.1 presents data on radionuclide fluxes for the period 1949–1956 (Degteva *et al.* 1994; Kryshev *et al.*, 1997, 1998) whereas Table 4.2 presents data on radionuclide fluxes for the period 1962–1994 compiled from different sources.

The fluxes of ^{90}Sr in 1962–1965 and 1976–1994 were given by JNRI (1997). Data on the annual concentration of ^{90}Sr and ^{137}Cs in Pershino and Muslumovo (Kryshev 1997; Kryshev *et al.*, 1998) were used to reconstruct the flux of radionuclides into the Techa River for the period 1957–1961. Data for 1966–1975 are missing; so, for this period the average of 1965 and 1976 (991 GBq/yr) was used. Data for 1977 were taken as 630 GBq/yr (Gedeonov *et al.*, 1998) instead of 15.5 GBq/yr from Table 4.2. Data on the annual concentration of ^{137}Cs in Muslumovo (27 km from the Mayak PA along the Techa River) and Pershino (162 km from the Mayak PA) (Kryshev, 1997, 1998) were also used to reconstruct the flux of ^{137}Cs in 1962–1994 (see Table 4.2). The distribution coefficients K_d^w and K_d^b in the RIVTOX model were $0.25\,m^3\,kg^{-1}$ and $0.8\,m^3\,kg^{-1}$ for ^{90}Sr and $3\,m^3\,kg^{-1}$ and $15\,m^3\,kg^{-1}$ for ^{137}Cs, respectively. The values of exchange rate were $a_{12}^w = 1\,da^{-1}$, $a_{12}^b = (100\,da)^{-1}$ for ^{90}Sr and $a_{12}^w - 1\,da^{-1}$, $a_{12}^b = (333\,da)^{-1}$ for ^{137}Cs, respectively.

Figures 4.11–4.17 present the contribution of the Mayak PA to ^{90}Sr and ^{137}Cs contamination of the Ob′ River system and the Kara Sea. In these figures the calculations are compared with ^{90}Sr measurements in Muslumovo and Pershino. As seen from Figure 4.11, the RIVTOX model reproduces observed concentrations of ^{90}Sr in Muslumovo and Pershino (Kryshev, 1997) quite well. Comparison between

Table 4.2. Assessment of radionuclide discharge into the Techa River in 1962–1994.

Year	^{90}Sr (GBq/yr)	^{137}Cs (GBq/yr)
1962	137	4,950
1963	914	—
1964	2,535	297
1965	1,690	—
1973	—	49.5
1976	292	—
1977	15.5	—
1978	996	42.9
1979	1,595	—
1980	797	—
1981	377	—
1982	203	—
1983	183	29.7
1984	288	—
1985	274	—
1986	348	—
1987	629	—
1988	539	49.5
1989	780	—
1990	356	16.5
1991	549	—
1992	512	—
1993	1,004	—
1994	907	—

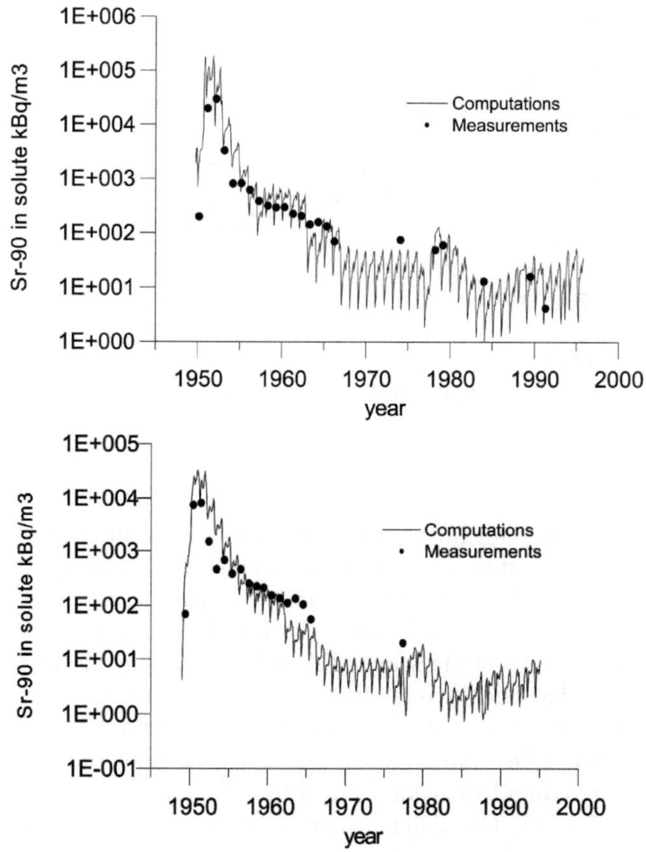

Figure 4.11. Simulated vs. observed (Kryshev, 1997) ^{90}Sr concentration in solute in Muslumovo (upper) and Pershino (lower).

the measured (Chumichev, 1995) and calculated concentration of ^{90}Sr in Salekhard in Figure 4.12 shows that after 1960, the Mayak PA input to river contamination was much less than global fallout. The observed ^{90}Sr concentration in the Lena River (Chumichev, 1995) is shown in the figure as a rough estimate of the global fallout effect. "Reconstructed" concentration in the Salekhard is represented in Figure 4.12 as the sum of calculated and measured concentrations in the Lena River. It is closer to the observed curve but the observed level of contamination is higher, which can be explained by the difference between the Ob′ and Lena River basins in terms of global fallout. The surface distribution of ^{90}Sr in solute and the concentration of ^{90}Sr in the bottom of Ob′ Bay in the summer of 1954 is presented in Figure 4.13. This corresponds to the period in 1953–1954 of maximal contamination of Ob′ Bay that exceeded 13,000 Bq m^3 in solute in the Ob′ delta. The figure also shows clean water from the Taz River that dilutes Ob′ water. RIVTOX/THREETOX–computed mean annual values of ^{90}Sr cross-sectionally averaged concentration in the mouth of Ob′ Bay are shown in Figure 4.14. Figure 4.15 shows the correlation between cross-sectionally averaged concentration of ^{90}Sr calculated by THREETOX in 1949–

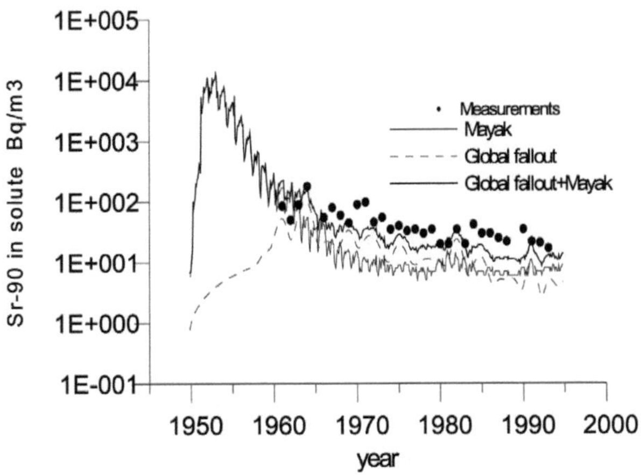

Figure 4.12. Simulated vs. observed (humichev, 1995) ^{90}Sr concentration in solute in Salekhard.

1956 and relevant values of the concentration calculated by RIVTOX. The figure suggests that in a relatively long and shallow estuary, the one-dimensional RIVTOX model can be applied to the entire estuary.

Figures 4.16–4.17 show the RIVTOX-calculated yearly averaged fluxes of ^{90}Sr and ^{137}Cs from the Mayak PA and the Ob′ Bay mouth to the Kara Sea. These results show that after strong initial contamination in the early 1950s, sediments in the Ob′ River were sources for secondary contamination of the river and estuary. The immense Ob′ estuary served as a reservoir of radionuclides with a large residence time. Around 15% and 79% of the ^{137}Cs and ^{90}Sr, respectively, released in 1949–1956 reached the Ob′ delta, whereas for the period 1949–1968 about 61% and 82% of released ^{137}Cs and ^{90}Sr, respectively, entered the estuary. However, only 1.3% and 64% of the ^{137}Cs and ^{90}Sr, respectively, released in 1949–1956 reached the Kara Sea, whereas for the period 1949–1968 about 13% and 76% of released ^{137}Cs and ^{90}Sr, respectively, entered the sea. These estimates do not include fluxes related to global atmospheric fallout, which is a dominant source of ^{137}Cs in the bottom sediments of Ob′ Bay (Sayles *et al.*, 1997).

Reconstruction of past contamination of the Yenisei River by the MCC nuclear plant

The RIVTOX–THREETOX model chain was customized for the Yenisei River to simulate the transport of radioactivity from the Krasnoyarsk MCC to the Arctic Ocean. The Yenisei River is described in detail in Section 2.1.3, and so we confine ourselves here to simply supplement that description with data pertinent to model customization. The main portion of river runoff is transported during the spring–summer period (i.e., more than 78% of the total volume of runoff occurs between May and September; Figure 4.18). Interannual variability in the mean annual values of Yenisei River runoff is less than for the Ob′ River, reaching 15% or even 17% from the mean value (Figure 4.19). The river is ice-bound for more than half the year, and ice dams can cause major flooding.

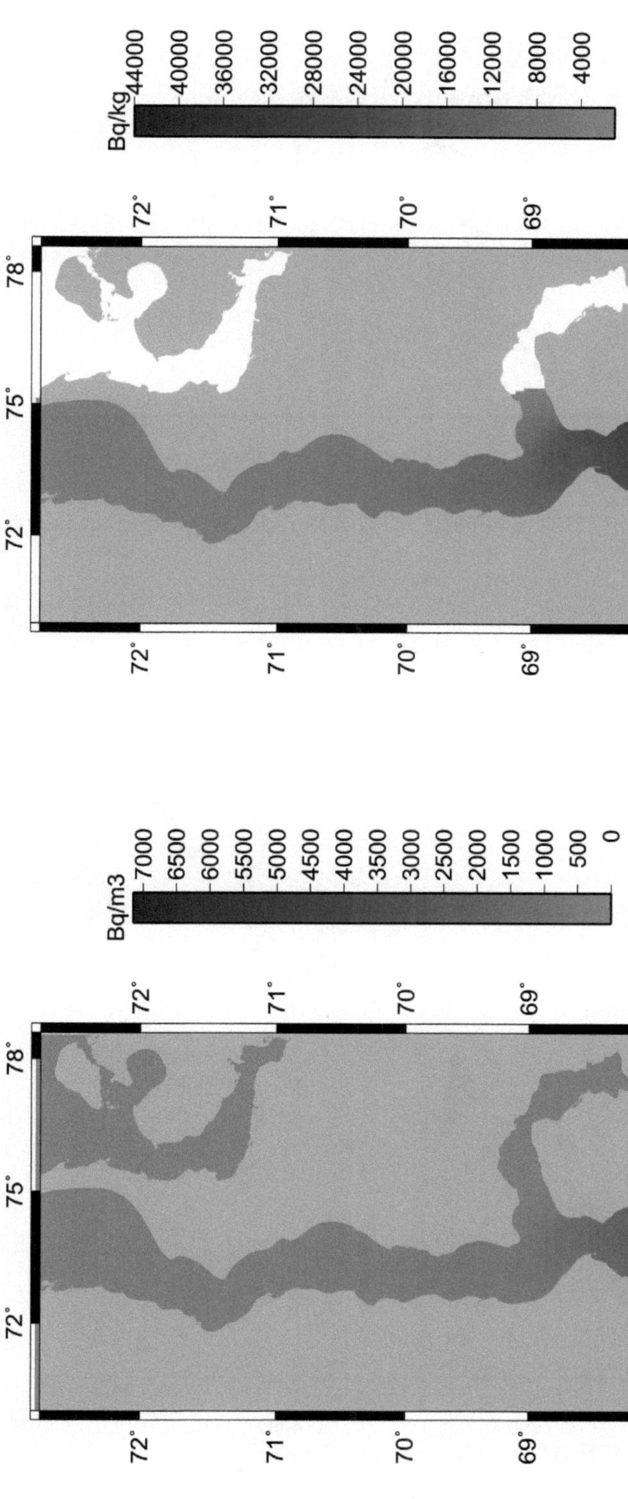

Figure 4.13. The simulated surface distribution of ^{90}Sr in solute (left) and concentration of ^{90}Sr in the bottom (right) of Ob′ Bay in summer 1954.

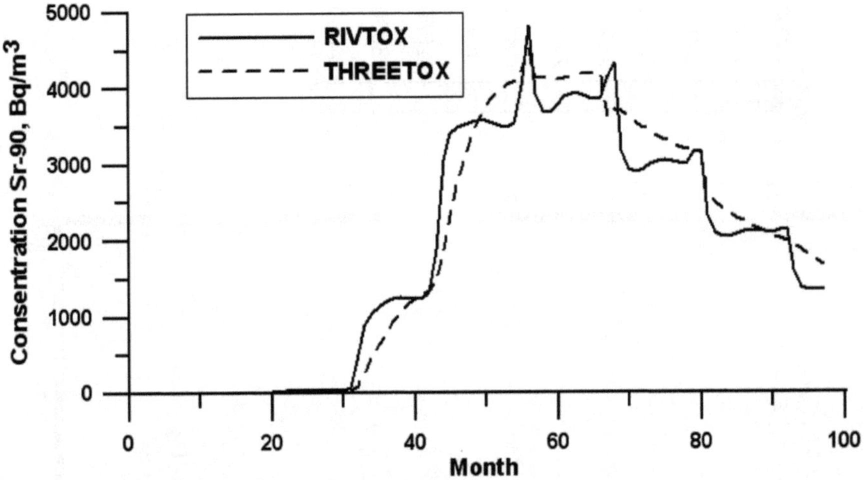

Figure 4.14. RIVTOX- and THREETOX-simulated ^{90}Sr concentration in solute in the mouth of Ob′ Bay.

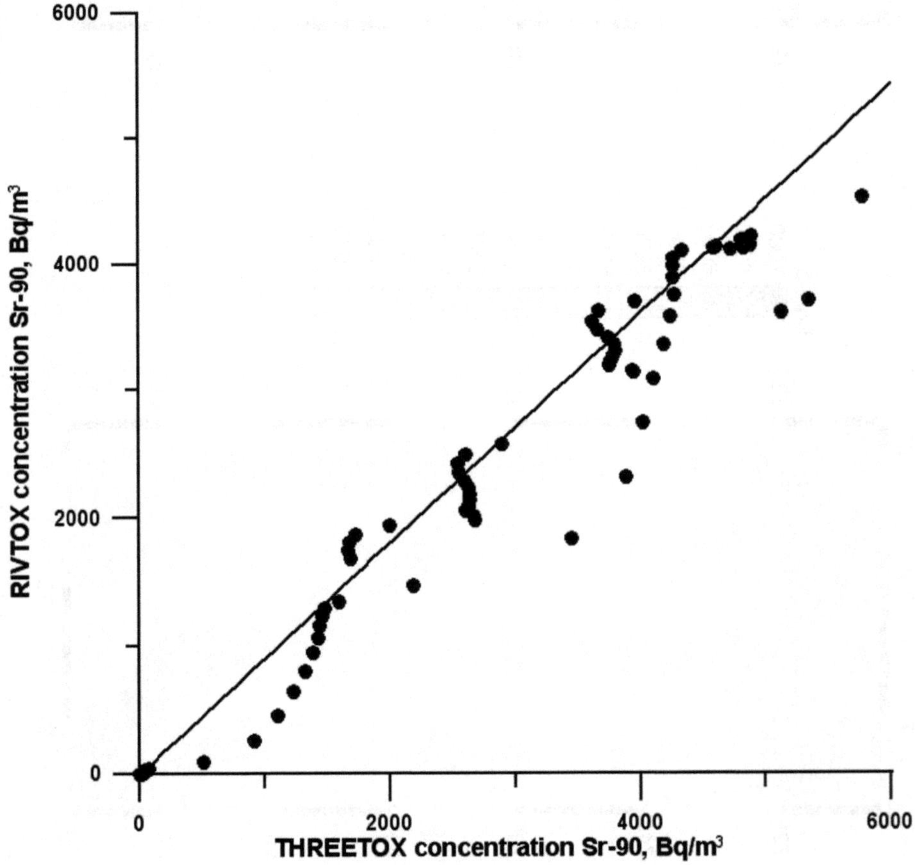

Figure 4.15. Correlation between cross-section–averaged flux of ^{90}Sr in the mouth of Ob′ Bay simulated by RIVTOX and THREETOX.

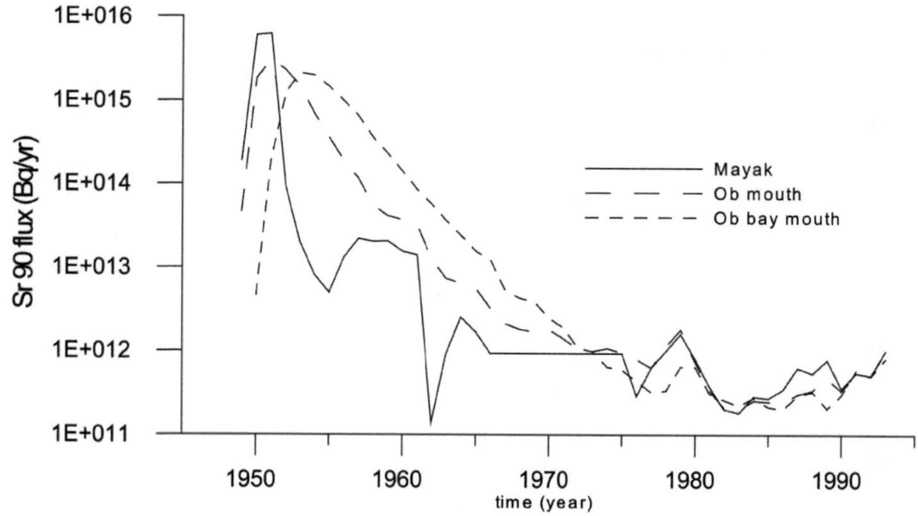

Figure 4.16. RIVTOX-simulated fluxes of ^{90}Sr from PA Mayak through the Ob′ River and Ob′ Bay 1949–1994.

The Yenisei River flows into Yenisei Gulf (Figure 4.18), which is about 225 km long (from Sopochnaya Karga cape to Dikson island), and has a surface area of 9,950 km^2 (Antonov, 1962). The width of the funnel-shaped gulf varies from 6 km to 7 km seaward from the Dorofeevsky cape to 120 km in the mouth of this estuary. The inner part of the estuary is separated by sill-formed sandbanks near Sopochnaya Karga. From the sill, depths gradually grow seaward reaching values of 30 m to 40 m. Fine silt is also found in the gulf. In contrast to Ob′ Bay, Yenisei Gulf has a clearly pronounced two-layer structure of water masses, with a sharp halocline and summer

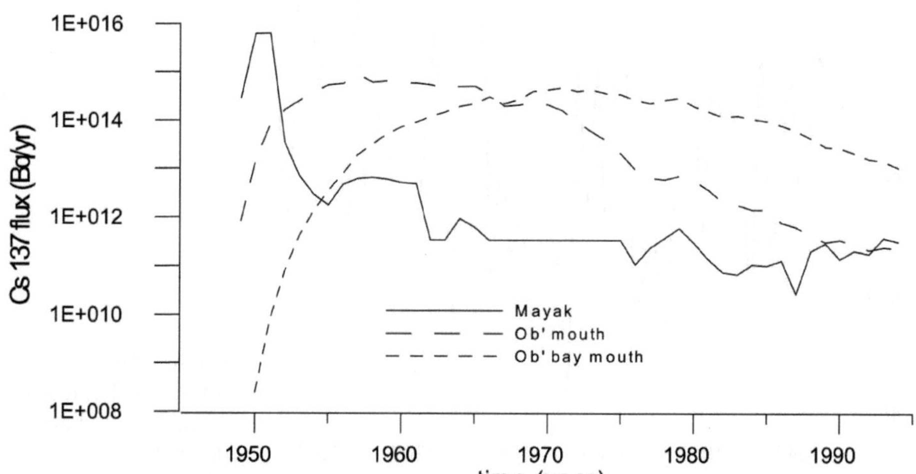

Figure 4.17. RIVTOX-simulated fluxes of ^{137}Cs from PA Mayak through the Ob′ River and Ob′ Bay 1949–1994.

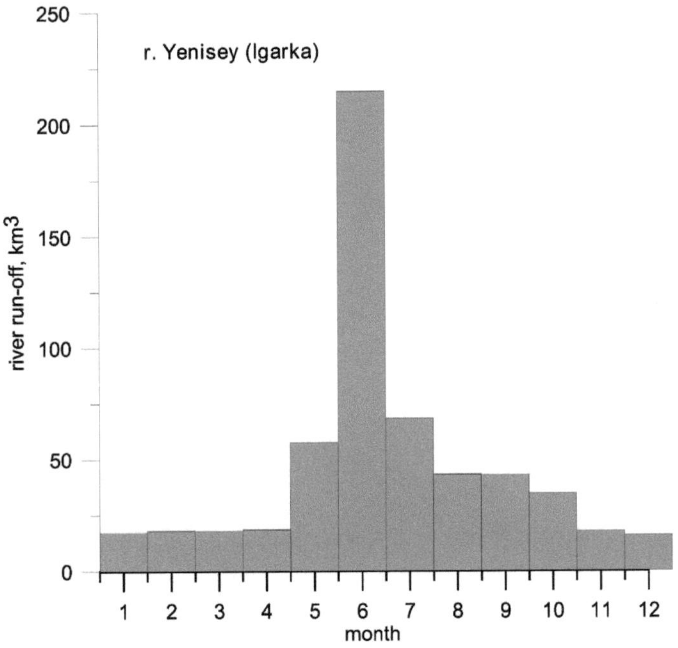

Figure 4.18. Seasonal variability of mean discharge of the Yenisei River, 1934–1995.

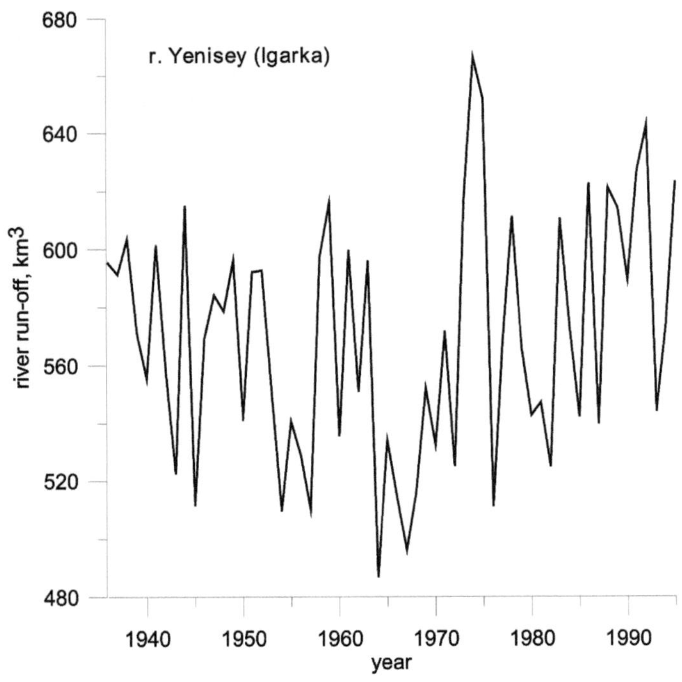

Figure 4.19. Interannual variability of annual discharge of the Yenisei River, 1934–1995.

thermocline. Vertical salinity and temperature gradients may be extremely strong and reach respective values of up to 5–10 per 10 cm and 1.5°C to 2.0°C per 10 cm (Harms *et al.*, 2001). A salt wedge (intrusion) reaches the sill and its position can change under local hydrological conditions. The two-layer structure also exists in winter; it shifts seaward only in June when river discharge is maximal.

The Yenisei River is usually divided into three parts (Akimenko *et al.*, 2001b): upper (from the river's source to the Tuba River), middle (from the confluence of the Tuba to the junction with the Angara), and lower Yenisei (from the Angara junction to the Kara Sea). The upper Yenisei is beyond the scope of this study, as it is located far upstream from the pollution source. Here, modeling the Yenisei River network includes that part of the Yenisei River from Bazaikha, which was taken as a "zero" point, to Dudinka near the beginning of the Yenisei Estuary. The zero point is selected based on the location of the main source of radioactive pollution for the Yenisei River system—the Krasnoyarsk MCC, situated 84 km downstream from Bazaikha, and the presence of the Bazaikha hydrological station. Observation points include Bazaikha, Yeniseysk, Podkamennaya Tunguska, and Igarka (Akimenko *et al.*, 2001b), which provide measurement data along the river from the Krasnoyarsk reservoir to Yenisei Gulf. The monthly river discharge values and parameters of water cross-sections for observation points from Akimenko *et al.* (2001b) were used for numerical simulation. River cross-sections at computational points were obtained by linear interpolation between observations. The river network is divided into seven branches to comply with the location of observation points and the main tributaries, and grid resolution is 1 km. The length of the simulated path along the Yenisei is 2,151 km.

To simulate circulation and radionuclide transport in Yenisei Gulf, the THREETOX model was customized in much the same way as in the Ob′ River case. The computational bathymetry map has 2′ resolution (Figure 4.20). The list of input meteorological data for THREETOX is the same as for Ob′ Bay. These data were processed from NCAR/NCEP reanalysis data; because of the relatively small area of the estuary, only one grid-point value of meteorological parameters was used. Yenisei River discharge, sediment concentration, and radionuclide concentration in water and sediments were provided by the RIVTOX model or by measurement data. Hydrological parameters at the mouth of the Yenisei estuary were adopted from available climatological data and expedition data. Temperature and salinity seasonal variations were reconstructed on the basis of surveys in 1977–1979 and KAREX-94. The amplitude of the dominant M_2 tide was prescribed to be 0.5 m, as given by Voinov (1999).

Validation of the one-dimensional hydrological river model was carried out on the basis of measurements taken along the Yenisei River (Akimenko *et al.*, 2001b). It was assumed that the size of particles in calculations was the same throughout the whole year and equal to a median value of 0.2 mm (Akimenko *et al.*, 2001b). The model describes the seasonal dynamics of discharge in the Yenisei well. Calculated level variations also agree with measurements, except for the Yeniseysk and the Igarka where the model underpredicts the maximum level (whereas mean discharge is predicted well). Comparison of calculated and measured discharge and the water

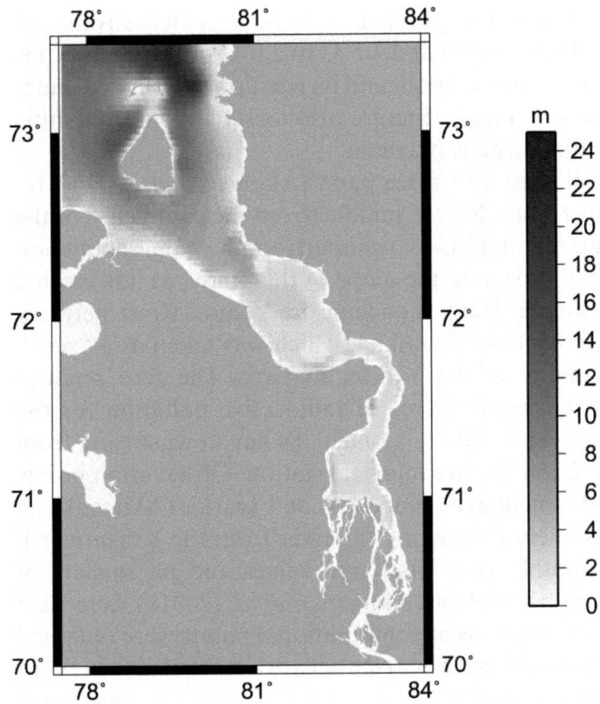

Figure 4.20. Bathymetry of the Yenisei Gulf.

level in Podkamennaya Tunguska for 1985 is given in Figure 4.21. Calculations generally agree with measurements well, given the assumptions that were made.

Calculations of circulation, temperature, and salinity were carried out for 1994. Simulation results agree with observations in the Yenisei estuary given by Antonov (1962), Solovieva and Antonova (1995), and Tretyakov (2001). The results of a comparison of simulations of surface temperature and salinity with measurements for September 1994 are given in Figure 4.22. Estuarine circulation is driven by Yenisei discharge, tidal forcing at the outer boundary of the estuary, and wind. Baroclinic effects are strong in the outer part of the estuary. Ice thickness in April ranges from 1.2 m to 0.7 m (Figure 4.23a), which is close to the value (1 m) given by Solovieva and Antonova (1995). Suspended sediment concentration in September varies between $0.002\,kg\,m^{-3}$ and $0.0001\,kg\,m^{-3}$ and decreases from the Yenisey delta to the estuary (Figure 4.23b).

Validation of the radionuclide transport submodel of RIVTOX was carried out on the basis of data (Kuznetsov *et al.*, 2001) on ^{90}Sr and ^{137}Cs contamination along the Yenisei River caused by KMCC liquid discharge in 1992. Figures 4.24–4.25 show calculated and measured ^{90}Sr and ^{137}Cs concentrations in solute. It should be mentioned that direct data on the discharge of radionuclides from the MCC were absent, and source characteristics were reconstructed using measured ^{90}Sr and ^{137}Cs concentrations at distances of 0.5 km, 5.5 km, and 15 km downstream of the point of discharge. Estimation comes up with a source capacity of $0.1056 \cdot 10^5$ Bq/s for ^{90}Sr

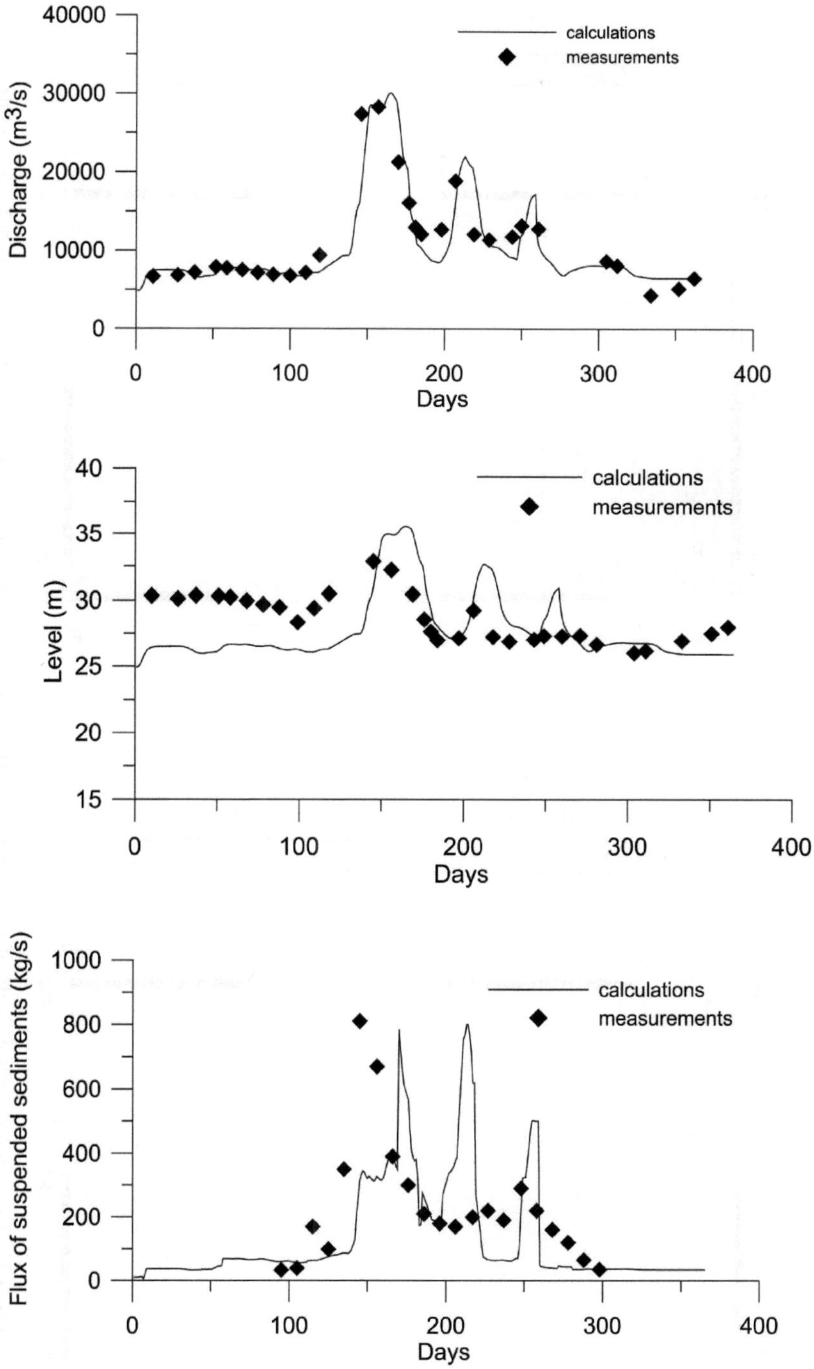

Figure 4.21. Simulated vs. observed Yenisei River discharge (upper), elevation (middle), and suspended sediment discharge (lower) in the mouth of the Podkamennaya Tunguska in 1985.

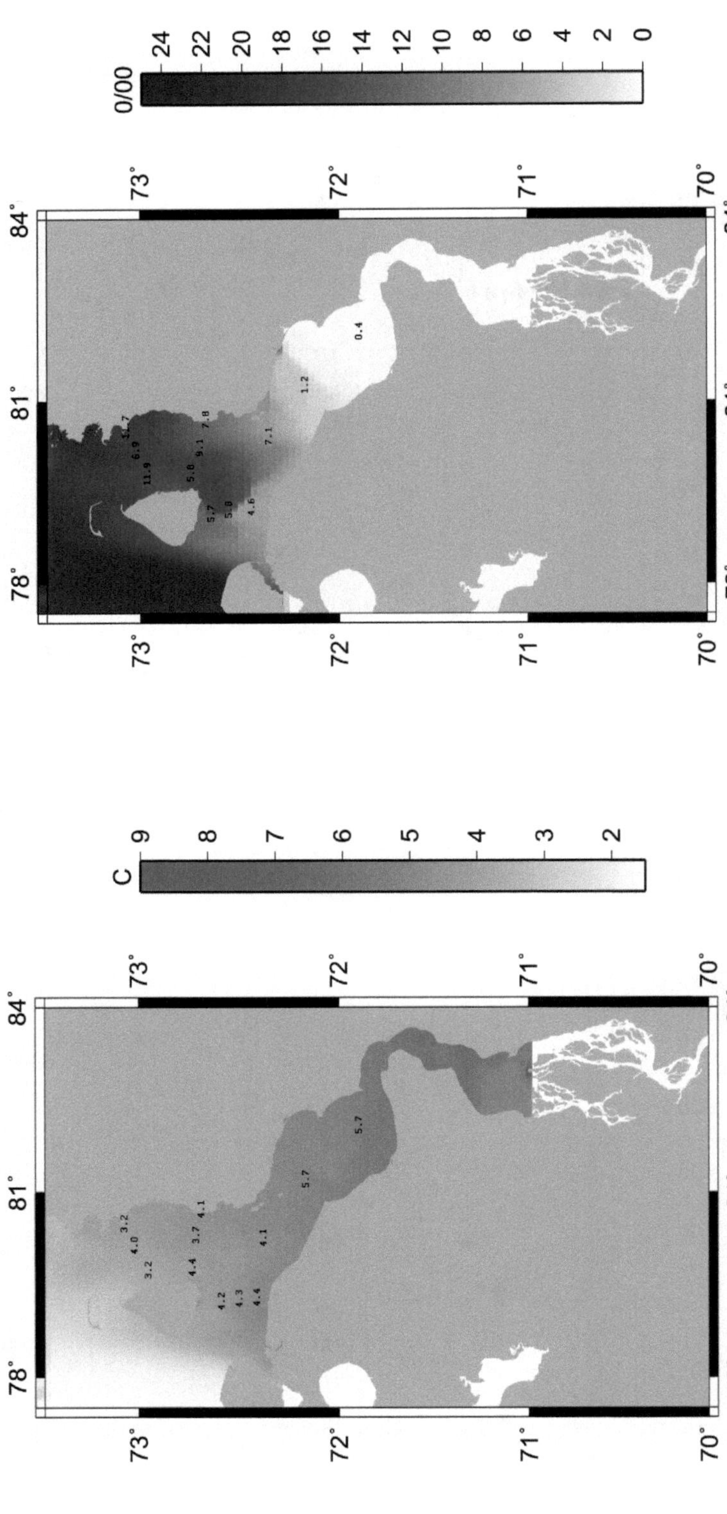

Figure 4.22. Computed surface fields of temperature (left) and salinity (right) vs. KAREX-94 survey data in Yenisei Gulf.

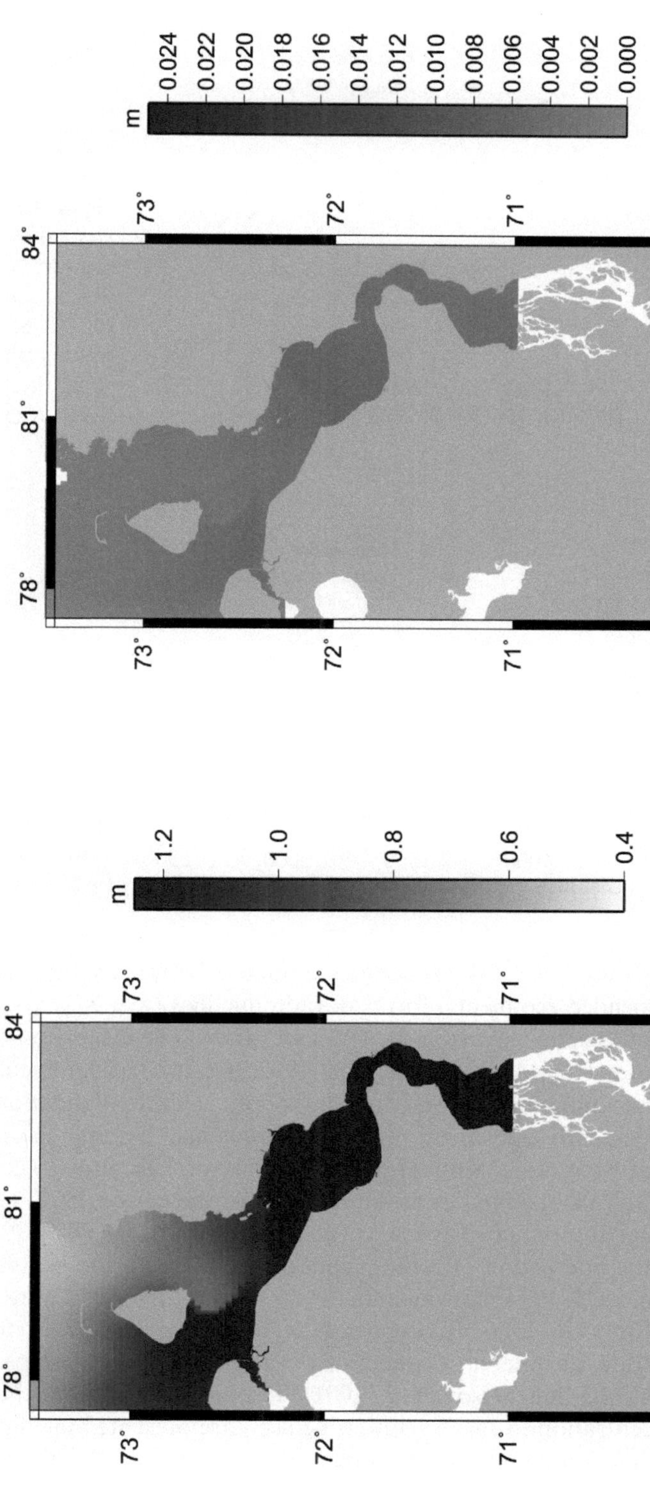

Figure 4.23. Computed ice thickness in April 1994 (left) and suspended sediment concentration in August 1994 (right) in Yenisei Gulf.

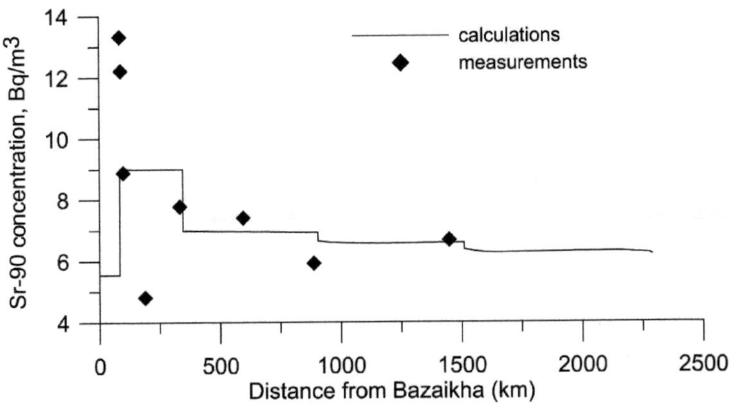

Figure 4.24. Simulated vs. measured (Kuznetsov *et al.*, 2001) ^{90}Sr concentration along the Yenisei River in 1992.

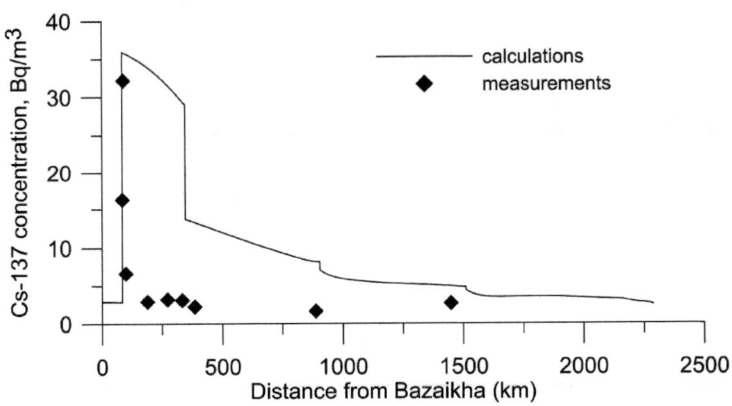

Figure 4.25. Simulated vs. measured (Kuznetsov *et al.*, 2001) concentration of ^{137}Cs along the Yenisei River in 1992.

and $0.6111 \cdot 10^4$ Bq/s for ^{137}Cs. All exchange parameters between soluble and solid (in bottom and suspended sediments) forms of radionuclides (K_{ds}, K_{db}, a_{12}, a_{21}, a_{13}, a_{31}) were chosen in the same way as in the Ob′ case study. The dispersion of radionuclides was calculated from August to October 1992, because measurement data of radionuclide concentrations correspond to this period of time. Validation results demonstrate a quite good agreement with measurements, bearing in mind the approximations that were done. Simulations of the flux of ^{90}Sr and ^{137}Cs through the Yenisei River/Gulf from activities at the KMCC for the period 1975–1995 were carried out using available data in Table 4.3. The data are only for 1975–1995. So, to take into account the interaction of radionuclides in solute with the bottom which was contaminated in 1959–1974, missing data for the period 1958–1975 were taken as the average data for 1976–1986. In Figure 4.26, temporal changes in the ^{137}Cs concentration in solute, at different distances from the KMCC, are compared with measurements (Table 4.3, Kuznetsov *et al.*, 2001). The measured (Chumichev, 1995) and simulated concentration in Igarka (lower Yenisei) are presented in Figure 4.27. This figure shows that contamination input from the KMCC into the Yenisei in the

Table 4.3. Assessment of radionuclide discharges into the Yenisei River in 1962–1994 (Kuznetsov *et al.*, 2001).

Year	^{90}Sr (GBq/yr)	^{137}Cs (GBq/yr)
1975	359	433
1976	266	10,508
1977	1,254	3752
1978	659	659
1979	740	592
1980	1,084	2,760
1981	1,014	1,225
1982	381	1,047
1983	744	1,177
1984	448	755
1985	185	2,109
1986	555	1,517
1987	367	1,072
1988	533	739
1989	360	778
1990	277	540
1991	200	434
1992	98	201
1993	52	47
1994	20	44
1995	35	38

1970s–1990s was much less than global fallout estimated using Lena River data. RIVTOX/THREETOX–computed values of ^{90}Sr cross-sectionally averaged concentration in the mouth of Ob′ Bay are shown in Figure 4.28. Figure 4.29 indicates the correlation between the cross-sectionally averaged concentration of ^{90}Sr calculated

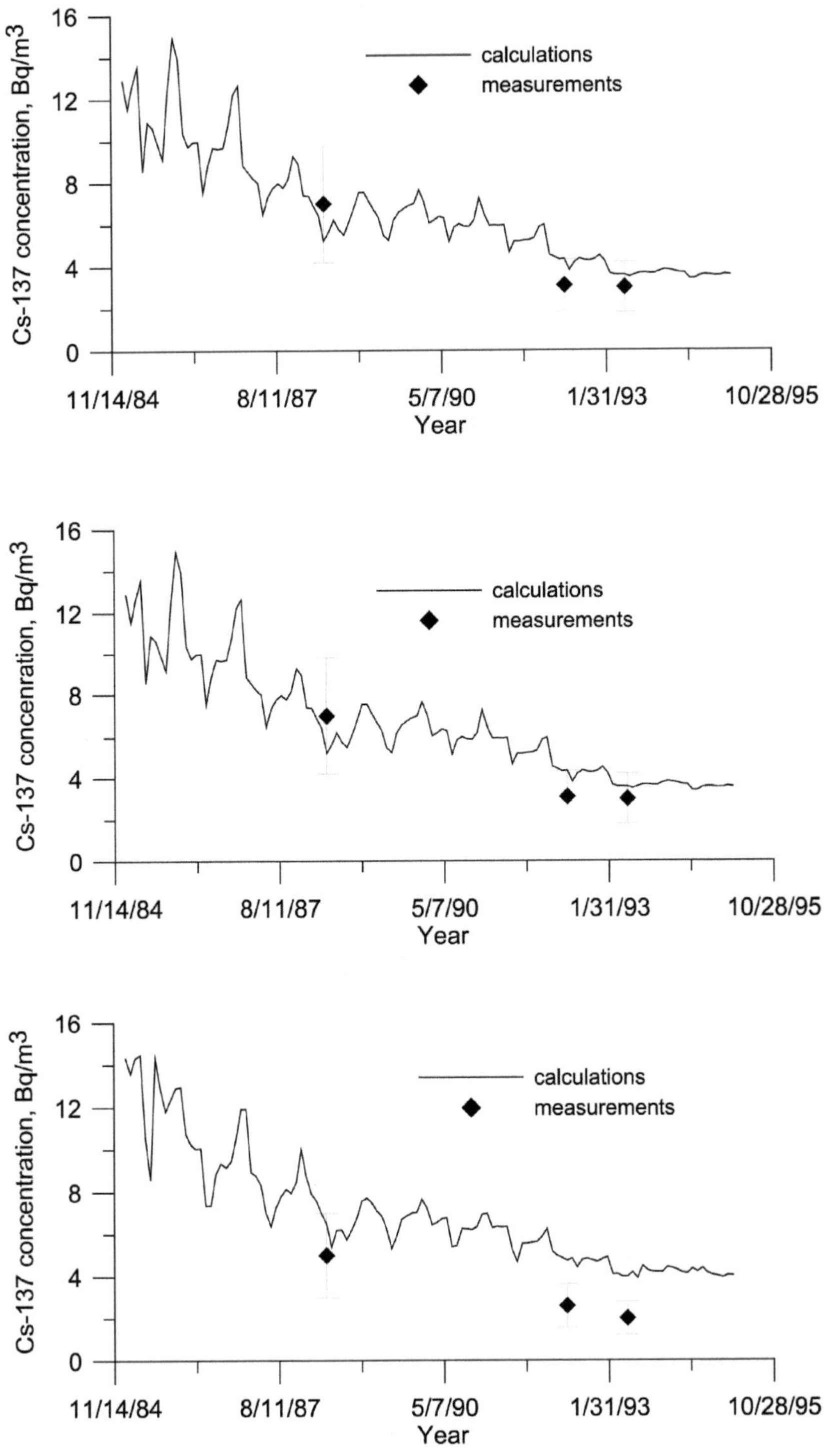

Figure 4.26. Temporal variability of ^{137}Cs concentration in solute in the Yenisei River at distances 250 km (upper), 510 km (middle), and 1,363 km (lower) from the MCC.

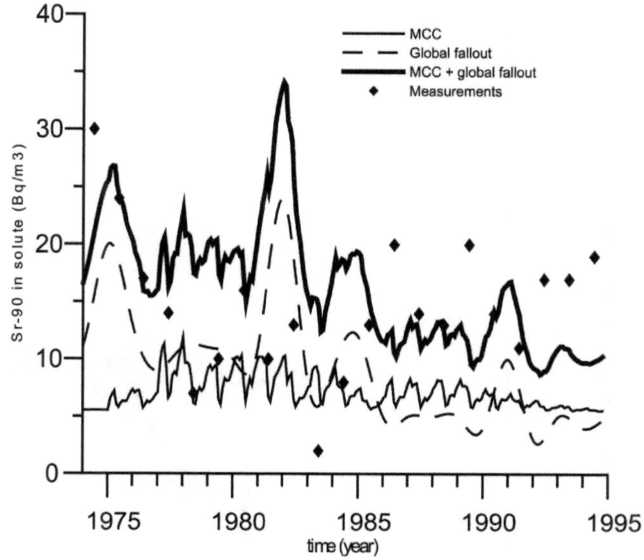

Figure 4.27. Simulated and observed (Chumichev, 1995) ^{90}Sr concentration in solute in the Igarka (1974–1995).

by THREETOX and relevant values of concentration calculated by RIVTOX. As follows from the figure, even in this deep estuary the one-dimensional RIVTOX model can be applied to simulate fuxes of radionuclide into the sea. The mean annual values of ^{90}Sr concentration in Yenisei Gulf are shown in Figure 4.30. Simulations and measurements show that the MCC was the main source of ^{137}Cs transported by Yenisei River water; however, the concentration of ^{137}Cs in the lower Yenisei does not exceed 8 Bq/m^3 in solute and 140 Bq/kg in bottom sediments.

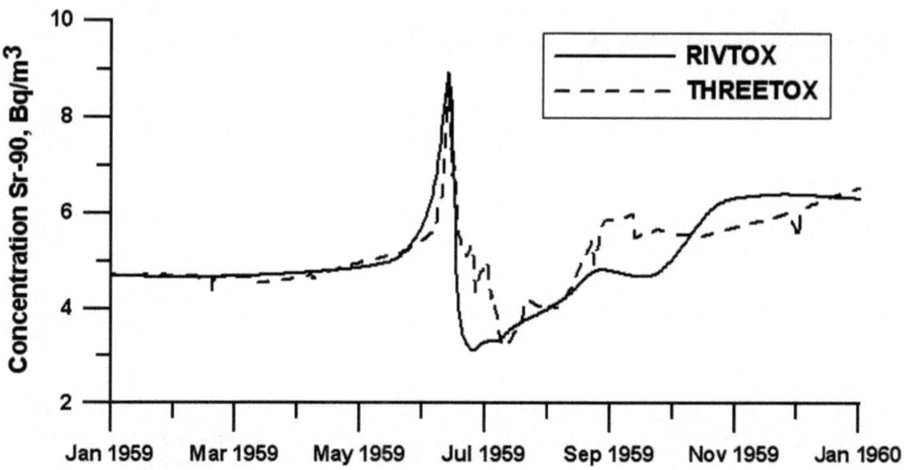

Figure 4.28. RIVTOX- and THREETOX-simulated ^{90}Sr concentration in solute in the mouth of Yenisei Gulf.

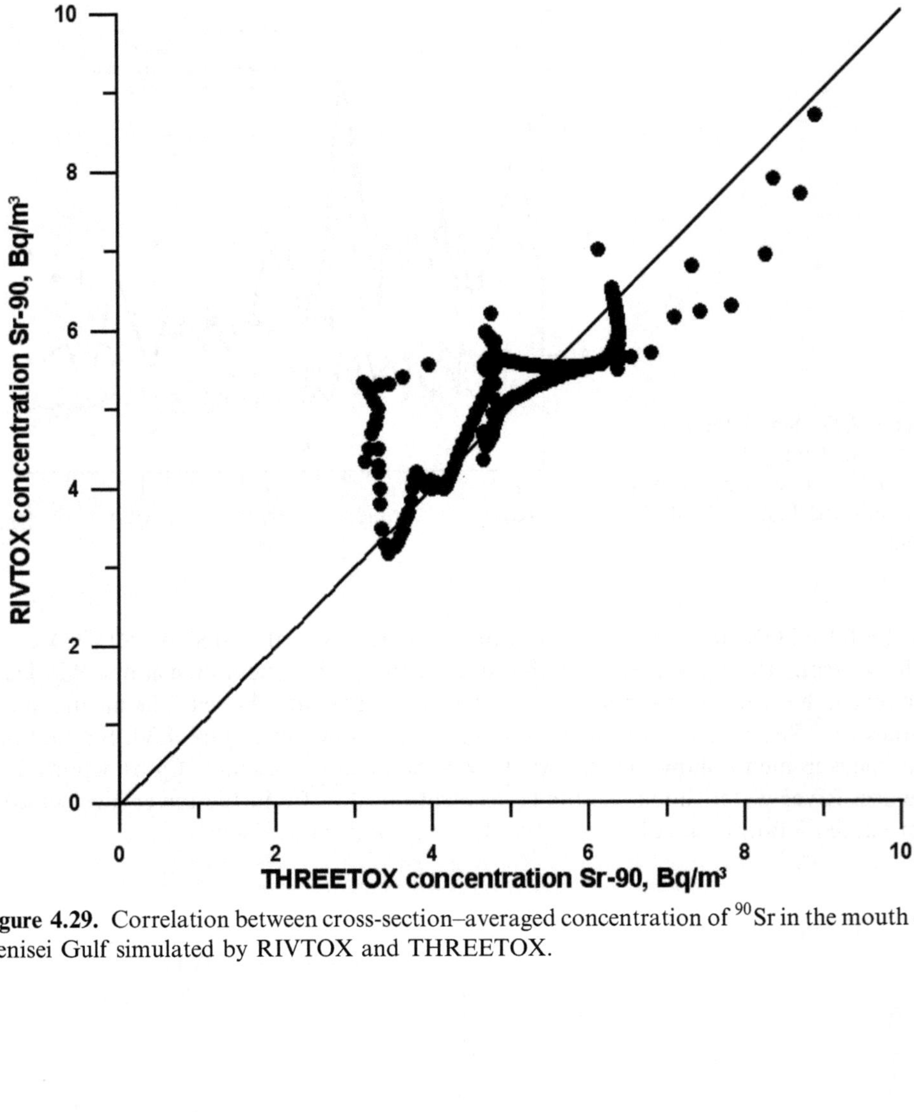

Figure 4.29. Correlation between cross-section–averaged concentration of ^{90}Sr in the mouth of Yenisei Gulf simulated by RIVTOX and THREETOX.

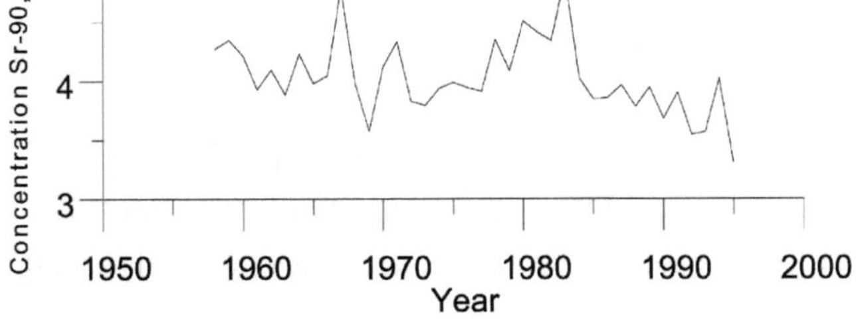

Figure 4.30. Simulated concentration of annually averaged ^{90}Sr in Yenisei Gulf (1958–1995).

4.1.2 Transport and dilution of radioactive waste and dissolved pollutants in the Kara Sea

Transport of strontium-90 from the Ob′ River from 1949 to 1965 (Historical Run 1)

Marine transport of ^{90}Sr originating from the Ob′ River was simulated in accordance with modeled Kara Sea current patterns driven by realistic atmospheric forcing fields. Source terms of ^{90}Sr and river discharge flux data at the mouth of the Ob′ estuary were supplied by river models.

For the purpose of modeling radionuclide dispersion, the model was expanded by using the passive tracer transport equation which is identical to the equation for heat–salt transport but coupled with the equation term describing the radioactive decay of the element. Here it is assumed that the radionuclide, in this case ^{90}Sr, is in dissolved form. The source of radioactive emission was the Mayak PA plant. To set the boundary condition at the Ob′ River, averaged monthly discharge of the river and modeled radionuclide ^{90}Sr transfer through the riverbed were used.

Charts describing Ob′ River discharge and ^{90}Sr concentration for the period from 1949 to 1965 are presented in Figures 4.31 and 4.32. During the period from 1951 to 1957, there was uncontrolled emission of nuclear waste to the river.

All the model runs cover the period from 1949 to 1965 and have the initial and boundary conditions described above. Model results are shown in Figure 4.33. Radionuclide transport follows the Kara Sea current system. Maximum strontium concentrations are obtained at the shallow-water shelf in 1954 and 1955. Radionuclides are carried farther to both the north boundary of the Kara Sea and the Kara Sea gate to the Pechora Sea. Some pollution reaches Vilkitsky Strait and enters the Laptev Sea. Because the ^{90}Sr half-life is 30 years, which considerably exceeds the

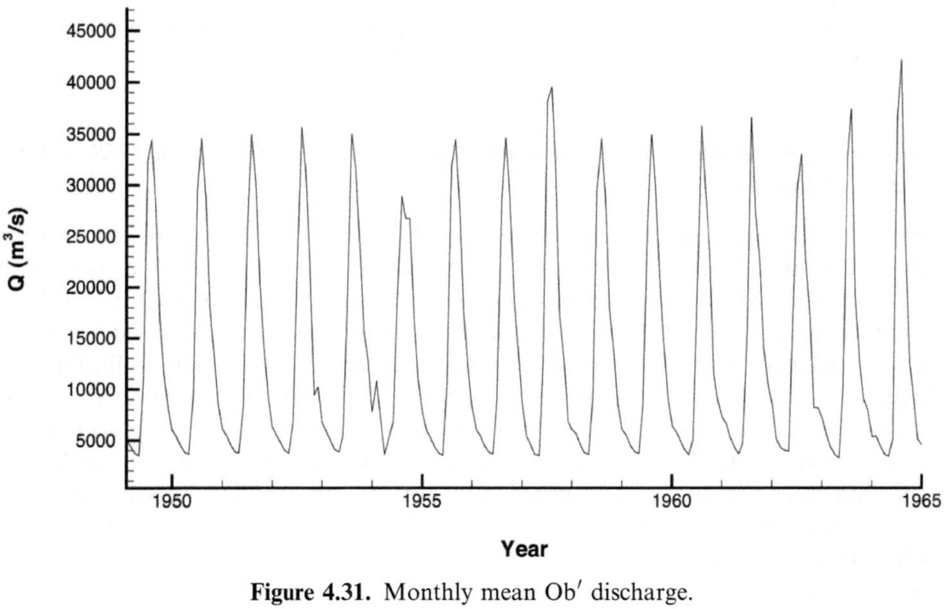

Figure 4.31. Monthly mean Ob′ discharge.

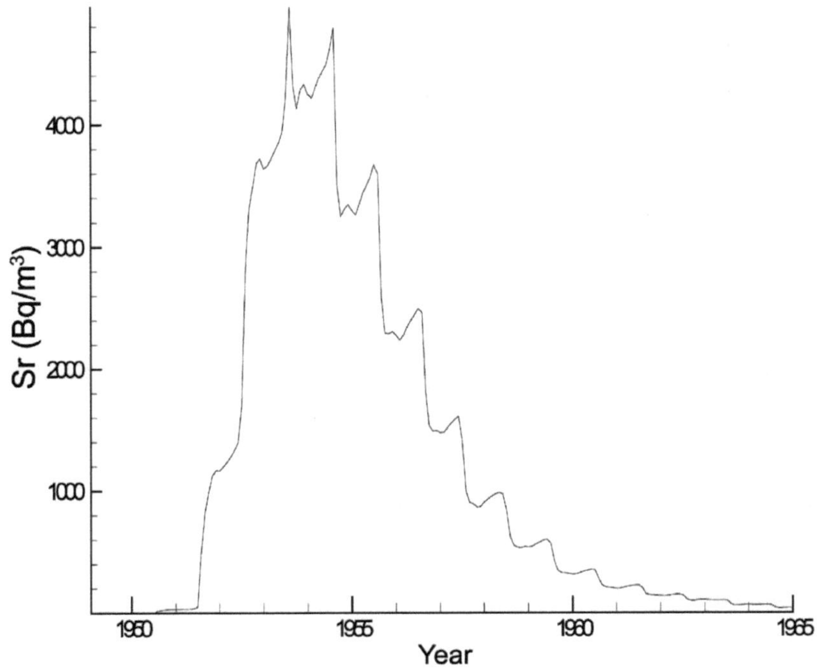

Figure 4.32. Monthly mean concentration of strontium-90.

source operation time and lifetime of water in the Kara Sea, the obtained results can be used in the interpretation of other dissolved pollutants. As for suspended solids, they precipitate mainly at Ob′ Bay or at the confluence of the Ob′ and Kara Sea.

This model run is a reconstruction of the real event and allows effects from the historical nuclear accident to be estimated successfully.

Transport of strontium-90 and cesium-137 from the Yenisei River from 1958 to 1993 (Historical Run 2)

Yenisei River runoff and radionuclide input data (both ^{90}Sr and ^{137}Cs) were estimated from radionuclide transport in the river and estuary system as simulated by river models. Marine transport/dispersion of radionuclides from the Yenisei River during the period 1958–1993 were simulated in accordance with modeled Kara Sea current patterns driven by realistic atmospheric forcing fields. The pollution discharge rate was spatially uniform with a stable distribution of, respectively, ^{90}Sr and ^{137}Cs during the years 1958–1993 (Figures 4.34 and 4.35, see color section for both). In contrast to the previous experiment (the Ob′ discharge of ^{90}Sr), pollution dispersion is less intensive. The circulation pattern initiated by the Ob′ discharge blocks westward dispersion from the Yenisei and thus limits the impact and spread of both radionuclides from the Yenisei. The main transport of radionuclides from the Yenisei is directed to the northeast into the Kara Sea.

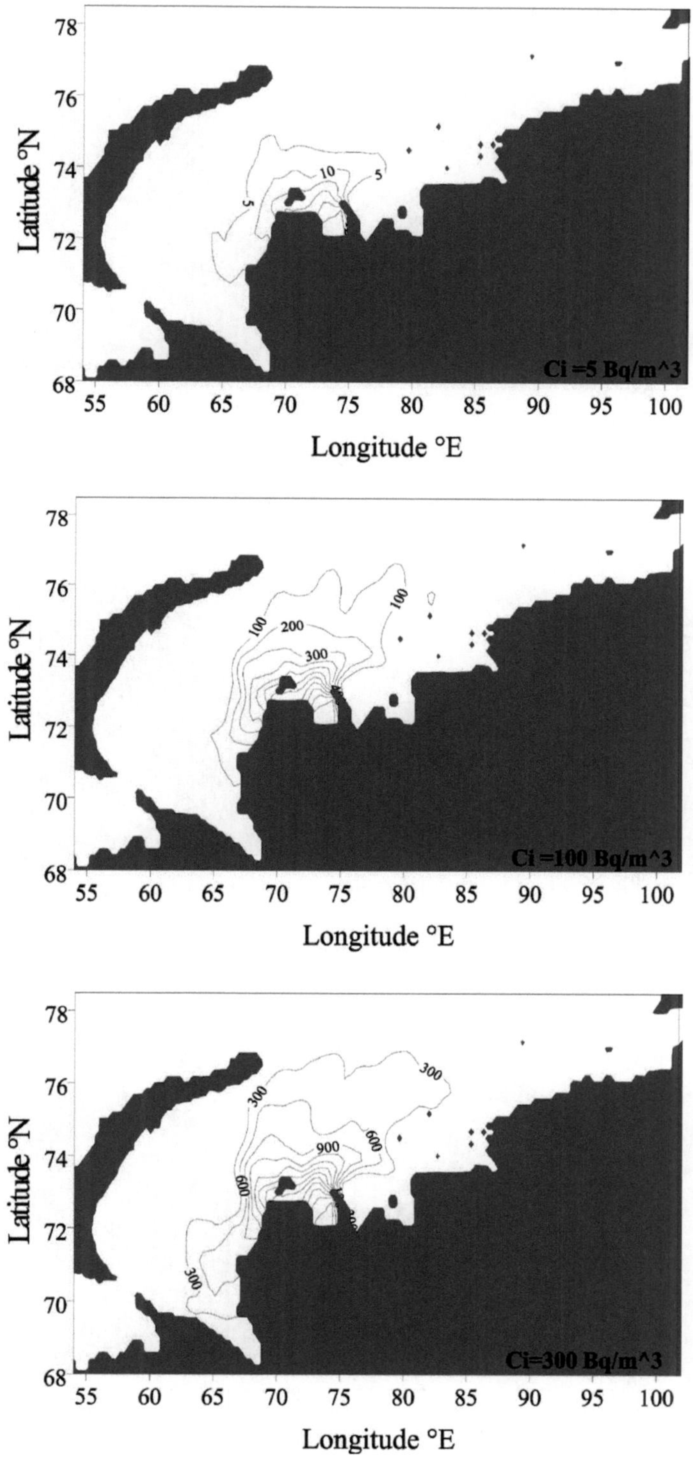

Figure 4.33. Simulated surface concentration of ^{90}Sr (Bq/m^3) due to Mayak release in January: (a) 1951, (b) 1952, (c) 1953.

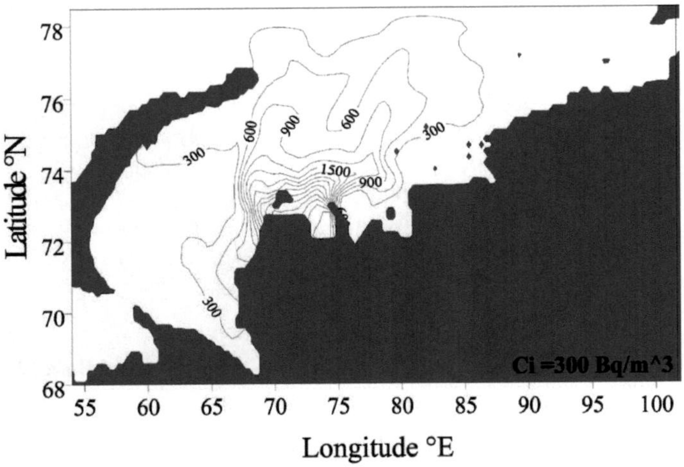

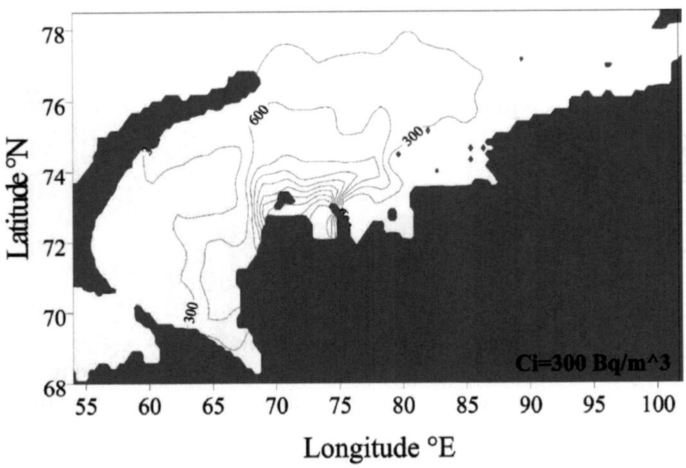

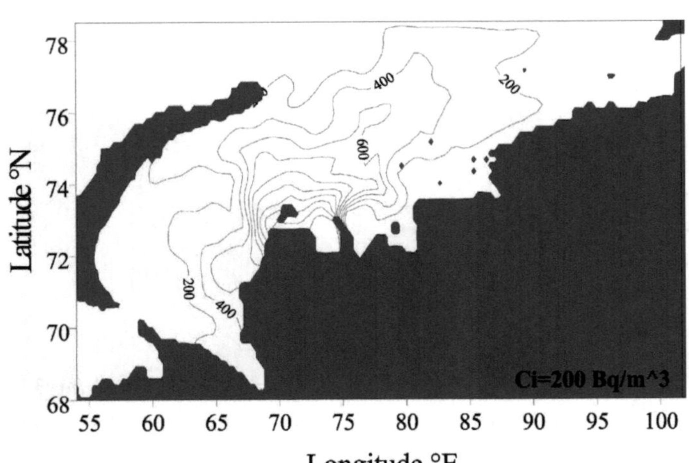

Figure 4.33 (*cont.*). Simulated surface concentration of ^{90}Sr (Bq/m^3) due to the Mayak PA release in January: (d) 1954, (e) 1955, (f) 1956.

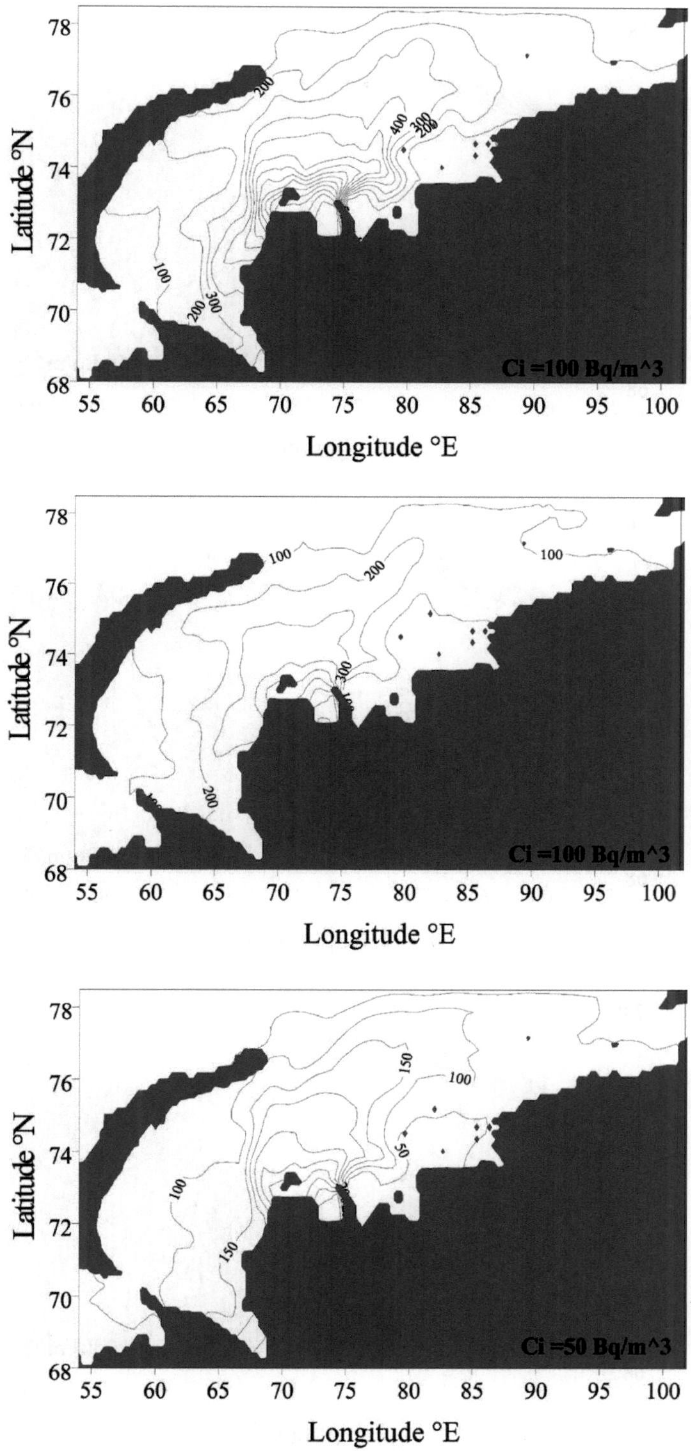

Figure 4.33 (*cont.*). Simulated surface concentration of ^{90}Sr (Bq/m^3) due to the Mayak PA release in January: (g) 1957, (h) 1958, (i) 1959.

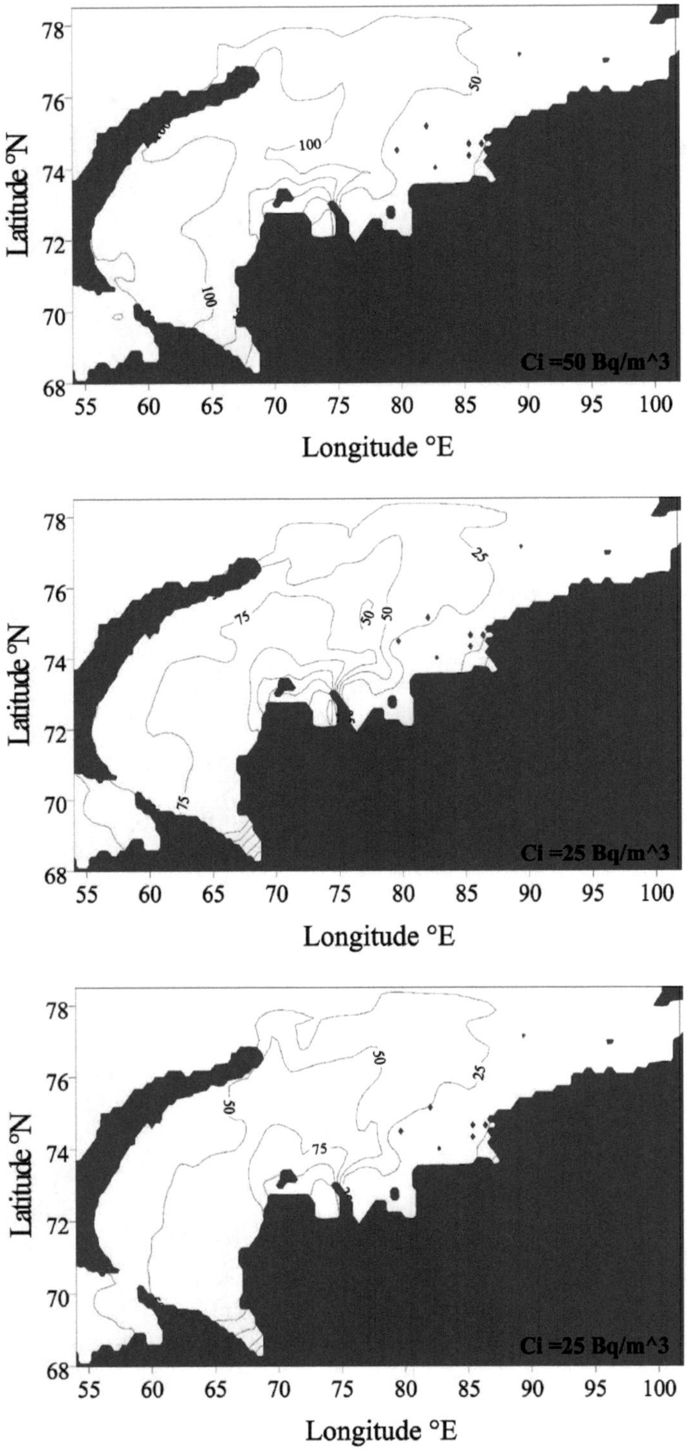

Figure 4.33 (*cont.*). Simulated surface concentration of ^{90}Sr (Bq/m^3) due to the Mayak PA release in January: (j) 1960, (k) 1961, (l) 1962.

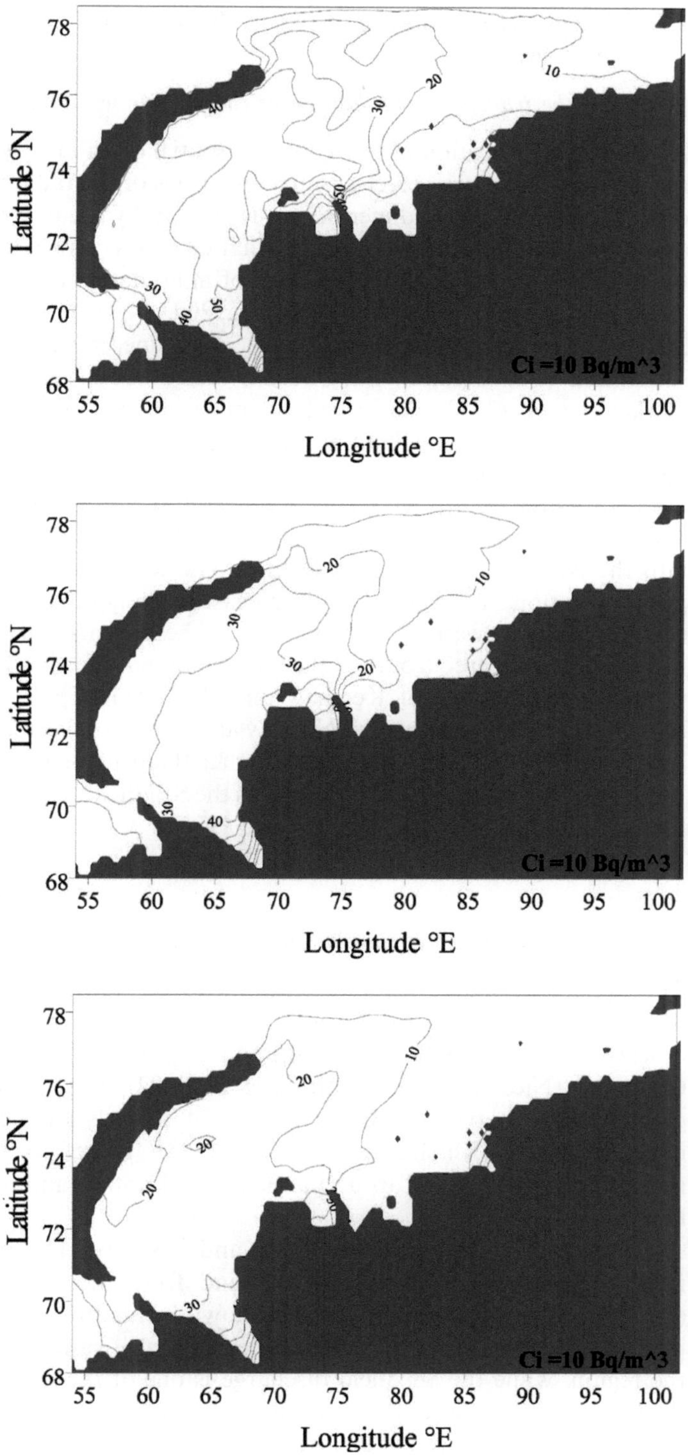

Figure 4.33 (*cont.*). Simulated surface concentration of ^{90}Sr (Bq/m^3) due to the Mayak PA release in January: (m) 1963, (n) 1964, and (o) 1965.

4.1.3 Transport and dilution of radioactive waste and dissolved pollutants from all sources

Modeling the transport and dilution of radioactive waste and dissolved pollutants

Accurate quantification of the temporal–spatial distributions of radionuclides can be achieved by combining precise source functions of the radionuclides, available *in situ* observations, and three-dimensional simulation by ocean general circulation models (OGCMs). It has been demonstrated that OGCMs can be powerful tools to simulate the temporal and spatial distributions of man-made radionuclides (e.g., Nies *et al.*, 1998; Gerland *et al.*, 2003; Tsumune *et al.*, 2003; Karcher *et al.*, 2004; Gao *et al.*, 2004, 2005; Orre *et al.*, 2007). For instance, Tsumune *et al.* (2003) performed a global simulation of ^{137}Cs and $^{239,240}Pu$ with the source of the atmospheric fallout, and Gerland *et al.* (2003) performed a simulation of ^{99}Tc with the source from the Sellafield Reprocessing Plant using a regional OGCM. Subsequently, Gao *et al.* (2004) used a global version of the Nansen Environmental and Remote Sensing Center version of MICOM (Bleck *et al.*, 1992) to examine the transport of ^{90}Sr and ^{137}Cs radionuclides as a result of atmospheric fallout and the Sellafield discharge into the North Atlantic–Arctic Oceans. The study showed that the simulated temporal evolution of surface ^{137}Cs is in reasonably good agreement with observations east of Scotland, west of Norway, and in the Barents Sea, and the transit time for radionuclides released from Sellafield to the Barents Sea is about 4–5 years, consistent with the observations of Kershaw and Baxter (1995). It was also shown that in Atlantic waters off the coast of Norway and in the southern Barents Sea, atmospheric fallout of ^{90}Sr and ^{137}Cs was much higher than the Sellafield release until the mid-1960s and from the early 1990s, whereas the Sellafield release was the main source for the two radionuclides in the 1970s and 1980s.

Setup of the radionuclide simulation (hindcast)

The important sources of radionuclides in the North Atlantic–Arctic region are nuclear bomb testing, the Chernobyl accident, and releases from the European reprocessing plants Sellafield (U.K.) and Cap de La Hague (France), as detailed in Chapter 1 (Nies *et al.*, 1998; Strand *et al.*, 2002). The Sellafield discharge has been greater that of Cap de La Hague in terms of quantity and impact for the majority of northern waters (Kershaw and Baxter, 1995). Therefore, the dispersion of radionuclides released from Cap de La Hague will not be addressed to the same degree here.

The temporal evolution of ^{137}Cs and ^{90}Sr over Denmark from the Sellafield release, from nuclear bomb testing, and from the Chernobyl accident in 1986 (S. Nielsen, pers. commun., 2001) is shown in Figure 4.36a, b. Atmospheric fallout provides the present background concentration of surface and subsurface waters of the region, while the Sellafield discharge is one of the major sources of radioactive contamination in the Arctic Ocean since the 1970s (Strand *et al.*, 1996). The Sellafield signal has been observed over the last two decades, making model evaluation of ^{137}Cs

distributions, in particular, possible. The temporal evolution of ^{99}Tc released from Sellafield and La Hague is shown in Figure 4.36c, d.

Furthermore, radioactive contaminants have been discharged from Russian nuclear facilities at high northern latitudes. Generally speaking, these discharges have played a less important role in overall radioactive contamination. However, Russian nuclear facilities that discharge into the Ob′ and Yenisei River systems and eventually reach the Arctic Ocean need to be considered since their discharges have been high historically (Strand *et al.*, 1998). There is concern about whether and how they have contaminated the Arctic Ocean, and whether and how potential accidents can lead to further contamination. To obtain the input flux of ^{90}Sr transported from the Ob′ and Yenisei Rivers, the one-dimensional river model (Zheleznyak *et al.*, 1992) and the three-dimensional estuary model (Margvelashvili *et al.*, 1997) were used along with output from the river model as input to the estuary model. Both the river model and estuary model have been validated against available observations. Finally, the simulated temporal distribution of ^{90}Sr concentration by the three-dimensional estuary model is used as the source function for the results presented here. Figure 4.37 shows the reconstructed time evolution of ^{90}Sr caused by river discharges at the mouths of the Ob′ and Yenisei Rivers into the Kara Sea (Johannessen *et al.*, 2003).

The latitudinal distribution of atmospheric ^{90}Sr deposition (UNSCEAR, 1982) makes it feasible to construct the latitudinal distribution of atmospheric fallout of ^{137}Cs, with the fallout of ^{137}Cs a factor 1.6 higher than that of ^{90}Sr (Strand *et al.*, 1998). Please note that the Chernobyl release in 1986, seen as the isolated peak in the atmospheric fallout in Figure 1.3, is not included in the simulations presented here. The initial fields of all the radionuclides are set to zero. The simulation starts in 1950 and ends at the present for global version simulations. Following the release history of ^{99}Tc, ^{99}Tc simulation using the regional model starts from 1975 and ends at the present day.

Results

The pathway of soluble radionuclides ^{137}Cs and ^{90}Sr from Sellafield to the Arctic has been summarized by Kershaw and Baxter (1995). Initially, the tracer is carried northward from the Irish Sea as a plume-like structure via the North Channel, and it then flows along the coast of Scotland into the North Sea. The tracer is transported northward before branching off northern Norway: one branch passes eastwards into the Barents Sea; the other passes through Fram Strait with the West Spitsbergen Current (WSC). The main surface-water return flow occurs with the East Greenland Current (EGC). The transit times for the Sellafield release have also been estimated based on available observations. It follows that the signal enters the Barents Sea after about 4 years. In Fram Strait, the transit time is about 5 years for Atlantic Water and 6–10 years for southward-flowing Arctic waters.

We now show an example of the simulated pathway of the Sellafield release from the 40 km version simulation (Figure 4.38, see color section). Results from the 80 km

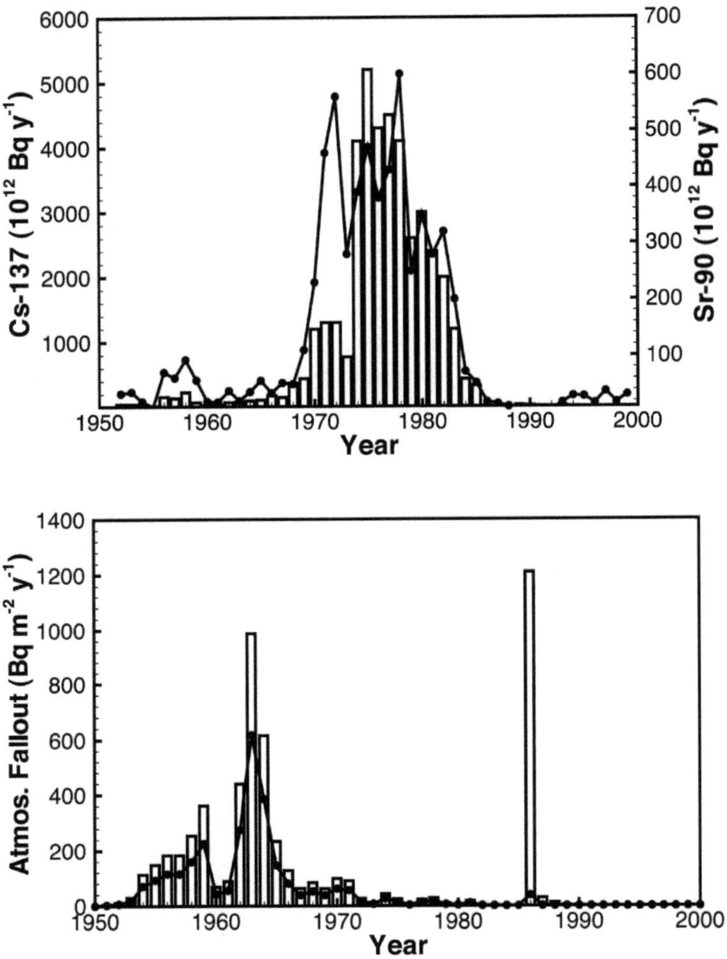

Figure 4.36. Time series of ^{137}Cs (bars) and ^{90}Sr (line) from (a) the Sellafield release (10^{12} Bq yr^{-1}), (b) atmospheric fallout (Bq m^{-2} yr^{-1}) over Denmark (S. Nielsen, pers. commun., 2001), (c) annual releases of ^{99}Tc (10^{12} Bq) from the reprocessing plant at Sellafield (black bars) and La Hague (gray bars), and (d) monthly releases of ^{99}Tc (10^{12} Bq) from the reprocessing plant at Sellafield during the 1990s. Parts (c), (d) are from fig. 2 in Orre *et al.* (2007).

and the regional model (not shown) are similar (Gao *et al.*, 2004; Orre *et al.*, 2007). Before extending northward with the North Atlantic Current (NAC), simulated Sellafield radionuclides circulate cyclonically in the North Sea. One branch enters the Barents Sea, and one branch heads towards Fram Strait following the West Spitsbergen Current (WSC). Simulated pathways of the Sellafield release to the Barents Sea are in general agreement with observations (Kershaw and Baxter, 1995) and broadly consistent with the known transport routes of Atlantic Water into the Nordic Seas: Initially, the tracer is carried northward from the Irish Sea as a plume-like structure via the North Channel. It then partly flows along the northern coast of Scotland into the North Sea (Kershaw and Baxter, 1995), and partly follows the shelf break northward off the coast of Norway (Orvik and Skagseth, 2003). Off northern Norway, one branch extends eastwards into the Barents Sea, and the other towards Fram Strait becoming the WSC. Near Fram Strait, some Atlantic Water

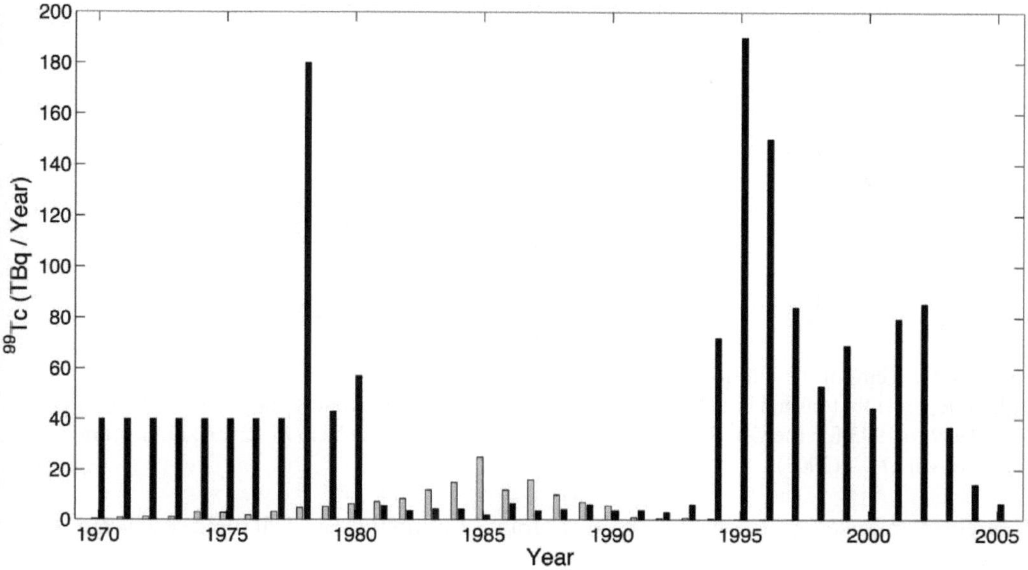

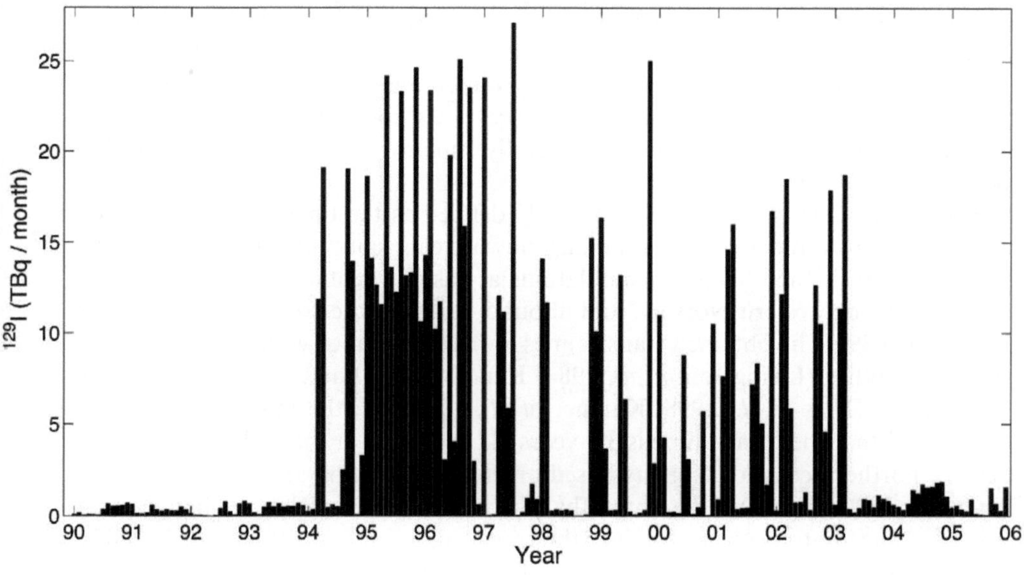

recirculates and continues southward, joining the EGC, whereas the remainder subducts and enters the Arctic Ocean under the upper, low-salinity Arctic surface water.

To quantitatively evaluate the simulated transport and mixing of the Sellafield discharge, a time series of the observed and simulated surface ^{137}Cs concentration east of Scotland (57.0–57.5°N, 1.5–2.0°W), west of Norway (59–61°N, 3.5–5.0°E), and in the southwestern Barents Sea region (71–72°N, 20–30°E) are shown in Figure

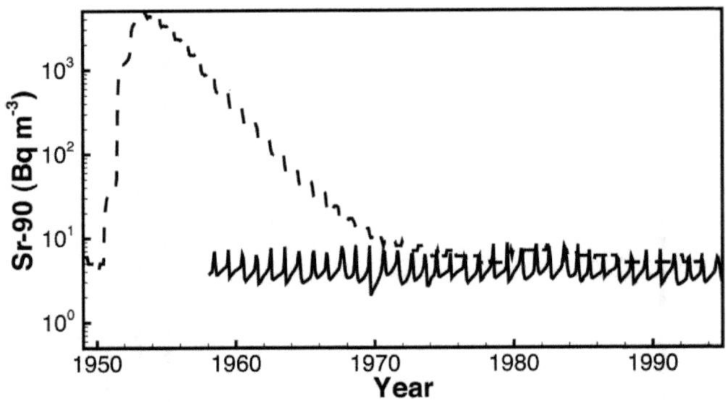

Figure 4.37. Temporal evolution of ^{90}Sr concentration ($Bq\,m^{-3}$) at the mouth of the Yenisei (solid line) and Ob′ (dashed line) Rivers in the Kara Sea caused by river discharge (monthly data from 1949 to 1993 for the Ob′ River and from 1958 to 1995 for the Yenisei River). Data from Johannessen *et al.* (2003).

4.39. Superimposed on the panels is a histogram of annual mean Sellafield discharge rates.

Atmospheric fallout clearly dominates surface ^{137}Cs and ^{90}Sr distributions in the mid-1960s in the three regions. However, from the late 1970s to the late 1980s, the Sellafield discharge heavily dominates. For the last 5–10 years of the simulation, the surface concentrations of ^{137}Cs and ^{90}Sr are governed mainly by atmospheric fallout. The temporal and spatial evolution of surface ^{137}Cs concentrations are in broad agreement with observations, and are generally superior to results from the 80 km model (thin solid line in Figure 4.39).

The transit time of a tracer is commonly defined as the time between observed peak release rates and the corresponding downstream concentration peak of the tracer. The transit time based on simulations at eastern Scotland is about 2 years, about 4 years for western Norway, and about 5 years for the southwestern Barents Sea (Figure 4.39). The obtained transit times are in accordance with earlier observation-based studies (Livingston *et al.*, 1984; Kershaw and Baxter, 1995), as well as model studies (Nies *et al.*, 1998; Karcher *et al.*, 2004). Furthermore, Brown *et al.* (2002) found that the transit time is 3.5 years (42 months) for the Sellafield signal to reach the northern coast of Norway based on cross-correlation analysis of observed ^{99}Tc in the 1990s. It should be noted that, because ocean circulation changes as a result of changed atmospheric forcing, the transit time can vary (Gao *et al.*, 2005). Therefore, the observation-based estimate of transit time is both spatially and temporally constrained by the limited observations. To acquire an accurate transit time, numerical simulation is needed based on a validated model. Here we show an example of the two-dimensional simulated apparent age in the 1980s from the 40 km model (Figure 4.40, see color section), which indicates that the Sellafield release takes 56 months to reach the Barents Sea.

For ^{99}Tc simulation using the regional model, as already mentioned, the simulated pathways and concentration of the Sellafield release are in broad agreement with observations. The radionuclides released from La Hague are initially

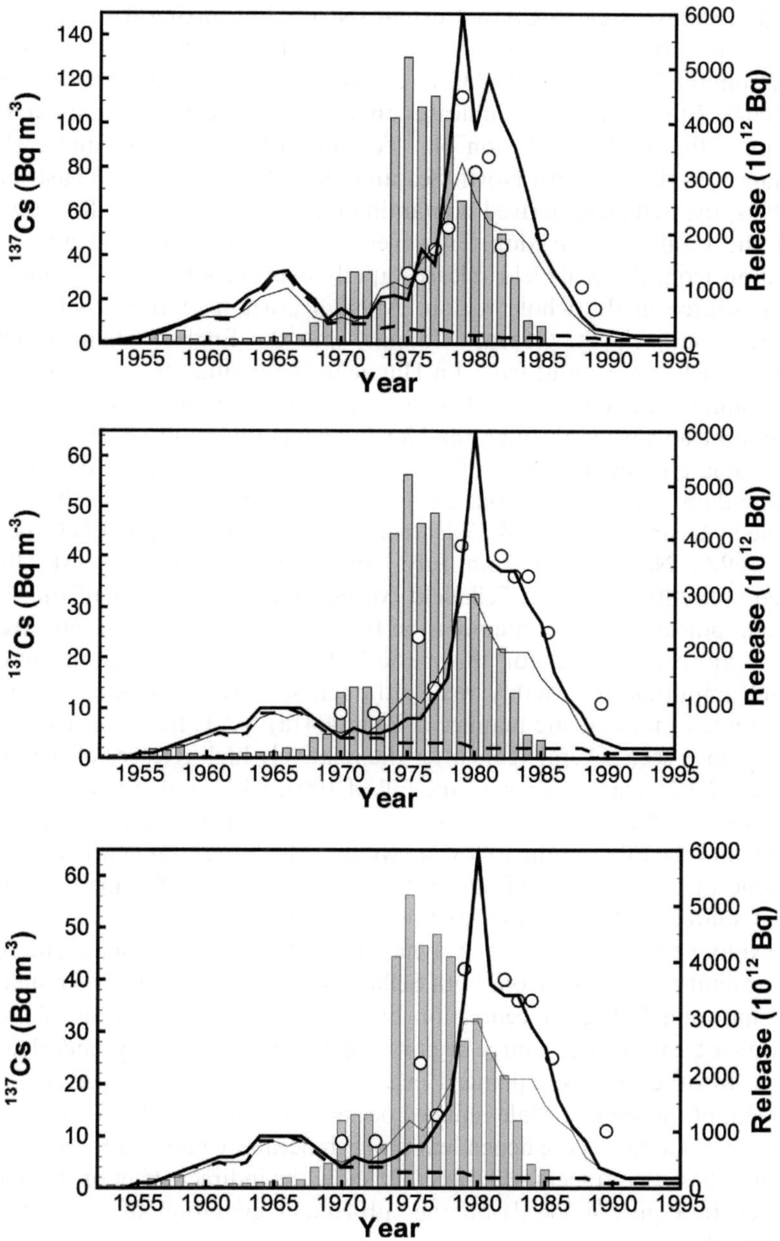

Figure 4.39. Time series of Sellafield release rates of ^{137}Cs (10^{12} Bq yr^{-1}; bars), observed ^{137}Cs surface concentration (Bq m^{-3}) from Kershaw and Baxter (1995) in circles, simulated surface concentration due to atmospheric fallout (dashed line), and the sum of concentrations from Sellafield release and atmospheric fallout (thick solid line) for (a) east of Scotland, (b) west of Norway, and (c) in the southwestern Barents Sea region. For comparison, the thin line corresponds to the thick solid line, but is taken from the 80 km resolution version of the model (Gao *et al.*, 2004). Note the different concentration scales on the panels.

entrained in the Norwegian Coast Current (NCC), but mixing with the Norwegian Coast Current (NwAC) along the Norwegian coast causes a net transport of radionuclides from the NCC to the NwAC, as can also be seen from observations (Gascard *et al.*, 2004). Further analysis indicates that the ^{99}Tc released from La Hague is significant in the total distribution of ^{99}Tc only in the late 1980s and early 1990s. For instance, in the southern North Sea and along the Norwegian coast during the early 1990s, the Sellafield-derived contamination of ^{99}Tc accounts for only 20% to 30% of the total concentration. However, in the late 1990s and afterward, the contribution from the Sellafield release rapidly overtakes that of La Hague as the dominant source in the whole region. A small portion of the Sellafield signal is transported south via St. Georges Channel into the English Channel. Therefore, due to the much lower input from La Hague in the period after 1990, the Sellafield signal dominates in the English Channel, even though the bulk of tracers from Sellafield are transported through the North Channel (Orre *et al.*, 2007). To quantitatively evaluate model behavior of ^{99}Tc simulation, a data–model comparison was made. The data cover two periods at two different locations along the Norwegian coast, one at Utsira (59.18°N, 4.53°E) from March 1986 to April 1993, the other at Hillesøy (69.65°N, 17.95°E) from July 1997 to December 2003. At Utsira, the simulated concentration from Sellafield corresponds roughly to half the total concentration, indicating that contamination from Sellafield and La Hague are equally important at this location for this period. The model captures the interannual trend and amplitude reasonably well, although seasonal variations are in general not represented in a realistic manner (Figure 4.41a). At Hillesøy an increase in the concentration is seen several years after the relatively high releases of the mid-1990s (Orre *et al.*, 2007), starting around the fall of 1997, with gradually decreasing levels starting in mid-2001. The observations show two distinct peaks, one in the winter 1999/2000 and the other in the following winter. These peaks are clearly seen and are comparable in time and amplitude in the OGCM as well. Sellafield is by far the dominant source at this location for this period (Figure 4.41b).

To explore the impact of ocean dynamics on tracer dispersion, idealized tracers with a constant release rate from both Sellafield and La Hague were simulated using the regional model. The concentration of idealized tracers from Sellafield and La Hague downstream of the source region is then only affected by underlying ocean dynamics. In other words, enhanced ocean transport gives a higher concentration downstream of the source; while reduced ocean transport gives less concentration. In contrast, the ^{99}Tc tracers are dependent on both ocean dynamics and the release rate. Comparing these two tracers will therefore show the indirect effect of the release rate (refer back to Figure 4.36c, d) on the subsequent evolution of concentrations at downstream locations.

Normalized concentration at Hillesøy of ^{99}Tc and idealized tracers from both Sellafield and La Hague are shown for the simulated time period in Figure 4.42. Apparently, variations in ^{99}Tc tracers from Sellafield follow the idealized tracers from Sellafield. However, some features can only be explained by the release rate. For instance, the two peak values at Hillesøy for the ^{99}Tc tracers already discussed are not as dominating for idealized tracers, hence only the release rate can explain them. The

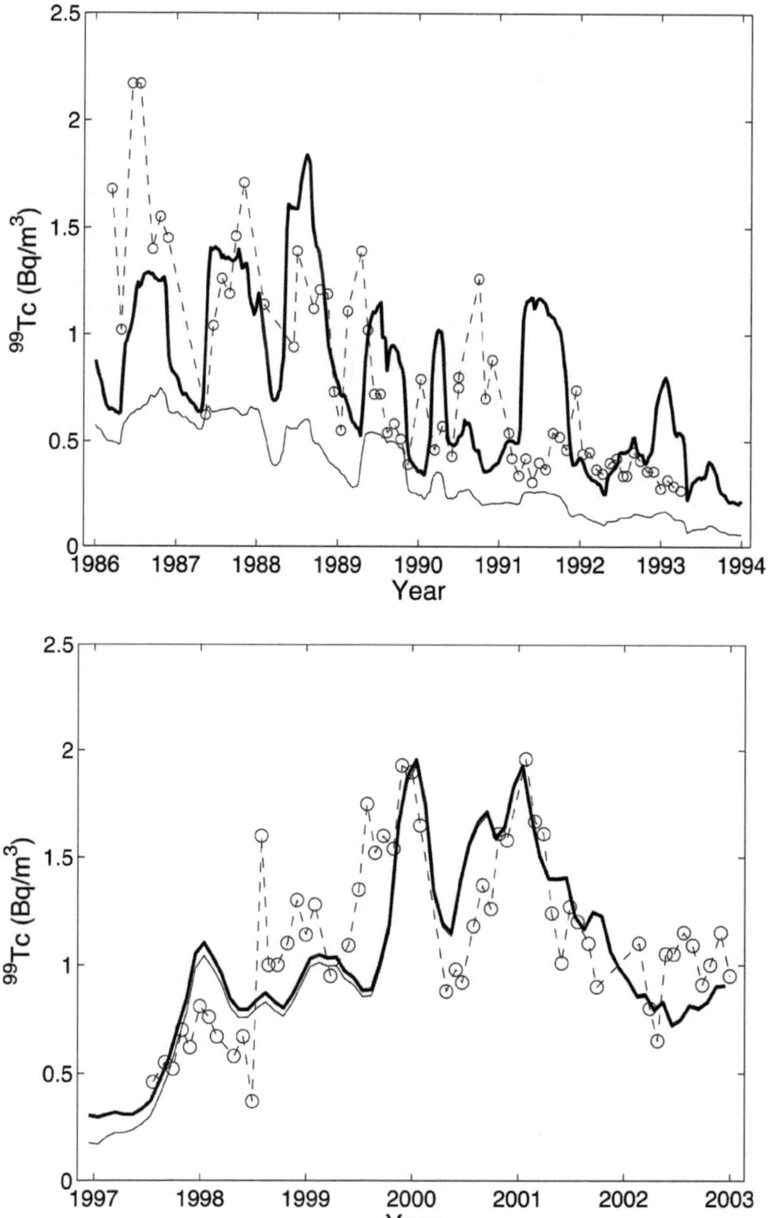

Figure 4.41. Concentration of ^{99}Tc (Bq m^{-3}) in seawater at Utsira (59.18°N, 4.53°E) off the southwest coast of Norway from observations (upper, open circles) and the surface of the OGCM (bold solid line) for the period 1986 to 1994. (Lower) ^{99}Tc concentration (Bq m^{-3}) in seawater at Hillesøy (69.65°N, 17.95°E) off the northwest coast of Norway from observations (open circles) and the surface of the OGCM (bold solid line) for the period 1997 to 2003. The simulated contribution from Sellafield alone is indicated by the thin solid line in both figures (figs 6 and 8 in Orre *et al.* (2007).

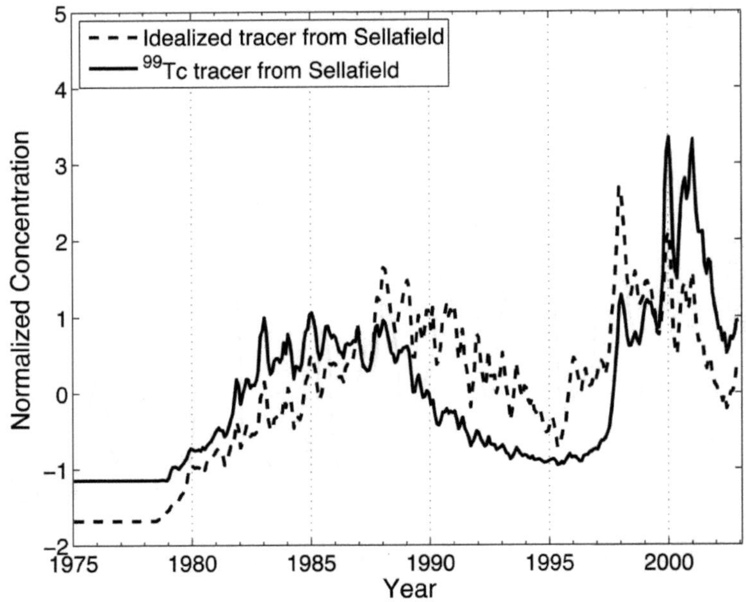

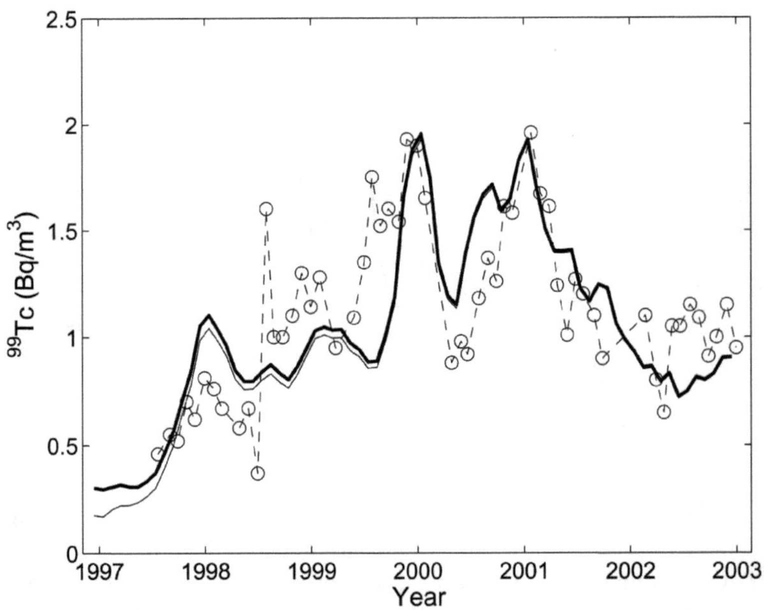

Figure 4.42. Simulated normalized concentration in the surface at Hillesøy of ^{99}Tc (solid line) and idealized tracers (dashed line) from (upper) Sellafield and (lower) La Hague. Mean value and standard deviation of ^{99}Tc from Sellafield is 0.49 and 0.43 Bq m^{-3}, respectively, and 0.16 and 0.16 Bq m^{-3} from La Hague. Normalized concentration of tracer C at station x_s is calculated as $(C(x_s, t) - \text{mean}(C(x_s, t)))/\text{std}(C(x_s, t))$ (after Fig. 9 in Orre *et al.* (2007).

variability in idealized tracers from Sellafield is much larger than that of ^{99}Tc in the late 1980s and early 1990s; this is most likely an effect of the relatively low input from Sellafield in this period. There was a steady decrease in the ideal tracer concentration from Sellafield at Hillesøy from 1988, followed by a rapid increase from mid-1995 and onwards. The flux of idealized tracers out of the Irish Sea is certainly an important factor in explaining these changes. Simulated tracer flux through the North Channel reached a record low value in 1992, accompanied by a record high tracer flux through the St. Georges Channel in the south. An anomalously high tracer flux through the North Channel was evident in the following three years: 1993, 1994, and 1995 (not shown). In addition, idealized tracers from La Hague exhibit substantial variations with a dominant seasonal cycle. ^{99}Tc tracers from La Hague are largely constrained by these variations, although the amplitude differs as a result of the varying release rate.

The following is an analysis of the ^{90}Sr released from atmospheric fallout, the Sellafield discharge, and discharges from Russian Arctic rivers. The temporal and spatial distributions of ^{90}Sr are diagnosed from 1955 to 1999. Here an example of the ^{90}Sr concentration distribution and ratios of the concentration caused by atmospheric fallout, the Sellafield release, and Ob′ River discharge to total surface ^{90}Sr concentration in 1955 is shown (Figure 4.43, see color section).

High concentrations of ^{90}Sr with values between several hundred to 3,000 Bq m^{-3} are found in the area close to the mouth of the Ob′ River in the Kara Sea in 1955. Concentrations below 5 Bq m^{-3} are found on the eastern and western sides of the North Atlantic and in the coastal waters off western Norway (Figure 4.43a, see color section). Atmospheric fallout dominates ^{90}Sr distribution in the North Atlantic and in the coastal waters off western Norway (Figure 4.43b), whereas Ob′ River discharge dominates ^{90}Sr distribution in the Arctic Ocean and the western side of the Nordic Seas (Figure 4.43d). The Sellafield discharge mainly influences the North Sea (Figure 4.43c).

In general, the radioactivity of ^{90}Sr declines dramatically from the 1950s to 1999. Ob′ River discharge had a strong influence on the Arctic Ocean during the 1950s and 1960s. However, atmospheric fallout is presently the dominant factor in the North Atlantic and Arctic region.

The spatial and temporal distributions of ^{90}Sr also indicate the pathways of the spread from different sources. The transport of ^{90}Sr from the Sellafield release is similar to that of simulated ^{137}Cs in Gao *et al.* (2004) and the observed estimate (Kershaw and Baxter, 1995). Radionuclides originating from the Ob′ River spread westward towards Fram Strait, thereafter they branch into two parts with one flowing southward with the EGC and the other continuously flowing westward following the northern coast of Greenland. Yenisei River discharge is mainly confined to the Kara Sea throughout the integration (not shown).

Simulated temporal evolution of surface ^{90}Sr concentration in the Kara Sea reproduces observed concentrations reasonably realistically (Figure 4.44, see color section). It follows that Ob′ River discharge dominates ^{90}Sr concentration before 1970 despite atmospheric fallout–induced ^{90}Sr reaching its peak value in the late

1960s. Between the mid-1980s and mid-1990s, atmospheric fallout, Ob′ River discharge, and the Sellafield release are nearly equally responsible for the surface concentration of ^{90}Sr. It is also evident that Yenisei River discharge plays a relatively minor role in the radioactivity of ^{90}Sr in the Kara Sea region.

4.2 SCENARIOS FOR POTENTIAL FUTURE RELEASES OF RADIOACTIVITY

Scenarios of hypothetical accidental radionuclide release into the environment from a range of sources are considered in this section, including estimates of the potential amount of radioactivity and release rate. Sources include the follwing three Russian radiochemical plants: the Mayak Production Association (PA), the Krasnoyarsk Mining and Chemical Combine (MCC), and the Tomsk Siberian Chemical Combine (SCC). These release scenarios together with a scenario for Sellafield are used in simulations by GMS of the spread of radioactivity in the Arctic for the present-day scenario and the $2 * CO_2$ global-warming scenario. Additionally, two hypothetical scenarios based on accidents onboard the submarines *Kursk* and *K-159* are considered. Simulation results are presented in Sections 4.3–4.5.

4.2.1 The Mayak PA scenario

In the Mayak PA accident scenario, the flux of ^{90}Sr from the Ob′ River to Ob′ Bay in the Kara Sea is simulated. A scenario in which a break in Reservoir-11's dam occurs was used (see Section 1.1.2). This scenario assumes a sudden and uncontrollable break in the dam of Reservoir-11 as a result of, for example, a strong flood, disastrous earthquake, or impact from an aircraft. Reservoir-11 parameters are: volume is 200E+06 (m^3), ^{90}Sr content in water in 1997 is 8,000 (Ci), and the ^{90}Sr content in water is 2,084−1,072 (Ci).

4.2.2 "Krasnoyarsk" scenario

The Krasnoyarsk MCC is a potential source of a sudden release of radioactive contamination into the Yenisei River. As at the Mayak PA there are special open-water storage reservoirs that collect liquid radioactive waste products that accumulate at MCC as a result of various technological operations. The function of these reservoirs is to hold the effluent so that short-lived radionuclides can decay and radioactive suspensions can be precipitated. The aqueous phase of radionuclides purified in special treating plants is placed at the bottom of the sea or is dumped in the Yenisei River. Here we consider the scenario of a sudden release of radionuclides from one such open-water storage reservoir; namely, Basin 365 (Waters *et al.*, 1999). It is assumed that a similar release of radionuclides in the Yenisei River may become possible at hypothetical damage to the dam that isolates the basin from the environment.

Basin 365 is an open-water storage reservoir located on the first super-floodplain terrace of the Yenisei River, approximately 100 m from the right bank of the Yenisei River (approximately 0.5 km upstream from Atamanovo) and 50 m above river level.

It was designed for the reception and interim storage of reactor emergency waters and off-grade non-processed wastewater from the radiochemical plant before they are sent to cleaning facilities. Isolation from groundwater is provided by an anti-filtration shield of clay, two asphalt layers on the bottom and slopes, and bottom and bank drainage systems for interception and leak detection in case of damage to the liners. The water in this basin is sent for further cleaning to nearby Pond 366 prior to its disposal. Bottom and bank drainage systems are designed to intercept and collect any leaks from the basin. Estimates of radionuclide inventories in the water and sediments in Basin 365 are shown in Table 4.4.

Based on the volume and surface area of the pond and assuming that the shape is generally triangular to represent a stream channel dammed at the lower end, the height of the resulting dam was estimated to be 8 m. The maximum flow rate from this pond was calculated using the broad-crested weir flow equation in the U.S. National Weather Service simplified dam break code, or SMPDBK (Wetmore and Fread, 1983). Using default parameters for earthen dams, the maximum rate of discharge is 525 m^3/s. The duration of discharge at this rate is approximately 6 minutes, which is considerably less than the 1-day time step used in the river transport model. Therefore, a 1-day pulse release rate is used in the calculations.

The data in Table 4.4 show that for radionuclides observed in both media over 99% of total activity is contained in the sediments. This also proves to be true for our

Table 4.4. Data for Basin 365 at the Krasnoyarsk MCC.

Parameter	*Value*
Water	
Volume (m^3)	204,000
Surface area (m^2)	53,000
Average depth (m)	3.8
Radionuclide content (GBq)	
^{60}Co	13
^{137}Cs	310
^{154}Eu	3
Total beta-activity	7,800
Sediments	
Total volume: solids, slurry, fluidized (m^3)	3,400
Average thickness (m)	0.064
Average porosity	0.4
Radionuclide content (GBq):	
^{60}Co	1,700
$^{103}Ru + ^{106}Ru$	5,800
^{137}Cs	42,000
^{154}Eu	520
^{238}U	11
^{239}Pu	2,000
Total beta-activity	2,5000–50,000

Table 4.5. Distribution coefficients (K_d) of radionuclides between the solid and water phases of the basin.

Radionuclide	K_d
^{137}Cs	10^3–10^4
^{90}Sr	$(2$–$8)10^1$
$^{239,240}Pu$	10^4–10^5
^{241}Am	10^4
^{152}Eu	10^4

determinations of distribution coefficient (K_d) radionuclides between the solid and liquid phases of the basin (Table 4.5).

Therefore, after a release from Basin 365, during transport basin water can be considered a relatively clean carrier of contaminated sediments. The particle-size distribution of basin sediments is unknown. Therefore, we assume that half the radionuclides are sorbed to silt and half to clay. This assumption balances deposition and washout between the silt and clay fractions.

According to an estimate by Waters *et al.* (1999), as a result of a sudden release of radionuclides from Basin 365, up to 37 TBq of radioactivity will enter the Yenisei River, with a ratio between separate radionuclides corresponding to the data given in Table 4.4.

The basic "depot" of radionuclides in the water system of the Yenisei River is floodland soils and bottom sediments. Bottom sediments with significant levels of radionuclides are extremely rare in the near zone of MCC influence (from the point of wastewater release from the MCC up to emptying of the Angara River into the Yenisei River). The content of radionuclides in these sediments is much less significant than those for floodland soils. It should be noted that Waters *et al.* (1999) proceeded from the assumption of identical efficiency of cesium and strontium receipt into floodland soil from the Yenisei River. However, we have proved experimentally that such an assumption is inaccurate: it is shown (Table 4.5) that the distribution coefficient for ^{137}Cs is more than an order of magnitude higher than that for ^{90}Sr. The data obtained here on ^{137}Cs and ^{90}Sr content in floodland soils are in agreement: in all samples, cesium content is an order of magnitude higher than strontium content.

Therefore, the initial data that can be used to model dissipation of radionuclide release from Basin 365 into the Yenisei River are as follows:

1. Suspension sedimentation rate (1.7–2.6×10^{-3} cm/s) and sediment formation rate (1 cm/yr, on average).
2. Radionuclide distribution coefficients [(2–8×10^4 g/L) and (1–4×10^6 g/L) for ^{137}Cs and $^{239,240}Pu$, respectively].

3. Radionuclide diffusion coefficients (1×10^{-8} cm^2/yr for ^{137}Cs).
4. Radionuclide migration forms: suspended, ^{239,240}Pu, ^{241}Am, ^{152,154}Eu (>90%), ^{137}Cs (>80%); soluble, ^{90}Sr (>80%).
5. Factor of radionuclide runoff from the catchment area (0.01–0.03)%/yr.
6. The rate of transfer of fixed forms of radionuclides due to regular sediment erosion/sedimentation is estimated at a distance of 1,500 km downstream from the MCC to be 20 km/yr to 50 km/yr.

At present, the floodlands of the Yenisei River (from the location of the MCC right up to where the Angara River joins the Yenisei River) have accumulated about 850 GBq of radionuclides. Comparison of this value with the amount of a hypothetical release shows that such a release is capable of increasing the total amount of radionuclides at this floodland site by 70 times.

4.2.3 "Tomsk" scenario

The Tomsk scenario involves radionuclide release from the SCC from a nonspecific hypothetical failure of one or more of the engineered containment systems on the site (Waters *et al.*, 1999). Due to the lack of necessary data, the assumption of a hypothetical sudden release (a worst case scenario) was accepted with a total radioactivity amount of 37 TBq (1,000 Ci) of ^{137}Cs and 37 TBq (1,000 Ci) of ^{90}Sr. Such a release is similar to that postulated by engineers at the MCC for evaluation of consequences to the Yenisei River, although it is significantly smaller than the total inventory of surface basins (ponds) at the SCC. However, there are significant uncertainties regarding these ponds. The release was assumed to enter the river primarily as contaminated soils and sediments via an overland flow and through the Ramashka River, which is used by the site to discharge processed water.

A significant difference between the Tom′ and Yenisei Rivers is the general absence of a large floodplain for Tom′ overflow. The river lies in a fairly narrow gorge with bluffs rising steeply to the east of the river and to a lesser extent on the western bank. The primary potential for floodplain-type deposition is on the islands in the Tom′ River. Due to the relative lack of a floodplain and the fact that much of the river channel is exposed during low flows, an overbank section was not modeled for the Tom′ River. This channel is assumed to be the contaminated region of interest.

Similar to the analysis of the MCC, 50% of radionuclides were assumed to be associated with silt sediments and 50% with clay sediments. Discharges around 4,000 m^3/s provide the most significant redistribution of contamination downstream from the release point within the reach of interest. At this rate of discharge, approximately 70% of the silt is expected to be retained here. The highest levels of contamination are expected to occur at the lowest discharges and at locations nearest the discharge point. Contamination densities of over 400 Bq/g may be possible. The highest doses resulting from a large and sudden release of radioactivity from the site into the river occur when the discharges are lowest, which is typically during the winter months. However, significant redistribution of contaminants would likely occur during such high-discharge conditions resulting from spring snowmelt, when the trap efficiency of the reach of interest for both silt and clay is low. Because spring flooding would result in a significant redistribution of radionuclides released to the Tom′, a relatively fast response may be necessary to remediate or contain a release prior to this flooding. This potential need for a relatively quick response highlights the urgency of developing contingency plans prior to such possible release events.

The upper level of soil contamination density for this release is limited to 3,700 Bq/kg so that the finer structure of the lower levels of contamination can be seen. Discharges around 4,000 m^3/s provide the most significant redistribution of contamination downstream of the release point within the reach of interest. At this rate of discharge, approximately 70% of the silt is expected to be retained in the reach. The highest levels of contamination are expected to occur at the lowest discharges and at locations nearest the discharge point. Contamination densities of over 400 Bq/g may be possible.

The results of this unit analysis indicate that the release of even a small fraction of the inventory known to exist in surface basins could result in locally severe contamination. Larger releases, of course, would result in an increase in both the magnitude and extent of contamination. The most important remedial activities that can be conducted in response to this scenario are likely to be those that focus on preventing large releases of radionuclides to the river. Contingency planning should be carried out to establish measures to prevent the spread of contamination from the Romashka to the Tom River system should such an event occur. However, because there are few available data on the characteristics of storage ponds, tanks, and subordinate objects within the SCC, no specific evaluations can be conducted at this time.

The estimates of bottom sediments of the Tom′ River (from the SCC to the confluence of the Tom′ and Ob′ Rivers) are 183.2 (existing stock of radionuclides in bottom sediments) and 33.3 TBq (prospective radionuclide stock in bottom sediments after a sharp release)—that is, the hypothetical release can increase the total inventory of radionuclides in bottom sediments in this river by 140 times.

4.2.4 "CO_2-doubling" scenario

The forcing fields under global warming are derived from an experiment performed using the Bergen Climate Model (BCM) (Furevik *et al.*, 2003), in which the present-day CO_2 level is assumed to increase at the rate of 1% per year until CO_2 is doubled.

4.2.5 "Submarine" scenarios

The simulation scenarios here are based on real accidents with the nuclear submarine *Kursk* that occurred in August 2000 and with the decommissioned nuclear submarine *K-159* on August 30, 2003. The *Kursk*, a nuclear submarine of the Northern Fleet of Russian Navy, sank on August 12, 2000 in the Barents Sea at 69°40′00N, 37°35′00E to the sea floor at a depth of 108 m, and at a distance of 150 miles from Murmansk. The submarine's displacement was around 14,000 tons surfaced and 24,000 tons submerged. It contained two pressurized water nuclear reactors OK-650b of 190 MW (FAS, 2000a). At the time of the accident, the vessel was participating in fleet exercises. The submarine was not armed with nuclear weapons. It apparently sank quickly, but it was reported that the reactors were shut down. All of the 118 crew members onboard perished. The boat was raised by a salvage operation in 2001. The reason for the disaster was an emergency situation in the submarine torpedo compartment. As a result of an explosion the front part of the pressure hull was seriously damaged. However, no enhanced radiation levels in the water or atmosphere were detected in measurements by the Russian and Norwegian authorities on August 14, 2000.

Information about the nuclear inventory of the *Kursk*'s reactors is still not publicly available. Therefore, we used information about the reactor onboard the *Komsomolets* (Høibråten *et al.*, 1997) which is of a similar type (OK-650b-3) to the two reactors onboard the *Kursk* (OK-650b). Information from the Kurchatov Institute about the *Komsomolets*'s reactor reveals that total activity was 29 PBq shortly after shutdown of the reactor (Høibråten *et al.*, 1997). This corresponds to permanent operation of the reactor for a period of 5 years. In all likehood this is right for the *Kursk*, which was commisioned in 1995. The ^{90}Sr inventory was 2,800 TBq, and for ^{137}Cs the amount was 3,100 TBq, shortly after shutdown of the *Komsomolets* reactor. Using these numbers the inventory of the two *Kursk* reactors can be estimated as 5,600 TBq for ^{90}Sr and 6,200 TBq for ^{137}Cs.

Only one hypothetical release scenario is presented in Section 4.5: instantaneous leakage. It was assumed the submarine was damaged during the attempt to lift it in the course of a salvaging operation on August 14, 2001. The reactor compartment disintegrates and an instantaneous release takes place. To simplify the problem, we considered only the effects of a release of ^{137}Cs. We adopted a value of 4,500 TBq, which is, of course, a strongly overestimated value. However, because the problem is linear the results of the calculation can also be used for other levels of activity using a multiplier. The month of August was chosen as the time of release, keeping in mind an appropriate time for the salvaging operation.

Another submarine, the *K-159*, sank August 30, 2003 in the Barents Sea to 238 m, when the vessel was being towed to a shipyard for dismantling, with a crew of nine and spent nuclear fuel in its reactors. Its displacement was around 3,065 tons surfaced and 4,075 tons submerged. It contained two pressurized water nuclear reactors VM-A of 70 MW (FAS, 2000b). It was assumed that the submarine potentially starts to leak radioactive material to the water from its location at 240 m depth in the Murmansk Fjord. The rate of leakage is constant and this leakage continues for 5 years.

4.3 ASSESSMENTS OF POTENTIAL ACCIDENTAL RELEASES FOR THE 21ST CENTURY

This section presents the potential radioactive contamination of the Ob′ and Yenisei Rivers and the flux of radionuclides from a hypothetical accidental radionuclide release into the environment from the following Russian radiochemical plants: the Mayak PA, the Krasnoyarsk MCC, and the Tomsk SCC. Five different situations (Scenarios I–V) of radionuclide pollution of the rivers for the present-day scenario and the $2 * CO_2$ scenario were simulated by a river–sea model chain to predict their spread in the Kara Sea, using the release scenarios from Section 4.2.

4.3.1 Potential radioactive contamination from rivers to the Kara Sea

In this subsection, potential radioactive contamination of the Ob′ and Yenisei Rivers and flux of radionuclides from a hypothetical accidental release from nuclear plants to the Kara Sea are simulated. The transport of radionuclides in rivers depends essentially on seasonal and interannual changes in discharge. Moreover, climate variations on decadal, multidecadal, and longer timescales can cause changes in river discharge. This effect can be important with respect to global warming and increasing CO_2 concentration. Therefore, the effect of discharge variability on seasonal and multidecadal scales is also considered and simulation results of potential radioactivity releases in the 20th and 21st centuries are compared.

River runoff in the 20th and 21st centuries

In Figures 4.19 and 4.31, variability in the annual runoff of the Ob′ and Yenisei are given for the period 1934–1995. The period 1960–1980 was chosen for its quasi-normal meteorological conditions. To estimate future changes in river runoff as a result of climate change, we used simulations made by the atmospheric general circulation model ECHAM of Ob′ and Yenisei River runoff in the 21st century under the $2 * CO_2$ scenario. Simulated river runoff in 1936–2099 under the $2 * CO_2$ scenario vs. observed discharges of the Ob′ and Yenisei rivers at Salekhard and Igarka, respectively, are shown in Figures 4.45–4.46. Whereas the simulated trend in Ob′ River runoff agrees with observations for 1936–1995, simulated values of Yenisei River runoff are offset from observations by about $159\,km^3/yr$. Therefore, the simulated runoff series was corrected accordingly and this corrected series was used for comparison with the period 2079–2099, which was chosen under climate change conditions ($2 * CO_2$). Observed annual runoff of the Ob′ and Yenisei rivers in 1960–1980 vs. that predicted by the ECHAM model in 2080–1999 under the $2 * CO_2$ scenario are shown in Figures 4.47–4.48. In both cases, mean runoff for the period 2080–1999 is greater than that for 1960–1979.

Climate simulations from the ECHAM model predict river discharges into large river mouths. However, discharges into river tributaries are necessary to simulate radionuclide transport through the river network (e.g., from the Mayak PA through the Techa–Iset–Tobol–Irtysh–Ob′ to the estuary). Therefore, a means of reconstructing river discharge into tributaries using river runoff and climate correlations had to be developed.

The mean annual value of river discharge in a river mouth is represented by Q; Q_i is mean annual discharge of the ith tributary, where $i = 1, m$ (m is the number of tributaries). Let us assume that $Q_i = \alpha_i Q$, where $\alpha_i \geq 0$. The climate values of

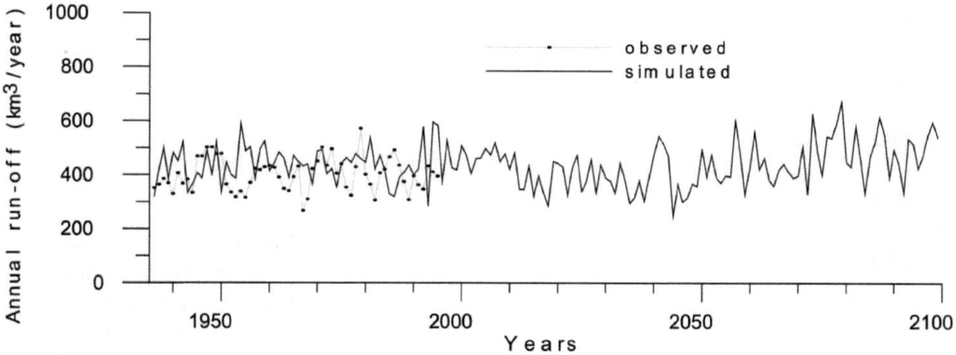

Figure 4.45. ECHAM model–simulated Ob′ River runoff in 1936–2099 with the $2 * CO_2$ scenario vs. observed discharge in Salekhard.

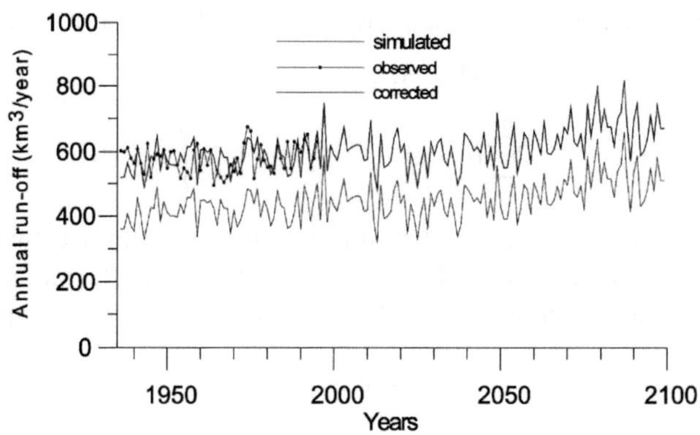

Figure 4.46. ECHAM model–simulated Yenisei River runoff in 1936–2099 with the $2 * CO_2$ scenario vs. observed discharge in Igarka. Yhe figure also shows the corrected ECHAM run.

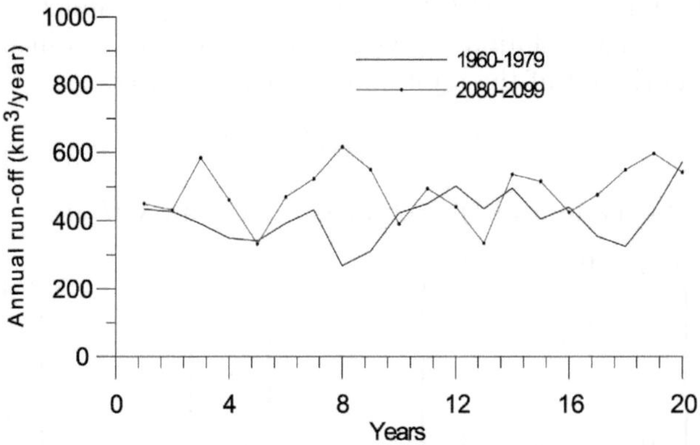

Figure 4.47. Observed 1960–1980 annual runoff of the Ob′ River vs. ECHAM model–predicted runoff from 2080 to 1999 with the $2 * CO_2$ scenario.

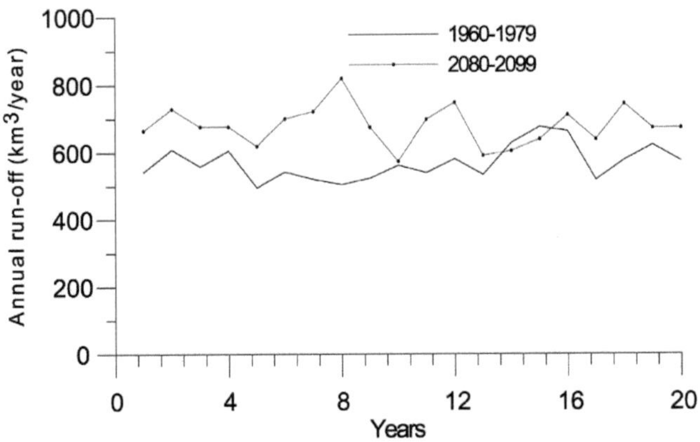

Figure 4.48. Observed 1960–1980 annual runoff the Yenisei River vs. ECHAM model–predicted runoff from 2080 to 1999 with the $2 * CO_2$ scenario.

discharges are $\bar{Q} = M, \bar{Q}_i = M_i$. Then the relevant relations for α_i are

$$\left.\begin{aligned} \alpha_i M = M_i, \qquad i = 1, m \\ \sum_1^m \alpha_i = 1 \end{aligned}\right\} \tag{4.1}$$

Hence, mean annual discharge into tributaries can be determined by solution of the system (4.1) about coefficients α_i. The system (4.1) is overdetermined, therefore the method of least squares can be used. Let us consider the system of equations written in matrix form

$$Hx = z \tag{4.2}$$

where

$$x = \begin{pmatrix} x_1 \\ \vdots \\ x_m \end{pmatrix}, \qquad z = \begin{pmatrix} z_1 \\ \vdots \\ z_n \end{pmatrix}$$

are vectors with dimension m and $n > m$, correspondingly; and H is a matrix with dimension $(n \times m)$. The solution of system (4.2) is vector $\hat{x}$, which minimizes the square of norm deviation of the right and left parts; that is

$$H\hat{x} \underset{x}{\longrightarrow} \min \|Hx - z\|^2 \tag{4.3}$$

For the system of equations (4.1), vectors x, z and matrix H are

$$H = \begin{pmatrix} M & 0 & \cdots & 0 \\ 0 & M & \cdots & 0 \\ \vdots & \vdots & & \vdots \\ 0 & 0 & \cdots & M \\ 1 & 1 & \cdots & 1 \end{pmatrix}, \qquad x = \begin{pmatrix} \alpha_1 \\ \vdots \\ \alpha_m \end{pmatrix}, \qquad z = \begin{pmatrix} M_1 \\ \vdots \\ M_m \\ 1 \end{pmatrix}$$

Considering that

$$\|Hx - z\|^2 = \sum_{i=1}^{m}(M_i - \alpha_i M)^2 + \left(1 - \sum_{i=1}^{m}\alpha_i\right)^2 \tag{4.4}$$

vector $(\alpha_1, \alpha_2, \ldots, \alpha_m)$, minimizing this square norm satisfies the equation set

$$Ax = \hat{z} \tag{4.5}$$

where A is a square nonsingular matrix having order $(m \times m)$; and $\hat{z}$ is a vector having order m written as

$$A = \begin{pmatrix} 1+M^2 & 1 & \cdots & 1 \\ 1 & 1+M^2 & \cdots & 1 \\ \vdots & \vdots & & \vdots \\ 1 & 1 & \cdots & 1+M^2 \end{pmatrix}, \quad \hat{z} = \begin{pmatrix} 1+M_1M \\ 1+M_2M \\ \vdots \\ 1+M_mM \end{pmatrix}$$

The linear equation set (4.5) was solved by the Gauss elimination method. Seasonal change of discharge into a river system can be calculated by the same procedure using seasonal variations of discharge for each tributary. Parameter values for such a reconstruction were calculated on the basis of 1960–1980 data whereas 1981–1990 data were used for validation. Observed and reconstructed mean year discharge of the Podkamennaya Tunguska in 1980–1990 are shown in Figure 4.49 and comparison of seasonal change between observed and reconstructed discharge of the Angara River in 1985 is shown in Figure 4.50.

Simulation of flux of radionuclides to the Kara Sea from a hypothetical accidental release from nuclear plants

The strong seasonal variability in river flow has an effect on radionuclide transport in the Ob′ and Yenisei River systems. Because of this, we consider scenarios of accidental releases from the Mayak PA, the SCC, and the MCC, listed in Table

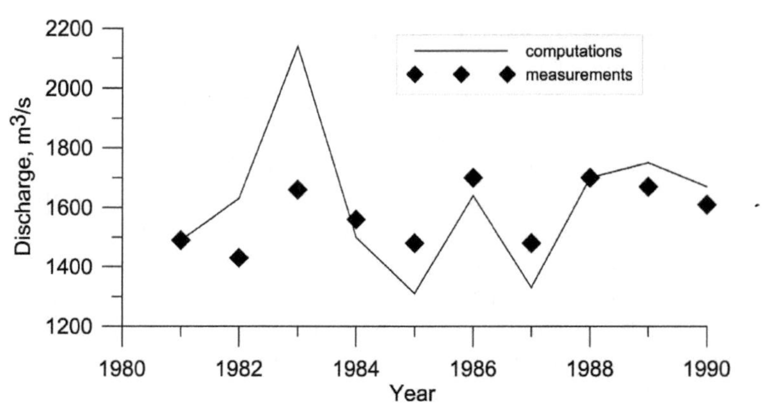

Figure 4.49. Observed vs. reconstructed mean year discharge of the Podkamennaya Tunguska River in 1980–1990.

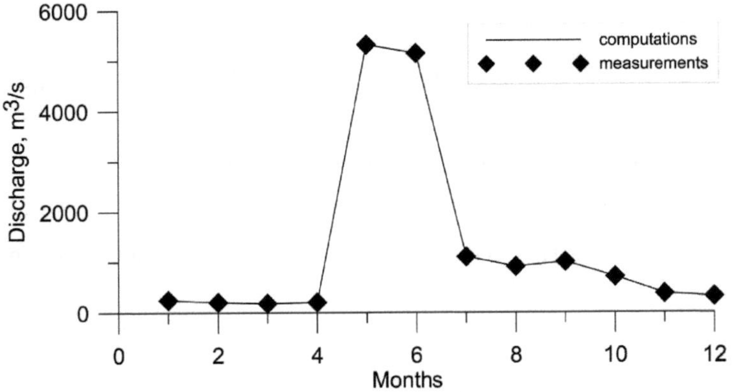

Figure 4.50. Observed vs. reconstructed discharge of the Podkamennaya Tunguska River in 1985.

4.6 for two seasons: winter and summer for the SCC and the MCC and spring and autumn for the Mayak PA. Spring and autumn were chosen for the Mayak PA because the small Techa River is completely frozen (down to the bottom) in winter. Results of a "quasi-normal" control run (QN run) can be compared with runs for the 21st century using CO_2-doubling forcing data to study the impact of global warming on radioactive spread ($2 * CO_2$ run).

Simulated discharges of the Ob′ River in 1960–1980 and 2080–2100 are given in Figure 4.51. The large interannual variability of Ob′discharge should be pointed out. Simulation results given in Section 4.1.1 showed that the flux of radionuclides calculated by the one-dimensional river model correlated well with calculations by the three-dimensional model. Therefore, we only present the simulation results of the Ob′ River and estuary by the one-dimensional RIVTOX model. Radionuclide concentration in the estuary mouth is given in Figures 4.52–4.54 for Scenarios I–III. Temporal evolution of the concentration of ^{90}Sr in the Ob′ estuary mouth for a hypothetical release from the Mayak PA is plotted in Figure 4.52, which shows the predominant transport to be the fraction of ^{90}Sr in solute. Transport through the river chain Techa–Iset–Tobol–Irtysh essentially depends on the season of release. Though the difference between release time (April and November) is 6 months, the difference in appearance of the maximum concentration of ^{90}Sr is around 1 year, which can be explained by the effect of high summer flooding in the Ob′ River system. The maxima of concentration in the mouth of the Ob′ estuary for the QN and $2 * CO_2$ scenarios are close and relatively low (100 Bq m^{-3}), although the forms of the curves differ. Figure 4.53 shows the concentration of ^{137}Cs in the Ob′ estuary mouth for a hypothetical release from the Mayak PA. As was the case in Section 4.1.1, most ^{137}Cs was deposited in river and estuary sediments and as such the effect of such a release was very low. Temporal evolution of the concentration of ^{90}Sr in the Ob′ estuary mouth for a hypothetical release from the SCC (Scenario III) is plotted in Figure 4.54. Due to the shorter pathway and higher velocities, radioactivity appears in the estuary mouth earlier than in Scenario I. As in Scenario I, the maxima of concentration in both the QN and $2 * CO_2$ cases are close, although once again the forms of the curves differ. Because of extremely high summer flooding in the Yenisei,

Table 4.6. Scenarios for a hypothetical accidental release from nuclear plants to the Kara Sea.

Scenario	*River*	*Scenario of hypothetical accident*	*Time of accident in QN scenario*	*Time of accident in $2 * CO_2$ scenario*	*Radionuclide*
1a	Ob′	Mayak, break in Reservoir-11's dam	April 5, 1967	April 5, 2084	^{90}Sr
1b	Ob′	Mayak, break in Reservoir-11's dam	November 5, 1967	November 5, 2084	^{90}Sr
2a	Ob′	SCC, break in the holding pond	January 5, 1967	January 5, 2084	^{137}Cs
2b	Ob′	SCC, break in the holding pond	June 4, 1967	June 4, 2084	^{137}Cs
3a	Ob′	SCC, break in the holding pond	January 5, 1967	January 5, 2084	^{90}Sr
3b	Ob′	SCC, break in the holding pond	June 4, 1967	June 4, 2084	^{90}Sr
4a	Yenisei	MCC, breaks in holding ponds No. 365 and No. 366	January 5, 1967	January 5, 2089	^{137}Cs
4b	Yenisei	MCC, breaks in holding ponds No. 365 and No. 366	June 3, 1967	June 3, 2089	^{137}Cs
5a	Yenisei	MCC, breaks in holding ponds No. 365 and No. 366	January 5, 1967	January 5, 2089	^{90}Sr
5b	Yenisei	MCC, breaks in holding ponds No. 365 and No. 366	June 3, 1967	June 3, 2089	^{90}Sr

the form of the curve of concentration in Scenarios IV and V is characterized by high and short spikes (Figure 4.55). Whereas the concentrations of ^{137}Cs and ^{90}Sr in these spikes are similar, the concentration of ^{137}Cs in sediments is some two orders of magnitude higher than ^{90}Sr (Figures 4.56 and 4.57). In total, as for the Ob′ River scenarios there is not much difference between the QN and $2 * CO_2$ cases.

4.3.2 Potential radioactive contamination in the Kara Sea

The climate change simulations in this section are done assuming a doubling of atmospheric CO_2 concentration and, accordingly, altering water fluxes in aquatic system (rivers, estuaries, and oceans).

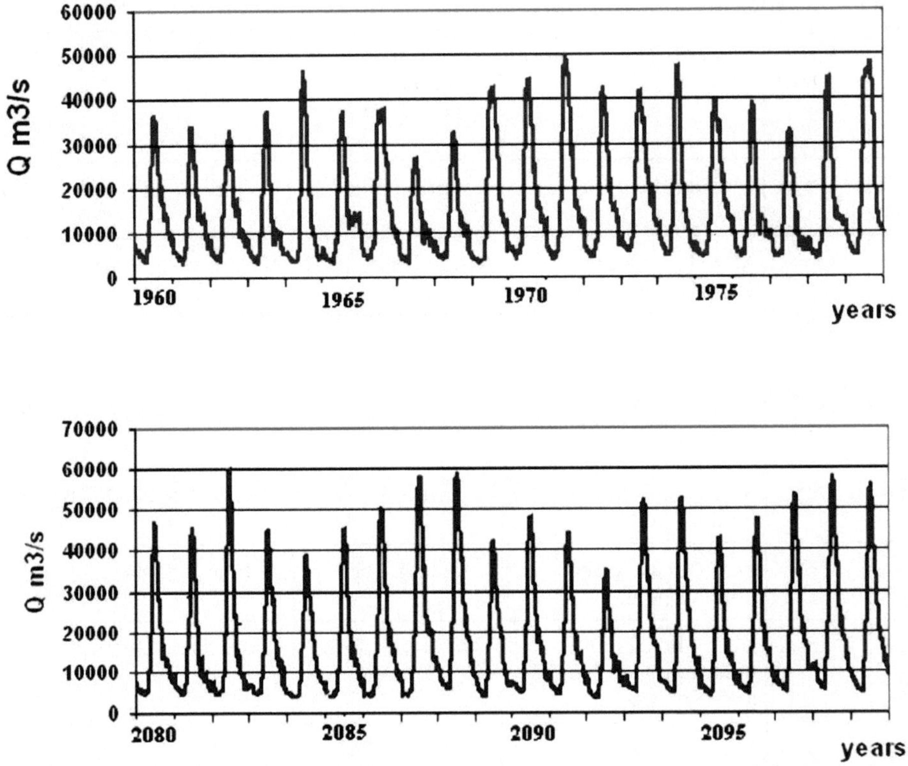

Figure 4.51. Discharge of the Ob′ River in 1960–1980 (upper) and 2080–2100 (lower).

Transport of ^{90}Sr from the Ob′ to the Kara Sea in 2084–2086 (Scenario I)

River runoff and radionuclide flux data were estimated from river and estuary simulations, taking into account $2 * CO_2$-induced climate changes, which accordingly influences precipitation and the river volume flux of water discharge. Radionuclide transport from the Ob′ River in 2084–2086 was driven by the simulated ocean circulation system of the Kara Sea. The only difference from previous experiments is that river runoff is changed and the source term of radionuclides is reduced as a result of their decay time.

Atmospheric circulation conditions and boundary conditions at the liquid boundary of the Kara Sea are the same as for 1948–2000. Therefore, the atmospheric impact and boundary conditions correspond to the 1984–1986 period. The Kara Sea circulation pattern was identical. The maximum radionuclide discharge was on April 5, 2084 according to this scenario. Simulated changes in the dispersal of ^{90}Sr concentration for the years 2085 to 2086 are shown in Figure 4.58 (see color section). The maximum concentration of 10 Bq/m^3 of radionuclide corresponds to January 2085 near the shelf adjoining Ob′ Bay. About a year later, in December 2086, the maximum concentration is half this value.

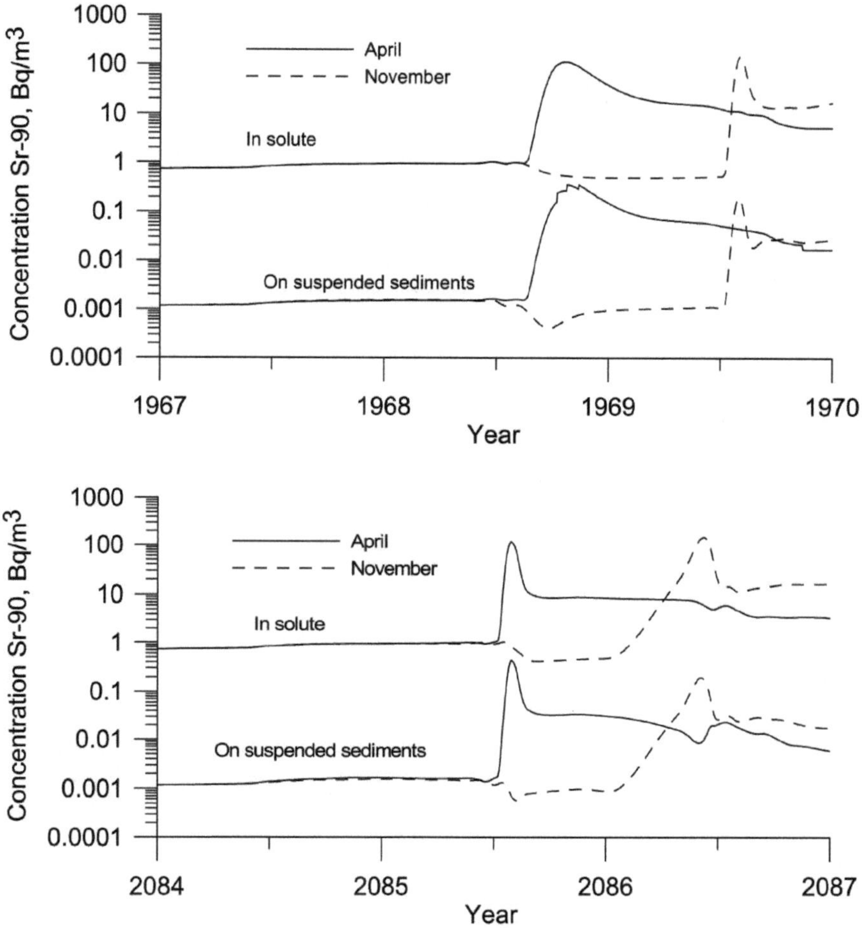

Figure 4.52. Calculated concentration of ^{90}Sr in the Ob′ estuary mouth for Scenario I.

Transport of ^{137}Cs from the Ob′ in 2084–2089 (Scenario II)

Simulations of the transport of ^{137}Cs from the Ob′ to the Kara Sea during 2084–2089 were driven by the ocean circulation system of the Kara Sea. As in the previous scenario, the only parameters affecting Kara Sea circulation and radionuclide transport were climatic alteration in river discharge and radionuclide concentrations. Atmospheric circulation conditions and boundary conditions for the Kara Sea's open boundaries were obtained from calculations for 1984–1989, as were the atmospheric impact and boundary conditions. River and estuary simulations indicate a peak (maximum) discharge of radionuclides which takes place on January 5, 2084. Figure 4.59 (see color section) shows the simulated distribution pattern of ^{137}Cs concentration in the Kara Sea for 2085–2089. The maximum spread of radionuclide according to the given scenario occurs in January 2088. By December 2086, radioactive pollutants have dispersed throughout the entire Kara Sea region.

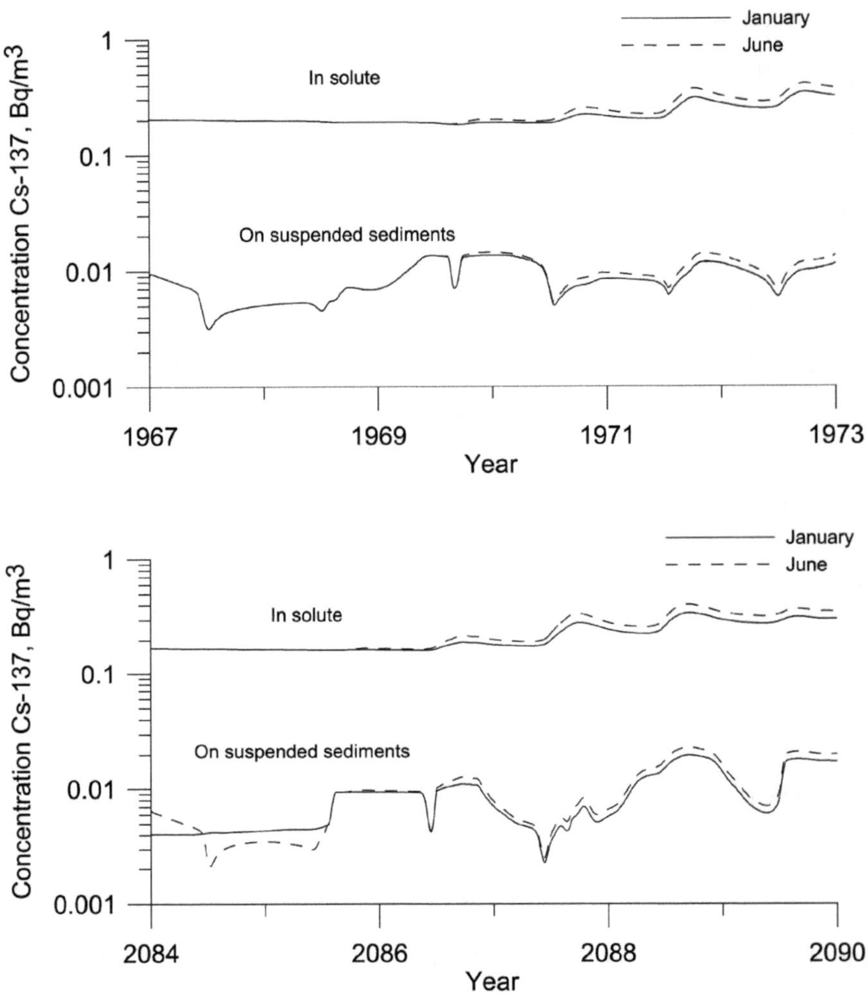

Figure 4.53. Calculated concentration of ^{137}Cs in the Ob′ estuary mouth for Scenario II: (upper) QN, and (lower) $2 * CO_2$.

Transport of ^{90}Sr from Ob′ to the Kara Sea during 2084–2086 (Scenario III)

Simulations of the transport of radionuclide ^{90}Sr from the Ob′ to the Kara Sea during 2084–2086 were driven by the ocean circulation system of the Kara Sea. As in the previous scenario the only parameter affecting Kara Sea circulation and radionuclide transport was slightly modified river discharge. Atmospheric circulation and boundary conditions on the open boundaries of the Kara Sea were obtained from calculations for the period 1984–1989. The peak (maximum) estuarine discharge of radionuclides took place on January 1, 2084. Simulated distribution of the ^{90}Sr concentration is shown in Figure 4.60 (see color section). The maximum radionuclide spread according to the given scenario occurs in January 2085. By December 2086,

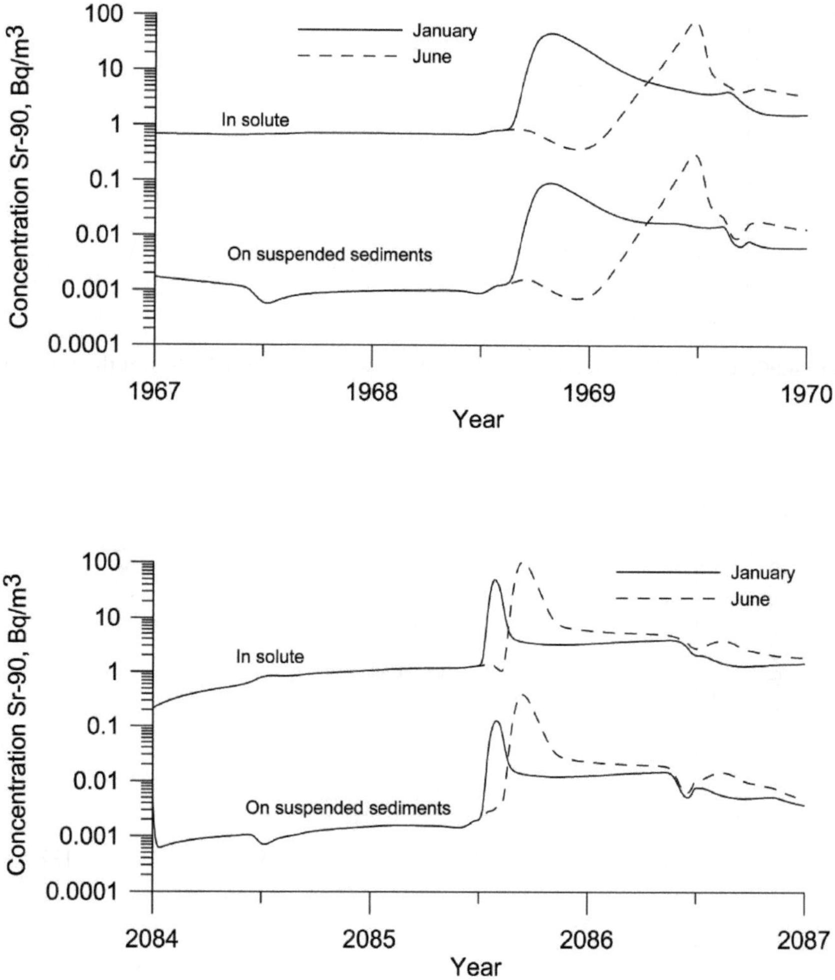

Figure 4.54. Calculated concentration of ^{90}Sr in the Ob′ estuary mouth ($2 * CO_2$ Scenario III): (upper) QN, and (lower) $2 * CO_2$.

radionuclear pollutants have dispersed throughout the entire Kara Sea region, and the maximum concentration near the Ob′ estuary has decreased tenfold.

^{137}Cs transport from the Yenisei during 2089–2094 (Scenario IV)

Simulated transport of ^{137}Cs from the Yenisei to the Kara Sea during 2089–2094 years was driven by the ocean circulation system of the Kara Sea. As in the previous scenario, the only parameter affecting Kara Sea circulation and radionuclide transport was slightly modified river discharge.

Atmospheric circulation conditions and boundary conditions at the Kara Sea's open boundary were obtained from calculations for 1989–1994. The peak (maximum)

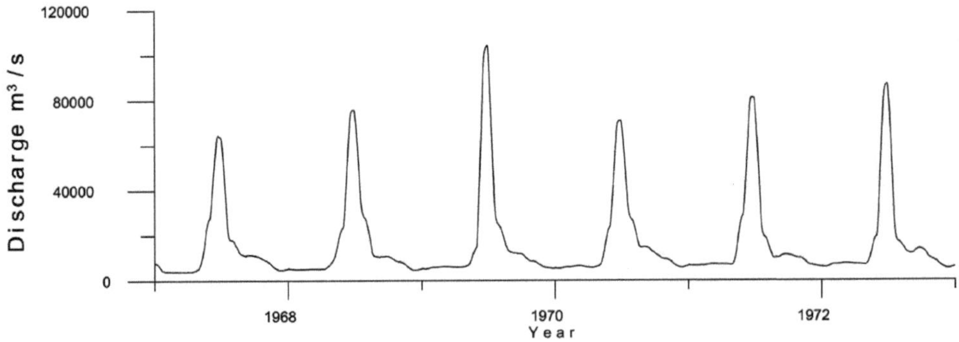

Figure 4.55. Calculated discharge of the Yenisei in the Yenisei Gulf mouth, 1967–1973.

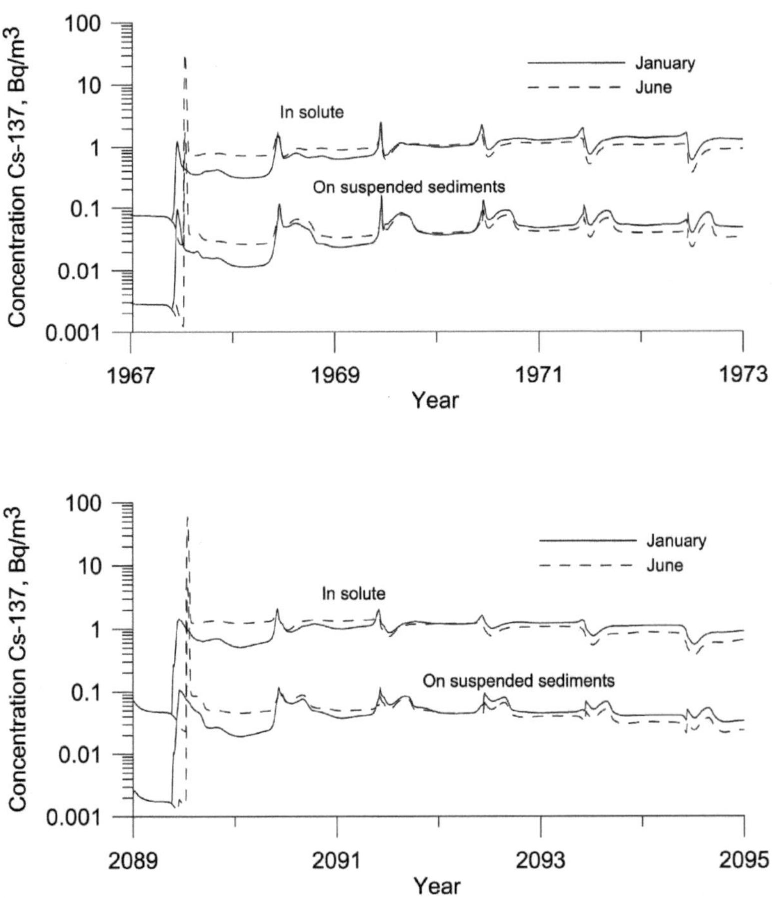

Figure 4.56. Calculated concentration of ^{137}Cs in the Yenisei Gulf mouth for Scenario IV: (upper) QN, and (lower) $2 * CO_2$.

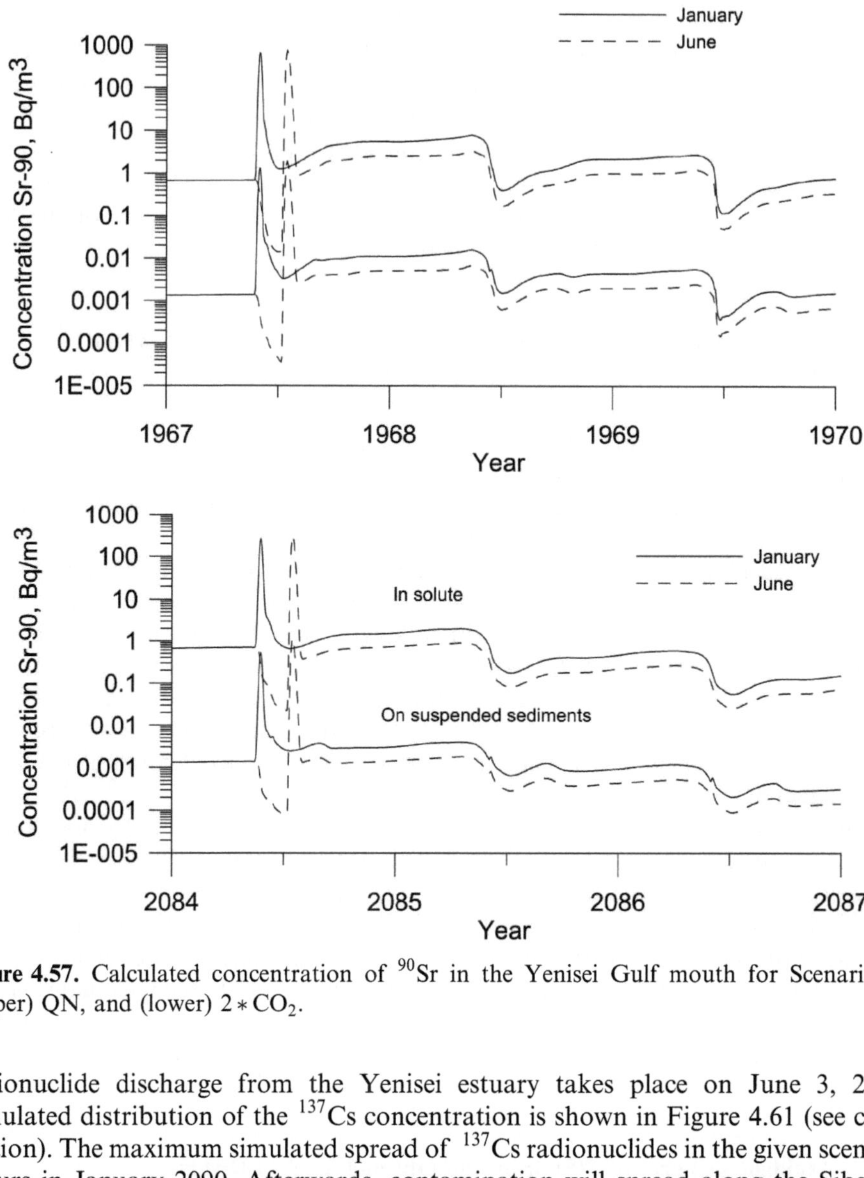

Figure 4.57. Calculated concentration of ^{90}Sr in the Yenisei Gulf mouth for Scenario V: (upper) QN, and (lower) $2 * CO_2$.

radionuclide discharge from the Yenisei estuary takes place on June 3, 2089. Simulated distribution of the ^{137}Cs concentration is shown in Figure 4.61 (see color section). The maximum simulated spread of ^{137}Cs radionuclides in the given scenario occurs in January 2090. Afterwards, contamination will spread along the Siberian coast to the northeast.

^{90}***Sr transport from the Yenisei for the 2084–2086 period (Scenario V)***

Simulated transport of ^{90}Sr from the Yenisei to the Kara Sea during the years 2084–2086 was performed according to the ocean circulation and currents of the Kara Sea. As in the previous scenario, the only parameter affecting Kara Sea circulation—and accordingly the transport of radionuclides—was a slightly modified river discharge. Atmospheric circulation conditions and boundary conditions at the Kara Sea's open boundary were obtained from hindcast calculations performed for the 1984–1986 time period. Peak (maximum) discharge of radionuclides takes place on January 5, 2084. Temporal evolution of the distribution of ^{90}Sr concentration is shown in Figure 4.62 (see color section). According to the simulations, the maximum spread of ^{90}Sr, given this river's input scenario, occurs one year after the release in January 2085. Afterwards the contaminated area advects along the coast of Siberia to the northeast with a gradual decrease in concentration.

4.4 TRANSPORT OF RADIOACTIVITY IN THE ARCTIC AND POSSIBLE IMPACT OF CLIMATE CHANGE

Ocean circulation and mixing processes govern the transport of all conservative substances in the ocean. Because the global climate system changes as a result of increasing greenhouse gases in the atmosphere, the dynamics and thermodynamics of the ocean will also change. The changed ocean climate will have consequences for the transport and mixing of radionuclides in the Atlantic–Arctic region. To assess the spread and mixing characteristics of accidentally released radionuclides under a global-warming scenario, twin experiments with the 80 km global version of MICOM, as used in Gao *et al.* (2004), have been performed with present-day forcing and with possible forcing resembling the future climate in the 21st century. Differences in transport and mixing between the two climate situations provide a scenario for possible changes in the transport routes and concentration distributions of radionuclides.

To simulate the ^{90}Sr originating from accidental releases in the Ob′ and Yenisei Rivers, twin-integrations were conducted. The baseline integration is similar to the hindcast simulation described above (i.e., a simulation under present climate conditions covering the time period from 1970 to 2000 using daily NCEP/NCAR forcing). For the global-warming scenario, daily atmospheric forcing fields were extracted from a $2 * CO_2$ simulation performed with the Bergen Climate Model (BCM) (Furevik *et al.* 2003; Sorteberg *et al.* 2005). Under $2 * CO_2$ forcing, the integration was performed twice using the same forcing to stabilize the surface water simulated.

4.4.1 Accident scenario of ^{90}Sr from the Ob′ and Yenisei Rivers

Hypothetical accidents were assumed in the simulation of accidental releases. In the Ob′ River, the scenario assumes a sudden and uncontrollable break in Reservoir-11's dam at the Mayak PA. In the Yenisei River, a hypothetical accident at the Mining and Chemical Combine (MCC) assumes an instantaneous release of ^{90}Sr into the Yenisei River. The river model is used to generate the temporal evolution of the ^{90}Sr concentration at Ob′ Bay and at the Yenisei Gulf of the Kara Sea. Maximum accidental releases take place in summer, with peak values of 116 Bq m^{-3} for the Ob′ River and 302 Bq m^{-3} for the Yenisei River.

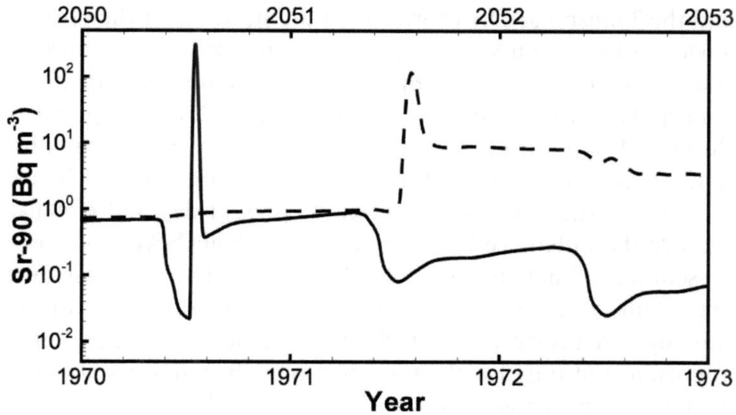

Figure 4.63. Accidental release of ^{90}Sr (Bq m^{-3}) from the the Ob′ River (dashed line) and the Yenisei River (solid line).

Figure 4.63 depicts a hypothesized accidental release of ^{90}Sr (Bq m^{-3}) from the Ob′ River (dashed line) and the Yenisei River (solid line). Data are from Johannessen *et al.* (2003). The x-axes indicate the time of the simulation under present-day (lower x-axis) and $2 * CO_2$ (upper x-axis) climate forcing.

Integration of the accidental release scenarios starts from 1970 with zero concentrations in seawater and ends in 2000 under present-day climate conditions, and from 2050 with zero concentrations in the sea and ends at 2080 under $2 * CO_2$ forcing.

4.4.2 Spread of accidentally released ^{90}Sr under present and $2 * CO_2$ warming scenarios

The short-term (5-year) surface spread of accidentally released ^{90}Sr from the Ob′ River is shown for both present-day (Figure 4.64a, see color section) and $2 * CO_2$ warming scenario runs (Figure 4.64b). It shows that ^{90}Sr spreads eastwards following the coast of Siberia and a large amount of accidentally released ^{90}Sr is confined to the Arctic Ocean in the $2 * CO_2$ warming scenario run, instead of being advected westward towards Fram Strait under the present-day climate situation. The spread of accidentally released ^{90}Sr from the Yenisei River (not shown) is similar to that of the Ob′: transport is also confined to the Arctic Ocean and the Nordic Seas are radioactivity-free in the $2 * CO_2$ warming scenario run within the first 5 years. For both Ob′ and Yenisei discharges, the peak values of surface ^{90}Sr concentration move out of the Kara Sea in the $2 * CO_2$ warming scenario run, instead of being blocked in the Kara Sea under present-day forcing. Consequently, vulnerable regions for potential accidental release of ^{90}Sr in the Ob′ and Yenisei Rivers in a future climate is along the coast of Siberia.

This finding strongly suggests that the high-latitude ocean circulation pattern is very different between the two forcing situations. The reason for modeled differences in the spatial distributions of Ob′ and Yenisei releases of ^{90}Sr follows the main surface circulation in the Arctic Ocean—namely, the strong clockwise Beaufort Gyre and the Transpolar Drift towards Fram Strait under the present-day climate (Figure 4.64c)—whereas under the global-warming scenario (Figure 4.64d) the circulation changes as follows: (1) Beaufort Gyre circulation is dramatically weakened, (2) the Transpolar Drift disappears, and (3) there is less transport from the Arctic Ocean through Fram Strait and along the EGC, with potentially major implications downstream in deepwater formation in the subpolar and Nordic Seas.

In fact, the difference between the present-day and global-warming simulations exceeds the difference between years with high-NAO and low-NAO forcing in the present-day climate (see fig. 8 in Gao *et al.*, 2004). This was the first time that such a drastic and fundamental change in Arctic Ocean circulation under global warming was suggested (Gao *et al.*, 2009). The significance of such a change would clearly be important not only for the transport of radioactivity and other pollutants, but also would have major geophysical significance. The robustness of modeling results and the underlying mechanisms for such a change will be explored in further research.

Long-term concentration characteristics are provided in Figure 4.65 (see color section). It follows that more of the released ^{90}Sr is confined to the Arctic Ocean in the global-warming run, and that the distribution is different in the Arctic Ocean with a change in the location of ^{90}Sr-enriched waters from the Kara Sea region under present-day forcing to the coastal, non-European sector in the global-warming run. It should be stressed that although there are large local differences in the spatial distribution of tracers, the displayed concentration levels are very low.

4.5 POTENTIAL TRANSPORT OF RADIOACTIVITY FROM SUBMARINE ACCIDENTS

The transport of radioactive materials by ship and the presence of naval submarines, ships, and icebreakers in the Arctic Ocean constitute a potential danger because of the large amount of radioactivity that could be accidentally released into the marine environment. The three latest accidents in the Norwegian and Barents Seas, involving the Russian submarines *Komsomolets*, *Kursk*, and *K-159*, have demonstrated the need to develop special tools for decision makers—tools that are be able to hindcast and forecast radioactive contamination in the accident area. These "submarine" scenarios are described in Section 4.2.5.

In this section, the simulation results of a hypothetical accidental release from sources in the Barents Sea are presented. Simulations were carried out to show the ability of the GMS to cope with environmental modeling associated with such scenarios. Two models have been applied: the first is the three-dimensional local hydrodynamic and radionuclide transport model THREETOX that was used for local-scale simulation of radionuclide dispersion as a result of a hypothetical release from the *Kursk*. The other model is the global version of MICOM. It was applied for large-scale and long-term simulations of radioactivity dispersal according to a hypothetical scenario of release from the *K-159*.

4.5.1 Local model simulations

The bottom topography of the Barents Sea is shown in Figure 4.66 (see color section). The area under consideration includes the Barents Sea, the White Sea, and the western part of the Kara Sea. Grid resolution in the horizontal was $24' \times 12'$ resulting in 75×103 nodes, with the nested $8' \times 4'$ grid resulting in 55×43 nodes. The vertical direction was resolved by 13 depth-adapted layers using logarithmic stretching for near-surface and near-bottom nodes. Daily wind, air temperature, humidity, cloudiness, and air pressure fields from NCEP/NCAR reanalysis were used as forcing in the simulations. The discharge rates of the Severnaya Dvina River and the Pechora River were obtained from Pavlov (1994). At open-sea boundaries we used simulation data from a regional model GMS. The initial values of the temperature and salinity fields were specified by using regional model GMS data. Simulations were carried out for the entire year (2001).

Information about the nuclear inventory of the *Kursk*'s reactors is still not publicly available. So, we used information about the reactor onboard the *Komsomolets* (Høibråten *et al.*, 1997), which is of a similar type (OK-650b-3) to the two reactors onboard the Kursk (OK-650b). Information from the Kurchatov Institute about the *Komsomolets* reactor reveals that total activity was 29 PBq shortly after shutdown of the reactor (Høibråten *et al.*, 1997). This corresponds to permanent operation of the reactor for a period of 5 years. In all likelihood this is essentially correct for the *Kursk*, which was commisioned in 1995. The ^{90}Sr inventory was 2,800 TBq, and for ^{137}Cs the amount was 3,100 TBq, shortly after shutdown of the *Komsomolets* reactor. After a few decades, ^{90}Sr and ^{137}Cs predominate. Using these numbers, the inventory of the two *Kursk* reactors can be estimated as 5,600 TBq for ^{90}Sr and 6,200 TBq for ^{137}Cs.

Only one hypothetical release scenario is presented here: instantaneous leakage. It is assumed that the *Kursk* is damaged during the attempt to lift it in the course of a salvaging operation on August 14, 2001. The reactor compartment disintegrates and an instantaneous release takes place. To simplify the problem, we considered only the effects of a release of ^{137}Cs. We adopted a value of 4,500 TBq, which is, of course, a strongly overestimated value. However, because the problem is linear, the calculation results can also be used for other levels of activity using a multiplier. The month of August was chosen as the time of release—an appropriate time for a salvaging operation.

Simulated concentrations of ^{137}Cs at the surface and at the near-bottom layer of the Barents Sea after 12, 27, 57, and 105 days since the release are shown in Figures 4.67 and 4.68 (see color section for both). The plume of contaminated water moves eastward and southward along the Kola Peninsula. The maximum surface concentration of ^{137}Cs in the plume diminishes gradually from a maximal value of 300 Bq/m^3 to 120 Bq/m^3. Maximum bottom contamination was around 3,000 Bq/kg. The bottom was a secondary source of water contamination, and the concentration here decreased threefold after 3.5 months. According to Kershow *et al.* (1997) the level of ^{137}Cs activity in water of the Barents Sea in 1994 was in the range 3 Bq/m^3 to 11 Bq/m^3. The levels of ^{137}Cs at the bottom were less than 10 Bq/kg (Foyn and Sværen,

1995). Therefore, long-term dilution resulted in decrease of the contamination level over the background level. However, if it is necessary to carry out a rescue operation in close proximity to a ship after radioactivity release (e.g., the crew of the *Komsomolets* spent some hours extinguishing a fire; some of the *Kursk*'s crew survived at least one day after an explosion in the bow compartment) then real-time tools should be used to estimate and forecast local radioactivity concentration and transport around the place of the accident.

4.5.2 Large-scale model simulations

In this study, the global version of MICOM (at 40 km resolution) in the North Atlantic/Nordic Seas is used to simulate the potential spread of radioactivity from the Russian submarine *K-159* (which sank on August 30, 2003). The model simulation is based on a hypothetical leak of radioactive material from the submarine at a depth of 240 m in the Murmansk Fjord. Ocean currents transport the material and it

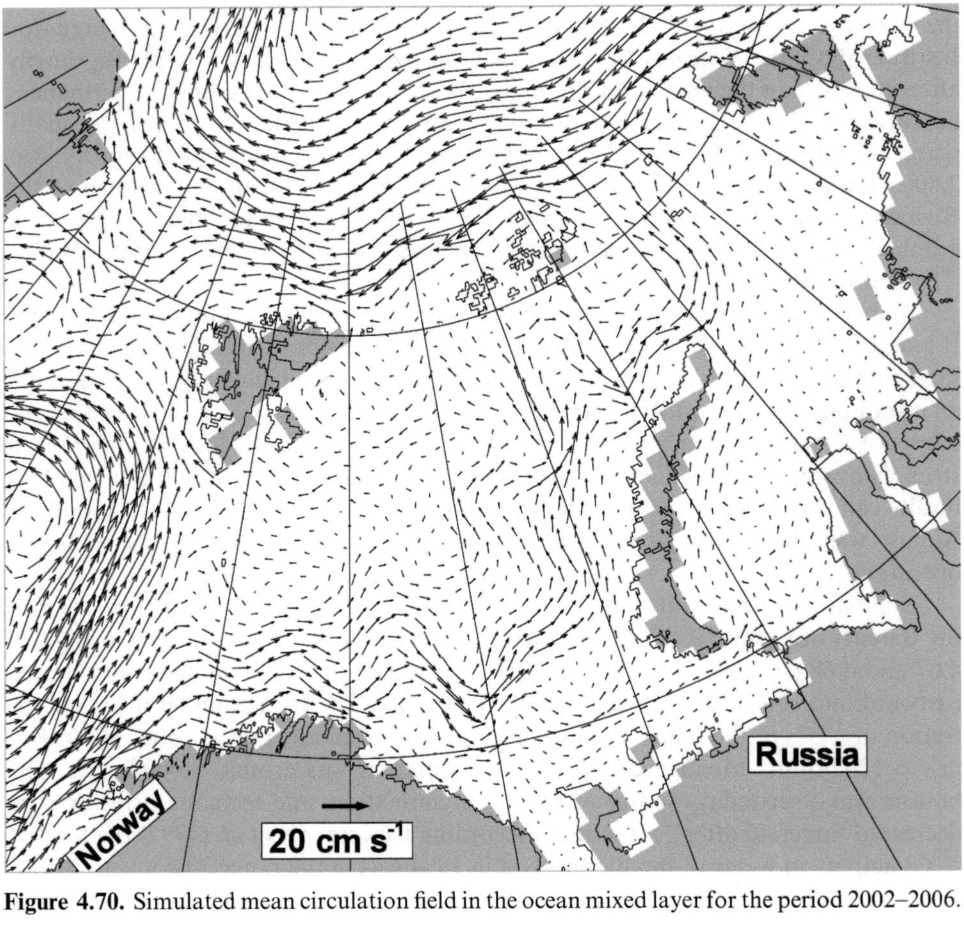

Figure 4.70. Simulated mean circulation field in the ocean mixed layer for the period 2002–2006.

mixes with water masses. The simulation shows relative concentrations from the source location and simulation is done for five consecutive years (2002–2006). Potential release from the *K-159* is assumed to be constant for 5 years.

Temporal evolution of the potential spread of radioactivity from the *K-159* is shown in Figure 4.69 (see color section). It is found that the spread of radioactivity moves eastward first, then northward following the western coast of Novaya Zemlya. The radioactivity level close to the northern tip of Novaya Zemlya is between 5% and 10% after constant release from the source over a period of 5 years. The spread of radioactivity naturally follows the simulated circulation pattern in the region of interest (Figure 4.70). Fish and spawn that periodically appear in the potentially contaminated water areas will be exposed to radioactivity from a possible leakage from the reactors of the sunken submarine. These results, together with more complete simulations, can be used in a risk analysis of actions that could be taken in response to such a leakage.

5

Studies of the spread of non-radioactive pollutants in the Arctic using the generic model system (GMS)

The Arctic Ocean is threatened with contamination not only from the spread of radionuclides (Chapters 1, 3, and 4) but also by other toxic pollutants—for example, persistent organic pollutants (POPs), petroleum hydrocarbons, and heavy metals (AMAP, 2004, 2009)—see also the latter sections in Chapter 1. Although the levels of many POPs have recently declined in the Arctic environment (AMAP, 2009), "legacy" POPs contaminate the Arctic largely as a result of past use and emissions, and emerging and current-use POPs have the potential to transport to and accumulate in the Arctic. Significant increases in oil exploration on Arctic shelf seas and its transportation are foreseen for the near future (AMAP, 2007). These activities will lead to increased risks of oil contamination of the cold Arctic environment, including ice-covered waters.

The generic model system (GMS)—see Chapter 3—which was developed to simulate radionuclide transport and its fate in Arctic rivers and seas, can be modified to simulate the spread of POPs, hydrocarbons, and other toxic non-radioactive pollutants. The realization of such GMS applications is considered in this chapter. Section 5.1 describes the approach for modeling non-radioactive pollutant transport, with the emphasis on POPs. Section 5.2 presents an example of GMS simulation of non-radioactive pollutant transport in the Arctic, focused on polychlorinated biphenyls (PCBs). Section 5.3 describes some aspects of modeling the spread of petroleum hydrocarbons in the marine environment.

5.1 APPROACH TO SIMULATION OF POLLUTANTS IN THE AQUATIC ENVIRONMENT

5.1.1 Persistent organic pollutants

Persistent organic pollutants (POPs) are organic compounds that, to a varying degree, resist photolytic, biological, and chemical degradation. POPs are often halogenated and characterized by low water solubility and high lipid solubility, leading to their bioaccumulation in fatty tissues (Ritter *et al.*, 1995). POPs are also noted for their semivolatility (i.e., that property of their physicochemical characteristics that permits these compounds to occur either in the vapor phase or adsorbed on atmospheric particles, thereby facilitating their long-range transport through the atmosphere).

The global threat of POPs was declared in the *UNEP Stockholm POP Convention* (2001) as chemicals that remain intact in the environment for long periods, become widely distributed geographically, and accumulate in the fatty tissue of humans and animals, thereby creating a serious threat for people and nature. Exposure to POPs can lead to a range of serious health effects. Given their long-range transport, no government acting alone can protect its citizens or its environment from POPs.

The following 12 POPs, known as the "Dirty Dozen" were initially included in the *UNEP Stockholm POP Convention* (2001): aldrin, chlordane, dichlorodiphenyl trichloroethane (DDT), dieldrin, endrin; heptachlor, hexachlorobenzene, mirex, polychlorinated biphenyls (PCBs) (see Section 5.2), polychlorinated dibenzo-*p*-dioxins, polychlorinated dibenzofurans, and toxaphene.

5.1.2 Basic processes and equations for modeling

The general principles of modeling water ecosystems and the basic approaches to parameterization of hydrodynamic, hydrophysical, hydrochemical, and hydrobiological processes used in water quality models of different levels of complexity are applicable in the construction of models for both radionuclide and non-radioactive pollutants (see, e.g., Onishi *et al.*, 1981; Schnoor, 1996; DiToro, 2000). The specificity of models of radionuclide migration is determined by inter-phase exchanges of (1) radionuclides located in dissolved form with radionuclides in bottom sediments of a water body, (2) radionuclides adhering to suspended sediment, and (3) radionuclides in hydrobionts. The basic processes determining pollutant transport in water systems used in GMS models are presented in Figure 5.1.

The equation for transport of pollutant diluted in water can be derived from the mass conservation equation. It can be expressed in terms of the advection–diffusion equation:

$$\frac{\partial C}{\partial t}+\frac{\partial}{\partial x_i}(U_i C)\left(\varepsilon_i\frac{\partial C}{\partial x_i}\right)+\lambda C+\sum_{j=1}^{n}R_j, \tag{5.1}$$

where C = concentration of the diluted contaminant;
t = time;
x_i = $(i=1,2,3)$ Cartesian co-ordinates;
U = velocity component for the relevant co-ordinate;
ε_i = diffusion/dispersion coefficient;
λ = decay rates for a contaminant;
$\sum_{j=1}^{n} R_j$ = sum of sink and/or source terms.

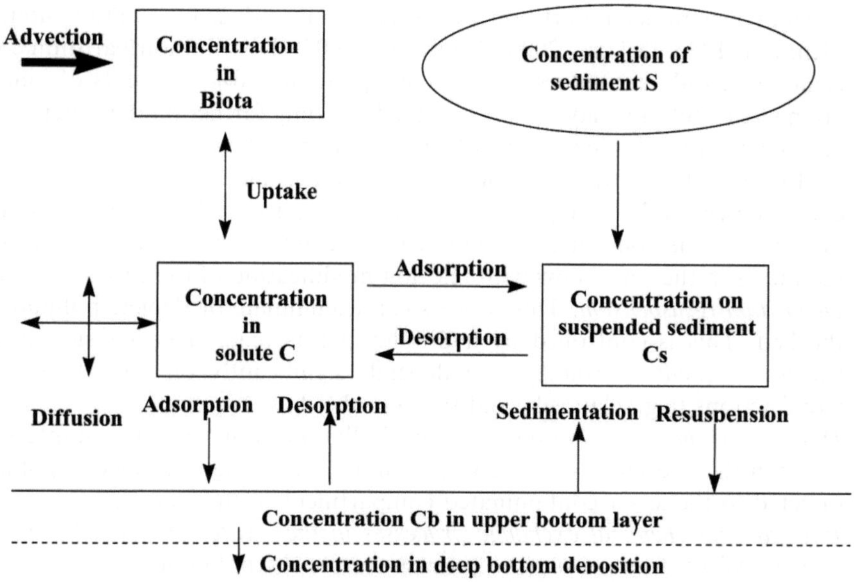

Figure 5.1. Flow diagram of key processes of pollutant transport in water bodies.

The usual rule of summing up recurring indices is used in Equation (5.1). For radionuclides, the last term in this equation describes the exchanges between solute, suspended sediments, and bottom depositions. For POPs and other non-radioactive pollutants, this sum can also include parameterization of the other physicochemical processes considered below. We shall briefly consider the basic processes (depicted in Figure 5.1):

(1) *Advective–diffusive transport of dissolved pollutants by water flow*. This consists of pollutant fluxes driven by currents (the second term on the left-hand side of Equation (5.1) and of fluxes driven by eddy diffusion (first term on the right-hand side of the equation). The currents and coefficients of eddy diffusion can be calculated from the hydrodynamic and turbulent transport equations of open flow before pollutant transport simulation.

(2) *Advective–diffusive transport of pollutants adhering to suspended sediments*. The process is determined by the quantity of suspended sediments carried by the flow and quantity of radionuclides adhering to sediments. Traditionally, the quantity of radionuclides in solid phase in the equilibrium state is determined by the distribution coefficient K_d (see, e.g., Onishi *et al.*, 1981; Codell *et al.*, 1982; Santschi *et al.*, 1989):

$$K_d = \frac{\text{equilibrium concentration of radionuclide in sediment (Bq/kg)}}{\text{equilibrium concentration of radionuclide in solute (Bq/m}^3\text{)}}$$

$$= \frac{C^s_{\text{equil}}}{C_{\text{equil}}} (\text{m}^3/\text{kg}) \tag{5.2}$$

A similar parameter–partition coefficient is introduced for POP modeling (Schnoor, 1996; DiToro, 2000; Wool *et al.*, 2005). Distribution/partition coefficients have different values for the same pollutants for sediments of different grain size (sand, silt, clay) and for biota, depending further on the water quality parameters and physicochemical forms of the pollutant.

In general, the transport of pollutants adhering to suspended sediments can be described by an equation similar to Equation (5.1), expressed as multiplication of the concentration adherent to sediments by the concentration of sediments in the water flow with relevant modification of the sink/source term.

(3) *Deposition/resuspension.* This process can accumulate or deplete pollutants in the bed. This is controlled by hydraulic factors (e.g., river/marine currents, waves, sediment transport), and depends significantly on typical sediment-size fractions (e.g., clay, silt, and sand) in the flow.
(4) *Diffusion at the water–bottom interface.* Pollutant diffusion between interstitial water in the bed and overlying water can create a long-term flux of pollutant directed to the lesser contaminated compartment of this system.
(5) *Physical and chemical exchange processes in the "water–suspended sediment" system.* Adsorption–desorption processes drive the pollutant transfer between water and suspended sediments. The approach used for parameterization of these processes in the GMS and some other contemporary pollutant modeling systems (e.g., WASP6) is based on implementation into the sink/source terms of formulas derived from the linear sorption isotherm with the exchange coefficient that characterizes the temporal scale of sorption. Then, pollutant in the solid phase approaches equilibrium at values defined by the partition/distribution coefficient.
(6) *Physical and chemical exchange processes in the "water–bottom sediment" system.* The main chemical mechanisms of radionuclide adsorption to and desorption from surface bed sediment are realized simultaneously with the above-mentioned diffusion to interstitial water. Parameterization of adsorption–desorption processes is based on the same methods as for the "water–suspended sediment" system. For pollutant transfer between the bottom and water column, the most important exchanges occur within the thin top layer of the river/marine bed.
(7) *Transfer processes between the upper bottom layer and deep buried sediments.* Through the thin top bed layer, pollutants can be accumulated into or depleted from the deeper bed sediment. These pollutants are later mixed within deeper riverbed layers by diffusion, bioturbation, and movements of bed sediment formation.
(8) *Biodegradation.* Biodegradation processes involve the uptake and subsequent decay and excretion of pollutants by aquatic biota and microorganisms. This process transfers pollutants from water to the bed sediments. Sometimes the water–bottom interface processes (diffusion, adsorption–desorption, biota fluxes) are combined under one definition—direct uptake processes.

The above processes should be taken into consideration when modeling radionuclide transport and its fate in river and marine systems. All of these processes are important for POP modeling; however, for some POPs, particulate organic carbon (POC) (i.e., the organic carbon fraction of suspended and bottom sediments) is the most important carrier of POP in solid matter. Therefore, a separate partition coefficient for POC can be used in POP modelling in addition to (or instead of) such a coefficient for bulk sediments—sand, silt, and clay.

The following processes should be considered, in addition to those mentioned above, for modeling POPs and other organic toxins (Schnoor, 1996; DiToro, 2000, Wool *et al.*, 2005):

(9) *Volatilization.* Volatilization is the movement of chemical across the air–water interface as the dissolved neutral concentration attempts to equilibrate with the gas-phase concentration. Equilibrium occurs when the partial pressure exerted by the chemical in solution equals the partial pressure of the chemical in the overlying atmosphere. The rate of exchange is proportional to the gradient between the dissolved concentration and the concentration in the overlying atmosphere and conductivity across the interface of the two fluids.

(10) *Photolysis.* Photolysis is the transformation or degradation of a compound that results directly from the adsorption of light energy. It is a function of the quantity and wavelength distribution of incident light, the light absorption characteristics of the compound, and the efficiency at which absorbed light produces a chemical reaction. Photolysis is classified into two types that are defined by the mechanism of energy absorption. Direct photolysis is the result of direct absorption of photons by the toxic chemical molecule. Indirect or sensitized photolysis is the result of energy transfer to the toxic chemical from some other molecule that has absorbed the radiation.

(11) *Hydrolysis.* Hydrolysis, or reaction of the chemical with water, is known to be a major pathway for the degradation of many toxic organics. Hydrolysis is a reaction in which cleavage of a molecular bond of the chemical and formation of a new bond with either the hydrogen or the hydroxyl component of a water molecule occurs.

(12) *Oxidation.* Chemical oxidation of organic toxicants in aquatic systems can be a consequence of interactions between free radicals and the pollutants. Free radicals can be formed as a result of photochemical reactions.

(13) *Ionization.* In an aquatic environment some chemicals may occur only in their neutral form while others may react with water molecules to form positively (cationic) or negatively (anionic) charged ions. These reactions are rapid and are generally assumed to be at local equilibrium. At equilibrium, the distribution of chemicals between the neutral and ionized species is controlled by the pH and temperature of the water and the ionization constants.

These analyses show that the modeling of POPs and other pollutants should be based on simulation of the same basic processes (1–8) as for radionuclides and other toxins. Different kinds of non-radioactive toxins can also be influenced by some supplementary processes (9–13) whose description can be included for the modifications of GMS models to optimize them for simulation of POPs and other non-radioactive toxins (Sections 5.1.3 and 5.2).

5.1.3 Modeling POP transport in the environment

Because of the global transfer of POPs through all components of the environment, "multimedia" POP models have been developed that integrate atmospheric, watershed, and ocean models at different levels of resolution.

An example of a spatially resolved regional multi-compartment model is MSCE-POP (Gusev *et al.*, 2005), whose ocean model is based on three-imensional equations similar to (5.1). Equilibrium sorption on suspended sediments is calculated via the partition coefficient for POC, and volatilization is parameterized. POP fallout fluxes are calculated from the atmospheric model, and POP transport with sea ice is taken into account.

Examples of box models on a global scale are (1) ChemRange, which represents the global circulation of chemicals in a one-dimensional loop consisting of soil, oceanic surface water, and tropospheric air, and (2) CliMoChem, which consists of latitudinal zones with different annual temperature cycles, land–sea fractions, and vegetation (Scheringer *et al.*, 2004). The ocean boxes in both models require setting up the degradation coefficients and partition coefficients.

The three-dimensional model FANTOM was recently used to study POP transport in the North Sea (Ilyina *et al.*, 2006). The hydrodynamics variables driving FANTOM were calculated using the HAMburg Shelf Ocean Model (HAMSOM) that was coupled with FANTOM. POP transformations in the sea were parameterized via partition coefficients for POC and degradation coefficients.

The modeling of POPs and other pesticide pollution has been strongly influenced by the approaches and methods of the parameterization of physicochemical processes presented in the U.S. Environmental Protection Agency (EPA) model WASP/TOXI (Ambrose *et al.*, 1993; Wool *et al.*, 2005). Such an approach can be applied to simulation of all sorts of toxic organic compounds in the water environment (Schnoor, 1996; Di Toro, 2000).

One of the most toxic and widespread POPs is polychlorinated biphenyls (PCBs). The behavior of PCBs in water is characterized by significant interactions with suspended and bottom sediments—in common with many radionuclides. The models describing PCB transport and its fate in natural waters thus have many common features with radionuclide transport models.

Models of the riverine transport of PCBs developed in recent decades are widely used in PCB assessment studies. An example is a model presented in Steinberg (1997), which includes simulation of such processes as hydrophobic sorption, volatilization, and sedimentation, and biological processes such as aerobic and anaerobic biodegradation.

Most of the models used for modeling the fate of PCBs in water bodies (QEA, 1999; BBL, 2000; LTI, 2000; TAMS, 2000) have used the WASP/TOXI modeling approach (Ambrose *et al.*, 1993). Sometimes small modifications to the WASP approach are applied to account for special features of PCB transport. However, the methods presented in WASP remain the basis for modeling PCBs and other POPs in the water environment. The basic formulas of this approach, which create the basis of the methods implemented in GMS models, are presented below. Table 5.1 provides a glossary of symbols, terms, and units.

Equilibrium sorption

Sorption reactions are usually fast relative to other environmental processes for PCBs, and equilibrium may be assumed (Ambrose, 1993; TAMS, 2000). Karickhoff (1984) showed that organic carbon is the principal sorbent compartment for hydrophobic organic chemicals such as PCBs in aquatic systems.

For environmentally relevant concentrations (less than 50% water solubility), equilibrium sorption varies linearly with dissolved chemical concentration (Karickhoff, 1984), or:

$$C'_s = K_{ps} C'_w \tag{5.3}$$

where C'_s (kg/kg$_s$) is the concentration of sorbed chemical on sediments of type "s"; and C'_w (kg/m^3_w) is the mass of dissolved chemical on the unit of volume of pure water. At equilibrium, then, distribution among the phases is controlled by partition coefficients K_{ps} (m^3/kg).

According to Ambrose (1993), dissolved chemicals in the water column and benthic segments interact with sediment particles and dissolved organic carbon (DOC) to form the following phases—dissolved, DOC-sorbed, and sediment-sorbed. The DOC phase is not considered here.

The phase concentrations C_w (kg/m^3), C_s (kg/m^3) are governed by the equilibrium partition coefficient:

$$K_{ps} = \frac{C'_s}{C'_w} = \frac{C_s n}{M_s C_w} \tag{5.4}$$

Total chemical concentration in water C (kg/m^3) is the sum of all the phase concentrations:

$$C = C'_w n + \sum_s C'_s M_s \tag{5.5}$$

Substituting Equation (5.3) by (5.5), factoring, and rearranging terms gives the mass fraction of dissolved concentration F_w:

$$F_w = \frac{C'_w n}{C} = \frac{n}{n + \sum_s K_{ps} M_s} \tag{5.6}$$

The values for partition coefficients can be obtained from laboratory experiments. For organic chemicals, laboratory studies have shown that the partition coefficient is related to the hydrophobicity of the chemical and the organic matter content of the sediment. Normalization of the partition coefficient by the organic carbon content of the sediment was shown in Ambrose (1993) to yield a coefficient, K_{oc} (the organic carbon partition coefficient), which is relatively independent of other sediment

Table 5.1. Glossary to Section 5.1.3, with symbols, variables, and units.

Symbol	*Variable or parameter*	*Units*
C'_s	Concentration of sorbed chemical on sediments of type "s"	kg/kg_s
C'_w	Mass of dissolved chemical in the unit of the volume of the pure water	kg/m^3_w
K_{ps}	Partition coefficient of the chemical on the sediment of the type "s"	m^3/kg
C_w	Concentration of dissolved chemical phase concentrations	kg/m^3
C_s	Concentration of sorbed chemical phase concentrations	kg/m^3
C	Total chemical concentration in water—dissolved + sorbed	kg/m^3
C_b	Concentration of PCBs in bottom sediments	kg/m^3
F_w	Mass fraction of dissolved concentration	—
K_{oc}	The organic carbon partition coefficient	—
f_{ocs}	Organic carbon fraction of sediment	—
K_{ow}	Octanol–water partition coefficient	—
Q	Water discharge	m^3/s
$\bar{Q}$	Average water discharge	m^3/s
K_v	Transfer rate	m/day
H	Water depth	m
C_a	Atmospheric concentration	kg/m^3
T_k	Water temperature	K
R	Universal gas constant	$Pa \cdot m^3/mole \cdot K$
H	Henry's law coefficient for the air–water partitioning of the chemical	$Pa \cdot m^3/mole$
R_L	Liquid phase resistance	s/m
K_L	Liquid phase transfer coefficient	m/s
R_G	Gas phase resistance	s/m
K_G	Gas phase transfer coefficient	m/s
K_a	Reaeration velocity	m/s
K_{v0}	Ratio of volatilization rate to reaeration rate	—
M_{O_2}	Molecular weight of the oxygen	g/mole

Symbol	*Variable or parameter*	*Units*
M_W	Molecular weight of the chemical	g/mole
U	Velocity of water	m/s
D_w	Diffusivity of oxygen in water	m^2/s
U_*	Shear velocity	m/s
W_{10}	Wind velocity at 10 m above the water surface	m/s
Cd	Drag coefficient	—
ρ_a	Density of air	kg/m^3
ρ_w	Density of water	kg/m^3
κ	Von Karman's constant	—
λ	Viscous sublayer thickness	—
Sc_a, Sc_w	Air and water Schmidt Numbers	—
K_0, k_a	Aerobic and anaerobic biodegradation rates	L/s
q_r, q_s	Resuspension and sedimentation velocities	m/s
A_{12}	Mass transfer velocity	m/s
Z_*	Depth of the bottom sediment layer	*M*

characteristics or geographic origin. PCBs are non-polar, hydrophobic compounds whose partition coefficients correlate quite well with the organic fraction of the sediment. Rao and Davidson (1980) and Karickhoff *et al.* (1979) developed empirical expressions relating equilibrium coefficients to laboratory measurements leading to fairly reliable methods of estimating appropriate values. The correlations used in WASP/TOXI are:

$$K_{ps0} = f_{\mathrm{ocs}} K_{\mathrm{oc}} \tag{5.7}$$

where K_{oc} (m_w^3/kg) is the organic carbon partition coefficient; and f_{ocs} is the organic carbon fraction of sediment.

The correlation of K_{oc} with the water solubility of the chemical or the octanol–water partition coefficient of the chemical has yielded successful predictive tools for incorporating the hydrophobicity of the chemical to estimate its partitioning. If the direct measurement of K_{oc} is not available, the following correlation with the octanol–water partition coefficient K_{ow} (m_w^3/ m_{oct}^3) is used:

$$\log K_{\mathrm{oc}} = a_0 + a_1 \log K_{\mathrm{ow}} \tag{5.8}$$

where a_0 and a_1 are typically considered to be log 0.6 and log 1.0, respectively.

For different PCB congeners, the value of K_{ow} can differ by an order of 2. Consequently, the mass-weighted value of the K_{ow} coefficient was used in TAMS (2000) to determine the "effective" octanol–water partition coefficient. The temperature dependence of K_{ow} was also used in TAMS (2000), in addition to the approach of Ambrose (1993). Determining the organic carbon fraction of sediment is a rather complicated task; its value should probably be measured. According to Ambrose (1993), the values of f_{oc} can range between 0.005 and 0.5. In TAMS (2000) it was found that the value of f_{oc} was correlated with flow discharge. The following empirical relation based on regression analyses of the available data in Hudson River was proposed there:

$$f_{oc} = 0.175\left(\frac{Q}{\bar{Q}}\right)^{-0.3687} \tag{5.9}$$

where Q (m^3/s) is water discharge at the given point; and $\bar{Q}$ (m^3/s) is average water discharge upstream of the given point. The corresponding values of f_{oc} varied from 0.08 to 0.22. In BBL (2000), measured values of f_{oc} were used varying from 0.01 to 0.1.

Volatilization

The rate of exchange through the water surface is proportional to the gradient between the dissolved concentration and concentration in the overlying atmosphere and conductivity across the interface of the two fluids.

According to Ambrose (1993), the dissolved concentration attempts to equilibrate with the gas-phase concentration, as given by:

$$\left.\frac{\partial C}{\partial t}\right|_{\text{vol}} = \frac{K_v}{h}\left(f_d C - \frac{C_a}{\dfrac{H}{RT_k}}\right) \tag{5.10}$$

where K_v (m/s) is the transfer rate; h (m) is the depth of water; C_a is atmospheric concentration, kg/m^3; R is the universal gas constant; T_k is water temperature, K; H is Henry's law coefficient for air–water partitioning of the chemical, Pa $\cdot$ m^3/mole. Equilibrium occurs when dissolved concentration equals partial pressure divided by Henry's Law constant.

In WASP/TOXI, the dissolved concentration of a chemical in a surface water column segment can volatilize at a rate determined by the two-layer resistance model. The two-resistance method assumes that two "stagnant films" are bounded on either side by well-mixed compartments. Concentration differences serve as the driving force for water layer diffusion. Pressure differences drive the diffusion for the air layer. From mass balance considerations, it is obvious that the same mass must pass through both films, thus the two resistances combine in series, so that conductivity is the reciprocal of total resistance:

$$K_v = (R_L + R_G)^{-1} = \left[K_L^{-1} + \left(K_G \frac{H}{RT_k}\right)^{-1}\right]^{-1} \tag{5.11}$$

where R_L (s/m) is liquid-phase resistance; K_L (m/s) is the liquid-phase transfer coefficient; R_G (s/m) is gas-phase resistance; and K_G (m/s) is the gas-phase transfer coefficient. Although this two-resistance method is rather simplified in its assumption of uniform layers, it has been shown to be as accurate as more complex models.

There have been a variety of methods proposed to compute liquid-phase (K_L) and gas-phase (K_G) transfer coefficients, several of which were included in WASP/TOXI. One option allows the input of an oxygen reaeration constant that is then adjusted to represent the liquid film transfer constant for the particular chemical. The rate (K_L) is computed from equation:

$$K_L = K_a K_{v0} \tag{5.12}$$

where K_a is the reaeration velocity, m/s; and K_{v0} is the ratio of volatilization rate to reaeration rate, which can be assumed to be (Ambrose, 1993):

$$K_{v0} = \sqrt{\frac{M_{O_2}}{M_w}} \tag{5.13}$$

where M_{O_2} is the molecular weight of the oxygen (32 g/mole); and M_w is the molecular weight of the chemical.

Approaches used in Steinberg (1997), BBL (2000), and TAMS (2000) to simulate the volatilization process are the same or simplified. The only difference is introduction of the temperature dependence of the Henry's Law constant in TAMS (2000). Note that application of the described approach requires knowledge of the concentration of PCBs in air C_a. When this is not measured, simplified approaches are usually used, such as in Steinberg (1997), with only the first term of Equation (5.11), including K_L.

Biodegradation

Biodegradation encompasses the broad and complex processes of enzymatic attack by organisms on organic chemicals. Bacteria, and to a lesser extent fungi, are the mediators of biological degradation in surface water systems. Two different types of biodegradation are usually recognized: aerobic and anaerobic.

In some studies (e.g., TAMS, 2000) these processes are neglected, whereas in Steinberg (1997) estimates of aerobic k_0 and anaerobic k_a biodegradation rates are given, together with their probability distributions. The correspondent average value for k_0 is $\sim 10^{-6}$ L/s and for k_a it is $\sim 10^{-9}$ L/s. From these estimates the order of the aerobic biodegradation rate is the same as that of the volatilization rate, and should be taken into account.

Exchange of PCBs between water and bottom sediments

There are two types of PCB exchange between water and bottom sediments; they can be considered separately. The first type of processes is settling and resuspension of sediment particles containing PCBs. These processes can be represented by introducing the velocities q_r (m/s) and q_s (m/s) of resuspension and settling, respectively.

During high-flow periods, sediment resuspension can be the dominant sediment–water transfer mechanism for PCBs. However, under low-flow conditions, flow-induced resuspension can be small relative to the other mechanisms causing sediment–water transfer of PCBs (BBL, 2000). Aside from flow-driven sediment erosion, the second type of potential mass transfer processes that influence the sediment–water transfer of PCBs includes (BBL, 2000):

(1) molecular diffusion of dissolved-phase or colloid-bound pore water PCBs;
(2) groundwater advection up through the sediment bed;
(3) hydrodynamically induced advective pumping through near-surface interstices in the sediment bed;
(4) biologically enhanced pore water and particulate transport within the sediment bed and at the sediment–water interface;
(5) physical disturbance from wind-driven waves, fish activity, or human activity;
(6) direct desorption from surface sediments to the water column.

The sum of these processes can be represented by the dissolved-phase mass transfer velocity a_{12} (m/s).

With these two kinds of processes affecting PCB exchange between water and bottom sediments, changes in total concentrations of PCBs in water and in bottom sediments due to these processes can be represented by the following equations, which coincide with the simplified versions of the relevant modules of the RIVTOX and THREETOX models:

$$\frac{\partial C}{\partial t} = \frac{q_r}{h} C_b - \frac{q_s}{h}(1 - F_w)C + \frac{a_{12}}{h}(C_B - F_w C) \tag{5.14}$$

$$\frac{\partial C_b}{\partial t} = \frac{q_s}{Z_*}(1 - F_w)C - \frac{q_r}{Z_*} C_b - \frac{a_{12}}{Z_*}(C_B - F_w C) \tag{5.15}$$

where C_b (kg/m^3) is the concentration of PCBs in bottom sediments; and Z_* (m) is the depth of the bottom sediment layer. A variety of approaches exist for determination of resuspension and sedimentation velocities; however, that is a topic for sediment transport modeling.

The magnitude of the mass transfer velocity a_{12} representing direct exchange of PCBs with the bottom can vary seasonally as a function of temperature and other climatic conditions. Biologically enhanced sediment–water transfer of PCBs is temperature-dependent due to increased biological activity during warm temperatures. PCB movement to groundwater may vary seasonally with the groundwater hydraulic gradient.

In the absence of any physical disturbance of the upper sediment layer, the exchange of PCBs between sediments and water takes place by molecular diffusion or Brownian diffusion. Valasaraj *et al.* (1997) estimated that mass transfer rates as a result of molecular diffusion should be of the order of 0.02 cm/da, assuming water diffusivity of 5.6×10^{-6} cm^2/s. Application of this mass transfer rate to pore-water concentrations of PCBs in the Kalamazoo River in TAMS (2000) results in a small mass flux from sediments to water that is insufficient to match observed water column concentrations under low-flow conditions.

This indicates that transfer mechanisms other than molecular diffusion are operative at rates sufficient to influence surface-water PCB concentrations under low-flow conditions. Valasaraj *et al.* (1997) estimated that a biodiffusion—bioturbation-induced mass transfer of pore-water chemicals—mass transfer rate would be approximately 12 cm/da. This estimate was established through calibration for the Kalamazoo River and through values from independent studies of the Upper Hudson River (QEA, 1999) and USEPA (TAMS *et al.*, 2000).

Thibodeaux (1996) presented a two-layer mass transfer resistance model to describe sediment–water interactions for hydrophobic contaminants that accounts for the effect of overlying water flow and bioturbation. This model relates mass transfer across the sediment–water interface to the waterside boundary layer and particle mixing velocities. The thickness of the boundary layer is governed by flow velocity. Lower flows permit a thicker boundary layer and cause more waterside resistance to entrainment of PCBs at the interface with the overlying water. This mechanism accounts for the lower mass transfer velocities observed for lakes as opposed to rivers. Sediment–side mass transfer resistance arises from the rate of chemical movement within the mixed layer due to biological mixing.

5.2 MODELING PCB SPREAD IN ARCTIC RIVERS AND COASTAL WATERS USING THE GMS

5.2.1 Modification of the models for simulation of PCBs

Because radionuclides are typical pollutants that adhere to sediments, the development of the radionuclide transport model within GMS required the elaboration of submodels that describe pollutant exchange in the "solute–suspended sediment–bottom deposition" system. Independent of model dimension and resolution, the GMS models RIVTOX and THREETOX describe the hydrodynamic parameters of flow, suspended sediment transport, and radionuclide transport, taking into account contaminant interaction with sediments, as described in Sections 5.1 and 3.4. Keeping in mind that RIVTOX and THREETOX include hydrodynamics and sediment transport models, modifications of the models for non-radioactive applications were required for sink–source terms.

The analyses presented in Section 5.1.2 indicate that the contemporary basis for simulation of PCB exchange processes in the water column is provided by approaches used in the WASP/TOXI model, taking into account the experience of PCB simulation by models as well (Steinberg, 1997; QEA, 1999; TAMS, 2000; BBL, 2000; LTI, 2000). The following model structure was accepted for RIVTOX and

THREETOX on the basis of this approach:

$$\frac{\partial C_a}{\partial t} + [\text{conv} + \text{dif}] = -\overbrace{k_{\text{vol}} F_w C_a}^{1} - \overbrace{k_0 F_w C_a}^{2} + \overbrace{\frac{q_b}{h} C_b}^{3} - \overbrace{\frac{q_s}{h}(1 - F_w) C_a}^{4} + \overbrace{\frac{a_{12}}{h} F_w \left(\frac{C_B}{n_B} - C_a\right)}^{6} \tag{5.16}$$

$$\frac{\partial C_b}{\partial t} + = \overbrace{\frac{q_s}{Z_*}(1 - F_w) C_a}^{3'} - \overbrace{\frac{q_r}{Z_*} C_b}^{4'} - \overbrace{k_a C_b}^{5} - \overbrace{\frac{a_{12}}{Z_*} F_w \left(\frac{C_B}{n_B} - C_a\right)}^{6} \tag{5.17}$$

where [conv + dif] = terms in the equations describing advective and diffusive transport of the pollutant (one-dimensional for RIVTOX and three-dimensional for THREETOX);

C_a (kg/m^3) = total PCB concentration in water in solute and on suspended sediments;

C_b (kg/m^3) = concentration of PCBs in bottom sediments;

k_{vol} (L/s) = volatilization coefficient;

k_0, k_a (L/s) = aerobic and anaerobic biodegradation (dechlorination) coefficients;

F_w = ratio of concentration in solute C_w to total concentration in water:

$$C_w = F_w C_a \tag{5.18}$$

q_b, q_s (m/s) = resuspension and sedimentation velocities;

h (m) = depth of the water column;

Z_* (m) = depth of the bottom sediment layer;

a_{12} (m/s) = mass transfer rate;

n_b = porosity of bottom sediments.

Terms 1–6 in Equations (5.16)–(5.17) describe 1 volatilization, 2 aerobic biodegradation, 3, 3′ sedimentation, 4, 4′ resuspension, 5 anaerobic biodegradation, 6 direct exchange of PCBs between water and bottom sediments. See Table 5.2 for parameters.

Sorption

$$F_w = 1/(1 + K_s S) \tag{5.19}$$

where S (kg/m^3) = concentration of sediments;

K_s (m^3/kg) = water–sediment partition coefficient of PCBs:

$$K_s = f_{\text{oc}} K_{\text{oc}} \tag{5.20}$$

f_{oc} = fraction of organic carbon in sediments ~0.15 (WASP);

K_{oc} (m^3/kg) = organic carbon partition coefficient, which according to WASP:

$$\log_{10} K_{\text{oc}} = \log_{10} K_{\text{ow}} - 0.22 \tag{5.21}$$

where K_{ow} = octanol–water partition coefficient is a table function dependent on the type of PCB congener.

Volatilization

According to Steinberg (1997):

$$k_{vol} = -\frac{K_l}{h}, \qquad K_l = \left(\frac{32}{M}\right)^{0.25}\left(\frac{D_o u}{h}\right)^{0.5} \tag{5.22}$$

where h = depth of water column;
M = molecular weight of PCBs;
D_o = molecular diffusivity of oxygen in the water;
u = stream velocity.

Table 5.2. Parameters for the model used here.

Symbol	*Parameter name*	*Parameter specification or equation*
K_{vol} (L/s)	Volatilization coefficient	$K_{vol} = -\frac{\sqrt{D_0 u}}{h\sqrt{h}}\left(\frac{32}{M}\right)^{0.25}$
D_0	Molecular diffusivity of the oxygen in water	$D_0 = 22\text{E-9}/M^{2/3} = 7.33\text{E-13}$
M	Molecular weight of PCBs	300
h (m)	Depth of the water column	
u (m/s)	Stream velocity	
F_w	Ratio of concentration in solute C_w to total concentration in water	$F_w = \frac{1}{1+K_s S}$
K_s (m^3/kg)	Water–sediment partition coefficient of PCBs	$K_s = f_{oc}K_{oc} = 25.47366$
f_{oc}	Fraction of organic carbon in sediments	0.15
K_{oc} (m^3/kg)	Organic carbon partition coefficient	$\log_{10} K_{oc} = \log_{10} K_{ow} - 0.22 = 169.8244$
K_{ow}	Weighted sum dependent on the composition of the PCB mixture	$10^{2.45} = 281.8383$
a_{12} (m/s)	Mass transfer rate	1.38E-6
k_0 (L/s)	Aerobic biodegradation	3E-6
k_a (L/s)	Anaerobic biodegradation	3E-9
n_B	Porosity of bottom sediments	0.6
Z_* (m)	Depth of bottom sediment layer	0.05

Biodegradation (dechlorination)

The processes of biodegradation were not included in the HudTox model (TAMS, 2000) nor in the KALSIM model (LTI, 2000). In the model used by Steinberg (1997) the following biodegradation parameters values were proposed based on calibration studies for the Hudson River:

$$k_0 \sim 3 * 10^{-6}\ \text{L/s} \quad \text{(aerobic biodegradation)}$$

$$k_a \sim 3 * 10^{-9}\ \text{L/s} \quad \text{(anaerobic biodegradation)}$$

Direct exchange of PCBs between water and bottom sediments

The following parameter values have been selected to simulate this process in (5.21)–(5.22):

$$a_{12} = 1.38 * 10^{-6}\ \text{m/s (LTI, 2000)}, \qquad n_b = 0.3$$

The parameters M (molecular weight) and K_{ow} are PCB-type dependent from the Hudson River study (TAMS, 2000) where it was proposed to use a weighted sum of them dependent on composition of the PCB mixture. The corresponding values applicable for such mixtures were chosen based on literature reviews. In particular, K_{ow} was chosen to be $10^{2.45}\ \text{m}^3/\text{kg}$.

5.2.2 GMS application to simulate the transport and fate of PCBs released in the Yenisei River and estuary

Here the one-dimensional river model RIVTOX and the three-dimensional model THREETOX—without modifying the numerical scheme, just the sink–source terms in Equations (5.17) and (5.18)—were applied to simulate the propagation of PCBs in the Yenisei River and estuary. The river part was simulated for an accidental release of PCBs into the river water at the city of Turukhansk near the Nizhnyaja Tunguska mouth for two scenarios of river discharge with the worst at 5,000 m^3/s (year 1974) and another at 3,500 m^3/s (year 1985) (Figure 5.1).

Simulated results of this artificial scenario (Figures 5.2–5.4) show that the model produces physically reasonable results, illustrating bottom contamination along the river due to interaction between PCBs propagated in dilute and contaminated sediments. The maximum PCB concentration in dilute reached the mouth of the river at the Yenisei estuary 5 days earlier in the high-discharge scenario (1985) than for the low-discharge scenario (1974) and had only half the PCB concentration at the mouth of the bay (Figure 5.4).

The application of THREETOX to simulate the fate of PCBs in the Yenisei estuary was also done for another case of PCB release into the water near Dudinka (Figures 5.5–5.6). The three-dimensional model describes the propagation of PCBs in solute and sediments creating the pattern of bottom sediment contamination (Figure 5.7).

It can be concluded that GMS models can be effectively modified to simulate POPs and other non-radioactive pollutants. It is shown that PCB transport is driven

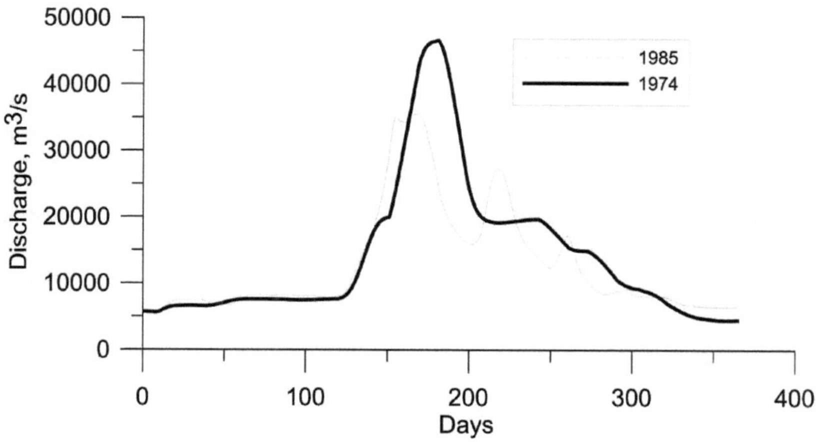

Figure 5.2. RIVTOX-simulated water discharges near Turukhansk for 1985 and 1974.

by the processes of pollutant exchange in the "solute–suspended sediment–bottom deposition" system, which are similar to radionuclide inter-phase exchange processes. Therefore, only the sink–source terms of the GMS models RIVTOX and THREETOX were changed, without needing replacement of the numerical algorithms, to simulate PCB transport in one-dimensional and three-dimensional resolution. Modified models were applied to simulate PCB spread in the Yenisei River and its estuary. The results are promising for efficient GMS model applications to simulate the spread of industrial chemical pollutants in the Arctic Basin.

5.3 MODELING PETROLEUM HYDROCARBON SPREAD USING THE GMS

5.3.1 Processes of oil spread in the marine environment

Significant increases in oil exploration in the seas of the Arctic Shelf and oil transportation are foreseen for the near future. These activities will lead to increased risks of oil contamination of the cold Arctic environment, including its ice-covered waters. Drifting sea ice can accumulate and transport oil over great distances. Growing concern over the impact of an accidental spillage in oil exploration fields and transport routes is a motivating factor for the development and implementation of decision support tools to evaluate oil spill response strategies, to provide environmental impact assessment, and to use in contingency planning and training.

The transport and fate of oil slicks are governed by a set of interrelated processes: advection by wind and currents, surface slick spread due to turbulent diffusion and gravitational and surface tension forces, mass transfer and changes in physico-chemical properties due to weathering processes, and interaction of oil with suspended and bottom sediments and the shoreline. Under calm-sea conditions, oil floats on the surface whereas in rough seas it is entrained by breaking waves. Wave breaking disintegrates and forces entrained oil in the form of droplets dispersed

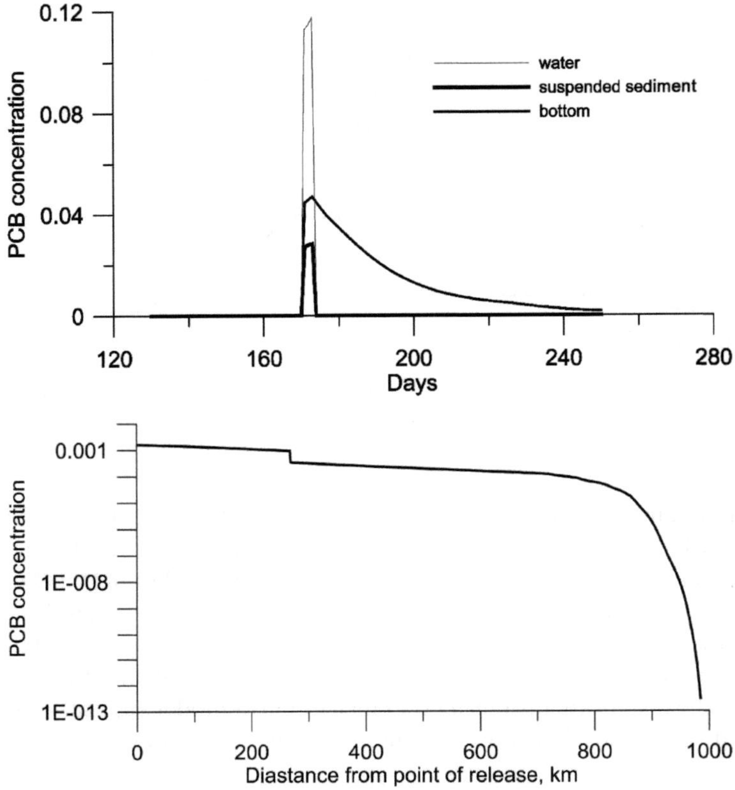

Figure 5.3. (a) PCB concentration in water, suspended sediment, and at the bottom near the point of release near Turukhansk (Nizhnyaja Tunguska mouth) for the 1985 river discharge scenario. (b) PCB concentration at the bottom along the Yenisei River from the point of release near Turukhansk (Nizhnyaja Tunguska mouth) for the 1985 river discharge scenario.

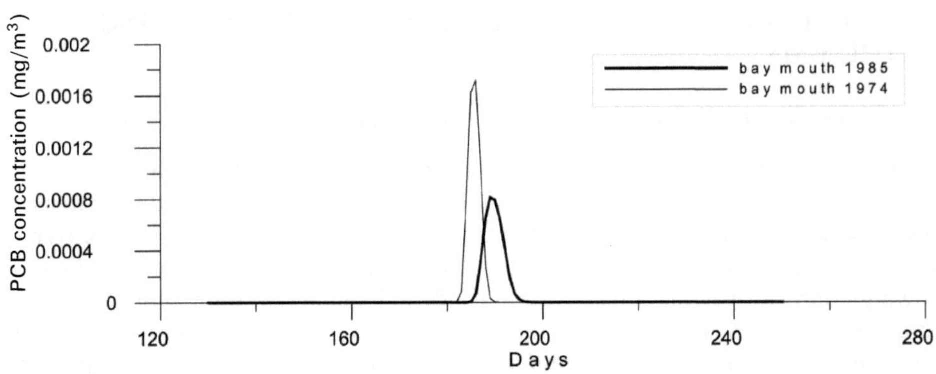

Figure 5.4. Influence of water discharge on the same PCB release near Turukhansk.

in water down to deep levels. The droplets can further break due to turbulence. The coalescence of large drops may cause the water in rising oil masses to become trapped and water-in-oil emulsions, often referred to as "chocolate mousse", to form. Both processes depend strongly on the hydrodynamics of breaking and oil features (density, viscosity, and interfacial tension). Low-viscosity crude oils disperse in rough

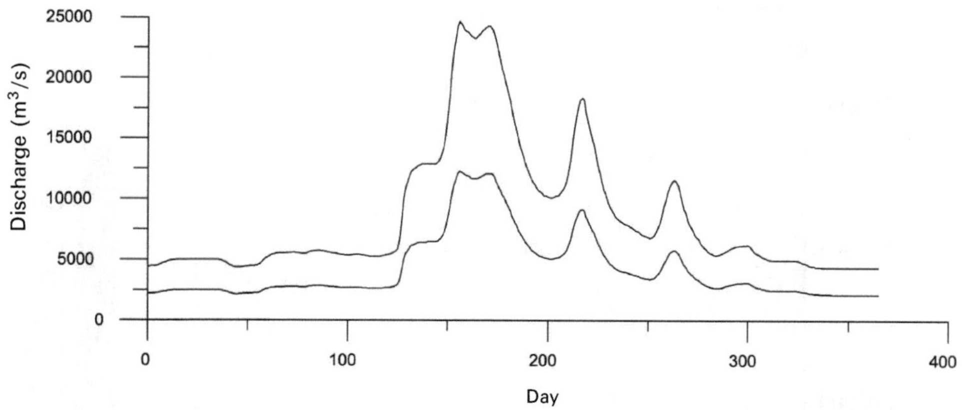

Figure 5.5. Simulated water discharge near Dudinka for the years 1985 and 1974.

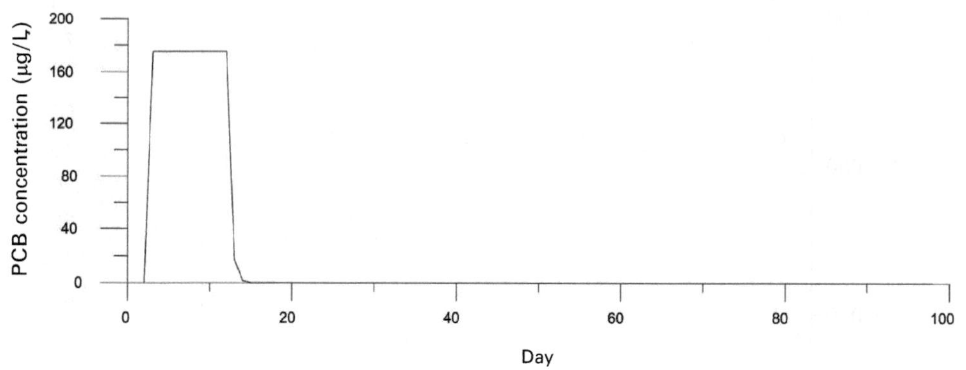

Figure 5.6. Simulated accidental release of PCBs to the Yenisei Estuary near Dudinka for the 1985 scenario.

seas whereas the increase of viscosity due to evaporation of more volatile components in the oil enhances formation of the emulsion. Significant quantities of air are also entrained in large breaking events. The trapped air disperses in bubble clouds. These bubbles rise through the water to the surface. They are an additional source of fine-scale turbulence near the surface and can entrain water in the slick.

Dispersed oil droplets are transported by currents and wave-induced Stokes drift. Both components of drift interact forming Langmuir circulation that contributes to the vertical and horizontal spread of oil. The shear of drift currents results in dispersed particles from the surface slick lagging behind. The slow rise of buoyant droplets at the surface causes the appearance of the thin "sheen" which trails the thickest part of the surface slick. In shallow areas the dispersed droplets interact with suspended particulates, which can lead to sedimentation of significant amounts of spilled oil. A well-known example is the "Braer" oil spill at the Shetland Islands in

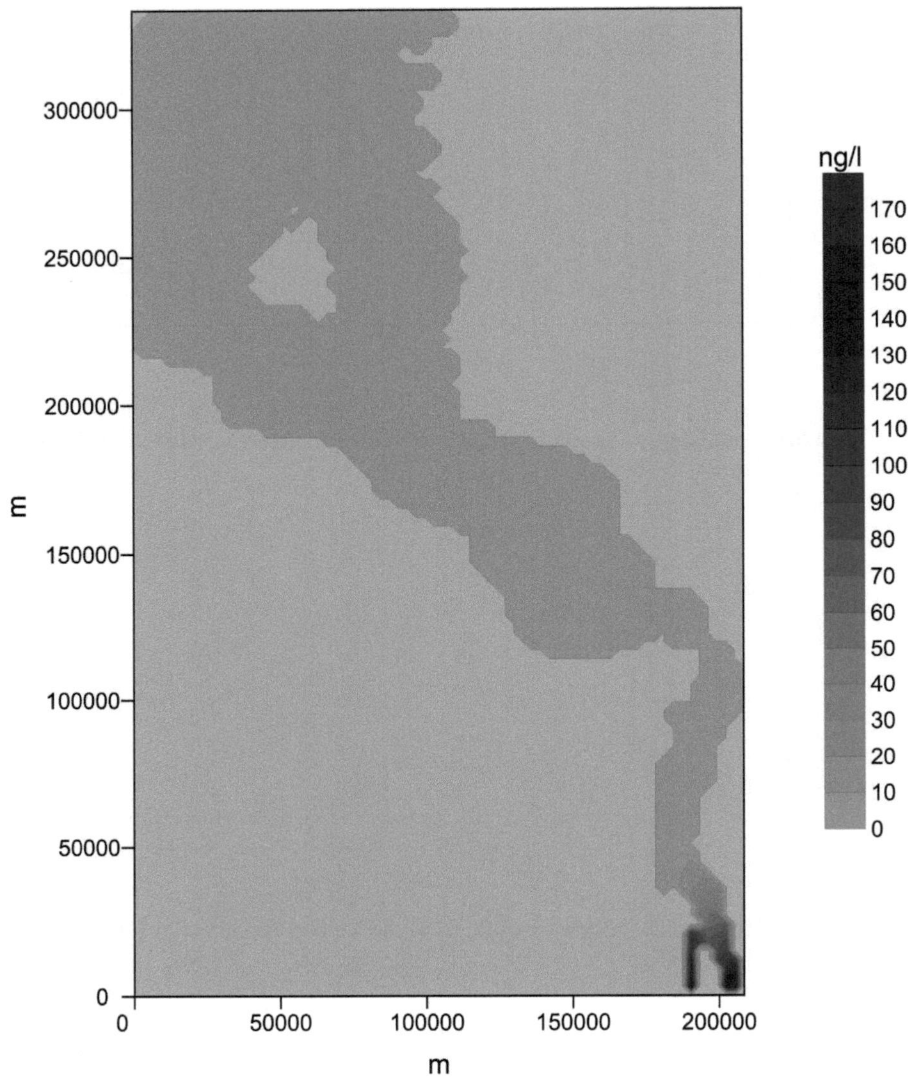

Figure 5.7. THREETOX-simulated PCB concentration in the Yenisei Estuary at 9 days after release near Dudinka near the surface.

1993. The released crude oil was dispersed by waves and currents into a water column and transported for a long distance. It was estimated that as much as 30% of the spilled oil was deposited at the bottom.

The presence of a sea ice cover fundamentally changes the processes of oil spread because oil can spread on the ice, under the ice, and in leads (e.g., Yapa and Weerasuriya, 1997). Ice concentration plays a crucial role in the intensity of weathering because the ice cover limits both spread and evaporation and results in wave damping (Reed *et al.*, 1999).

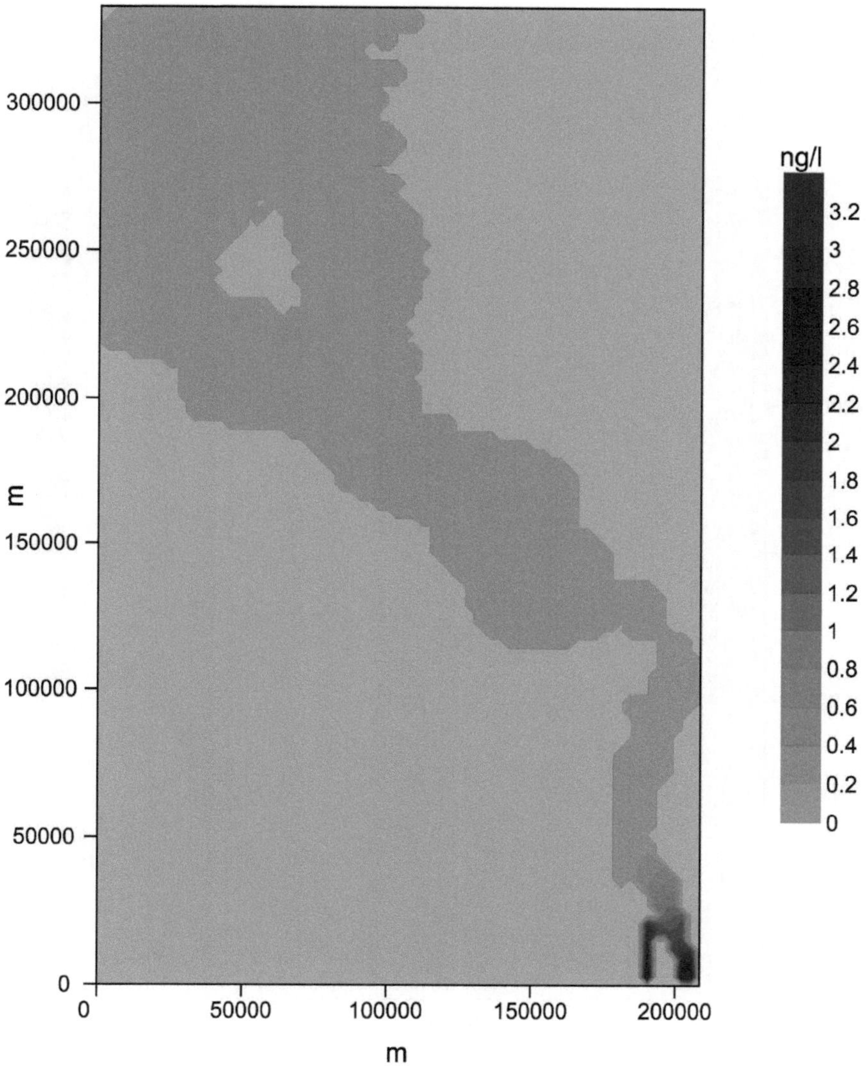

Figure 5.7 (*cont.*). THREETOX-simulated PCB concentration in the Yenisei Estuary at 9 days after release near Dudinka near the bottom.

5.3.2 Modeling oil spread in the marine environment

Modeling petroleum hydrocarbon spread, which is different in that the pollutant is transported mainly in surface slicks at the initial stages of oil spills, is considered in this section. The results of laboratory and field experiments over the past 30 years have advanced our understanding of oil transport and fate processes and have formed the basis for development of the models that describe oil spills—see the state-of-the-art reviews of ASCE (1996) and Reed *et al.* (1999). However, despite

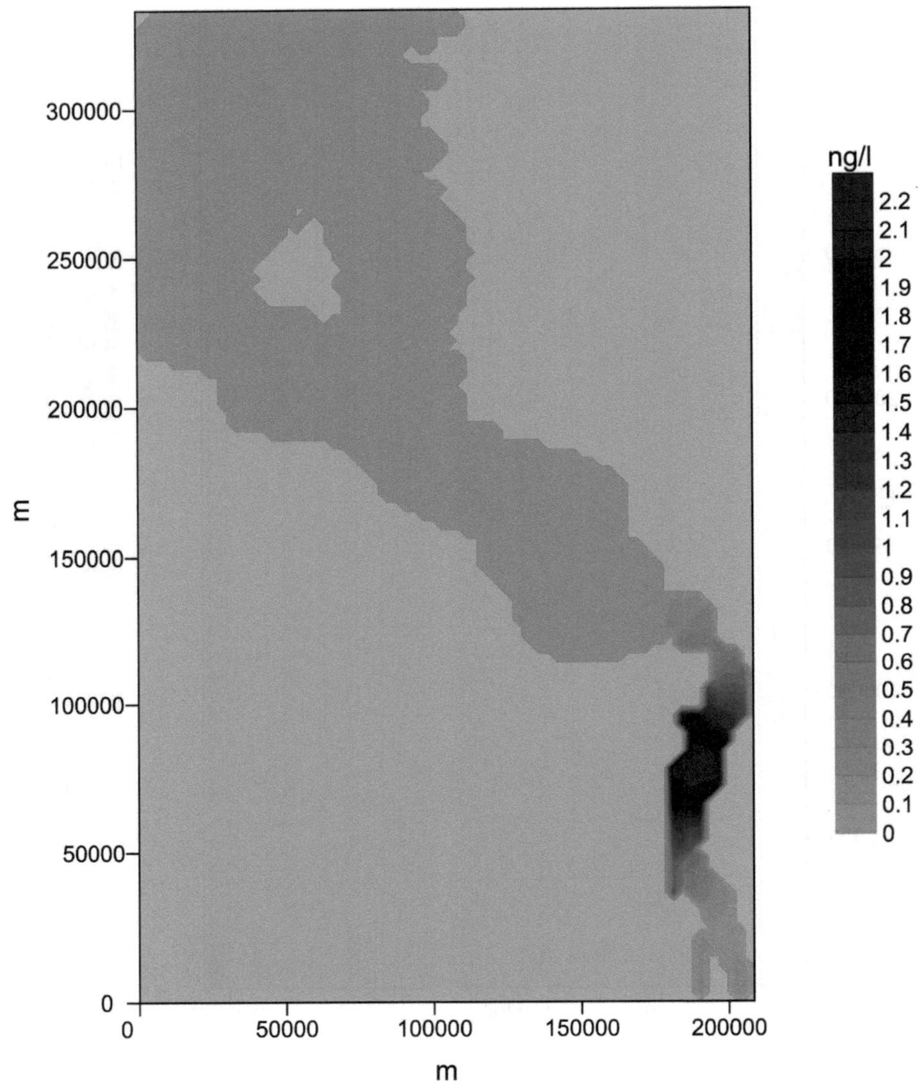

Figure 5.7 (*cont.*). THREETOX-simulated PCB concentration in the Yenisei Estuary at 25 days after release near Dudinka near the surface.

the complexity of the oil spill model, the accuracy of such model predictions has been seriously constrained by limitations in the availability of forecasts of marine and atmospheric environment data for the spill site. Presently, however, increasing computational resources are allowing oil spill models to be coupled with models of weather, circulation, and wave forecasts. Oil spill models can be considered as an integral part of operational oceanographic systems, and the efficiency of such models strongly depends on the increasing accuracy of wave and circulation models. The recently developed modeling system for simulation of an accidental oil spill

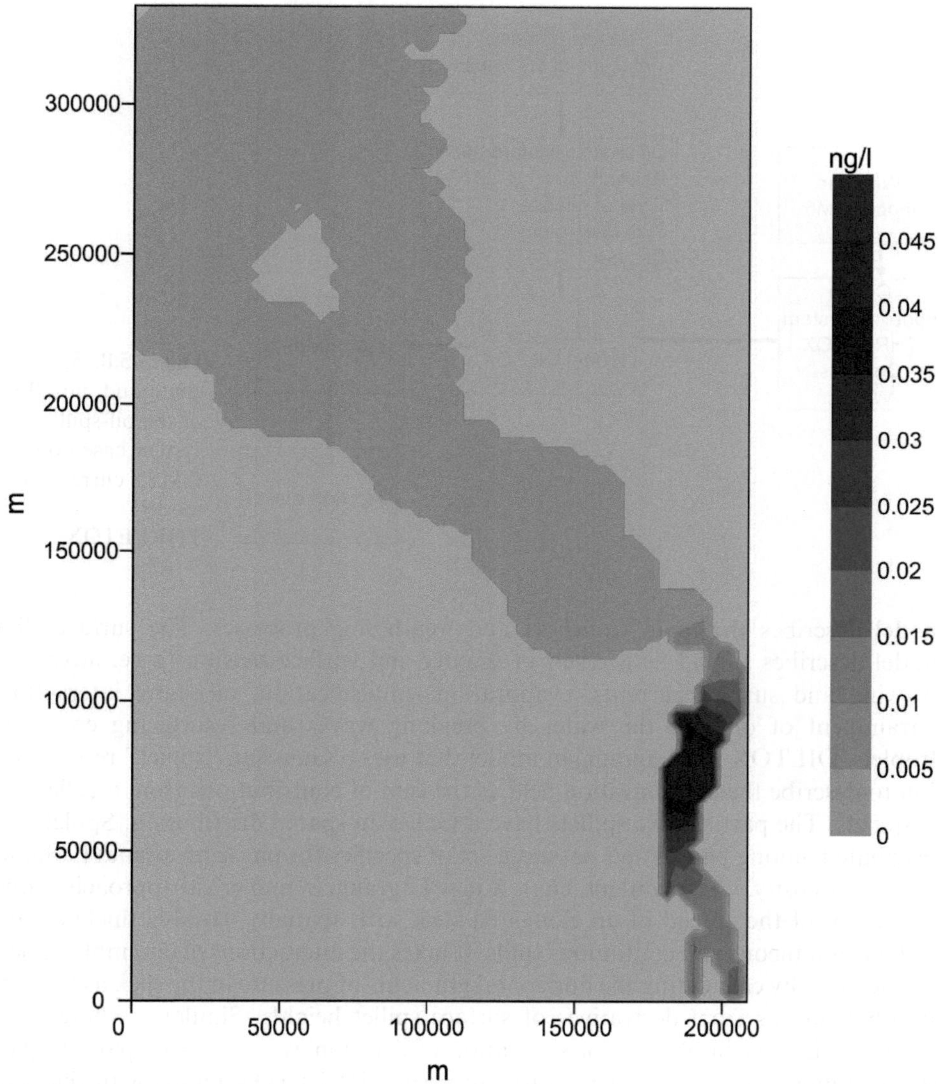

Figure 5.7 (*cont.*). THREETOX-simulated PCB concentration in the Yenisei Estuary at 25 days after release near Dudinka near the bottom.

(Brovchenko *et al.*, 2003) includes the three-dimensional oil-spill-and-fate model OILTOX, the circulation-and-sediment-transport model THREETOX, and the wave forecast model WAVEWATCH III (Tolman, 1999) coupled with the weather forecast model MM5 (Grell *et al.*, 1994), shown in Figure 5.7.

OILTOX is a model that simulates oil transport and fate in five interactive phases: oil on surface, oil in water, oil on bottom, oil on suspended sediments, and oil at shoreline. It is being developed to support a response to accidental oil spillages in the Black and Azov Seas and large river reservoirs in the Ukraine. The

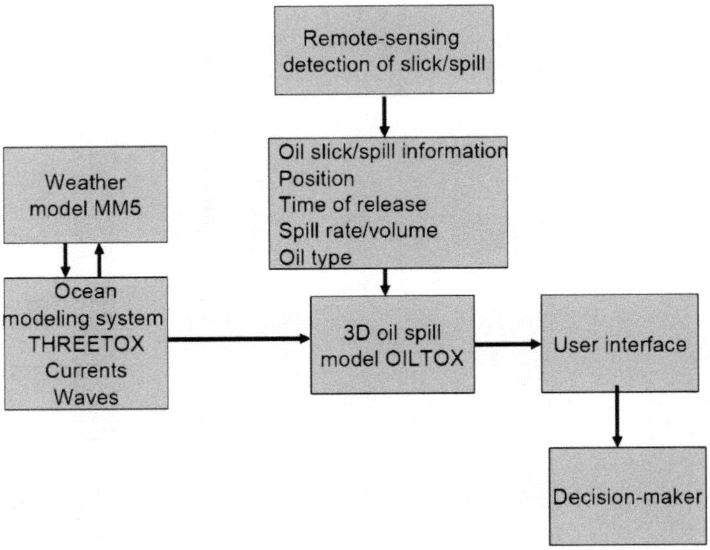

Figure 5.8. Model chain and data flow of the oil-spill system based on the GMS's current flow model THREETOX.

model describes the main transport and weathering processes. The surface slick model describes spread as a result of gravity and surface tension force, advection by wind and surface currents, evaporation, emulsification, oil–shore interaction, entrainment of oil into the water by breaking waves and resurfacing entrained droplets. OILTOX is a Lagrangian model that uses a Gaussian "spillet" representation to describe the concentration field as the sum of contributions from a collection of spillets. The particle-like spillets have a Gaussian spatial distribution. Spillets are distributed among phases and possess a set of specifics for phase properties (volume, density, viscosity, water content, etc.). A new Lagrangian numerical approach allows simulation of the spread of an elongated slick with spatially variable thickness for both instantaneous and continuous spills. It takes the interactions of ensemble spillets into account by calculating the horizontal gradients of pressure in the slick as the sum of contributions from derivatives of surface spillet heights. Similar methods were developed in plasma physics for simulation of collisionless systems by particle–particle methods. By coupling the hydrodynamical THREETOX model with the two-equation turbulence model, the dispersion model allows simulation of oil concentration with depth. Horizontal and vertical turbulent diffusion processes are simulated using Lagrangian stochastic modeling techniques based on the random walk method for Gaussian spillets. Algorithms of evaporation, emulsification, partitioning, sedimentation, and oil–shore interaction are based on state-of-the-art published research.

The modeling system has been adapted to the Black Sea basin. A hypothetical scenario of a spillage of 75,000 tons of crude oil on the northwestern shelf of the Black Sea on September 25, 2002 at a distance of around 50 km from the Odessa oil terminal was considered. The oil slick drifted to a shallow area due to storm winds and surface currents. Oil from the slick was entrained by breaking waves at a rate calculated by WAVEWATCH III wave parameters. The oil droplets were trans-

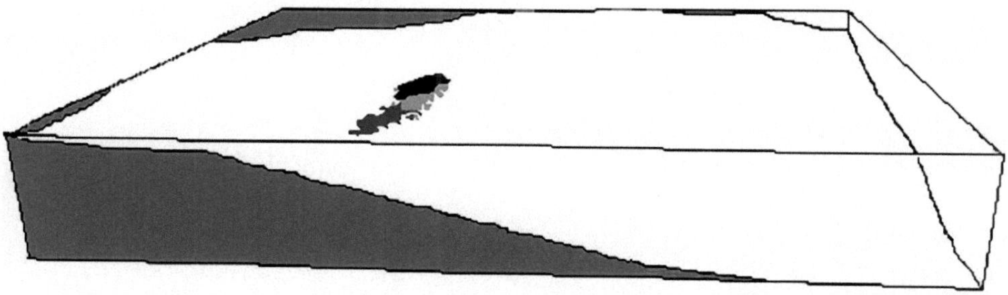

Figure 5.9. Three-dimensional representation of the oil spill at the northwestern shelf of the Black Sea. The surface slick, oil-in-water, and oil-on-bottom concentrations are shown in black, light gray, and dark gray, respectively.

ported by currents, dispersed by turbulence, interacted with suspended and bottom sediments, and gradually resurfaced by buoyancy forces, forming a thin film (sheen) behind the thick slick area. The resulting three-dimensional distribution of oil concentration in three phases (surface, water, and bottom) is shown in Figure 5.8. These results demonstrate the sort of modeling capabilities that can be applied, with further modification, to oil spread in the Arctic marine environment in the future.

6

Assessment and input to risk management

This chapter presents an assessment of the risk of radiological impact on humans as a consequence of the potential release of radioactive material. Although it is clearly beyond the scope of this chapter to provide a comprehensive risk assessment of all potential environmental and human impacts from all scenarios of radioactive releases in Arctic marine and terrestrial realms, we are able to focus on one major set of risks. These are risks to humans associated with potential releases along the major Siberian rivers—the Ob′ and Yenesei—including an assessment of how global warming may affect the consequences. Section 6.1 is an introduction to the assessment, while various scenarios are described in Section 6.2. Section 6.3 describes how dose models are formulated and implemented. The results of risk assessment modeling are provided in Section 6.4. Section 6.5 presents a summary and major conclusions.

6.1 INTRODUCTION

6.1.1 Purpose, endpoints, and philosophy

The purpose of this assessment is (1) to evaluate the radiological impact on humans from potential accidental releases of large amounts of radioactive material located at sites along the Russian rivers Ob′ and Yenisei and transported toward the Arctic Ocean, and (2) to estimate the impact of global warming on radiological consequences.

The endpoints of the assessment are radiological risks to individuals living along the Ob′ and Yenisei and on the coastlines of the Arctic Ocean—risks arising from hypothetical accidental releases of radioactivity. Radiological risks are evaluated at their maximum annual value and assumed to be proportional to radiation doses received by individuals as a result of exposure to radioactive substances transported downstream by the rivers.

Assumptions made for the assessment are generally taken to be realistic rather than conservative. This includes assumptions used to predict river transport of radionuclides from the storage sites of radioactive waste to locations of human occupancy, and assumptions about human lifestyles when estimating radiation doses.

6.1.2 Source term characteristics

Source terms considered for the assessment are nuclear waste disposals from the three biggest combines of the Russian nuclear industry: Chelyabinsk-65, the Mayak Production Association (PA); Krasnoyarsk-26, the Mining and Chemical Combine (MCC); and Tomsk-7, the Siberian Chemical Combine (SCC). Nuclear enterprises and the characteristics of waste disposals are described in detail in Chapter 1.

6.1.3 Environmental characteristics

The Mayak PA is located in the closed town of Ozersk in the Southern Urals, situated approximately 15 km east of the town of Kyshtym and 70 km north of the city of Chelyabinsk. During operation of the plant from 1949 to 1956, large amounts of radioactive waste were discharged into the nearby Techa River and since 1951 into Lake Karachay. Spring floods have contaminated large areas of the Asanov Swamp in the upper reaches of the Techa River with nuclear waste. Between 1951 and 1966, a system of dams was constructed to contain the waste by creating artificial reservoirs. The Techa River connects—via the Iset, Tobol, and Irtysh Rivers—the Ob′ River which flows to the Kara Sea in the Arctic. The downstream distance from the Mayak PA to the outlet of the Ob′ River in the Arctic Ocean is approximately 2,800 km.

The Siberian Chemical Combine (SCC) at Tomsk-7 is located on the Tom River near the small city of Seversk about 15 km northwest of the city of Tomsk. The Tom River is a tributary to the Ob′ River, which flows into the Kara Sea.

The Krasnoyarsk Mining and Chemical Combine (MCC) is located at the town of Zheleznogorsk (formerly known as Krasnoyarsk-26) on the Yenisei River some 60 km northeast of the city of Krasnoyarsk. Powerful seasonal floods control the hydrological features of the Yenisei River. As a result, floodplain deposits are the cause of long-posed environmental problems since the MCC started to operate. The floodplain soils along with bottom sediments are the most contaminated environmental compartments influenced by the MCC. These deposits occupy considerable areas on many islands and along the riverbanks. The Yenisei River then flows to the Kara Sea. The geographical and hydrological characteristics of these river systems are detailed in Chapter 2.

6.1.4 Time frames and societal assumptions

The assessment focuses on maximum annual radiation doses received by individuals affected by hypothetical releases of radioactivity from nuclear waste sites. The time frames involved for accident scenarios thus cover the time needed for transport of contaminated water and sediments from waste disposals to locations where humans become exposed, typically a few years from the time of release.

To evaluate the impact of global warming, a doubling of atmospheric CO_2 concentrations is assumed to occur over a century, after which accident scenarios are assumed to release radionuclides from waste disposals.

Societal conditions and human lifestyles are assumed to remain unchanged during the time frames considered for the assessment.

6.2 SCENARIOS

6.2.1 Source term scenarios

The accident scenario for the Mayak PA enterprise involves assuming a major break in the dam of Reservoir No. 11, which is the last barrier between the Mayak PA's reservoirs and the Techa River. The event triggering the accident could be a strong flood, an earthquake, or the crash of a large airplane into the dam. The accident scenario stipulates that Reservoir-11's inventory of ^{90}Sr (approximately 300 TBq) is transferred to the Techa River instantaneously.

The accident scenario for the Siberian Chemical Combine (SCC) at Tomsk-7 is based on a nonspecific hypothetical failure of nuclear waste containment systems on the site (e.g., break in the dam of a holding pond), resulting in releases of 37 TBq of ^{90}Sr and 37 TBq of ^{137}Cs. The radionuclides are assumed to enter the Tom River primarily with contaminated soils and sediments via an overland flow and through the Ramashka River, which is used at the site to discharge process water.

The accident scenario for the Krasnoyarsk Mining and Chemical Combine (MCC) assumes a hypothetical failure of nuclear waste containment systems—in this case open-water storage basins, Reservoirs No. 365 and No. 366. Instantaneous releases of ^{90}Sr and ^{137}Cs are assumed to occur in much the same way as those for the SCC scenario.

Scenarios for accidental releases are described in more detail in Chapter 4.

6.2.2 Climate scenarios

Due to the significant seasonal variation in water flow of the large rivers, release scenarios are selected to represent both low-flow conditions (autumn/winter) and high-flow conditions (spring/summer). The variability in water flows between season of the Ob′ and Yenisei Rivers is very significant and described in further detail in Chapter 2. The year 1967 was chosen as a reference for hydrological situations in the rivers.

Predicted climatic changes due to a doubling of atmospheric CO_2 concentrations over a century provide the input for environmental transport models, which have used these assumptions for another set of calculations of accident scenarios. The years of 2084 and 2089 were chosen as references for this situation. Thus, it is possible to directly compare radiological risks with and without climatic changes.

6.3 FORMULATION AND IMPLEMENTATION OF DOSE MODELS

The methodology used to assess radiation doses to individuals is based on predicted concentrations of radionuclides in water and sediments, the transfer of radionuclides through food chains, and human lifestyles. Best estimate values of model parameters are used throughout the calculations.

The assessment focused on annual average individual doses to adults, and dose models are therefore based on assumptions and parameter values that provide annual average doses across seasons. Thus, annual average concentrations in water and sediment have been used as input for the dose assessment. The models used to predict environmental transfer and dispersion are capable of providing data with a much higher time resolution, although this was not considered of relevance for the assessment.

Human occupancy and farming practices along contaminated riverbanks and coastlines give rise to human exposure to radiation and transfer of radionuclides to agricultural products. Dose models consider the transfer of radionuclides through food chains and a range of pathways of human exposure including external exposure from radionuclides in the soil and internal exposure from the intake of radionuclides through inhalation and ingestion.

Radionuclides in soil cause external radiation aboveground and contamination of vegetation by transfer from soil to plant, and contaminated pastures give rise to contamination of animal produce. These processes are accounted for by model parameters, the values of which have been determined from experiments and field studies. Concentration factors are used to calculate the transfer of radionuclides from soil to vegetation, as well as the transfer from water to biota. The transfer of radionuclides to milk and meat is calculated using parameters that give the fractions of radioactivity found per liter of milk and kilogram of meat relative to the daily intakes of radioactivity by the animals. External dose rates are calculated from average radionuclide concentrations in the soil and the energy of radiation emitted (EC, 1990). Dose factors give the relationship between activity ingested or inhaled and the corresponding committed effective radiation dose (EC, 1996). The values of environmental transfer parameters are recommended by the International Atomic Energy Agency (IAEA, 1985, 1994) and are given in Table 6.1 together with other parameter values.

Radiation doses are determined for individuals who are most exposed, so-called critical groups. The assumptions for these individuals involve annual intakes of food and time spent outdoors, which include inadvertent ingestion of soil and inhalation of resuspended airborne soil particles. Data on human food consumption and occupancy are derived from lifestyle surveys carried out in Brodokalmak on the Techa River (Aarkrog *et al.*, 2000). Lifestyle assumptions are listed in Table 6.2.

Concentrations of radionuclides in vegetation and animal produce are calculated from predicted concentrations in river water and riverbanks. Drinking water intake is assumed to be 1% river water.

The doses estimated are thus the sum of committed effective doses due to the ingestion of radionuclides in a year, and the external effective radiation dose received

Table 6.1. Parameter values (concentration factors, transfer factors, dose factors, and other) used for calculating the transfer of ^{137}Cs and ^{90}Sr from soil to vegetation and from water to biota and sediments (IAEA, 1985, 1994).

Item	*Parameter*	^{137}Cs	^{90}Sr	*Unit*
Grass	CF	0.24	1.7	Bq/kg per Bq/kg
Green vegetables	CF	0.46	3	Bq/kg per Bq/kg
Root vegetables	CF	0.17	0.26	Bq/kg per Bq/kg
Grain	CF	0.026	0.21	Bq/kg per Bq/kg
Milk	F_m	0.0079	0.0028	Bq/L per Bq/da
Beef	F_f	0.05	0.008	Bq/kg per Bq/da
Freshwater fish	CF	2,000	60	Bq/kg per Bq/kg
Marine fish	CF	100	2	Bq/kg per Bq/kg
Crustaceans	CF	30	2	Bq/kg per Bq/kg
Mollusks	CF	30	1	Bq/kg per Bq/kg
Sediment, coastal	CF	3,000	1,000	Bq/kg per Bq/L
Mean gamma-ray energy	E_γ	0.564	1.7E-6	MeV
Grass intake, dairy cow			16	kg/da
Grass intake, beef cattle			7	kg/da
Dose factor, ingestion	DF_{ing}	1.3E-8	2.8E-8	Sv/Bq
Dose factor, inhalation	DF_{inh}	3.9E-8	1.6E-7	Sv/Bq

in a year as defined by the International Commission on Radiological Protection (ICRP, 1990) for comparison with annual dose limits. Committed effective doses due to internal radiation are integrated over 50 years for adults. Therefore, doses are not the doses actually received in a year but rather those that individuals are committed to.

6.4 RESULTS

Radiation doses to individuals living in settlements on the rivers downstream from the release points have been calculated. For each accident scenario, three settlement locations were considered: one location relatively close to the point of release, one

Table 6.2. Assumptions on human habits used to calculate radiation doses for adults from ingestion and inhalation.

Item	*Value*	*Unit*
Water	800	kg/yr
Green vegetables	50	kg/yr
Root vegetables	120	kg/yr
Grain products	150	kg/yr
Milk and milk products	100	kg/yr
Meat	50	kg/yr
Freshwater fish	40	kg/yr
Marine fish	30	kg/yr
Crustaceans	10	kg/yr
Mollusks	10	kg/yr
Soil ingestion, inadvertent	0.04	kg/yr
Inhalation rate	1	m^3/h
Soil concentration in air	1E-07	kg/m^3
Outdoor occupancy	1,000	h/yr

Table 6.3. Locations for critical groups for calculation of doses.

Accident scenario	*Near release point*	*River mouth*	*River estuary*
Mayak PA	Brodokalmak	Ob′ River mouth/ Ob′ delta (Salekhard)	Ob′ estuary mouth
Tomsk-7 SCC	Tomsk-7	Ob′ River mouth/ Ob′ delta (Salekhard)	Ob′ estuary mouth
Krasnoyarsk-26 MCC	Krasnoyarsk-26	Yenisei River mouth/ Yenisei delta	Yenisei estuary mouth

location at the mouth of the river, and one location on the estuary of the river at the interface to the Kara Sea.

The locations at which doses to critical groups are calculated are listed by accident scenario in Table 6.3. Tomsk-7 on the Tom River and Krasnoyarsk-26

Table 6.4. Accident scenarios considered for the assessment.

Accident site	*Time of release*	*Radionuclides released*
Mayak PA	April 1967 November 1967 April 2084 November 2084	^{90}Sr
Tomsk-7 SCC	January 1967 June 1967 January 2084 June 2084	^{90}Sr and ^{137}Cs
Krasnoyarsk-26 MCC	January 1967 June 1967 January 2089 June 2089	^{90}Sr and ^{137}Cs

on the Yenisei River are located close to the assumed points of release, whereas Brodokalmak on the Techa River is located some 110 km from the Mayak PA.

Input data for dose calculations comprising time series of concentrations of radionuclides in water and sediments at these locations were calculated for the accident scenarios. To take the effect of seasonality into account, these input data were calculated for accidents occurring for both low-flow and high-flow conditions in the rivers. Furthermore, to illustrate the effect on risk assessment of climate change due to a doubling of atmospheric CO_2 concentrations over a century, the calculations were repeated using environmental data for the years 2084–2089 on precipitation and temperature to force the river models. To facilitate conclusions about the effects due to climate change, the latter calculations were made with the same source terms as during 1967, disregarding radioactive decay. The full set of scenarios is listed in Table 6.4 and the time series of concentrations predicted in water and sediments are given in Appendix A.

From the time series of annual average radionuclide concentrations in water and sediments, those years showing the highest concentrations were chosen to calculate doses to individuals in critical groups. The resulting total annual doses are shown in Tables 6.5–6.7 and Figures 6.1–6.3.

For the Mayak PA accident scenario, the maximum annual doses to individuals in Brodokalmak are predicted to be in the range 10 mSv to 11 mSv, in the Ob′ delta in the range 4 μSv to 8 μSv and in the Ob′ estuary in the range 4 μSv to 6 μSv. Seasonal variations of doses are small for all three locations, in the range 2% to 7% with doses higher in April than in November. The variation of doses between years is significant only for the Ob′ delta, with doses in 2084 some 30% lower than in 1967. The variations are calculated as differences relative to mean values.

For the Tomsk-7 SCC accident scenario, the maximum annual doses to individuals in Tomsk-7 are predicted in the range 0.09 mSv to 0.13 mSv, in the

Table 6.5. Total maximum annual doses to individuals ($Sv\,yr^{-1}$) from accident scenarios of a break in Reservoir-11's dam at Mayak. Doses are shown at three downstream sites (Brodokalmak, Ob′ delta, and Ob′ estuary) by time of accident.

Location	*April 1967*	*November 1967*	*April 2084*	*November 2084*
Brodokalmak	1.0E-02	9.8E-03	1.1E-02	1.0E-02
Ob′ delta	8.3E-06	7.4E-06	7.1E-06	3.7E-06
Ob′ estuary	5.8E-06	4.3E-06	5.1E-06	4.6E-06

Table 6.6. Total maximum annual doses to individuals ($Sv\,yr^{-1}$) from accident scenarios of a break in a waste-holding pond at the Siberian Chemical Combine (SCC) at Tomsk. Doses are shown at three downstream sites (Tomsk, Ob′ delta, and Ob′ estuary) by time of accident.

Location	*January 1967*	*June 1967*	*January 2084*	*June 2084*
Ob′ delta	1.0E-05	3.8E-05	8.7E-06	1.7E-05
Ob′ estuary	2.8E-06	4.2E-06	2.6E-06	4.1E-06
Tomsk	1.3E-04	1.3E-04	8.9E-05	8.9E-05

Table 6.7. Total maximum annual doses to individuals ($Sv\,yr^{-1}$) from accident scenarios of a break in waste Holding Ponds No. 365 and No. 366 at the Mining and Chemical Combine (MCC) at Krasnoyarsk. Doses are shown at three downstream sites (Krasnoyarsk-26, Yenisei delta, and Yenisei estuary) by time of accident.

Location	*January 1967*	*June 1967*	*January 2089*	*June 2089*
Krasnoyarsk-26	7.3E-04	7.3E-04	4.3E-04	4.3E-04
Yenisei delta	7.2E-06	1.2E-05	5.7E-06	1.3E-05
Yenisei estuary	6.6E-06	9.0E-06	3.4E-06	6.0E-06

Ob′ delta in the range 9 μSv to 38 μSv, and in the Ob′ estuary in the range 3 μSv to 4 μSv. The seasonal variation of doses for the Tomsk-7 location is nil, but 29% for the Ob′ delta and 10% for the Ob′ estuary, with higher doses in June than in January. The variation of doses between years is 33% for Tomsk, 45% for the Ob′ delta, and insignificant for the Ob′ estuary with doses lower in 2084 than in 1967.

For the Krasnoyarsk-26 MCC accident scenario, the maximum annual doses to individuals in Krasnoyarsk-26 are predicted in the range 0.4 mSv to 0.7 mSv, in the Yenisei delta in the range 6 μSv to 13 μSv, and in the Yenisei estuary in the range

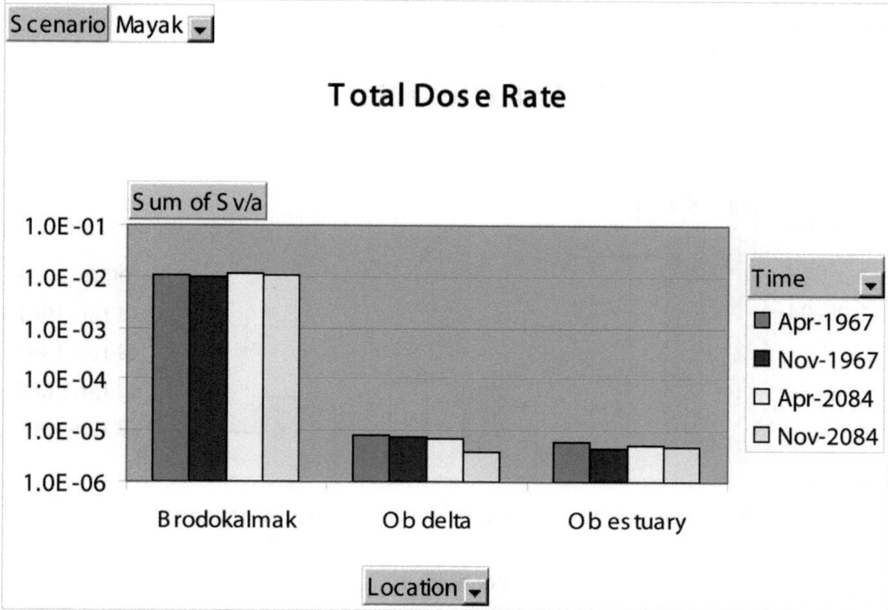

Figure 6.1. Total maximum annual doses to individuals (Sv y^{-1}) from accident scenarios of a break in the reservoir-11 dam at Mayak.

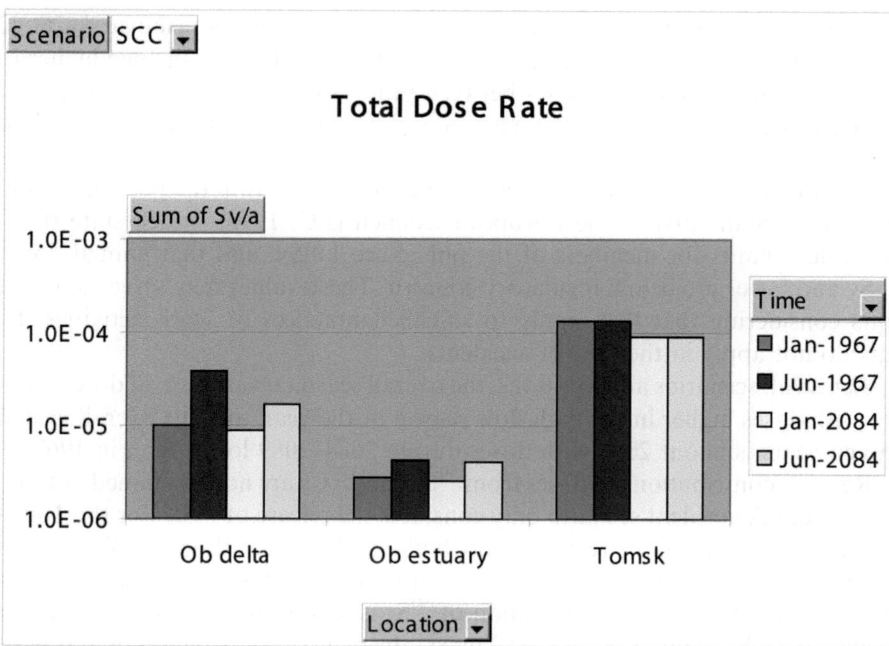

Figure 6.2. Total maximum annual doses to individuals (Sv y^{-1}) from accident scenarios of a break in a waste-holding pond at the Tomsk SCC.

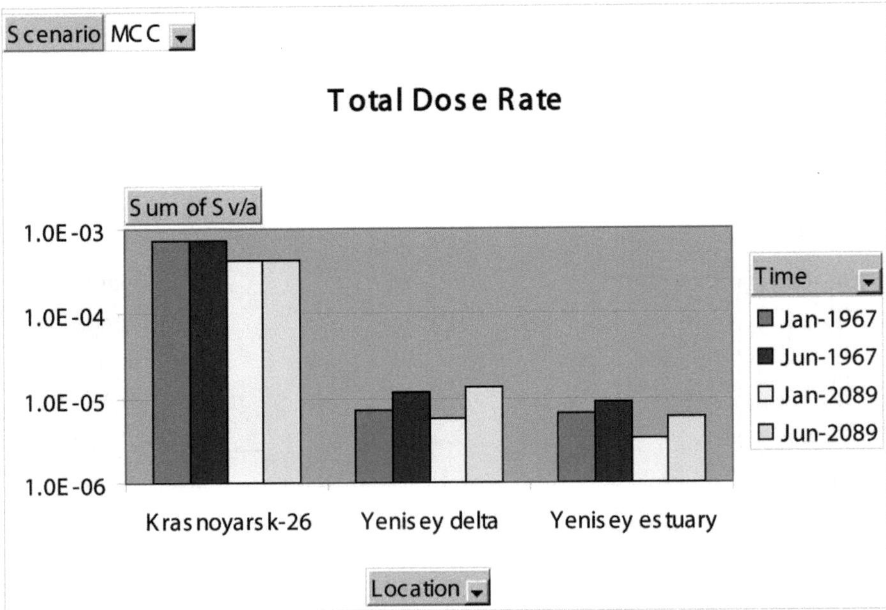

Figure 6.3. Total maximum annual doses to individuals ($Sv\,y^{-1}$) from accident scenarios of a break in waste-holding ponds Nos. 365 and 366 at the Krasnoyarsk MCC.

3 μSv to 9 μSv. Seasonal variation between doses is nil for Krasnoyarsk-26, 12% for the Yenisei delta, and 8% for the Yenisei estuary with doses in June higher than in January. The variation of doses between years is 40% at Krasnoyarsk-26, nil at the Yenisei delta, and 40% at the Yenisei estuary with doses lower in 2089 than in 1967.

Radiological risks from the doses may be considered from the perspective of the Basic Safety Standards of the European Council (EC, 1996), which state that the annual dose limits for members of the public are 1 mSv, and that annual doses of 10 μSv and lower are below regulatory concern. These values may serve as reference points considering that they apply to so-called practices or work activities. Dose limits do not apply in the case of accidents.

Across all scenarios and locations, the overall seasonal variation of doses is about 10% with doses higher in the high-flow season of the year, and the overall variation between years is about 20% with doses during 2084–2089 lower than in 1967.

Relative contributions to doses from ^{90}Sr and ^{137}Cs are now examined as follows. The Mayak PA accident scenario only considers the release of ^{90}Sr. For the Tomsk-7 SCC accident scenario, the relative contributions of ^{90}Sr are shown in Table 6.8 and Figure 6.4. Relative contributions of ^{137}Cs to the total dose are complementary to those of ^{90}Sr. The relative contribution of ^{90}Sr is seen to increase with downstream distance from the point of release. At Tomsk, the relative contribution of ^{90}Sr is 11%, at the Ob′ delta in the range 37% to 50%, and in the Ob′ estuary in the range 82% to 87%. This is due to the fact that cesium for chemical reasons is bound much more

Table 6.8. Relative contribution of ^{90}Sr to doses from accident scenarios of a break in a waste-holding pond at the Siberian Chemical Combine (SCC) at Tomsk. Contributions are shown at three downstream sites (Tomsk, Ob′ delta, and Ob′ estuary) by time of accident.

Location	*January 1967*	*June 1967*	*January 2084*	*June 2084*
Ob′ delta	0.37	0.50	0.40	0.38
Ob′ estuary	0.82	0.86	0.82	0.87
Tomsk	0.11	0.11	0.11	0.11

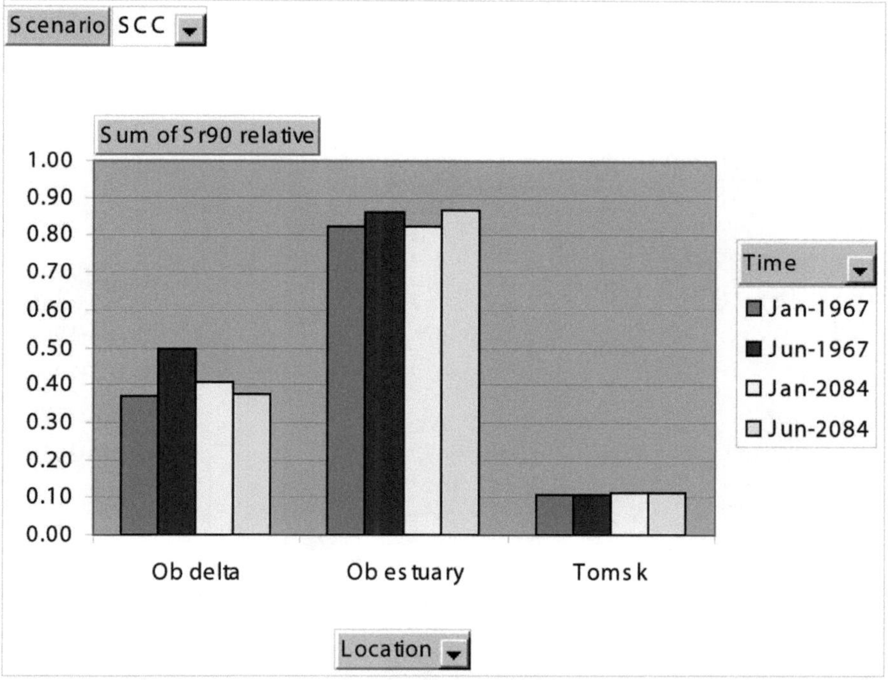

Figure 6.4. Relative contribution of ^{90}Sr to doses from accident scenarios of a break in a waste-holding pond at the Tomsk SCC.

strongly to minerals in the river sediments than strontium, thus causing a gradual depletion with downstream distance of ^{137}Cs relative to ^{90}Sr.

For the Krasnoyarsk-26 accident scenario, relative contributions of ^{90}Sr are shown in Table 6.9 and Figure 6.5. Again we find that relative contributions of ^{90}Sr increase with downstream distance from the point of release. At Krasnoyarsk-26 the relative contribution of ^{90}Sr is 2%, at the Yenisei delta it is in the range 10% to 63%, and in the Yenisei estuary it is in the range 25% to 73%.

The doses calculated for individuals sorted by radionuclide and exposure pathway are detailed in Appendix B. For the Mayak PA accident scenarios, ^{90}Sr

Table 6.9. Relative contribution of ^{90}Sr to doses from accident scenarios of a break in waste-holding ponds No. 365 and No. 366 at the Mining and Chemical Combine (MCC) at Krasnoyarsk. Contributions are shown at three downstream sites (Krasnoyarsk-26, Yenisei delta, and Yenisei estuary) by time of accident.

Location	*January 1967*	*June 1967*	*January 2089*	*June 2089*
Krasnoyarsk-26	0.02	0.02	0.02	0.02
Yenisei delta	0.63	0.54	0.23	0.10
Yenisei estuary	0.69	0.73	0.40	0.25

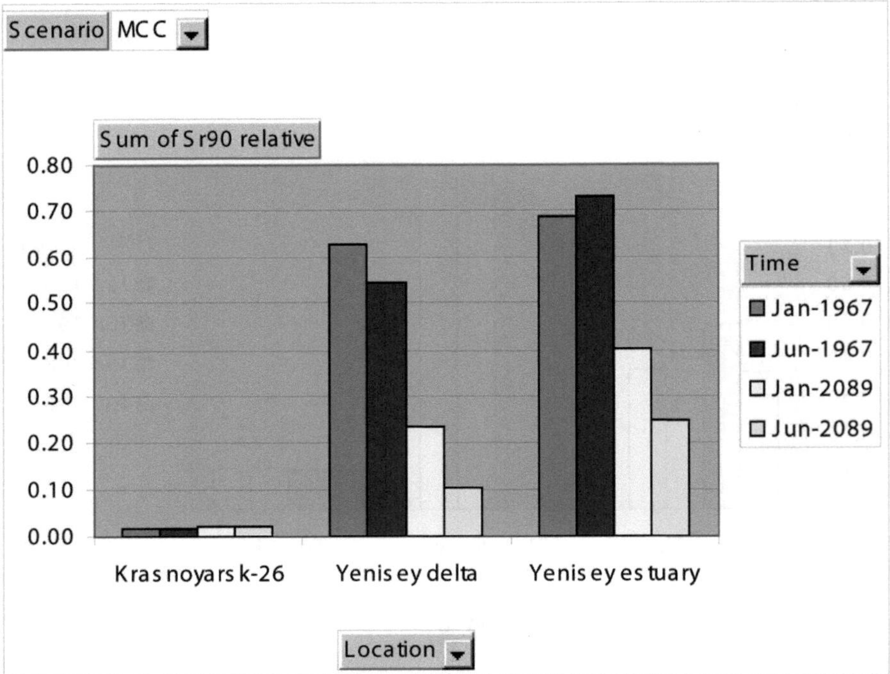

Figure 6.5. Relative contribution of ^{90}Sr to doses from accident scenarios of a break in waste-holding ponds Nos. 365 and 366 at the Krasnoyarsk MCC.

is the only radionuclide considered, and ingestion of fish is the predominant exposure pathway for all locations except for the Ob′ estuary, where ingestion of grain in some cases predominates.

For the Tomsk-7 SCC accident scenarios, ^{137}Cs is the dominating radionuclide at the Tomsk-7 location. At the Ob′ delta ^{137}Cs dominates in most cases, and at the Ob′ estuary ^{90}Sr is the predominant radionuclide. Ingestion of fish is the predominant exposure pathway for all locations and all accident scenarios.

For the Krasnoyarsk-26 accident scenarios, ^{137}Cs is the predominant

Table 6.10. Predominant radionuclides and exposure pathways across all accident scenarios.

Accident scenario	*Location*	*Nuclide*	*Pathway*
Mayak PA	Brodokalmak	^{90}Sr	Ingestion of fish
	Ob′ delta	^{90}Sr	Ingestion of fish
	Ob′ estuary	^{90}Sr	Ingestion of grain/fish
Tomsk-7 SCC	Tomsk-7	^{137}Cs	Ingestion of fish
	Ob′ delta	$^{137}Cs/^{90}Sr$	Ingestion of fish
	Ob′ estuary	^{90}Sr	Ingestion of fish
Krasnoyarsk-26	Krasnoyarsk-26	^{137}Cs	External exposure
	Yenisei delta	$^{90}Sr/^{137}Cs$	Ingestion of fish
	Yenisei estuary	$^{90}Sr/^{137}Cs$	Ingestion of fish

radionuclide at the Krasnoyarsk-26 location where external exposure is the predominant exposure pathway. At the Yenesei delta and estuary locations the results show that ^{90}Sr dominates for the scenarios in 1967 while ^{137}Cs dominates in 2089; ingestion of fish is the dominating exposure pathway.

These results are summarized across accident scenarios in Table 6.10.

6.5 CONCLUSIONS

A radiological risk assessment has been carried out to evaluate the potential impact on humans of hypothetical accidental releases of radioactive material from waste disposal sites on the Russian rivers Ob′ and Yenisei and transported downstream to the Kara Sea in the Arctic Ocean. Furthermore, the potential impact of global warming on radiological consequences has been evaluated.

The source terms considered for assessment are hypothetical accidental releases of radioactive material from nuclear waste disposal sites located at the three biggest combines of the Russian nuclear industry: Chelyabinsk-65, the Mayak Production Association; Krasnoyarsk-26, Mining and Chemical Combine (MCC); and Tomsk-7, the Siberian Chemical Combine (SCC). The accident scenarios assume hypothetical failures of waste containment systems such as a break in dams of holding ponds due to anthropogenic or natural events involving, for example, airplane crashes, floods, or earthquakes. The radionuclides considered for the accident scenarios are ^{90}Sr and ^{137}Cs.

Radiological risks to individuals of critical groups living on the Ob′ and Yenisei rivers and on the river estuaries bordering the coastlines of the Kara Sea are estimated. Lifestyles of the critical group are chosen to represent those individuals that are most exposed to radioactive material. Radiological risks are evaluated at their maximum annual values and assumed to be proportional to radiation doses from exposure to radioactive material. The locations of critical groups closest to the discharge points are Brodokalmak on the Techa River downstream from the Mayak PA, the Tomsk-7 site on the Tom River, and Krasnoyarsk-26 on the Yenisei River. Further critical groups are located at the deltas and estuaries of the Ob′ and Yenisei Rivers.

Accident scenarios at the Mayak PA are predicted to give rise to radiation doses of 10 mSv to 11 mSv at Brodokalmak, 4 μSv to 8 μSv at the Ob′ River delta, and 4 μSv to 6 μSv at the Ob′ estuary. The most important exposure pathway at Brodokalmak is ingestion of ^{90}Sr in freshwater fish. The most important pathways at the Ob′ delta and estuary are ingestion of ^{90}Sr in fish, grain products, and green vegetables. Variations in doses as a result of seasonal timing of the accident are small, in the range of 2% to 7% with doses higher in April than in November.

Accident scenarios at the Tomsk SCC are predicted to cause radiation doses in the range 0.09 mSv to 0.13 mSv at Tomsk-7, 9 μSv to 38 μSv at the Ob′ delta, and 3 μSv to 4 μSv at the Ob′ estuary. Cesium-137 is the predominant radionuclide at the Tomsk-7 location. At the Ob′ delta ^{137}Cs is the predominant radionuclide in most cases, and at the Ob′ estuary ^{90}Sr is predominant. Ingestion of fish is the dominating exposure pathway for all locations and accident scenarios. Variability in doses as a result of seasonal timing of the accident are zero for the Tomsk-7 site, 29% for the Ob′ delta, and 10% for the Ob′ estuary, with doses higher in June than in January.

Accident scenarios at the Krasnoyarsk MCC are predicted to cause radiation doses in the range 0.4 mSv to 0.7 mSv at Krasnoyarsk-26, 6 μSv to 13 μSv at the Yenisei delta, and 3 μSv to 9 μSv at the Yenisei estuary. Cesium-137 is the dominating radionuclide at the Krasnoyarsk-26 location, where external exposure is the dominating exposure pathway. At the Yenesei delta and estuary locations, results show that ^{90}Sr dominates for the scenarios in 1967 while ^{137}Cs dominates in 2089; ingestion of fish is the dominating exposure pathway. Variations in doses as a result of seasonal timing of the accident are zero for the Krasnoyarsk-26 site, 12% for the Yenisei delta, and 8% for the Yenisei estuary with doses higher in June than in January.

The relative contributions of ^{90}Sr to radiation doses increase with downstream distance from the point of release. At Tomsk the relative contribution of ^{90}Sr is 11%, at the Ob′ delta it is in the range 37% to 50%, and in the Ob′ estuary it is in the range 82% to 87%. This is due to the fact that cesium (for chemical reasons) is bound much more strongly to minerals in river sediments than strontium, thus causing a gradual depletion with downstream distance of ^{137}Cs relative to ^{90}Sr.

Across all scenarios and locations, overall seasonal variation in doses is about 10% with doses higher in the high-flow season of the year, and overall variation between years is about 20% with doses during 2084–2089 lower than in 1967. The impact from global warming on radiological risks is thus estimated to be relatively minor and to have the effect of reducing the risks about 20% as a result of increased precipitation and reduced concentrations of radionuclides in the rivers.

Radiological risks from estimated doses do not give rise to great concern, not even to critical groups located near the points of discharge. The highest annual doses are predicted at 10 mSv to 11 mSv at the Brodokalmak site with annual doses at all other sites below 1 mSv. Doses may be considered from the perspective of the Basic Safety Standards of the European Council (EC, 1996), which state that annual dose limits for members of the public are 1 mSv, and that annual doses of 10 μSv and lower are below regulatory concern. It is important to note, however, that these values may serve as reference points considering that they apply to ongoing activities; dose limits do not apply in the case of accidents.

Appendix A

Time series of annual average concentrations of radionuclides in water and sediments by accident scenario and location used for dose calculations

Table A.1. Scenario 1a, break in Reservoir-11's dam at the Mayak PA, April 1967. Annual average concentrations of ^{90}Sr in water solution C ($Bq\,m^{-3}$), sediments CS ($Bq\,kg^{-1}$), and suspended sediment loads S ($kg\,m^{-3}$).

	Brodokalmak			*Ob′ River mouth (Salekhard)*			*Ob′ estuary mouth*		
Year	*C* (Bq/m^3)	*CS* (Bq/kg)	*S* (kg/m^3)	*C* (Bq/m^3)	*CS* (Bq/kg)	*S* (kg/m^3)	*C* (Bq/m^3)	*CS* (Bq/kg)	*S* (kg/m^3)
1967	82,515	21,684	1.00E-03	49.82	21.93	6.02E-03	0.84	0.67	1.99E-03
1968	1,084	1,995	9.98E-04	2.76	1.5	7.61E-03	17.72	20.55	2.29E-03
1969	109	382	8.56E-04	0.83	0.74	1.97E-03	10.69	8.64	4.13E-03
1970	34.48	62.98	1.01E-03	0.32	0.26	2.00E-03	3.28	2.81	2.24E-03
1971	11.97	21.77	1.03E-03	0.1	0.09	1.83E-03	1.28	1.02	2.00E-03
1972	4.15	7.52	7.33E-04	0.04	0.03	1.86E-03	0.51	0.4	1.89E-03
1973	1.44	2.6	1.02E-03	0.01	0.01	1.77E-03	0.22	0.17	2.08E-03
1974	0.49	0.87	9.83E-04	6.49E-03	5.23E-03	1.84E-03	8.03E-02	6.18E-02	1.85E-03
1975	0.17	0.3	9.98E-04	2.05E-03	1.74E-03	1.77E-03	3.72E-02	2.96E-02	2.00E-03

Table A.2. Scenario 1b, break in Reservoir-11's dam at the Mayak PA, November 1967. Annual average concentrations of ^{90}Sr in water solution C ($Bq\,m^{-3}$), sediments CS ($Bq\,kg^{-1}$), and suspended sediment loads S ($kg\,m^{-3}$).

	Brodokalmak			*Ob' River mouth (Salekhard)*			*Ob' estuary mouth*		
Year	*C* (Bq/m^3)	*CS* (Bq/kg)	*S* (kg/m^3)	*C* (Bq/m^3)	*CS* (Bq/kg)	*S* (kg/m^3)	*C* (Bq/m^3)	*CS* (Bq/kg)	*S* (kg/m^3)
1967	77,682	20190	1.00E-03	0.32	0.18	6.02E-03	0.84	0.67	1.99E-03
1968	1,212	2845	9.98E-04	54.27	16.87	7.61E-03	0.86	0.65	2.29E-03
1969	143.58	279.3	8.56E-04	2.13	1.87	1.97E-03	21.05	12.88	4.13E-03
1970	49.86	90.58	1.01E-03	0.77	0.63	2.00E-03	7.94	7.09	2.24E-03
1971	17.29	31.29	1.03E-03	0.25	0.21	1.83E-03	3.08	2.46	2.00E-03
1972	5.99	10.81	7.33E-04	0.09	0.08	1.86E-03	1.18	0.94	1.89E-03
1973	2.07	3.73	1.02E-03	0.03	0.02	1.77E-03	0.54	0.43	2.08E-03
1974	0.72	1.25	9.83E-04	0.01	0.01	1.84E-03	0.19	0.15	1.85E-03
1975	0.25	0.43	9.98E-04	0.01	0	1.77E-03	0.09	0.07	2.00E-03

Table A.3. Scenario 1a, break in Reservoir-11's dam at the Mayak PA, April 2084. Annual average concentrations of ^{90}Sr in water solution C ($Bq\,m^{-3}$), sediments CS ($Bq\,kg^{-1}$), and suspended sediment loads S ($kg\,m^{-3}$).

	Brodokalmak			*Ob' River mouth (Salekhard)*			*Ob' estuary mouth*		
Year	*C* (Bq/m^3)	*CS* (Bq/kg)	*S* (kg/m^3)	*C* (Bq/m^3)	*CS* (Bq/kg)	*S* (kg/m^3)	*C* (Bq/m^3)	*CS* (Bq/kg)	*S* (kg/m^3)
2084	84,455	23,884.48	1.00E-03	39.6	19.7	1.22E-02	0.85	6.80E-01	2.00E-03
2085	1,119.8	2,169.59	7.63E-04	1.78	9.55E-01	1.31E-02	17.02	17.39	3.41E-03
2086	145.5	504.61	7.33E-05	0.6	6.19E-01	1.17E-03	5.35	4.19	4.03E-03
2087	68.49	108.09	9.62E-05	2.08E-01	1.27E-01	9.24E-03	1.83	1.5	1.89E-03
2088	70.63	68.34	7.63E-04	8.03E-02	4.84E-02	1.04E-02	7.19E-01	5.85E-01	4.65E-03
2089	8.72	30.08	7.33E-05	3.92E-02	3.89E-02	1.57E-03	3.55E-01	2.74E-01	4.29E-03
2090	4.07	5.88	9.62E-05	1.32E-02	6.55E-03	1.27E-02	1.79E-01	1.43E-01	2.00E-03
2091	4.23	3.3	7.63E-04	5.04E-03	2.68E-03	1.43E-02	6.63E-02	4.94E-02	4.28E-03
2092	5.02E-01	1.72	7.33E-05	2.46E-03	2.04E-03	1.59E-03	3.41E-02	2.61E-02	4.15E-03

Table A.4. Scenario 1b, break in Reservoir-11's dam at the Mayak PA, November 2084. Annual average concentrations of ^{90}Sr in water solution C ($Bq\,m^{-3}$), sediments CS ($Bq\,kg^{-1}$), and suspended sediment loads S ($kg\,m^{-3}$).

	Brodokalmak			*Ob' River mouth (Salekhard)*			*Ob' estuary mouth*		
Year	*C* (Bq/m³)	*CS* (Bq/kg)	*S* (kg/m³)	*C* (Bq/m³)	*CS* (Bq/kg)	*S* (kg/m³)	*C* (Bq/m³)	*CS* (Bq/kg)	*S* (kg/m³)
2084	80,960	21,325.68	1.00E-03	0.28	0.12	1.21E-02	0.85	0.68	2.00E-03
2085	1,487.51	3,428.22	7.63E-04	37.48	5.29	1.31E-02	0.73	0.5	3.41E-03
2086	209.87	727.73	7.33E-05	2.08	2.14	1.17E-03	22.61	13.87	4.03E-03
2087	73.29	134.33	9.62E-05	0.67	0.35	9.24E-03	5.29	4.22	1.89E-03
2088	83.79	207.09	7.63E-04	0.27	0.16	1.04E-02	2.23	1.59	4.65E-03
2089	12.4	42.86	7.33E-05	0.11	0.09	1.57E-03	1.09	0.79	4.29E-03
2090	5.79	8.32	9.62E-05	0.04	0.02	1.27E-02	0.52	0.43	2.00E-03
2091	6.03	11.83	7.63E-04	0.01	0.01	1.43E-02	0.19	0.14	4.28E-03
2092	0.71	2.45	7.33E-05	0.01	0	1.59E-03	0.09	0.07	4.15E-03

Table A.5. Scenario 2a, break in waste-holding pond at the Siberian Chemical Combine (SCC) at Tomsk, January 1967. Annual average concentrations of ^{137}Cs in water solution C ($Bq\,m^{-3}$), sediments CS ($Bq\,kg^{-1}$), and suspended sediment loads S ($kg\,m^{-3}$).

	Tomsk			*Ob' River mouth (Salekhard)*			*Ob' estuary mouth*		
Year	*C* (Bq/m³)	*CS* (Bq/kg)	*S* (kg/m³)	*C* (Bq/m³)	*CS* (Bq/kg)	*S* (kg/m³)	*C* (Bq/m³)	*CS* (Bq/kg)	*S* (kg/m³)
1967	84.23	730.5	5.29E-02	3.83	56.3	5.81E-03	0.2	3.03	1.74E-03
1968	0	0		2.54	33.92	8.25E-03	0.19	2.91	1.94E-03
1969	0	0		1.87	29.87	1.92E-03	0.19	2.8	4.23E-03
1970	0	0		1.1	17.14	1.89E-03	0.2	2.99	2.62E-03
1971	0	0		0.55	8.95	1.73E-03	0.25	3.79	2.27E-03
1972	0	0		0.34	5.39	1.84E-03	0.29	4.51	1.98E-03

Table A.6. Scenario 2a, break in waste-holding pond at the Siberian Chemical Combine (SCC) at Tomsk, June 1967. Annual average concentrations of ^{137}Cs in water solution C (Bq m^{-3}), sediments CS (Bq kg^{-1}), and suspended sediment loads S (kg m^{-3}).

	Tomsk			*Ob′ River mouth (Salekhard)*			*Ob′ estuary mouth*		
Year	*C* (Bq/m³)	*CS* (Bq/kg)	*S* (kg/m³)	*C* (Bq/m³)	*CS* (Bq/kg)	*S* (kg/m³)	*C* (Bq/m³)	*CS* (Bq/kg)	*S* (kg/m³)
1967	84.23	730.5	5.29E-02	6.93	273.69	5.81E-03	0.2	3.03	1.74E-03
1968	0	0		2.04	24.98	8.25E-03	0.19	2.91	1.94E-03
1969	0	0		1.28	20.21	1.92E-03	0.19	2.85	4.23E-03
1970	0	0		0.7	10.74	1.89E-03	0.22	3.26	2.62E-03
1971	0	0		0.33	5.32	1.73E-03	0.29	4.43	2.27E-03
1972	0	0		0.19	3.01	1.84E-03	0.35	5.28	1.98E-03

Table A.7. Scenario 2a, break in waste-holding pond at the Siberian Chemical Combine (SCC) at Tomsk, January 2084. Annual average concentrations of ^{137}Cs in water solution C (Bq m^{-3}), sediments CS (Bq kg^{-1}), and suspended sediment loads S (kg m^{-3}).

	Tomsk			*Ob′ delta (Salekhard)*			*Ob′ estuary (mouth)*		
Year	*C* (Bq/m³)	*CS* (Bq/kg)	*S* (kg/m³)	*C* (Bq/m³)	*CS* (Bq/kg)	*S* (kg/m³)	*C* (Bq/m³)	*CS* (Bq/kg)	*S* (kg/m³)
2084	56.44	474.21	5.29E-02	3.62	33.87	1.71E-02	0.17	2.01	2.08E-03
2085	0	0		1.94	19.22	1.58E-02	0.16	1.97	3.39E-03
2086	0	0		1.52	22.24	1.25E-03	0.17	2.11	4.33E-03
2087	0	0		0.62	5.72	1.03E-02	0.22	2.72	1.69E-03
2088	0	0		0.34	3.14	1.34E-02	0.29	3.50	4.56E-03
2089	0	0		0.39	4.74	2.20E-03	0.29	3.62	3.69E-03

Table A.8. Scenario 2b, break in waste-holding pond at the Siberian Chemical Combine (SCC) at Tomsk, June 2084. Annual average concentrations of ^{137}Cs in water solution C ($Bq\,m^{-3}$), sediments CS ($Bq\,kg^{-1}$), and suspended sediment loads S ($kg\,m^{-3}$).

	Tomsk			*Ob' delta (Salekhard)*			*Ob' estuary (mouth)*		
Year	*C* (Bq/m^3)	*CS* (Bq/kg)	*S* (kg/m^3)	*C* (Bq/m^3)	*CS* (Bq/kg)	*S* (kg/m^3)	*C* (Bq/m^3)	*CS* (Bq/kg)	*S* (kg/m^3)
2084	56.44	474.21	5.29E-02	5.67	115.44	1.71E-02	0.17	2.02	1.74E-03
2085	0	0		1.43	13.76	1.58E-02	0.16	1.98	3.14E-03
2086	0	0		0.99	14.34	1.25E-03	0.19	2.29	4.33E-03
2087	0	0		0.38	3.48	1.03E-02	0.26	3.17	1.69E-03
2088	0	0		0.20	1.82	1.34E-02	0.34	4.09	4.56E-03
2089	0	0		0.22	2.62	2.20E-03	0.34	4.2	3.69E-03

Table A.9. Scenario 3a, break in waste-holding pond at the Siberian Chemical Combine (SCC) at Tomsk, January 1967. Annual average concentrations of ^{90}Sr in water solution C ($Bq\,m^{-3}$), sediments CS ($Bq\,kg^{-1}$), and suspended sediment loads S ($kg\,m^{-3}$).

	Tomsk			*Ob' River mouth (Salekhard)*			*Ob' estuary (mouth)*		
Year	*C* (Bq/m^3)	*CS* (Bq/kg)	*S* (kg/m^3)	*C* (Bq/m^3)	*CS* (Bq/kg)	*S* (kg/m^3)	*C* (Bq/m^3)	*CS* (Bq/kg)	*S* (kg/m^3)
1967	216.76	0	5.29E-02	16.87	11.43	5.81E-03	0.68	0.54	1.74E-03
1968	0	0		0.43	0.17	8.25E-03	7.89	7.8	1.94E-03
1969	0	0		0.11	0.09	1.92E-03	3.76	2.82	4.23E-03

Table A.10. Scenario 3b, break in waste-holding pond at the Siberian Chemical Combine (SCC) at Tomsk, June 1967. Annual average concentrations of ^{90}Sr in water solution C ($Bq\,m^{-3}$), sediments CS ($Bq\,kg^{-1}$), and suspended sediment loads S ($kg\,m^{-3}$).

	Tomsk			*Ob' River mouth (Salekhard)*			*Ob' estuary (mouth)*		
Year	*C* (Bq/m^3)	*CS* (Bq/kg)	*S* (kg/m^3)	*C* (Bq/m^3)	*CS* (Bq/kg)	*S* (kg/m^3)	*C* (Bq/m^3)	*CS* (Bq/kg)	*S* (kg/m^3)
1967	216.76	0	5.29E-02	28.73	74.82	5.81E-03	0.68	0.54	1.74E-03
1968	0	0		0.35	0.19	8.25E-03	0.71	0.56	1.94E-03
1969	0	0		0.09	0.08	1.92E-03	13.93	12.12	4.23E-03

Table A.11. Scenario 3a, break in waste-holding pond at the Siberian Chemical Combine (SCC) at Tomsk, January 2084. Annual average concentrations of ^{90}Sr in water solution C (Bq m^{-3}), sediments CS (Bq kg^{-1}), and suspended sediment loads S (kg m^{-3}).

	Tomsk			*Ob′ delta (Salekhard)*			*Ob′ estuary (mouth)*		
Year	*C* (Bq/m³)	*CS* (Bq/kg)	*S* (kg/m³)	*C* (Bq/m³)	*CS* (Bq/kg)	*S* (kg/m³)	*C* (Bq/m³)	*CS* (Bq/kg)	*S* (kg/m³)
2084	145.22	0	5.29E-02	14.36	11.43	1.71E-02	0.76	0.59	2.08E-03
2085	0	0		0.31	0.17	1.58E-02	8.09	7.08	3.39E-03
2086	0	0		0.09	0.09	1.25E-03	2.24	1.74	4.33E-03

Table A.12. Scenario 3b, break in waste-holding pond at the Siberian Chemical Combine (SCC) at Tomsk, June 2084. Annual average concentrations of ^{90}Sr in water solution C (Bq m^{-3}), sediments CS (Bq kg^{-1}), and suspended sediment loads S (kg m^{-3}).

	Tomsk			*Ob′ delta (Salekhard)*			*Ob′ estuary (mouth)*		
Year	*C* (Bq/m³)	*CS* (Bq/kg)	*S* (kg/m³)	*C* (Bq/m³)	*CS* (Bq/kg)	*S* (kg/m³)	*C* (Bq/m³)	*CS* (Bq/kg)	*S* (kg/m³)
2084	145.22	0	5.29E-02	26.37	21.09	1.71E-02	0.76	0.59	2.08E-03
2085	0	0		0.28	0.15	1.58E-02	14.30	11.43	3.39E-03
2086	0	0		0.08	0.08	1.25E-03	3.47	2.77	4.33E-03

Table A.13. Scenario 4a, break in waste-holding ponds No. 365 and No. 366 at the Mining and Chemical Combine (MCC) at Krasnoyarsk, January 1967. Annual average concentrations of ^{137}Cs in water solution C (Bq m^{-3}), sediments CS (Bq kg^{-1}), and suspended sediment loads S (kg m^{-3}).

	Krasnoyarsk-26			*Yenisei River mouth*			*Yenisei estuary (mouth)*		
Year	*C* (Bq/m³)	*CS* (Bq/kg)	*S* (kg/m³)	*C* (Bq/m³)	*CS* (Bq/kg)	*S* (kg/m³)	*C* (Bq/m³)	*CS* (Bq/kg)	*S* (kg/m³)
1967	0.58	16,690	4.28E-03	1.62	25.51	4.19E-03	0.43	8.03	3.34E-03
1968	0	0		1.82	17.64	1.18E-02	0.76	12.26	3.91E-03
1969	0	0		1.79	18.64	7.04E-03	1	14.95	3.82E-03
1970	0	0		1.62	11.54	1.63E-02	1.23	18.17	3.81E-03
1971	0	0		1.11	8.88	1.05E-02	1.2	16.65	3.81E-03
1972	0	0		0.72	4.39	1.34E-02	1.1	14.51	3.71E-03

Table A.14. Scenario 4b, break in waste-holding ponds No. 365 and No. 366 at the Mining and Chemical Combine (MCC) at Krasnoyarsk, June 1967. Annual average concentrations of ^{137}Cs in water solution C ($Bq\,m^{-3}$), sediments CS ($Bq\,kg^{-1}$), and suspended sediment loads S ($kg\,m^{-3}$).

	Krasnoyarsk-26			*Yenisei River mouth*			*Yenisei estuary (mouth)*		
Year	*C* (Bq/m^3)	*CS* (Bq/kg)	*S* (kg/m^3)	*C* (Bq/m^3)	*CS* (Bq/kg)	*S* (kg/m^3)	*C* (Bq/m^3)	*CS* (Bq/kg)	*S* (kg/m^3)
1967	0.58	16,690	4.28E-03	3.33	43.44	4.19E-03	1.43	21.72	3.34E-03
1968	0	0		1.68	14.69	1.18E-02	0.93	13.94	3.91E-03
1969	0	0		1.27	12.17	7.04E-03	0.97	13.67	3.82E-03
1970	0	0		0.93	6.14	1.63E-02	1.03	14.48	3.81E-03
1971	0	0		0.55	4.16	1.05E-02	0.89	11.9	3.81E-03
1972	0	0		0.32	1.83	1.34E-02	0.74	9.53	3.71E-03

Table A.15. Scenario 4a, break in waste-holding ponds No. 365 and No. 366 at the Mining and Chemical Combine (MCC) at Krasnoyarsk, January 2089. Annual average concentrations of ^{137}Cs in water solution C ($Bq\,m^{-3}$), sediments CS ($Bq\,kg^{-1}$), and suspended sediment loads S ($kg\,m^{-3}$).

	Krasnoyarsk-26			*Yenisei delta*			*Yenisei estuary (mouth)*		
Year	*C* (Bq/m^3)	*CS* (Bq/kg)	*S* (kg/m^3)	*C* (Bq/m^3)	*CS* (Bq/kg)	*S* (kg/m^3)	*C* (Bq/m^3)	*CS* (Bq/kg)	*S* (kg/m^3)
2089	0.43	9950.17	5.28E-03	2.72	35.85	6.92E-03	0.69	12.58	3.42E-03
2090	0	0		1.94	15.29	2.38E-02	1.05	16.44	3.68E-03
2091	0	0		1.28	9.13	2.53E-02	1.18	17.33	3.78E-03
2092	0	0		1.04	10.83	7.13E-03	1.23	18.05	3.45E-03
2093	0	0		0.56	3.82	1.78E-02	1.06	15.03	3.49E-03
2094	0	0		0.30	1.94	2.06E-02	0.84	11.55	3.55E-03

Table A.16. Scenario 4b, break in waste-holding ponds No. 365 and No. 366 at the Mining and Chemical Combine (MCC) at Krasnoyarsk, June 2089. Annual average concentrations of ^{137}Cs in water solution C ($Bq\,m^{-3}$), sediments CS ($Bq\,kg^{-1}$), and suspended sediment loads S ($kg\,m^{-3}$).

	Krasnoyarsk-26			*Yenisei delta*			*Yenisei estuary (mouth)*		
Year	*C* (Bq/m³)	*CS* (Bq/kg)	*S* (kg/m³)	*C* (Bq/m³)	*CS* (Bq/kg)	*S* (kg/m³)	*C* (Bq/m³)	*CS* (Bq/kg)	*S* (kg/m³)
2089	0.43	9,950.17	5.28E-03	5.29	149.67	6.92E-03	2.36	48.21	3.42E-03
2090	0	0		1.24	8.99	2.38E-02	1.29	18.76	3.68E-03
2091	0	0		0.72	4.83	2.53E-02	1.12	15.73	3.78E-03
2092	0	0		0.54	5.44	7.13E-03	1.03	14.59	3.45E-03
2093	0	0		0.27	1.78	1.78E-02	0.81	11.13	3.49E-03
2094	0	0		0.14	0.85	2.06E-02	0.59	7.89	3.55E-03

Table A.17. Scenario 5a, break in waste-holding ponds No. 365 and No. 366 at the Mining and Chemical Combine (MCC) at Krasnoyarsk, January 1967. Annual average concentrations of ^{90}Sr in water solution C ($Bq\,m^{-3}$), sediments CS ($Bq\,kg^{-1}$), and suspended sediment loads S ($kg\,m^{-3}$).

	Krasnoyarsk-26			*Yenisei River mouth*			*Yenisei estuary (mouth)*		
Year	*C* (Bq/m³)	*CS* (Bq/kg)	*S* (kg/m³)	*C* (Bq/m³)	*CS* (Bq/kg)	*S* (kg/m³)	*C* (Bq/m³)	*CS* (Bq/kg)	*S* (kg/m³)
1967	175.18	0	4.28E-03	25.49	12.43	4.19E-03	22.24	13.67	3.34E-03
1968	0	0		1.29	0.45	1.18E-02	2.38	1.43	3.91E-03
1969	0	0		0.39	0.17	7.04E-03	0.8	0.47	3.82E-03

Table A.18. Scenario 5b, break in waste-holding ponds No. 365 and No. 366 at the Mining and Chemical Combine (MCC) at Krasnoyarsk, June 1967. Annual average concentrations of ^{90}Sr in water solution C ($Bq\,m^{-3}$), sediments CS ($Bq\,kg^{-1}$), and suspended sediment loads S ($kg\,m^{-3}$).

	Krasnoyarsk-26			*Yenisei River mouth*			*Yenisei estuary (mouth)*		
Year	*C* (Bq/m³)	*CS* (Bq/kg)	*S* (kg/m³)	*C* (Bq/m³)	*CS* (Bq/kg)	*S* (kg/m³)	*C* (Bq/m³)	*CS* (Bq/kg)	*S* (kg/m³)
1967	175.18	0	4.28E-03	27.2	19.86	4.19E-03	26.74	21.32	3.34E-03
1968	0	0		0.65	0.23	1.18E-02	1.02	0.59	3.91E-03
1969	0	0		0.2	0.09	7.04E-03	0.35	0.2	3.82E-03

Table A.19. Scenario 5a, break in waste-holding ponds No. 365 and No. 366 at the Mining and Chemical Combine (MCC) at Krasnoyarsk, January 2089. Annual average concentrations of ^{90}Sr in water solution C ($Bq\,m^{-3}$), sediments CS ($Bq\,kg^{-1}$), and suspended sediment loads S ($kg\,m^{-3}$).

	Krasnoyarsk-26			*Yenisei delta*			*Yenisei estuary (mouth)*		
Year	*C* (Bq/m³)	*CS* (Bq/kg)	*S* (kg/m³)	*C* (Bq/m³)	*CS* (Bq/kg)	*S* (kg/m³)	*C* (Bq/m³)	*CS* (Bq/kg)	*S* (kg/m³)
2089	128.51	0	5.28E-03	9.88	2.98	6.92E-03	9.01	3.43	3.42E-03
2090	0	0		0.26	0.08	2.38E-02	0.65	0.43	3.68E-03
2091	0	0		0.07	0.02	2.53E-02	0.20	0.13	3.78E-03

Table A.20. Scenario 5b, break in waste-holding ponds No. 365 and No. 366 at the Mining and Chemical Combine (MCC) at Krasnoyarsk, June 2089. Annual average concentrations of ^{90}Sr in water solution C ($Bq\,m^{-3}$), sediments CS ($Bq\,kg^{-1}$), and suspended sediment loads S ($kg\,m^{-3}$).

	Krasnoyarsk-26			*Yenisei delta*			*Yenisei estuary (mouth)*		
Year	*C* (Bq/m³)	*CS* (Bq/kg)	*S* (kg/m³)	*C* (Bq/m³)	*CS* (Bq/kg)	*S* (kg/m³)	*C* (Bq/m³)	*CS* (Bq/kg)	*S* (kg/m³)
2089	128.51	0	5.28E-03	10.18	3.07	6.92E-03	10.01	3.61	3.42E-03
2090	0	0		0.12	0.04	2.38E-02	0.29	0.18	3.68E-03
2091	0	0		0.04	0.01	2.53E-02	0.09	0.06	3.78E-03

Appendix B

Doses to individuals in critical groups from all accident scenarios given by radionuclide and exposure pathway

The results are given in tables and and corresponding column charts (figures)

Table B.1. Maximum annual doses from radionuclides and pathways to individuals ($Sv\,yr^{-1}$) at Brodokalmak from Accident Scenario 1a: break in Reservoir-11's dam at the Mayak PA in April 1967.

Sum of Sv/yr	*Nuclide*		
Pathway	^{137}Cs	^{90}Sr	*Grand total*
External	0.0E+00	1.1E-09	1.1E-09
Fish	0.0E+00	5.5E-03	5.5E-03
Grain	0.0E+00	1.9E-03	1.9E-03
Green veg	0.0E+00	1.8E-03	1.8E-03
Meat	0.0E+00	2.9E-04	2.9E-04
Milk	0.0E+00	4.6E-04	4.6E-04
Root veg	0.0E+00	3.8E-04	3.8E-04
Soil ingestion	0.0E+00	2.4E-06	2.4E-06
Soil inhalation	0.0E+00	3.5E-08	3.5E-08
Water	0.0E+00	1.8E-05	1.8E-05
Grand total	*0.0E+00*	*1.0E-02*	*1.0E-02*

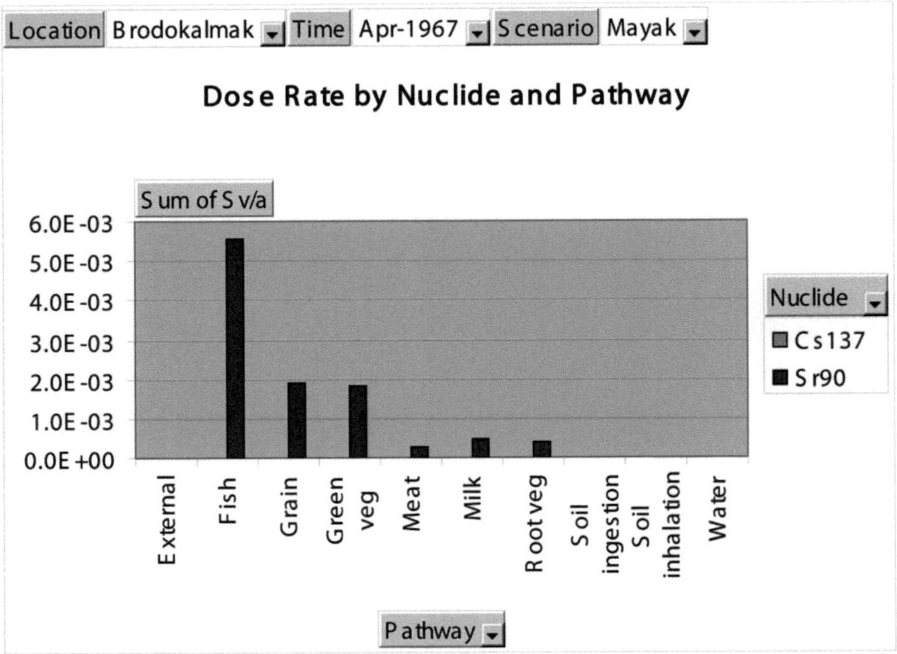

Table B.2. Maximum annual doses from radionuclides and pathways to individuals (Sv yr^{-1}) at the Ob′ River mouth (Salekhard) from Accident Scenario 1a: break in Reservoir-11's dam at the Mayak PA in April 1967.

Sum of Sv/yr	*Nuclide*		
Pathway	^{137}Cs	^{90}Sr	*Grand total*
External	0.0E+00	1.1E-12	1.1E-12
Fish	0.0E+00	3.3E-06	3.3E-06
Grain	0.0E+00	1.9E-06	1.9E-06
Green veg	0.0E+00	1.8E-06	1.8E-06
Meat	0.0E+00	2.9E-07	2.9E-07
Milk	0.0E+00	4.7E-07	4.7E-07
Root veg	0.0E+00	3.8E-07	3.8E-07
Soil ingestion	0.0E+00	2.5E-09	2.5E-09
Soil inhalation	0.0E+00	3.5E-11	3.5E-11
Water	0.0E+00	1.1E-08	1.1E-08
Grand total	*0.0E+00*	*8.3E-06*	*8.3E-06*

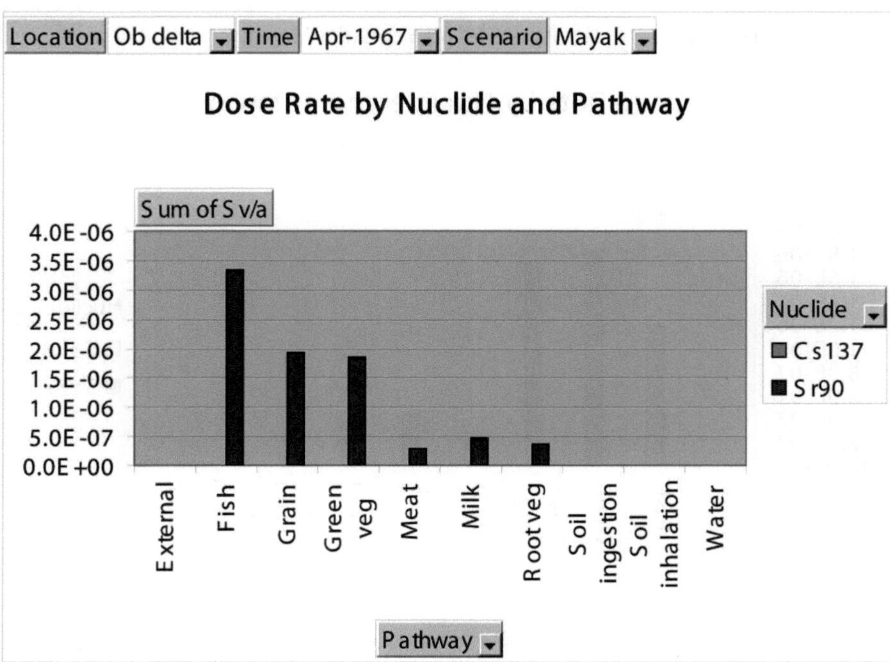

Table B.3. Maximum annual doses from radionuclides and pathways to individuals ($Sv\,yr^{-1}$) at the Ob′ estuary mouth from Accident Scenario 1a: break in Reservoir-11's dam at the Mayak PA in April 1967.

Sum of Sv/yr	*Nuclide*		
Pathway	^{137}Cs	^{90}Sr	*Grand total*
External	0.0E+00	1.0E-12	1.0E-12
Fish	0.0E+00	1.2E-06	1.2E-06
Grain	0.0E+00	1.8E-06	1.8E-06
Green veg	0.0E+00	1.7E-06	1.7E-06
Meat	0.0E+00	2.7E-07	2.7E-07
Milk	0.0E+00	4.4E-07	4.4E-07
Root veg	0.0E+00	3.6E-07	3.6E-07
Soil ingestion	0.0E+00	2.3E-09	2.3E-09
Soil inhalation	0.0E+00	3.3E-11	3.3E-11
Water	0.0E+00	4.0E-09	4.0E-09
Grand total	*0.0E+00*	*5.8E-06*	*5.8E-06*

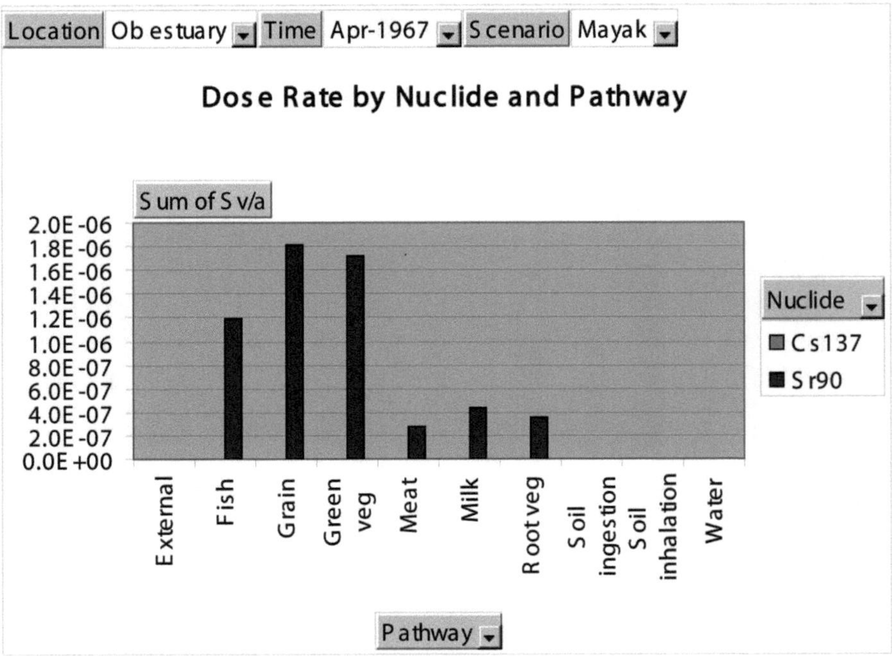

Table B.4. Maximum annual doses from radionuclides and pathways to individuals ($Sv\,yr^{-1}$) at Brodokalmak from Accident Scenario 1b: break in Reservoir-11's dam at the Mayak PA in November 1967.

Sum of Sv/yr	*Nuclide*		
Pathway	^{137}Cs	^{90}Sr	*Grand total*
External	0.0E+00	1.0E-09	1.0E-09
Fish	0.0E+00	5.2E-03	5.2E-03
Grain	0.0E+00	1.8E-03	1.8E-03
Green veg	0.0E+00	1.7E-03	1.7E-03
Meat	0.0E+00	2.7E-04	2.7E-04
Milk	0.0E+00	4.3E-04	4.3E-04
Root veg	0.0E+00	3.5E-04	3.5E-04
Soil ingestion	0.0E+00	2.3E-06	2.3E-06
Soil inhalation	0.0E+00	3.2E-08	3.2E-08
Water	0.0E+00	1.7E-05	1.7E-05
Grand total	*0.0E+00*	*9.8E-03*	*9.8E-03*

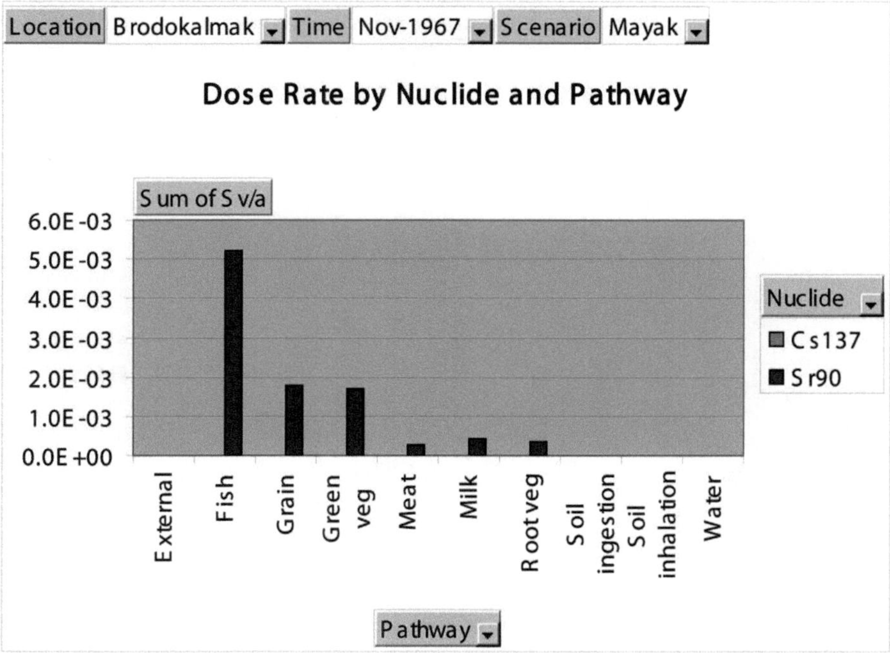

Table B.5. Maximum annual doses from radionuclides and pathways to individuals (Sv yr^{-1}) at the Ob′ River mouth from Accident Scenario 1b: break in Reservoir-11's dam at the Mayak PA in November 1967.

Sum of Sv/yr	*Nuclide*		
Pathway	^{137}Cs	^{90}Sr	*Grand total*
External	0.0E+00	8.6E-13	8.6E-13
Fish	0.0E+00	3.6E-06	3.6E-06
Grain	0.0E+00	1.5E-06	1.5E-06
Green veg	0.0E+00	1.4E-06	1.4E-06
Meat	0.0E+00	2.2E-07	2.2E-07
Milk	0.0E+00	3.6E-07	3.6E-07
Root veg	0.0E+00	2.9E-07	2.9E-07
Soil ingestion	0.0E+00	1.9E-09	1.9E-09
Soil inhalation	0.0E+00	2.7E-11	2.7E-11
Water	0.0E+00	1.2E-08	1.2E-08
Grand total	*0.0E+00*	*7.4E-06*	*7.4E-06*

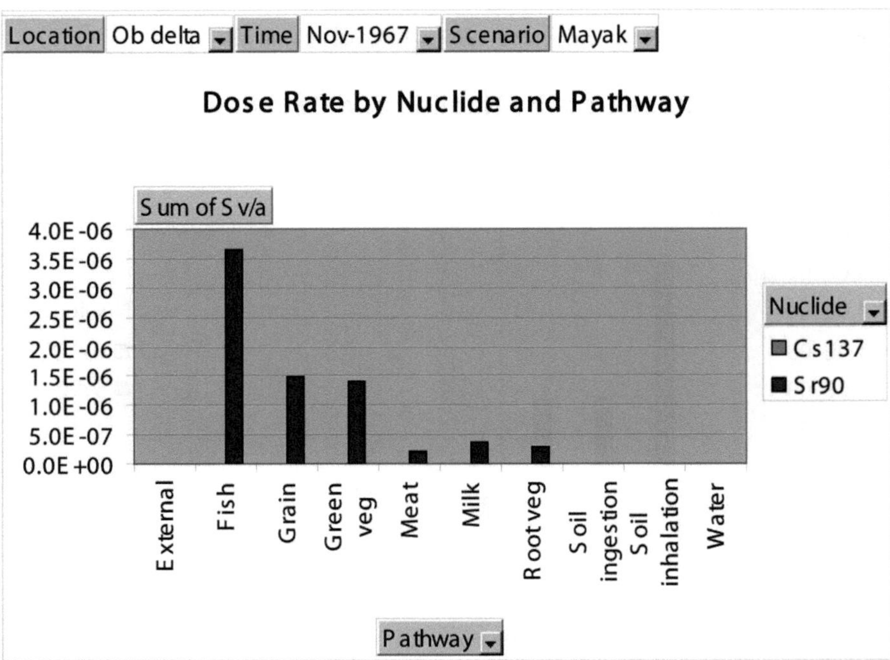

Table B.6. Maximum annual doses from radionuclides and pathways to individuals ($Sv\,yr^{-1}$) at the Ob′ estuary mouth from Accident Scenario 1b: break in Reservoir-11's dam at the Mayak PA in November 1967.

Sum of Sv/yr	*Nuclide*		
Pathway	^{137}Cs	^{90}Sr	*Grand total*
External	0.0E+00	6.6E-13	6.6E-13
Fish	0.0E+00	1.4E-06	1.4E-06
Grain	0.0E+00	1.1E-06	1.1E-06
Green veg	0.0E+00	1.1E-06	1.1E-06
Meat	0.0E+00	1.7E-07	1.7E-07
Milk	0.0E+00	2.7E-07	2.7E-07
Root veg	0.0E+00	2.3E-07	2.3E-07
Soil ingestion	0.0E+00	1.4E-09	1.4E-09
Soil inhalation	0.0E+00	2.1E-11	2.1E-11
Water	0.0E+00	4.7E-09	4.7E-09
Grand total	*0.0E+00*	*4.3E-06*	*4.3E-06*

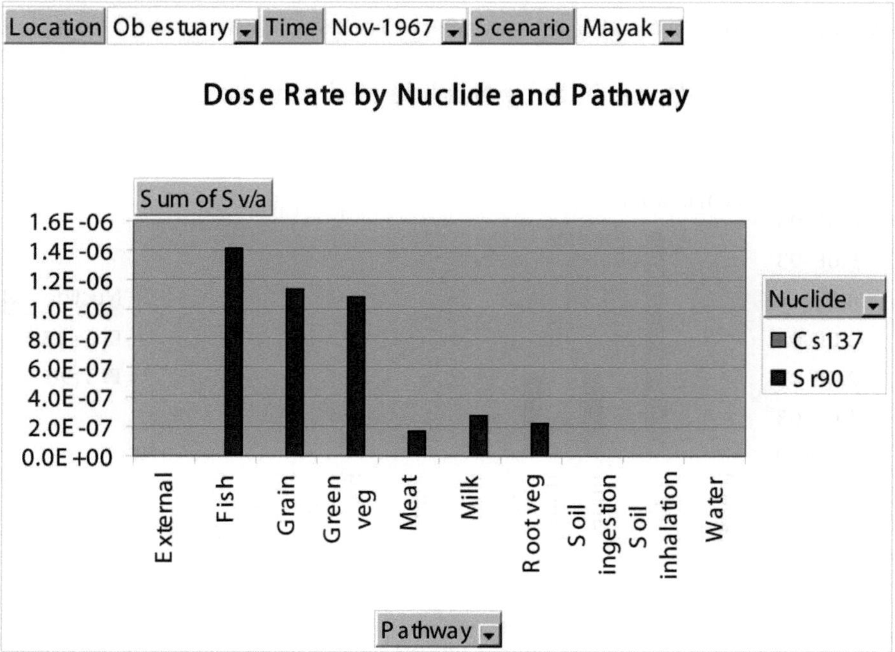

Table B.7. Maximum annual doses from radionuclides and pathways to individuals (Sv yr^{-1}) at Brodokalmak from Accident Scenario 1a: break in Reservoir-11's dam at the Mayak PA in April 2084.

Sum of Sv/yr	*Nuclide*		
Pathway	^{137}Cs	^{90}Sr	*Grand total*
External	0.0E+00	1.2E-09	1.2E-09
Fish	0.0E+00	5.7E-03	5.7E-03
Grain	0.0E+00	2.1E-03	2.1E-03
Green veg	0.0E+00	2.0E-03	2.0E-03
Meat	0.0E+00	3.2E-04	3.2E-04
Milk	0.0E+00	5.1E-04	5.1E-04
Root veg	0.0E+00	4.2E-04	4.2E-04
Soil ingestion	0.0E+00	2.7E-06	2.7E-06
Soil inhalation	0.0E+00	3.8E-08	3.8E-08
Water	0.0E+00	1.9E-05	1.9E-05
Grand total	*0.0E+00*	*1.1E-02*	*1.1E-02*

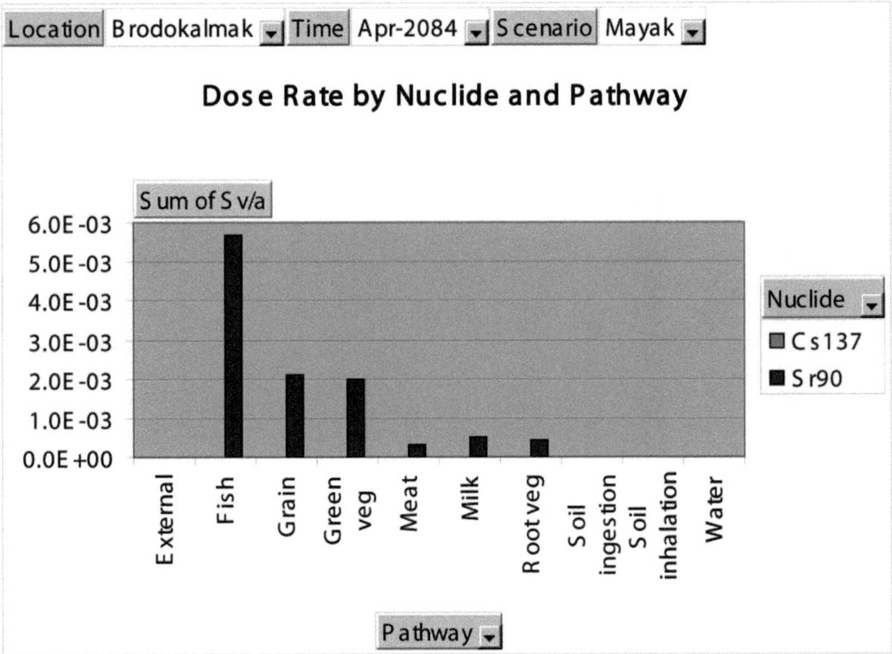

Table B.8. Maximum annual doses from radionuclides and pathways to individuals ($Sv\,yr^{-1}$) at the Ob′ River mouth (Salekhard) from Accident Scenario 1a: break in Reservoir-11's dam at the Mayak PA in April 2084.

Sum of Sv/yr	*Nuclide*		
Pathway	^{137}Cs	^{90}Sr	*Grand total*
External	0.0E+00	1.0E-12	1.0E-12
Fish	0.0E+00	2.7E-06	2.7E-06
Grain	0.0E+00	1.7E-06	1.7E-06
Green veg	0.0E+00	1.7E-06	1.7E-06
Meat	0.0E+00	2.6E-07	2.6E-07
Milk	0.0E+00	4.2E-07	4.2E-07
Root veg	0.0E+00	3.4E-07	3.4E-07
Soil ingestion	0.0E+00	2.2E-09	2.2E-09
Soil inhalation	0.0E+00	3.2E-11	3.2E-11
Water	0.0E+00	8.9E-09	8.9E-09
Grand total	*0.0E+00*	*7.1E-06*	*7.1E-06*

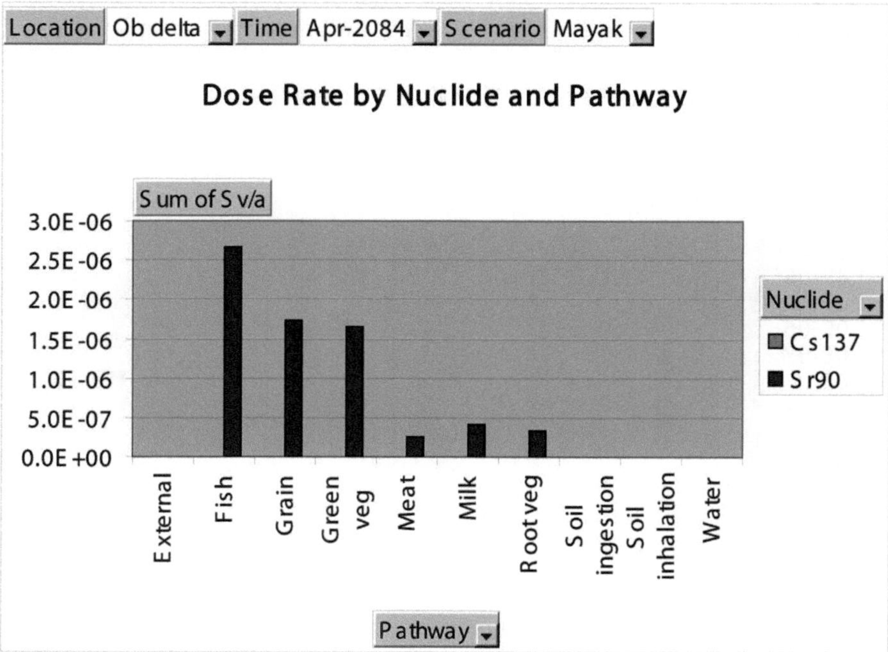

Table B.9. Maximum annual doses from radionuclides and pathways to individuals (Sv yr^{-1}) at the Ob′ estuary mouth from Accident Scenario 1a: break in Reservoir-11's dam at the Mayak PA in April 2084.

Sum of Sv/yr	*Nuclide*		
Pathway	^{137}Cs	^{90}Sr	*Grand total*
External	0.0E+00	8.9E-13	8.9E-13
Fish	0.0E+00	1.1E-06	1.1E-06
Grain	0.0E+00	1.5E-06	1.5E-06
Green veg	0.0E+00	1.5E-06	1.5E-06
Meat	0.0E+00	2.3E-07	2.3E-07
Milk	0.0E+00	3.7E-07	3.7E-07
Root veg	0.0E+00	3.0E-07	3.0E-07
Soil ingestion	0.0E+00	1.9E-09	1.9E-09
Soil inhalation	0.0E+00	2.8E-11	2.8E-11
Water	0.0E+00	3.8E-09	3.8E-09
Grand total	*0.0E+00*	*5.1E-06*	*5.1E-06*

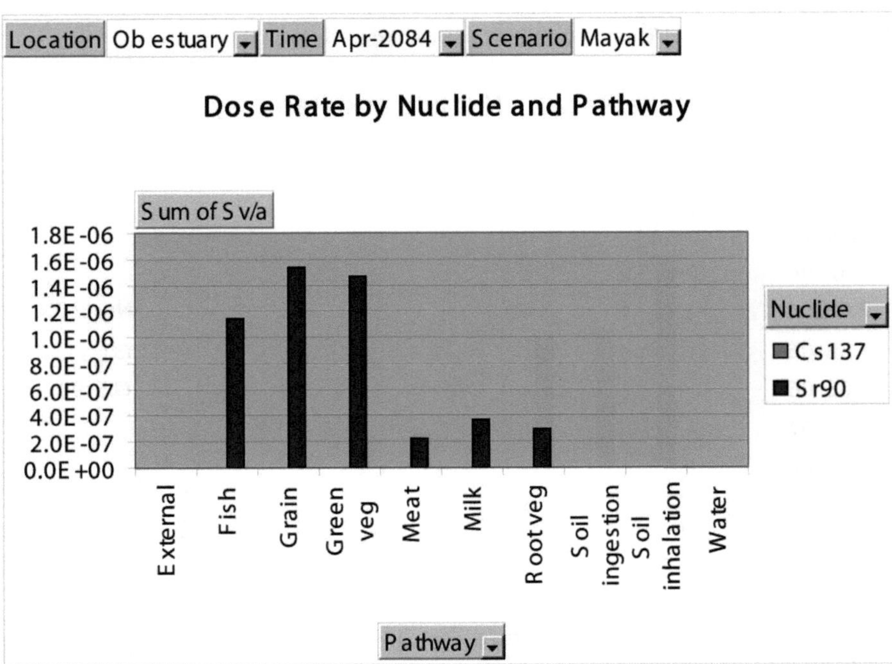

Table B.10. Maximum annual doses from radionuclides and pathways to individuals ($Sv\,yr^{-1}$) at Brodokalmak from Accident Scenario 1b: break in Reservoir-11's dam at the Mayak PA in November 2084.

Sum of Sv/yr	*Nuclide*		
Pathway	^{137}Cs	^{90}Sr	*Grand total*
External	0.0E+00	1.1E-09	1.1E-09
Fish	0.0E+00	5.4E-03	5.4E-03
Grain	0.0E+00	1.9E-03	1.9E-03
Green veg	0.0E+00	1.8E-03	1.8E-03
Meat	0.0E+00	2.8E-04	2.8E-04
Milk	0.0E+00	4.5E-04	4.5E-04
Root veg	0.0E+00	3.7E-04	3.7E-04
Soil ingestion	0.0E+00	2.4E-06	2.4E-06
Soil inhalation	0.0E+00	3.4E-08	3.4E-08
Water	0.0E+00	1.8E-05	1.8E-05
Grand total	*0.0E+00*	*1.0E-02*	*1.0E-02*

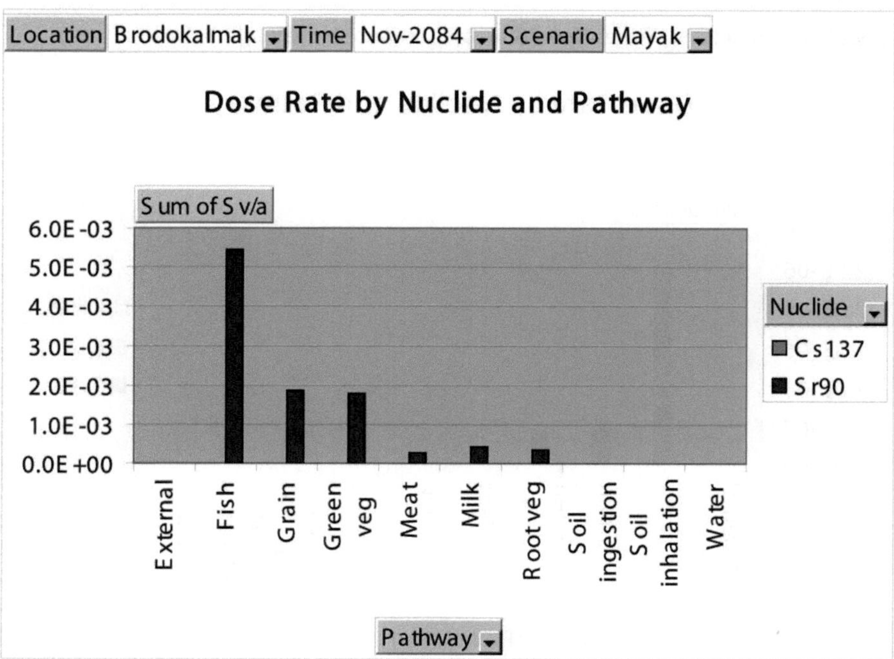

Table B.11. Maximum annual doses from radionuclides and pathways to individuals (Sv yr^{-1}) at the Ob′ River mouth (Salekhard) from Accident Scenario 1b: break in Reservoir-11's dam at the Mayak PA in November 2084.

Sum of Sv/yr	*Nuclide*		
Pathway	^{137}Cs	^{90}Sr	*Grand total*
External	0.0E+00	2.7E-13	2.7E-13
Fish	0.0E+00	2.5E-06	2.5E-06
Grain	0.0E+00	4.7E-07	4.7E-07
Green veg	0.0E+00	4.4E-07	4.4E-07
Meat	0.0E+00	7.1E-08	7.1E-08
Milk	0.0E+00	1.1E-07	1.1E-07
Root veg	0.0E+00	9.2E-08	9.2E-08
Soil ingestion	0.0E+00	5.9E-10	5.9E-10
Soil inhalation	0.0E+00	8.5E-12	8.5E-12
Water	0.0E+00	8.4E-09	8.4E-09
Grand total	*0.0E+00*	*3.7E-06*	*3.7E-06*

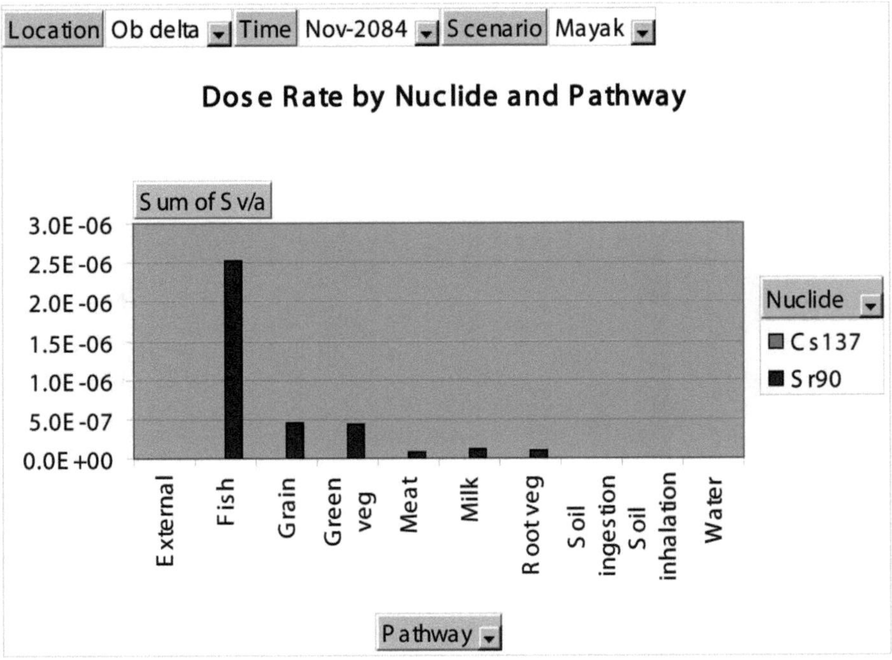

Table B.12. Maximum annual doses from radionuclides and pathways to individuals ($Sv\,yr^{-1}$) at the Ob′ estuary mouth from Accident Scenario 1b: break in Reservoir-11's dam at the Mayak PA in November 2084.

Sum of Sv/yr	*Nuclide*		
Pathway	^{137}Cs	^{90}Sr	*Grand total*
External	0.0E+00	7.1E-13	7.1E-13
Fish	0.0E+00	1.5E-06	1.5E-06
Grain	0.0E+00	1.2E-06	1.2E-06
Green veg	0.0E+00	1.2E-06	1.2E-06
Meat	0.0E+00	1.8E-07	1.8E-07
Milk	0.0E+00	3.0E-07	3.0E-07
Root veg	0.0E+00	2.4E-07	2.4E-07
Soil ingestion	0.0E+00	1.6E-09	1.6E-09
Soil inhalation	0.0E+00	2.2E-11	2.2E-11
Water	0.0E+00	5.1E-09	5.1E-09
Grand total	*0.0E+00*	*4.6E-06*	*4.6E-06*

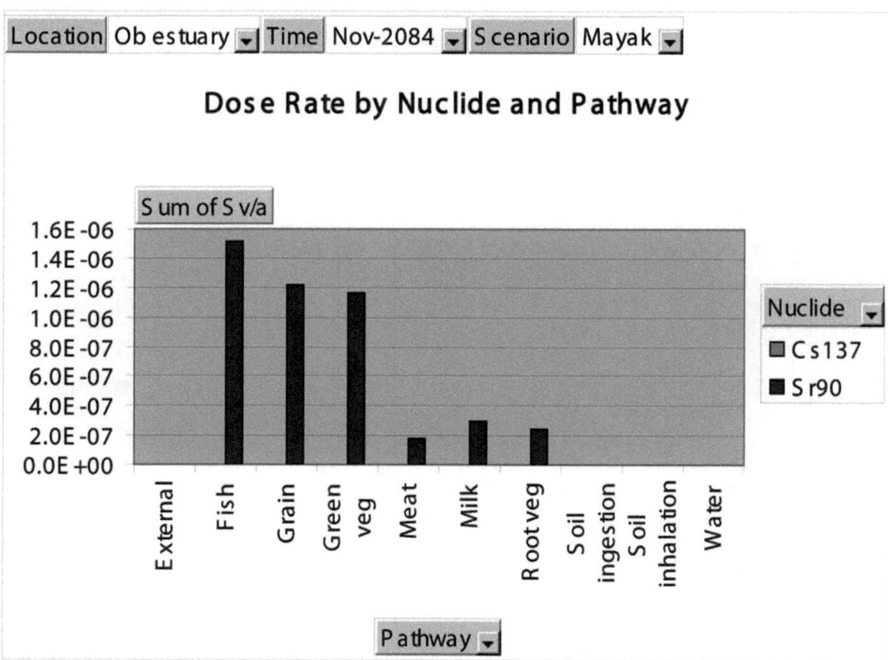

Table B.13. Maximum annual doses from radionuclides and pathways to individuals (Sv yr^{-1}) at Tomsk from Accident Scenarios 2a and 3a: break in a waste-holding pond at the Siberian Chemical Combine (SCC) at Tomsk in January 1967.

Sum of Sv/yr	*Nuclide*		
Pathway	^{137}Cs	^{90}Sr	*Grand total*
External	1.2E-05	0.0E+00	1.2E-05
Fish	8.8E-05	1.5E-05	1.0E-04
Grain	3.7E-06	0.0E+00	3.7E-06
Green veg	4.4E-06	0.0E+00	4.4E-06
Meat	4.0E-06	0.0E+00	4.0E-06
Milk	2.9E-06	0.0E+00	2.9E-06
Root veg	3.9E-06	0.0E+00	3.9E-06
Soil ingestion	3.8E-08	0.0E+00	3.8E-08
Soil inhalation	2.8E-10	0.0E+00	2.8E-10
Water	8.8E-09	4.9E-08	5.7E-08
Grand total	*1.2E-04*	*1.5E-05*	*1.3E-04*

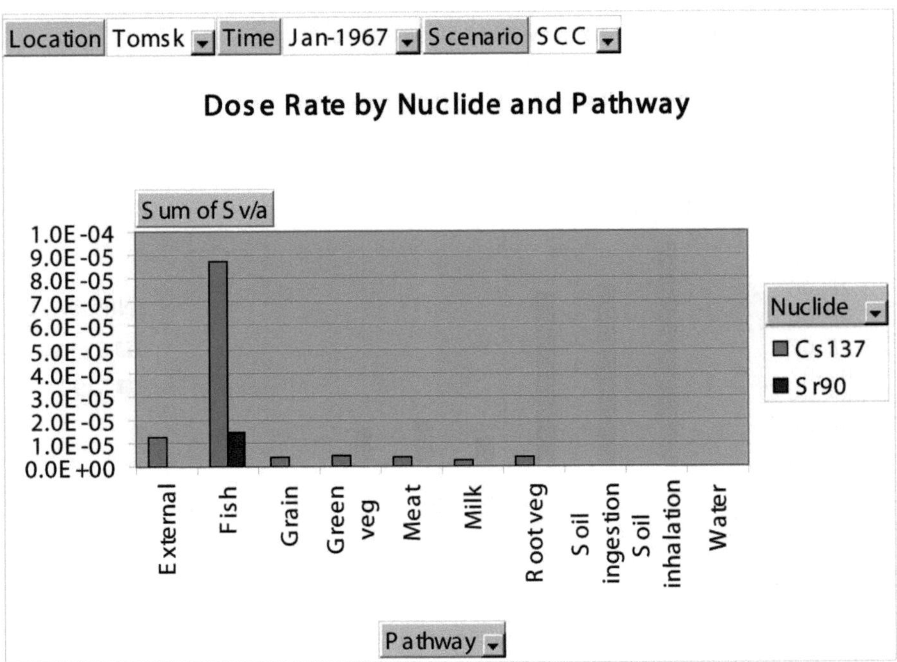

Table B.14. Maximum annual doses from radionuclides and pathways to individuals (Sv yr^{-1}) at the Ob′ River mouth (Salekhard) from Accident Scenarios 2a and 3a: break in a waste-holding pond at the Siberian Chemical Combine (SCC) at Tomsk in January 1967.

Sum of Sv/yr	*Nuclide*		
Pathway	^{137}Cs	^{90}Sr	*Grand total*
External	9.5E-07	5.8E-13	9.5E-07
Fish	4.0E-06	1.1E-06	5.1E-06
Grain	2.9E-07	1.0E-06	1.3E-06
Green veg	3.4E-07	9.6E-07	1.3E-06
Meat	3.1E-07	1.5E-07	4.6E-07
Milk	2.2E-07	2.4E-07	4.7E-07
Root veg	3.0E-07	2.0E-07	5.0E-07
Soil ingestion	2.9E-09	1.3E-09	4.2E-09
Soil inhalation	2.2E-11	1.8E-11	4.0E-11
Water	4.0E-10	3.8E-09	4.2E-09
Grand total	*6.4E-06*	*3.7E-06*	*1.0E-05*

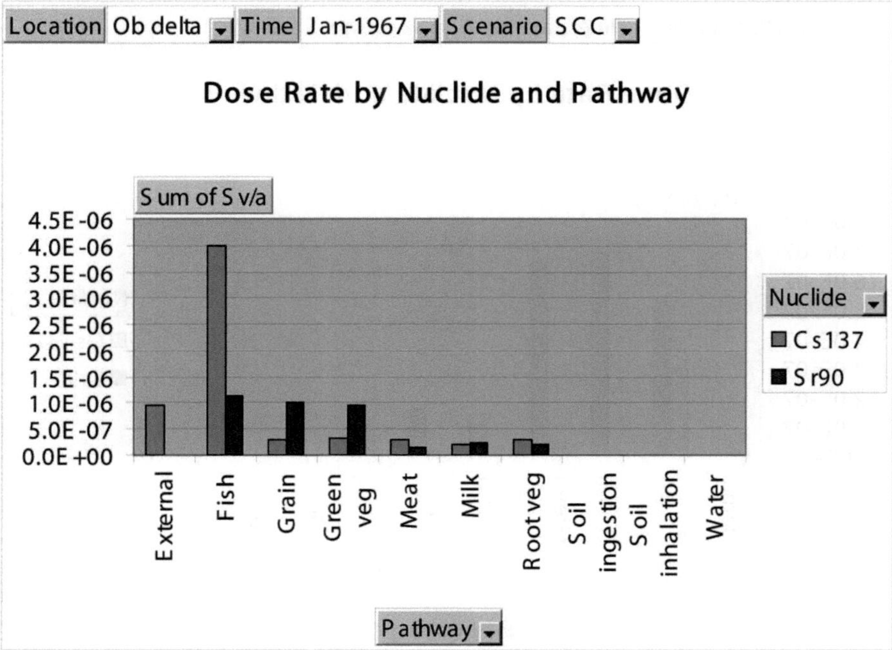

Table B.15. Maximum annual doses from radionuclides and pathways to individuals ($Sv\,yr^{-1}$) at the Ob′ estuary mouth from Accident Scenarios 2a and 3a: break in a waste-holding pond at the Siberian Chemical Combine (SCC) at Tomsk in January 1967.

Sum of Sv/yr	*Nuclide*		
Pathway	^{137}Cs	^{90}Sr	*Grand total*
External	7.6E-08	4.0E-13	7.6E-08
Fish	3.0E-07	5.3E-07	8.3E-07
Grain	2.3E-08	6.9E-07	7.1E-07
Green veg	2.7E-08	6.6E-07	6.8E-07
Meat	2.5E-08	1.0E-07	1.3E-07
Milk	1.8E-08	1.7E-07	1.8E-07
Root veg	2.4E-08	1.4E-07	1.6E-07
Soil ingestion	2.3E-10	8.7E-10	1.1E-09
Soil inhalation	1.8E-12	1.2E-11	1.4E-11
Water	3.0E-11	1.8E-09	1.8E-09
Grand total	*4.9E-07*	*2.3E-06*	*2.8E-06*

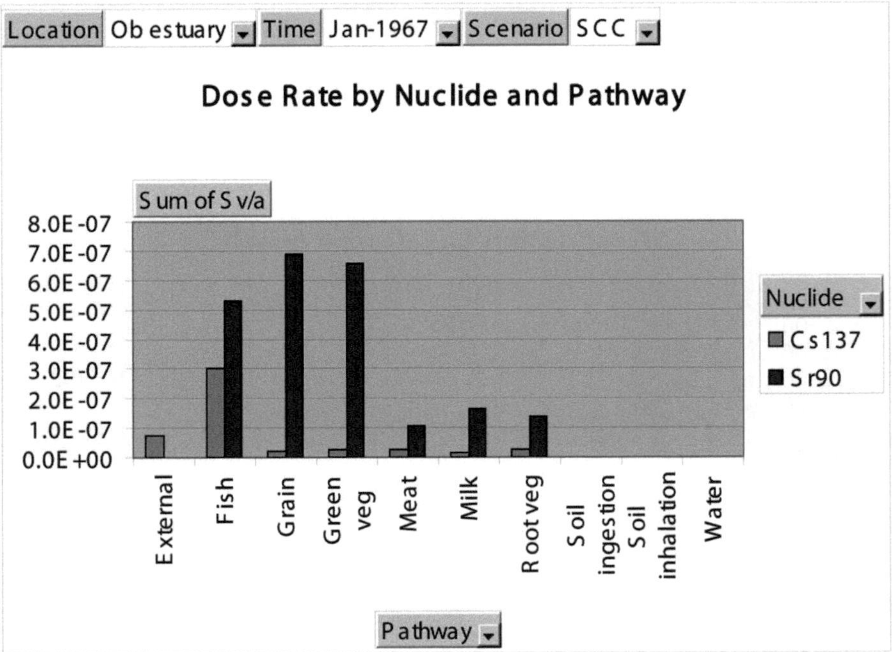

Table B.16. Maximum annual doses from radionuclides and pathways to individuals ($Sv\,yr^{-1}$) at Tomsk from Accident Scenarios 2b and 3b: break in a waste-holding pond at the Siberian Chemical Combine (SCC) at Tomsk in June 1967.

Sum of Sv/yr	*Nuclide*		
Pathway	^{137}Cs	^{90}Sr	*Grand total*
External	1.2E-05	0.0E+00	1.2E-05
Fish	8.8E-05	1.5E-05	1.0E-04
Grain	3.7E-06	0.0E+00	3.7E-06
Green veg	4.4E-06	0.0E+00	4.4E-06
Meat	4.0E-06	0.0E+00	4.0E-06
Milk	2.9E-06	0.0E+00	2.9E-06
Root veg	3.9E-06	0.0E+00	3.9E-06
Soil ingestion	3.8E-08	0.0E+00	3.8E-08
Soil inhalation	2.8E-10	0.0E+00	2.8E-10
Water	8.8E-09	4.9E-08	5.7E-08
Grand total	*1.2E-04*	*1.5E-05*	*1.3E-04*

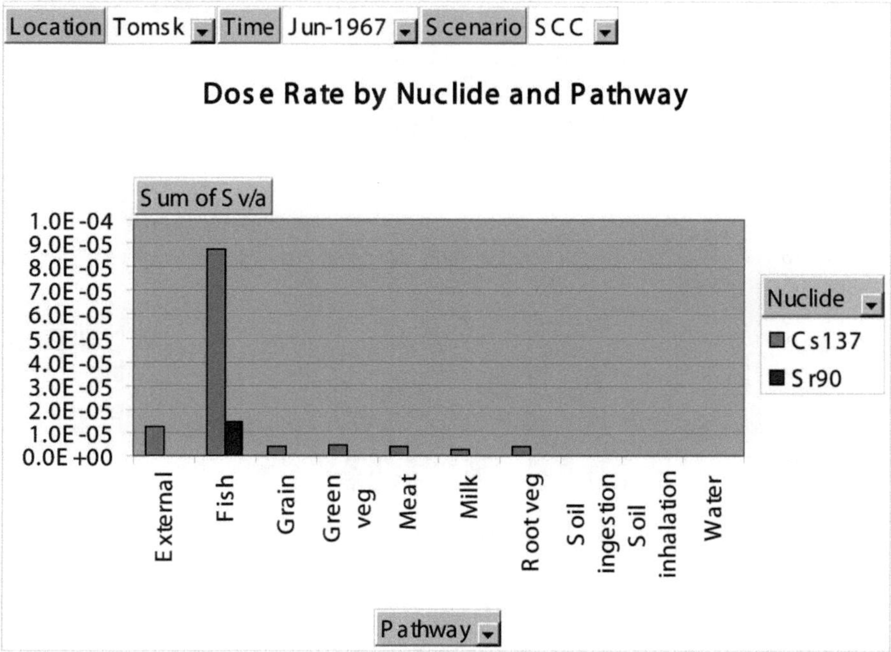

Table B.17. Maximum annual doses from radionuclides and pathways to individuals ($Sv\,yr^{-1}$) at the Ob′ River mouth (Salekhard) from Accident Scenarios 2a and 3a: break in a waste-holding pond at the Siberian Chemical Combine (SCC) at Tomsk in June 1967.

Sum of Sv/yr	*Nuclide*		
Pathway	^{137}Cs	^{90}Sr	*Grand total*
External	4.6E-06	3.8E-12	4.6E-06
Fish	7.2E-06	1.9E-06	9.1E-06
Grain	1.4E-06	6.6E-06	8.0E-06
Green veg	1.6E-06	6.3E-06	7.9E-06
Meat	1.5E-06	1.0E-06	2.5E-06
Milk	1.1E-06	1.6E-06	2.7E-06
Root veg	1.5E-06	1.3E-06	2.8E-06
Soil ingestion	1.4E-08	8.4E-09	2.3E-08
Soil inhalation	1.1E-10	1.2E-10	2.3E-10
Water	7.2E-10	6.4E-09	7.2E-09
Grand total	*1.9E-05*	*1.9E-05*	*3.8E-05*

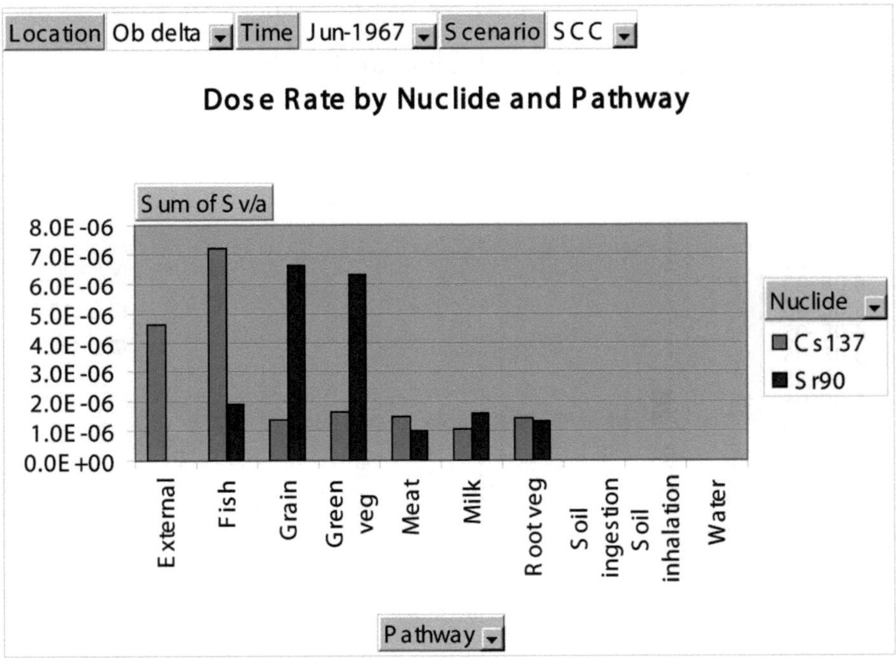

Table B.18. Maximum annual doses from radionuclides and pathways to individuals (Sv yr^{-1}) at the Ob′ estuary mouth from Accident Scenarios 2b and 3b: break in a waste-holding pond at the Siberian Chemical Combine (SCC) at Tomsk in June 1967.

Sum of Sv/yr	*Nuclide*		
Pathway	^{137}Cs	^{90}Sr	*Grand total*
External	8.9E-08	6.2E-13	8.9E-08
Fish	3.6E-07	9.4E-07	1.3E-06
Grain	2.7E-08	1.1E-06	1.1E-06
Green veg	3.2E-08	1.0E-06	1.0E-06
Meat	2.9E-08	1.6E-07	1.9E-07
Milk	2.1E-08	2.6E-07	2.8E-07
Root veg	2.8E-08	2.1E-07	2.4E-07
Soil ingestion	2.7E-10	1.4E-09	1.6E-09
Soil inhalation	2.1E-12	1.9E-11	2.1E-11
Water	3.6E-11	3.1E-09	3.2E-09
Grand total	*5.9E-07*	*3.7E-06*	*4.2E-06*

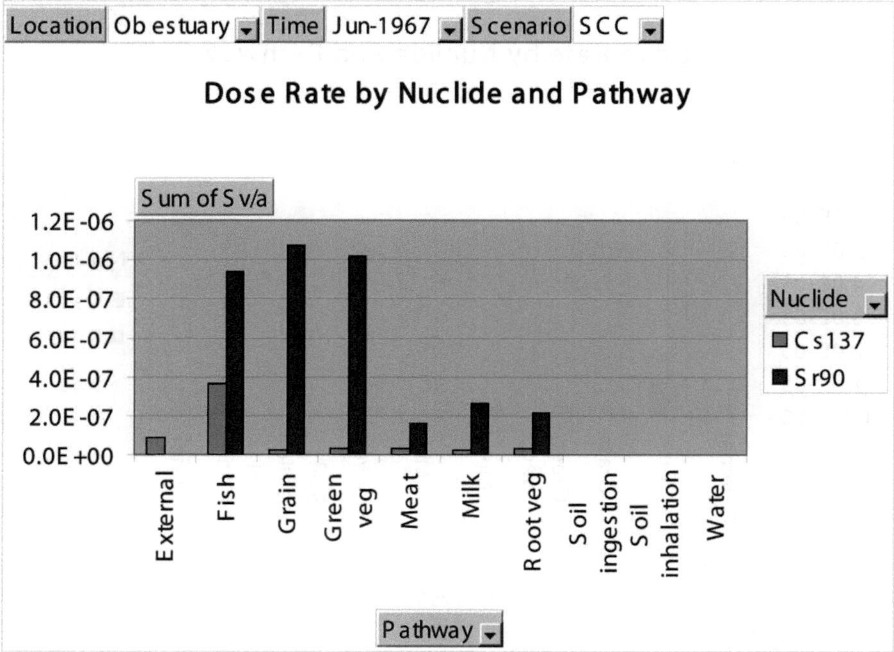

Table B.19. Maximum annual doses from radionuclides and pathways to individuals ($Sv\,yr^{-1}$) at Tomsk from Accident Scenarios 2a and 3a: break in a waste-holding pond at the Siberian Chemical Combine (SCC) at Tomsk in January 2084.

Sum of Sv/yr	*Nuclide*		
Pathway	^{137}Cs	^{90}Sr	*Grand total*
External	8.0E-06	0.0E+00	8.0E-06
Fish	5.9E-05	9.8E-06	6.8E-05
Grain	2.4E-06	0.0E+00	2.4E-06
Green veg	2.8E-06	0.0E+00	2.8E-06
Meat	2.6E-06	0.0E+00	2.6E-06
Milk	1.9E-06	0.0E+00	1.9E-06
Root veg	2.5E-06	0.0E+00	2.5E-06
Soil ingestion	2.5E-08	0.0E+00	2.5E-08
Soil inhalation	1.8E-10	0.0E+00	1.8E-10
Water	5.9E-09	3.3E-08	3.8E-08
Grand total	*7.9E-05*	*9.8E-06*	*8.9E-05*

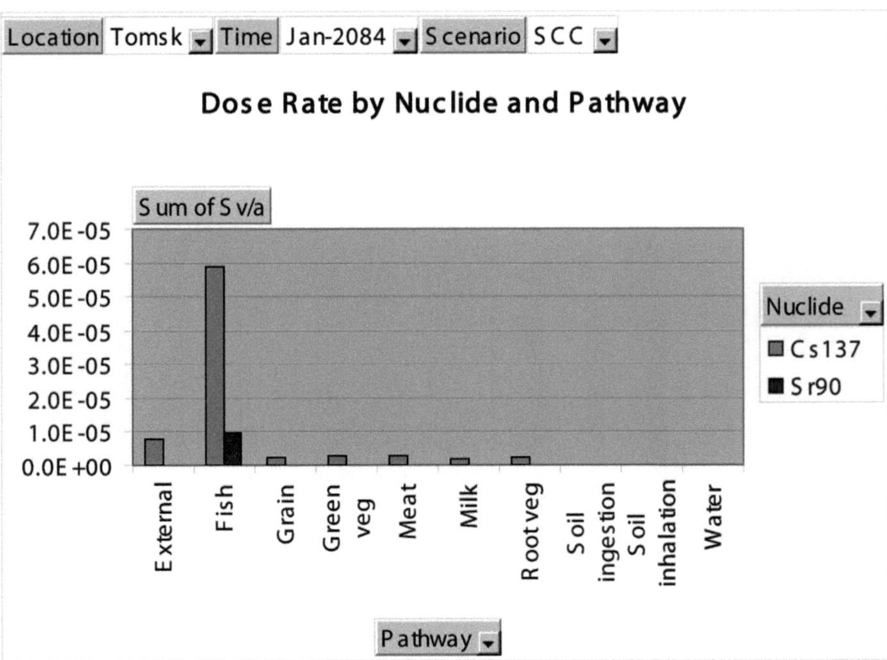

Table B.20. Maximum annual doses from radionuclides and pathways to individuals (Sv yr^{-1}) at the Ob′ River mouth (Salekhard) from Accident Scenarios 2a and 3a: break in a waste-holding pond at the Siberian Chemical Combine (SCC) at Tomsk in January 2084.

Sum of Sv/yr	*Nuclide*		
Pathway	^{137}Cs	^{90}Sr	*Grand total*
External	5.7E-07	5.8E-13	5.7E-07
Fish	3.8E-06	9.6E-07	4.7E-06
Grain	1.7E-07	1.0E-06	1.2E-06
Green veg	2.0E-07	9.6E-07	1.2E-06
Meat	1.8E-07	1.5E-07	3.4E-07
Milk	1.3E-07	2.4E-07	3.8E-07
Root veg	1.8E-07	2.0E-07	3.8E-07
Soil ingestion	1.8E-09	1.3E-09	3.0E-09
Soil inhalation	1.3E-11	1.8E-11	3.1E-11
Water	3.8E-10	3.2E-09	3.6E-09
Grand total	*5.2E-06*	*3.5E-06*	*8.7E-06*

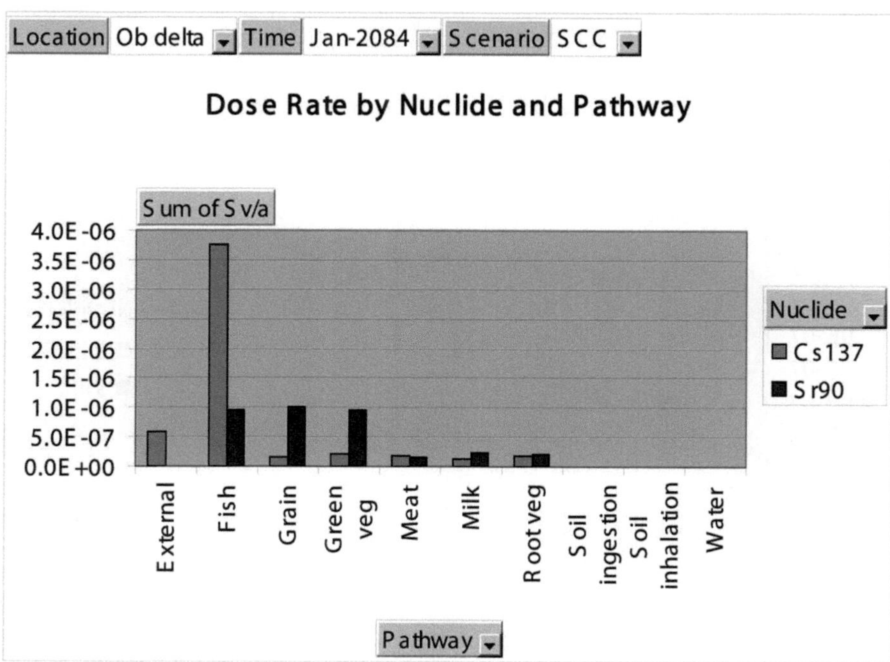

Table B.21. Maximum annual doses from radionuclides and pathways to individuals (Sv yr^{-1}) at the Ob′ estuary mouth from Accident Scenarios 2a and 3a: break in a waste-holding pond at the Siberian Chemical Combine (SCC) at Tomsk in January 2084.

Sum of Sv/yr	*Nuclide*		
Pathway	^{137}Cs	^{90}Sr	*Grand total*
External	6.1E-08	3.6E-13	6.1E-08
Fish	3.0E-07	5.4E-07	8.5E-07
Grain	1.8E-08	6.2E-07	6.4E-07
Green veg	2.2E-08	5.9E-07	6.2E-07
Meat	2.0E-08	9.4E-08	1.1E-07
Milk	1.4E-08	1.5E-07	1.7E-07
Root veg	1.9E-08	1.2E-07	1.4E-07
Soil ingestion	1.9E-10	7.9E-10	9.8E-10
Soil inhalation	1.4E-12	1.1E-11	1.3E-11
Water	3.0E-11	1.8E-09	1.8E-09
Grand total	*4.6E-07*	*2.1E-06*	*2.6E-06*

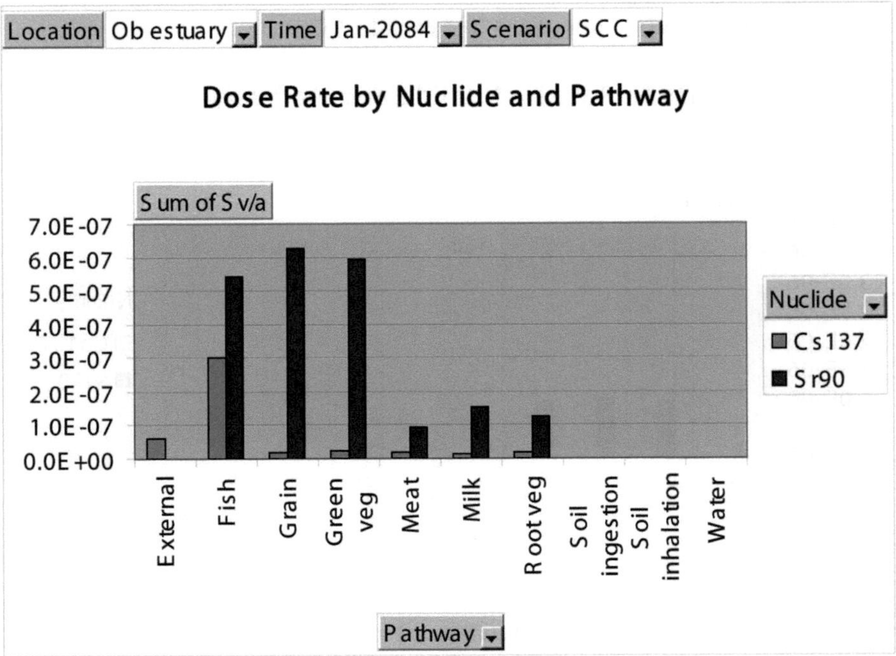

Table B.22. Maximum annual doses from radionuclides and pathways to individuals (Sv yr^{-1}) at Tomsk from Accident Scenarios 2b and 3b: break in a waste-holding pond at the Siberian Chemical Combine (SCC) at Tomsk in June 2084.

Sum of Sv/yr	*Nuclide*		
Pathway	^{137}Cs	^{90}Sr	*Grand total*
External	8.0E-06	0.0E+00	8.0E-06
Fish	5.9E-05	9.8E-06	6.8E-05
Grain	2.4E-06	0.0E+00	2.4E-06
Green veg	2.8E-06	0.0E+00	2.8E-06
Meat	2.6E-06	0.0E+00	2.6E-06
Milk	1.9E-06	0.0E+00	1.9E-06
Root veg	2.5E-06	0.0E+00	2.5E-06
Soil ingestion	2.5E-08	0.0E+00	2.5E-08
Soil inhalation	1.8E-10	0.0E+00	1.8E-10
Water	5.9E-09	3.3E-08	3.8E-08
Grand total	*7.9E-05*	*9.8E-06*	*8.9E-05*

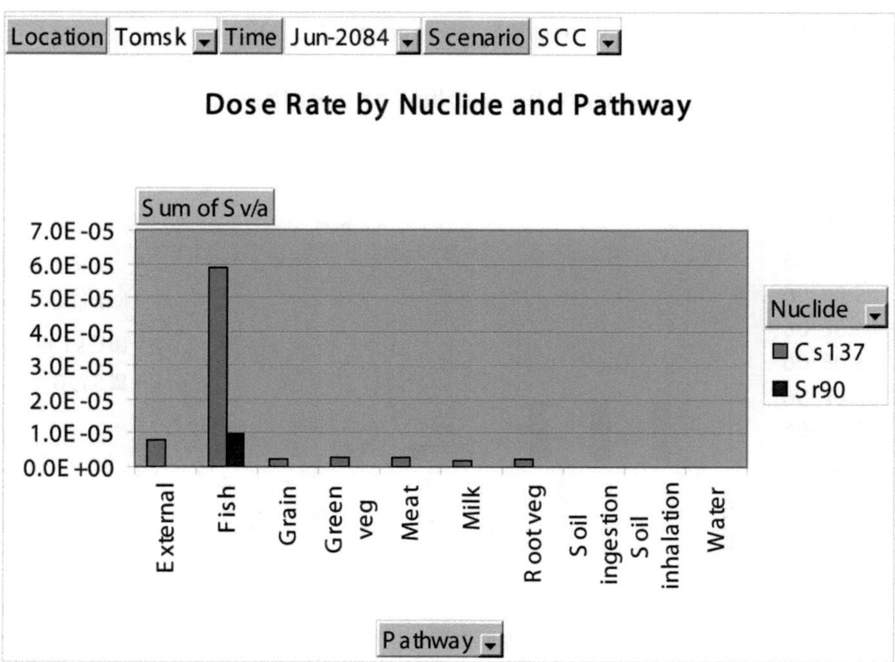

Table B.23. Maximum annual doses from radionuclides and pathways to individuals (Sv yr^{-1}) at the Ob′ River mouth (Salekhard) from Accident Scenarios 2b and 3b: break in a waste-holding pond at the Siberian Chemical Combine (SCC) at Tomsk in June 2084.

Sum of Sv/yr	*Nuclide*		
Pathway	^{137}Cs	^{90}Sr	*Grand total*
External	2.0E-06	1.1E-12	2.0E-06
Fish	5.9E-06	1.8E-06	7.7E-06
Grain	5.9E-07	1.9E-06	2.4E-06
Green veg	6.9E-07	1.8E-06	2.5E-06
Meat	6.3E-07	2.8E-07	9.1E-07
Milk	4.6E-07	4.5E-07	9.0E-07
Root veg	6.1E-07	3.7E-07	9.8E-07
Soil ingestion	6.0E-09	2.4E-09	8.4E-09
Soil inhalation	4.5E-11	3.4E-11	7.9E-11
Water	5.9E-10	5.9E-09	6.5E-09
Grand total	*1.1E-05*	*6.5E-06*	*1.7E-05*

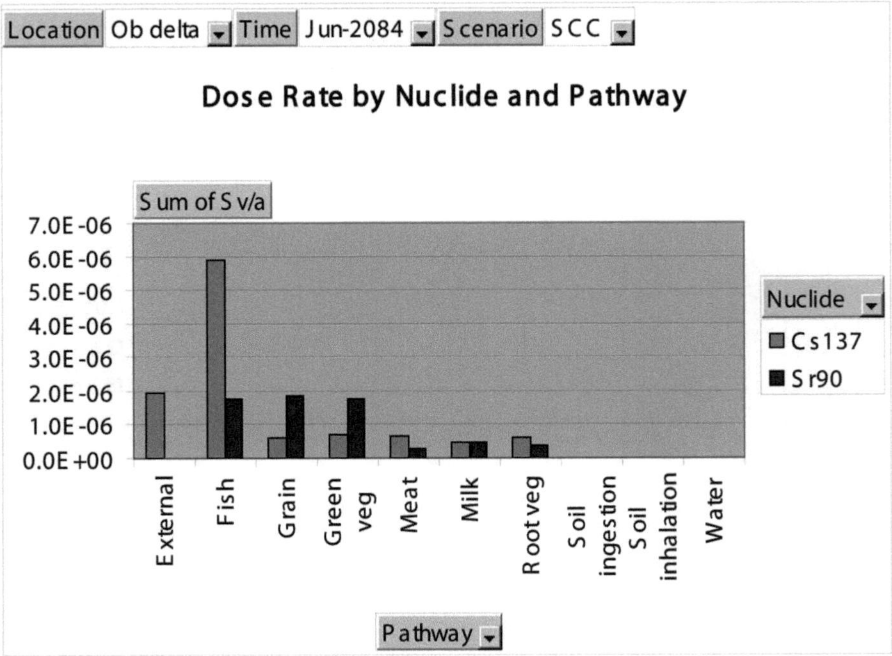

Table B.24. Maximum annual doses from radionuclides and pathways to individuals (Sv yr^{-1}) at the Ob′ estuary mouth from Accident Scenarios 2b and 3b: break in a waste-holding pond at the Siberian Chemical Combine (SCC) at Tomsk in June 2084.

Sum of Sv/yr	*Nuclide*		
Pathway	^{137}Cs	^{90}Sr	*Grand total*
External	7.1E-08	5.8E-13	7.1E-08
Fish	3.5E-07	9.6E-07	1.3E-06
Grain	2.1E-08	1.0E-06	1.0E-06
Green veg	2.5E-08	9.6E-07	9.9E-07
Meat	2.3E-08	1.5E-07	1.8E-07
Milk	1.7E-08	2.4E-07	2.6E-07
Root veg	2.2E-08	2.0E-07	2.2E-07
Soil ingestion	2.2E-10	1.3E-09	1.5E-09
Soil inhalation	1.6E-12	1.8E-11	2.0E-11
Water	3.5E-11	3.2E-09	3.2E-09
Grand total	*5.3E-07*	*3.5E-06*	*4.1E-06*

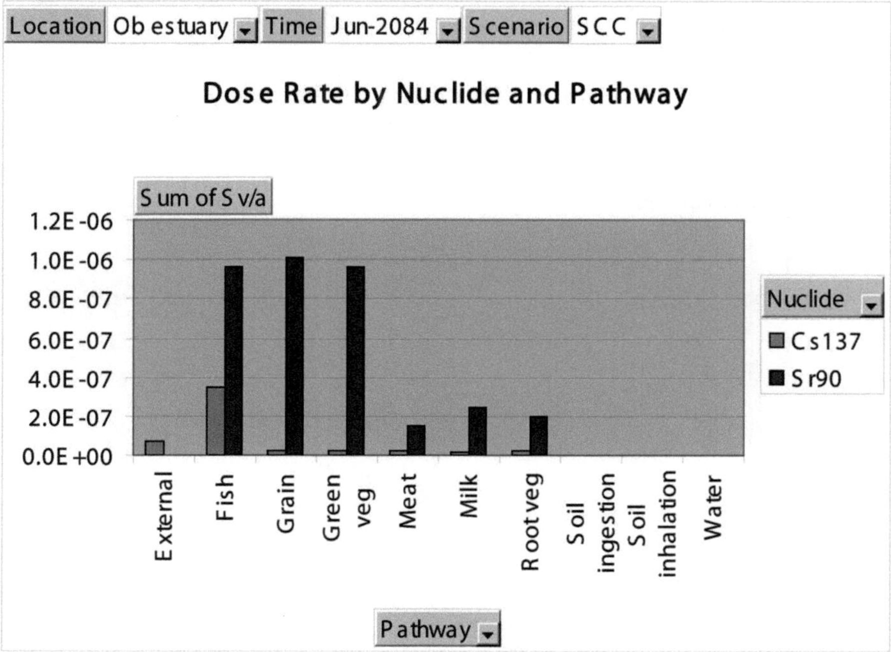

Table B.25. Maximum annual doses from radionuclides and pathways to individuals ($Sv\,yr^{-1}$) at Krasnoyarsk from Accident Scenarios 4a and 5a: break in waste-holding ponds No. 365 and No. 366 at the Mining and Chemical Combine (MCC) at Krasnoyarsk in January 1967.

Sum of Sv/yr	*Nuclide*		
Pathway	^{137}Cs	^{90}Sr	*Grand total*
External	2.8E-04	0.0E+00	2.8E-04
Fish	6.0E-07	1.2E-05	1.2E-05
Grain	8.5E-05	0.0E+00	8.5E-05
Green veg	1.0E-04	0.0E+00	1.0E-04
Meat	9.1E-05	0.0E+00	9.1E-05
Milk	6.6E-05	0.0E+00	6.6E-05
Root veg	8.9E-05	0.0E+00	8.9E-05
Soil ingestion	8.7E-07	0.0E+00	8.7E-07
Soil inhalation	6.5E-09	0.0E+00	6.5E-09
Water	6.0E-11	3.9E-08	3.9E-08
Grand total	*7.1E-04*	*1.2E-05*	*7.3E-04*

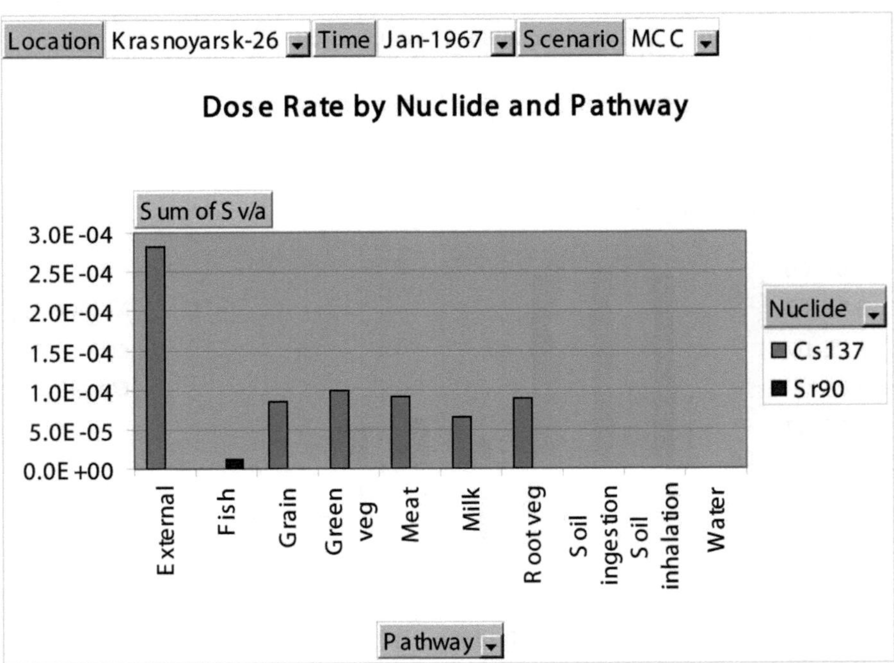

Table B.26. Maximum annual doses from radionuclides and pathways to individuals (Sv yr^{-1}) at the Yenisei River mouth from Accident Scenarios 4a and 5a: break in waste-holding ponds No. 365 and No. 366 at the Mining and Chemical Combine (MCC) at Krasnoyarsk in January 1967.

Sum of Sv/yr	*Nuclide*		
Pathway	^{137}Cs	^{90}Sr	*Grand total*
External	3.2E-07	6.3E-13	3.2E-07
Fish	1.9E-06	1.7E-06	3.6E-06
Grain	9.5E-08	1.1E-06	1.2E-06
Green veg	1.1E-07	1.0E-06	1.2E-06
Meat	1.0E-07	1.7E-07	2.7E-07
Milk	7.4E-08	2.7E-07	3.4E-07
Root veg	9.9E-08	2.2E-07	3.2E-07
Soil ingestion	9.7E-10	1.4E-09	2.4E-09
Soil inhalation	7.3E-12	2.0E-11	2.7E-11
Water	1.9E-10	5.7E-09	5.9E-09
Grand total	2.7E-06	4.5E-06	7.2E-06

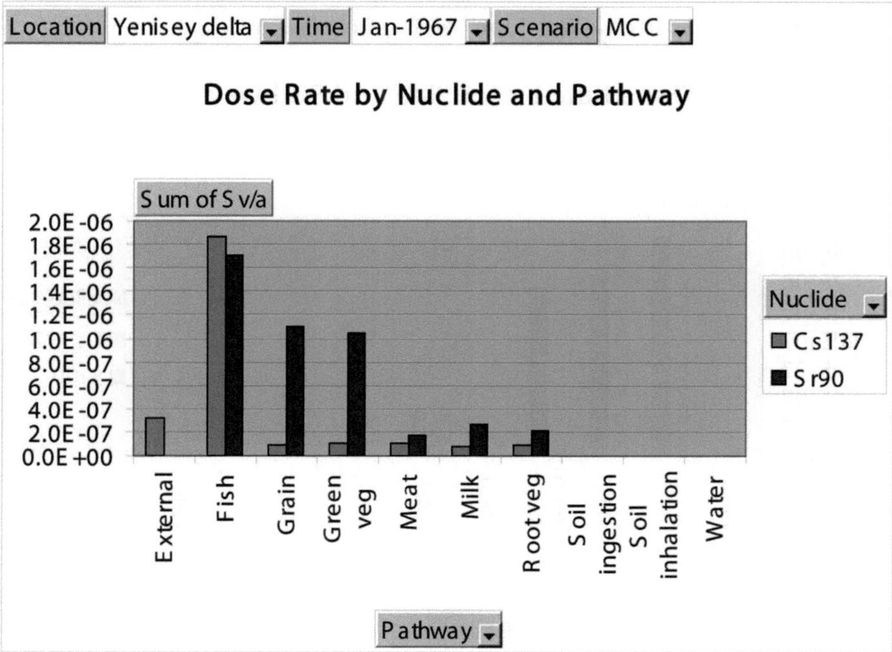

Table B.27. Maximum annual doses from radionuclides and pathways to individuals (Sv yr^{-1}) at the Yenisei estuary mouth from Accident Scenarios 4a and 5a: break in waste-holding ponds No. 365 and No. 366 at the Mining and Chemical Combine (MCC) at Krasnoyarsk in January 1967.

Sum of Sv/yr	*Nuclide*		
Pathway	^{137}Cs	^{90}Sr	*Grand total*
External	3.1E-07	7.0E-13	3.1E-07
Fish	1.3E-06	1.5E-06	2.8E-06
Grain	9.2E-08	1.2E-06	1.3E-06
Green veg	1.1E-07	1.1E-06	1.3E-06
Meat	9.9E-08	1.8E-07	2.8E-07
Milk	7.2E-08	2.9E-07	3.6E-07
Root veg	9.6E-08	2.4E-07	3.4E-07
Soil ingestion	9.4E-10	1.5E-09	2.5E-09
Soil inhalation	7.1E-12	2.2E-11	2.9E-11
Water	1.3E-10	5.0E-09	5.1E-09
Grand total	2.1E-06	4.6E-06	6.6E-06

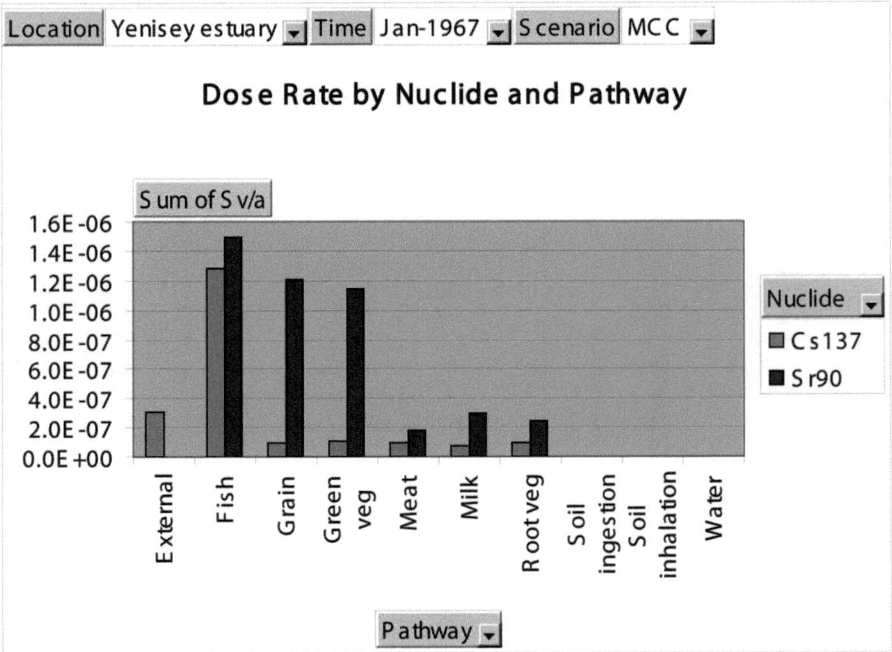

Table B.28. Maximum annual doses from radionuclides and pathways to individuals (Sv yr^{-1}) at Krasnoyarsk from Accident Scenarios 4a and 5a: break in waste-holding ponds No. 365 and No. 366 at the Mining and Chemical Combine (MCC) at Krasnoyarsk in June 1967.

Sum of Sv/yr	*Nuclide*		
Pathway	^{137}Cs	^{90}Sr	*Grand total*
External	2.8E-04	0.0E+00	2.8E-04
Fish	6.0E-07	1.2E-05	1.2E-05
Grain	8.5E-05	0.0E+00	8.5E-05
Green veg	1.0E-04	0.0E+00	1.0E-04
Meat	9.1E-05	0.0E+00	9.1E-05
Milk	6.6E-05	0.0E+00	6.6E-05
Root veg	8.9E-05	0.0E+00	8.9E-05
Soil ingestion	8.7E-07	0.0E+00	8.7E-07
Soil inhalation	6.5E-09	0.0E+00	6.5E-09
Water	6.0E-11	3.9E-08	3.9E-08
Grand total	*7.1E-04*	*1.2E-05*	*7.3E-04*

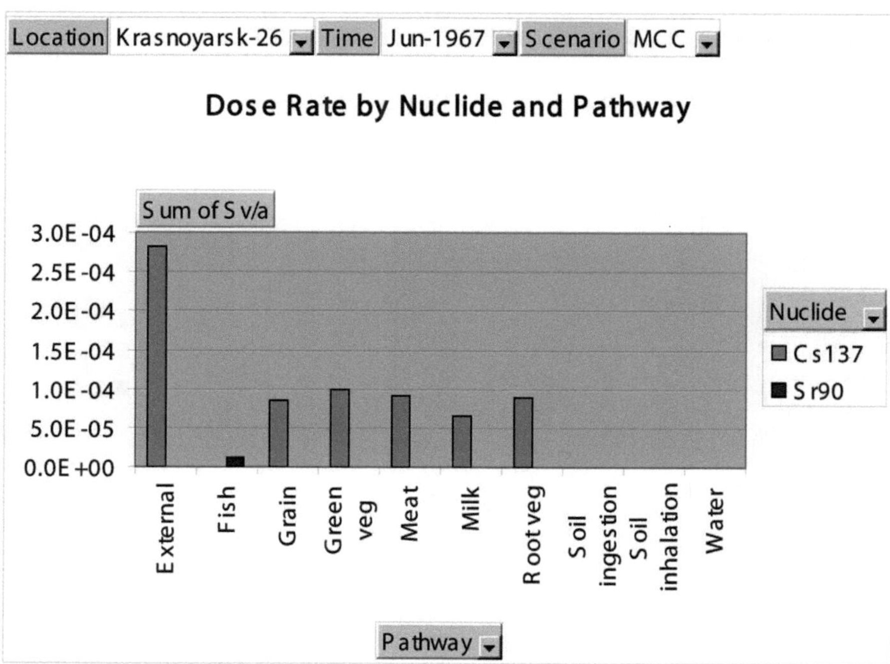

Table B.29. Maximum annual doses from radionuclides and pathways to individuals (Sv yr^{-1}) at the Yenisei River mouth from Accident Scenarios 4a and 5a: break in waste-holding ponds No. 365 and No. 366 at the Mining and Chemical Combine (MCC) at Krasnoyarsk in June 1967.

Sum of Sv/yr	*Nuclide*		
Pathway	^{137}Cs	^{90}Sr	*Grand total*
External	7.4E-07	1.0E-12	7.4E-07
Fish	3.5E-06	1.8E-06	5.3E-06
Grain	2.2E-07	1.8E-06	2.0E-06
Green veg	2.6E-07	1.7E-06	1.9E-06
Meat	2.4E-07	2.6E-07	5.0E-07
Milk	1.7E-07	4.2E-07	5.9E-07
Root veg	2.3E-07	3.5E-07	5.8E-07
Soil ingestion	2.3E-09	2.2E-09	4.5E-09
Soil inhalation	1.7E-11	3.2E-11	4.9E-11
Water	3.5E-10	6.1E-09	6.4E-09
Grand total	*5.3E-06*	*6.3E-06*	*1.2E-05*

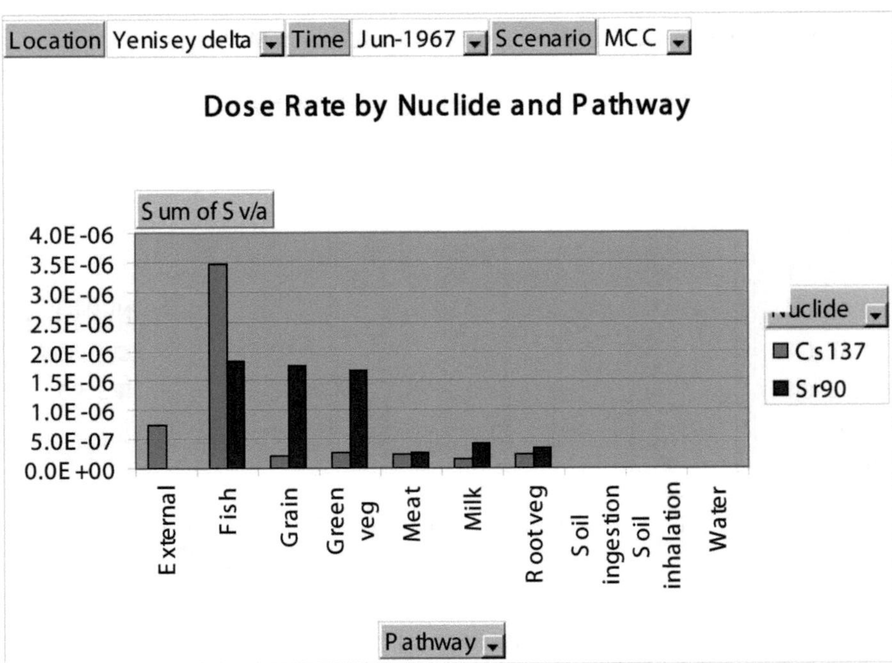

Table B.30. Maximum annual doses from radionuclides and pathways to individuals (Sv yr^{-1}) at the Yenisei estuary mouth from Accident Scenarios 4a and 5a: break in waste-holding ponds No. 365 and No. 366 at the Mining and Chemical Combine (MCC) at Krasnoyarsk in June 1967.

Sum of Sv/yr	*Nuclide*		
Pathway	^{137}Cs	^{90}Sr	*Grand total*
External	3.7E-07	1.1E-12	3.7E-07
Fish	1.5E-06	1.8E-06	3.3E-06
Grain	1.1E-07	1.9E-06	2.0E-06
Green veg	1.3E-07	1.8E-06	1.9E-06
Meat	1.2E-07	2.8E-07	4.0E-07
Milk	8.6E-08	4.5E-07	5.4E-07
Root veg	1.2E-07	3.7E-07	4.9E-07
Soil ingestion	1.1E-09	2.4E-09	3.5E-09
Soil inhalation	8.5E-12	3.4E-11	4.3E-11
Water	1.5E-10	6.0E-09	.1E-09
Grand total	*2.4E-06*	*6.6E-06*	*9.0E-06*

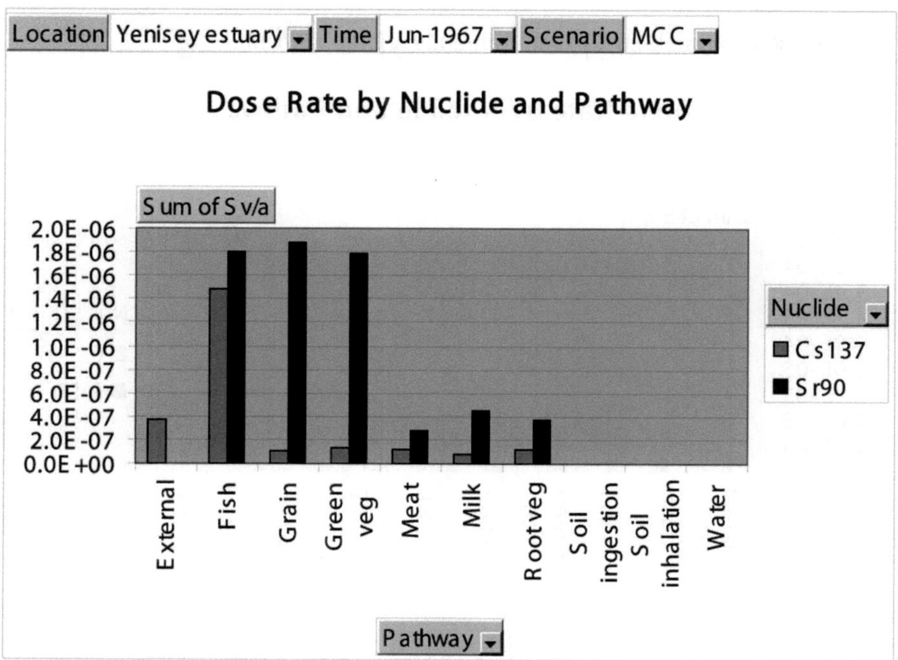

Table B.31. Maximum annual doses from radionuclides and pathways to individuals (Sv yr^{-1}) at Krasnoyarsk from Accident Scenarios 4a and 5a: break in waste-holding ponds No. 365 and No. 366 at the Mining and Chemical Combine (MCC) at Krasnoyarsk in January 2089.

Sum of Sv/yr	*Nuclide*		
Pathway	^{137}Cs	^{90}Sr	*Grand total*
External	1.7E-04	0.0E+00	1.7E-04
Fish	4.5E-07	8.6E-06	9.1E-06
Grain	5.0E-05	0.0E+00	5.0E-05
Green veg	6.0E-05	0.0E+00	6.0E-05
Meat	5.4E-05	0.0E+00	5.4E-05
Milk	3.9E-05	0.0E+00	3.9E-05
Root veg	5.3E-05	0.0E+00	5.3E-05
Soil ingestion	5.2E-07	0.0E+00	5.2E-07
Soil inhalation	3.9E-09	0.0E+00	3.9E-09
Water	4.5E-11	2.9E-08	2.9E-08
Grand total	*4.3E-04*	*8.7E-06*	*4.3E-04*

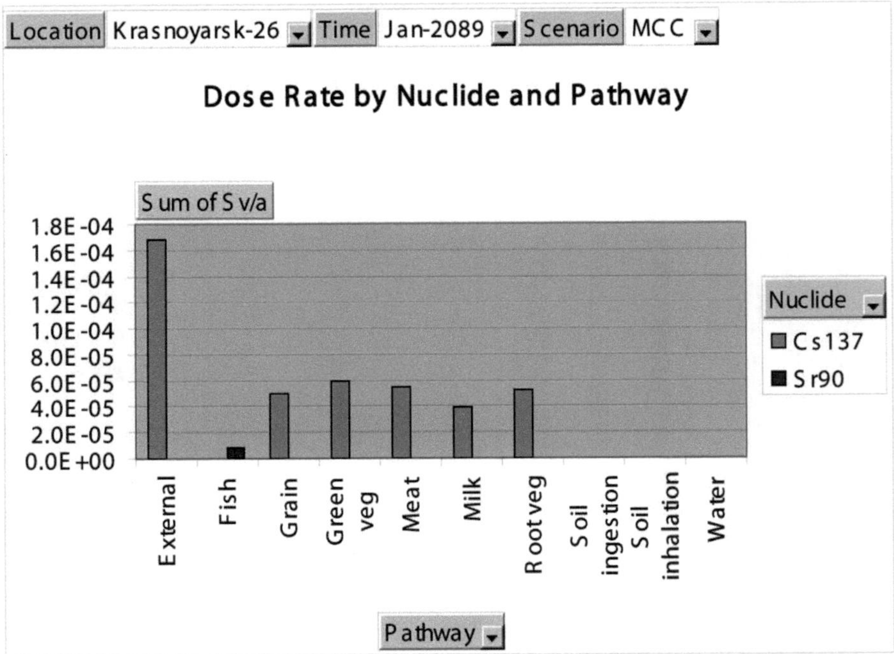

Table B.32. Maximum annual doses from radionuclides and pathways to individuals (Sv yr^{-1}) at the Yenisei River mouth from Accident Scenarios 4a and 5a: break in waste-holding ponds No. 365 and No. 366 at the Mining and Chemical Combine (MCC) at Krasnoyarsk in January 2089.

Sum of Sv/yr	*Nuclide*		
Pathway	^{137}Cs	^{90}Sr	*Grand total*
External	6.1E-07	1.5E-13	6.1E-07
Fish	2.8E-06	6.6E-07	3.5E-06
Grain	1.8E-07	2.6E-07	4.4E-07
Green veg	2.1E-07	2.5E-07	4.6E-07
Meat	2.0E-07	4.0E-08	2.4E-07
Milk	1.4E-07	6.4E-08	2.0E-07
Root veg	1.9E-07	5.2E-08	2.4E-07
Soil ingestion	1.9E-09	3.3E-10	2.2E-09
Soil inhalation	1.4E-11	4.8E-12	1.9E-11
Water	2.8E-10	2.2E-09	2.5E-09
Grand total	4.4E-06	1.3E-06	5.7E-06

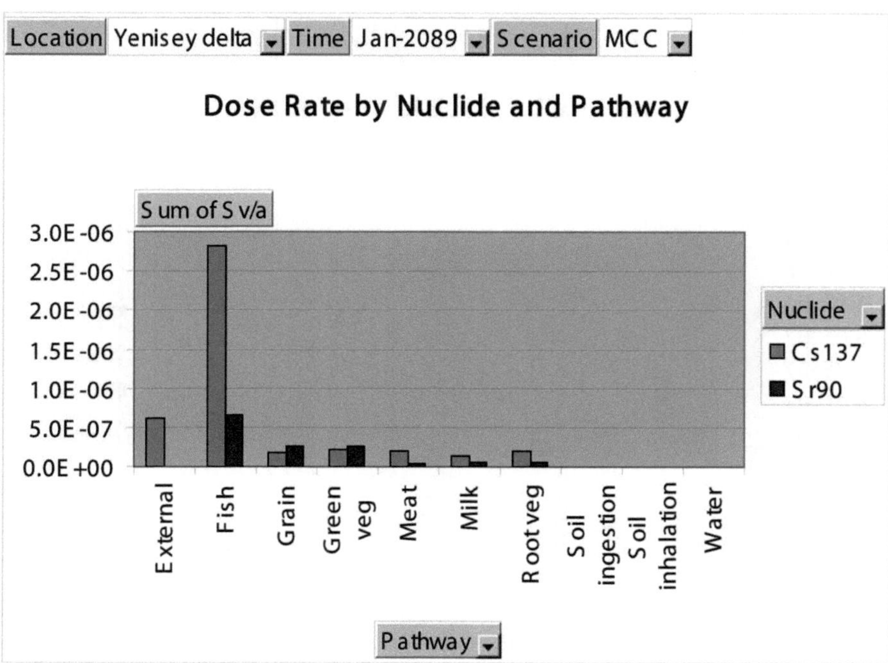

Table B.33. Maximum annual doses from radionuclides and pathways to individuals (Sv yr^{-1}) at the Yenisei estuary mouth from Accident Scenarios 4a and 5a: break in waste-holding ponds No. 365 and No. 366 at the Mining and Chemical Combine (MCC) at Krasnoyarsk in January 2089.

Sum of Sv/yr	*Nuclide*		
Pathway	^{137}Cs	^{90}Sr	*Grand total*
External	3.1E-07	1.7E-13	3.1E-07
Fish	1.3E-06	6.1E-07	1.9E-06
Grain	9.2E-08	3.0E-07	3.9E-07
Green veg	1.1E-07	2.9E-07	4.0E-07
Meat	9.9E-08	4.6E-08	1.4E-07
Milk	7.1E-08	7.3E-08	1.4E-07
Root veg	9.6E-08	6.0E-08	1.6E-07
Soil ingestion	9.4E-10	3.8E-10	1.3E-09
Soil inhalation	7.0E-12	5.5E-12	1.3E-11
Water	1.3E-10	2.0E-09	2.1E-09
Grand total	*2.1E-06*	*1.4E-06*	*3.4E-06*

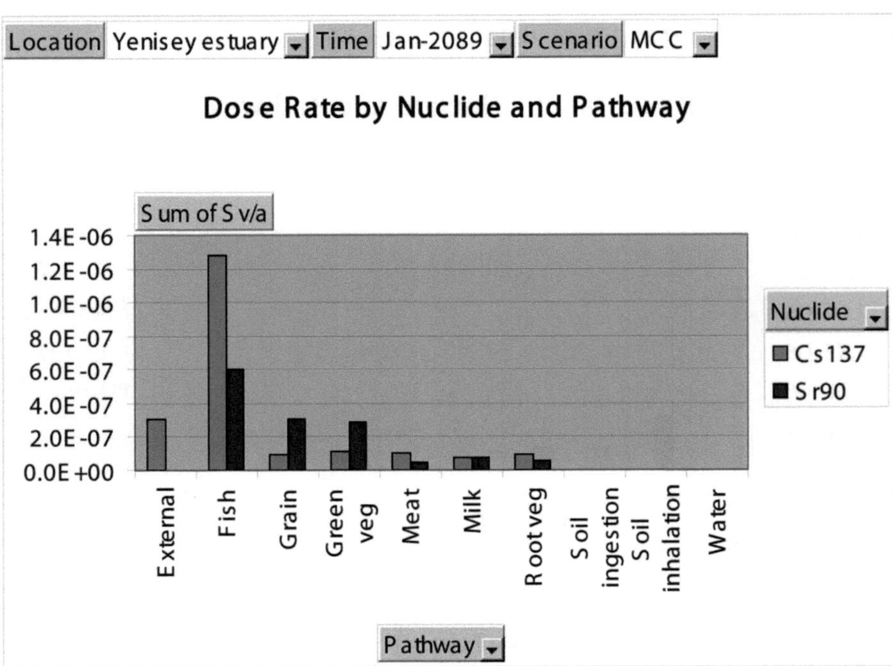

Table B.34. Maximum annual doses from radionuclides and pathways to individuals (Sv yr^{-1}) at Krasnoyarsk from Accident Scenarios 4a and 5a: break in waste-holding ponds No. 365 and No. 366 at the Mining and Chemical Combine (MCC) at Krasnoyarsk in June 2089.

Sum of Sv/yr	*Nuclide*		
Pathway	^{137}Cs	^{90}Sr	*Grand total*
External	1.7E-04	0.0E+00	1.7E-04
Fish	4.5E-07	8.6E-06	9.1E-06
Grain	5.0E-05	0.0E+00	5.0E-05
Green veg	6.0E-05	0.0E+00	.0E-05
Meat	5.4E-05	0.0E+00	5.4E-05
Milk	3.9E-05	0.0E+00	3.9E-05
Root veg	5.3E-05	0.0E+00	5.3E-05
Soil ingestion	5.2E-07	0.0E+00	5.2E-07
Soil inhalation	3.9E-09	0.0E+00	3.9E-09
Water	4.5E-11	2.9E-08	2.9E-08
Grand total	*4.3E-04*	*8.7E-06*	*4.3E-04*

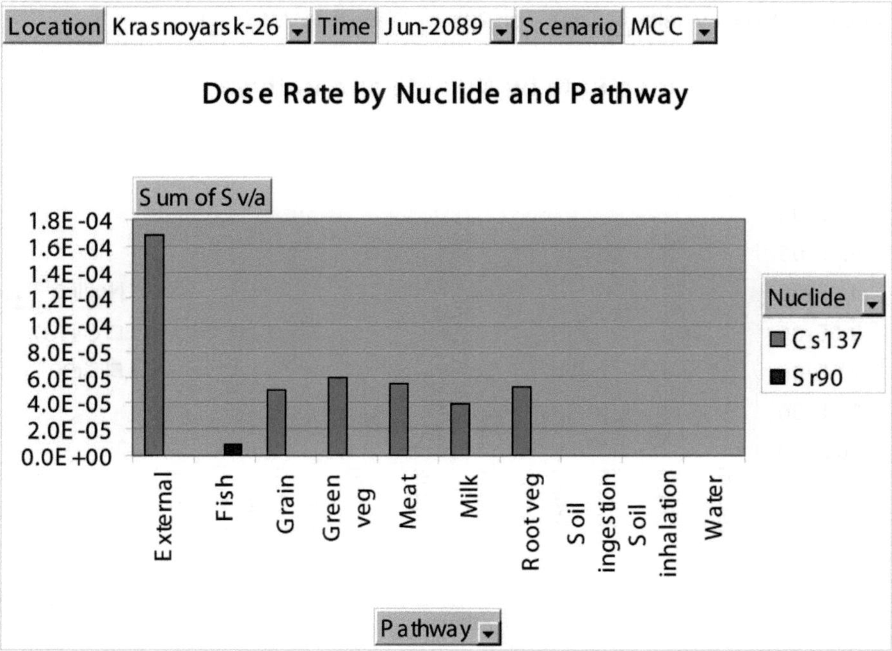

Table B.35. Maximum annual doses from radionuclides and pathways to individuals (Sv yr^{-1}) at the Yenisei River mouth from Accident Scenarios 4a and 5a: break in waste-holding ponds No. 365 and No. 366 at the Mining and Chemical Combine (MCC) at Krasnoyarsk in June 2089.

Sum of Sv/yr	*Nuclide*		
Pathway	^{137}Cs	^{90}Sr	*Grand total*
External	2.5E-06	1.6E-13	2.5E-06
Fish	5.5E-06	6.8E-07	6.2E-06
Grain	7.6E-07	2.7E-07	1.0E-06
Green veg	9.0E-07	2.6E-07	1.2E-06
Meat	8.2E-07	4.1E-08	8.6E-07
Milk	5.9E-07	6.5E-08	6.6E-07
Root veg	7.9E-07	5.4E-08	8.5E-07
Soil ingestion	7.8E-09	3.4E-10	8.1E-09
Soil inhalation	5.8E-11	4.9E-12	6.3E-11
Water	5.5E-10	2.3E-09	2.8E-09
Grand total	*1.2E-05*	*1.4E-06*	*1.3E-05*

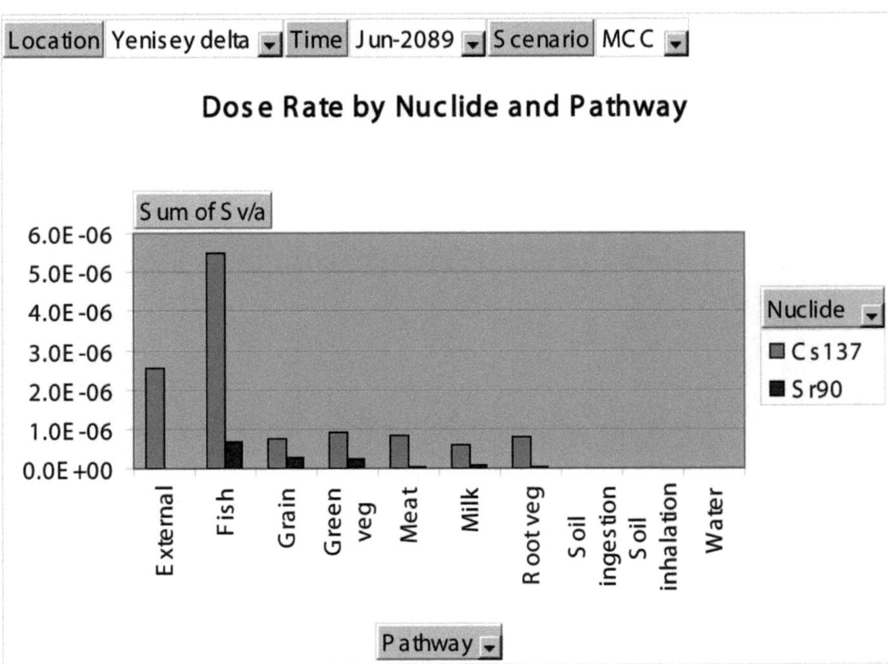

Table B.36. Maximum annual doses from radionuclides and pathways to individuals ($Sv\,yr^{-1}$) at the Yenisei estuary mouth from Accident Scenarios 4a and 5a: break in waste-holding ponds No. 365 and No. 366 at the Mining and Chemical Combine (MCC) at Krasnoyarsk in June 2089.

Sum of Sv/yr	*Nuclide*		
Pathway	^{137}Cs	^{90}Sr	*Grand total*
External	8.2E-07	1.8E-13	8.2E-07
Fish	2.5E-06	6.7E-07	3.1E-06
Grain	2.4E-07	3.2E-07	5.6E-07
Green veg	2.9E-07	3.0E-07	5.9E-07
Meat	2.6E-07	4.8E-08	3.1E-07
Milk	1.9E-07	7.7E-08	2.7E-07
Root veg	2.6E-07	6.3E-08	3.2E-07
Soil ingestion	2.5E-09	4.0E-10	2.9E-09
Soil inhalation	1.9E-11	5.8E-12	2.5E-11
Water	2.5E-10	2.2E-09	2.5E-09
Grand total	*4.5E-06*	*1.5E-06*	*6.0E-06*

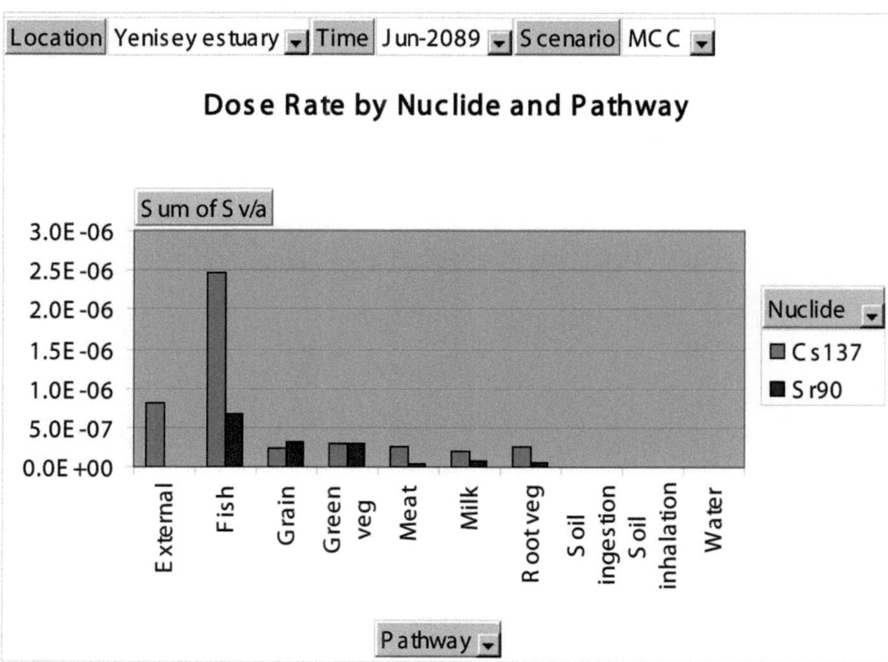

Afterword

Radioactivity in the Nordic Seas and the Arctic is a major concern because contamination can persist for long periods and spread over great distances, primarily through river and marine pathways, but also in the atmosphere. Pathways in the terrestrial environment can lead to high exposures to biota and people. Therefore, the ability to understand and predict the spread of radioactivity and other pollution (e.g., persistent organic pollutants—POPs) through the environment is extremely important, complex, and challenging. This requires a multi-scale, multi-model system that can deal with flows in rivers (including sediment), shelf seas, and the deep Arctic Ocean and Nordic Seas. The Generic Model System (GMS) put forth here in this book represents a means to do so—the GMS is thus one of the key results of the decade of studies upon which this book is based.

As described in Chapters 1 and 2, there are many potential sources of radioactive pollution in both the terrestrial and marine environment). In parts of the Arctic, there is a high density of sources of radionuclides. The risk of accidents combined with the vulnerability of the Arctic environment to radioactive contamination raises the necessity for continuous efforts to understand and reduce risks.

There has been significant progress in recent years regarding what can be done to reduce the risks of radioactive contamination from some of the potential sources, largely as a result of national and international actions (AMAP, 2009). Many of these potential sources are located in northern and northwestern Russia, including obsolete nuclear icebreakers and the nuclear submarines of the Russian Northern Fleet, the majority of which are being safely decommissioned. In the Arctic marine environment, discharges to the sea from the reprocessing of spent nuclear fuel at Sellafield (U.K.) and Cap de la Hague (France) have been a major source of radioactive contamination, mainly ^{137}Cs but also other radionuclides. As shown in the RADARC studies presented here and in other studies, these contaminants can be transported over long distances; they follow the Atlantic Current to the Norwegian

Sea and farther along the Norwegian coast into the Barents Sea on a pathway that takes 4–5 years, eventually reaching the Arctic Ocean.

The releases of radionuclides and the levels of radiation—as well as POPs—observed in the Arctic and Nordic Seas have decreased in recent years. However, much remains to be done to reduce, predict, and manage radiation risks. As we describe and model in Chapters 3 and 4, the potential spread of radionuclides from releases at sources such as the Russian Nuclear Complex (notably those situated along the Ob′ and Yenisei Rivers) and the Sellafield and Cap de la Hague plants, can be far-reaching and long-lasting. Moreover, there are new potential sources of radioactive pollution; for example, Russian plans to build floating nuclear power plants in the Russian Arctic Seas, and non-radioactive pollution from increased hydrocarbon exploitation and marine transport in the Arctic, as anticipated from projections of further reductions in Arctic sea ice in summer (Johannessen *et al.* 2004; IPCC, 2007; Johannessen, 2008).

Anthropogenic climate change is likely to affect the transport of contaminants in the marine environment in the 21st century. Changes in temperature and atmosphere–ocean circulation and exchanges are expected to alter the pathways and residence time of pollutants in the marine environment. A new finding put forth in Chapter 4 of this book which is potentially important—both geophysically and in terms of pollution transport—is the fundamental change in the large-scale circulation of the Arctic Ocean and adjacent Nordic Seas. Here we show that the modeled circulation pattern is very different from the present. For example, the reason for modeled differences in the spatial distributions of the Ob′ and Yenisei releases of ^{90}Sr are changes in major surface circulation features in the Arctic Ocean; namely, the strong clockwise Beaufort Gyre is dramatically weakened, the Transpolar Drift towards Fram Strait disappears, and there is reduced transport from the Arctic Ocean through Fram Strait. This will cause the contaminants that enter the Arctic marine environment to stay in the Arctic much longer than present. Moreover, anthropogenic climate change is expected to increase the risk of non-radioactive pollution, through increased hydrocarbon resource exploitation and transportation (Chapter 5).

What then of the risks of future releases of radioactive pollution and their transport to the Arctic and Nordic Seas? In the past, the resulting exposure to biota and people in the Arctic has been relatively low (AMAP, 2009) because of the wide dispersion and massive dilution of these contaminants in water. In Chapter 6 we assess the risk of releases of radionuclides from point sources near the Ob′ and Yenisei Rivers, and find that in general the effects should be moderate, except in the case of a serious accident. The modeling tools put forth in this book can be very useful in predicting the spread and effects of such pollution events in the future—a future where climate change can play a major role in the fate of the spread of pollution.

References

Aarkrog, A. (1993) Radioactivity in polar regions: Main sources. In: *Environmental Radioactivity in the Arctic and Antarctic, Proceedings of the International Conference on Environmental Radioactivity in the Arctic and Antarctic, Kirkenes*, P. Strand and E. Holm (Eds.), pp. 15–36.

Aarkrog, A.; Simmonds, J.; Strand, P.; Christensen, G; and Salbu, B. (2000) *Radiological Assessment of Past, Present and Potential Sources to Environmental Contamination in the Southern Urals and Strategies for Remedial Measures (SUCON)*, Risø-R-1243(EN). Risø National Laboratory, Roskilde, Denmark.

Achman, D. R.; Brownawell, B. J; and Zhang, L. (1996) Exchange of polychlorinated biphenyls between sediment and water in the Hudson River Estuary. *Estuaries*, **19**(4), 950–965.

Ackers, P; and White, R. (1973) Sediment transport: New approach and analyses. *Journal of the Hydraulic Division ASCE*, **99**, 2041–2060.

Aibulatov, N. A. (1993) Radioactive graves in the Kara Sea. *Zemlya i Vselennaya*, **4**, 28–33 [in Russian].

Akers, D. W.; Tolman, E. L.; Kuan, P.; Golden, D. W; and Nishio, M. (1989). Three Mile Island Unit 2 fission product inventory estimates. *Nuclear Technology*, **87**, 205–213.

Akimenko, T. A.; Zakharova, E. A.; and Kouraev, A.V. (2001) *Hydrology of the Yenisei River System*, Working Paper. INTAS Project 97-3127.

Aldahan, A.; Alfimov, V.; and Possnert, G. (2007) ^{129}I anthropogenic budget: Major sources and sinks. *Applied Geochemistry*, **22**, 606–618.

Alfimov, V.; Aldahan, A.; and Possnert, G. (2004a) Tracing water masses with ^{129}I in the western Nordic Seas in early spring 2002. *Geophysical Research Letters*, **31**(L19305).

Alfimov, V.; Aldahan, A.; Possnert, G.; and Winsor, P. (2004b) Anthropogenic iodine-129 in seawater along a transect from the Norwegian Coastal Current to the North Pole. *Marine Pollution Bulletin*, **49**, 1097–1104.

Ali, M.; Beaumont, H. M.; Dutton, L. M. C.; Handy, B. J.; Hosty, M.; Parsons, A. F.; Nielsen, S. P.; Sivintsev, Yu.; Lystsov V.; Yefimov, E. I.; Sazukina, T.; Zheleznyak, M.; Maderich, V; and Margvelashvili, N. (1997) *Evolution of the Radiological Situation around the Nuclear Reactors with Spent Fuel which Have Been Scuttled in the Kara Sea*, Final

Report of CEC Study Contract B-6340/95/000879/MAR/C3. NNC Ltd., Warrington, U.K.

Amante, C.; and Eakins, B. W. (2009) *ETOPO1 1 Arc-Minute Global Relief Model: Procedures, Data Sources and Analysis*, NOAA Technical Memo. NESDIS NGDC-24. National Oceanic and Atmospheric Administration, Silver Springs, MD, 19 pp.

AMAP (1997) *Arctic Pollution Issues: A State of the Arctic Environment Report*. Arctic Monitoring and Assessment Program, Oslo.

AMAP (1998) *Arctic Pollution Issues*, Chapter 8: *Radioactivity*. Arctic Monitoring and Assessment Program, Oslo, 188 pp.

AMAP (2004) *AMAP Assessment 2002: Persistent Organic Pollutants (POPs) in the Arctic*. Arctic Monitoring and Assessment Program, Oslo, 310 pp.

AMAP (2009) *Arctic Pollution 2009*. Arctic Monitoring and Assessment Program, Oslo. 83 pp.; ISBN 978-82-7971-050-9.

Ambrose, R. B. *et al.* (1993) *The Water Quality Analyses Simulation Program, WASP5—Part A: Model Documentation*. Environmental Research Laboratory, Office of Research and Development, USEPA, Athens, GA.

Anon. (1987) *Guidelines for Calculating Derived Release Limits for Radioactive Material in Airborne and Liquid Effluents for Normal Operation of Nuclear Facilities*, CAN/CSAN288.1M87. Canadian Standards Association, Toronto, Ontario, Canada, 69 pp.

Anon. (1994) *Methodology for Evaluating the Radiological Consequences of Radioactive Effluents Released in Normal Operations*. National Radiological Protection Board, Chilton, U.K.

Antonov, V. S. (1957) Distribution of river water in the Arctic Seas. *Proceedings AARI*, **208**(2), 25–52 [in Russian].

Antonov, V. S. (1962) Yenisei (a hydrological–navigation description). *Proceedings AARI*, 99 pp. [in Russian].

Antonov, V. S. (1970) Rivers of Siberia and Arctic seas. *Problems of the Arctic and Antarctic*, **36/37**, 142–152 [in Russian].

Antonov, V. S.; and Maslayeva N. G. (1965) *The Ob′ River Lower Reaches and Mouth (A Hydrological and Navigation Overview)*. Gidrometeoizdat, 82 pp.

Arakawa, A.; and Lamb, V. (1977) Computational design of the basic processes of the UCLA General Circulation Model. *Methods of Computational Physics*, **17**, 174–265.

Ariathurai, R.; and Krone, R. B. (1976) Finite element model for cohesive sediment transport. *Journal of the Hydraulic Division ASCE*, **104**, HY2, 323–328.

ASCE (1996) State-of-art review of modeling transport and fate of oil spills, ASCE Committee on Modeling Oil Spills. *Journal of Hydraulic Engineering*, **122**, 594–609.

Atlas of the Arctic (1985) Moscow, 204 pp. [in Russian].

Atlas of the North Polar Region Water Balance (1996) Hydrometeoizdat, St. Petersburg, 82 pp. [in Russian].

Atlas of the Oceans. Appendix: Terminology, Definitions, Reference Tables (1980) Main Administration for Navigation and Oceanography of the U.S.S.R. Ministry for Defense, Leningrad, 156 pp. [in Russian].

Atlas of the Oceans. Arctic Ocean (1980) U.S.S.R. Navy, Leningrad, 189 pp. [in Russian].

Atom without "Classified" Stamp: Viewpoints (1992) A. Emelianenkov and V. Popov (Eds.). Moscow [in Russian].

Azarova, S. (2001) Gone with the wind. *Oil and Gas Vertical*, **13**, 81 84 [in Russian].

Bambulyak, A.; and Frantsen, B. (2005) *Oil Transportation from the Russian Part of the Barents Region: The State as of January 2005*. Svanhovd Ecological Center, Svanhovd, 91 pp. [in Russian].

Ban′ko, Yu. (2002) Oil gates of the Arctic have opened wider. *Oil of Russia*, **12**. 90–92 [in Russian].

Barron, N.; Kara, A. B.; Martin, P. J.; Rhodes, R. C.; and Smedstad, L. F. (2006) Formulation, implementation and examination of vertical coordinate choices in the Global Navy Coastal Ocean Model (NCOM). *Ocean Modelling*, **11**, 347–375.

Baskakov, G. A; and Spaikher, A. O. (1978) The boundaries and the main morphometric characteristics of the Siberian shelf. *Proceedings AARI*, **349**, 76–83 [in Russian].

BBL (2001) PCB in the Kalamazoo River: Update for decision makers. *Latest Findings for Sediment, Surface Water, and Fish at Portage Creek/Kalamazoo River Superfund Site Kalamazoo and Allegan Counties, Michigan*, Report. Allied Paper/Blasland, Bouck & Lee, Syracuse, NY.

Beckers, J.-M. (1991) Application of a 3D model to the Western Mediterranean. *Journal of Marine Systems*, **1**, 315–332.

Beloyarsk NPP (*http://decomatom.org.ru/?q=node/42*).

Belyshev, A. P.; Klevantsov, Yu. P.; and Rozhkov, V. A. (1983) *Probability Analysis of Sea Currents*. Gidrometeoizdat, Leningrad, 264 pp. [in Russian].

Benes, P.; and Cernik, M. (1990) *Model Calculations and Experimental Analysis of Transport of Radiocesium in the System Wastewater Channel, Dudvah (in Czech Republic)*, Report. Faculty of Nuclear Sciences and Physical Engineering, Czech Technical University, Prague.

Benes, P.; Cernik, M.; and Slavik, O. (1994) Modelling of migration of ^{137}Cs accidentally released into a small river. *Journal of Environmental Radioactivity*, **22**, 279–293.

Bentsen, M.; Evensen, G.; Drange, H.; and Jenkins, A. D. (1999) Coordinate transformation on a sphere using conformal mapping. *Monthly Weather Review*, **127**, 2733–2740.

Berezkin, Vs. A. (1945) *General Current Pattern of the Arctic Ocean and the Adjacent Seas*, The Reports of the Jubilee Session of the Arctic Institute (celebrating 25 years of scientific activity of the Arctic Institute). Izdatelstvo Glavsevmorputi, Leningrad, 20 pp. [in Russian].

Berezkin, Vs. A.; and Ratmanov, G. Ye. (1940) *General Current Pattern of the Arctic Ocean and the Adjacent Seas*. Hydrographic Administration of the Navy, Leningrad, 10 pp. [in Russian].

Bijker, E. W. (1968) *Some Considerations about Scales for Coastal Models with Movable Bed*, Publ. No 50. Delft Hydrographical Laboratory.

Biotesting and Forecasting of Variability of the Aquatic Ecosystems with Anthropogenic Pollution (2003) G. G. Matishov (Ed.). KNC RAN, Apatity, 468 pp. [in Russian].

Blackadar, A. (1979) High resolution models of the planetary boundary layer. In: *Advances in Environmental Science and Engineering*, J. R. Pfafflin and E. N. Ziegler (Eds.). Gordon & Breach, New York, pp. 50–85.

Bleck, R. (1978) Finite difference equations in generalized vertical coordinates. Part I: Total energy conservation. *Beiträge zur Physik der Atmosphäre*, **51**, 360–372.

Bleck, R.; and Smith, L. T. (1990) A wind-driven isopycnic coordinate model of the North and Equatorial Atlantic Ocean. 1: Model development and supporting experiments. *Journal of Geophysical Research*, **95**, 3273–3285.

Bleck, R.; Rooth, C.; Hu, D.; and Smith, L. T. (1992) Salinity-driven thermohaline transients in a wind- and thermohaline-forced isopycnic coordinate model of the North Atlantic. *Journal of Physical Oceanography*, **22**, 1486–1515.

Blindheim, J.; and Østerhus, S. (2005) The Nordic Seas, main oceanographic features. In: *The Nordic Seas: An Integrated Perspective*, H. Drange, T. Dokken, T. Furevik, R. Gerdes, and W. Berger (Eds.). American Geophysical Union, Washington, D.C., pp. 11–38.

Blumberg, A. F.; and Kantha, L. H. (1985) Open boundary condition for circulation models. *Journal of Hydraulic Engineering*, **111**, 237–255.

Blumberg, A. F.; and Mellor, G. L. (1987) A description of a three-dimensional coastal ocean circulation model. In: *Three-Dimensional Coastal Ocean Models*, N. Heaps (Ed.). American Geophysical Union, Washington, D.C., pp. 1–16.

Bobylev, L.; Kondratyev, K. Ya; and Johannessen, O. M. (2003) *Arctic Environment Variability in the Context of Global Change*. Springer/Praxis, Heidelberg, Germany/ Chichester, U.K., 470 pp.

Bobylev, K.; Plekhov, V.; Kuznetsov, Yu.; Tishkov, V.; and Stepanov, A. (2000) The influence of the land-based sources of radionuclides on radioactive contamination of the Kara Sea. *Third Russian Conference on Radiochemistry, Radiochemistry-2000*, Thesis of Reports. St. Petersburg, Russia.

Bobylev, K.; Volkov, V. A.; Johannessen, O. M.; Pettersson, L. H; and Stepanov, A. (2002) The Geographical Information System (GIS) application for study, assessments and forecasting of a radionuclide distribution processes in the Kara Sea. *Proceedings VIII St. Petersburg International Conference: Regional Informatics—2002* [in Russian].

Bohmer, N. (1994) Atomsvineri fra Storbritannia. *Bellona Magasin*, June/July, 30–31 [in Norwegian].

Booth, R. S. (1975). *A System Analysis Model for Calculating Radionuclide Transport between Receiving Water and Bottom Sediments*, ORNL-TM-4751. Oak Ridge National Laboratory, 37 pp.

Borzilov, V. A.; Sedunov, Yu. S.; Novittsky, M. A. *et al.* (1989). Forecasting of secondary radioactive contamination of the rivers within a 30 kilometer zone of the Chernobyl NPP. *Meteorologia i Gidrologia*, **2**, 5–13 [in Russian].

Boundaries of the Oceans and the Seas (1960) Izdatelstvo Glavsevmorputi, Leningrad, 402 pp. [in Russian].

Bradley, D. J. (1997) *Third International Conference on Environmental Radioactivity in the Arctic, Tromsø, Norway, June 1–5*.

Bradley, D. J.; and Jenquin, U. P. (1995) Radioactive inventories and sources for contamination of the Kara Sea. In: *Environmental Radioactivity in the Arctic*, P. Strand and A. Cooke (Eds.), Oslo, Norway, pp. 51–56.

Brovchenko, I.; and Maderich, V. (2005) Oil dispersion by breaking waves and currents. Modeling of transport of spilled oil in wind and wave driven sea. *Sea Technology*, **46**(4), 17–21.

Brovchenko, I.; Kuschan, A.; Maderich, V.; Shliakhtun, M.; Koshebutsky, V.; and Zheleznyak, M. (2003) Model for oil spill simulation in the Black Sea. *Proc. Third Int. Conf. Oil Spills, Oil Pollution and Remediation, September 16–18*. Bogazici University, Istanbul, pp. 101–112.

Brown, J.; Iospje, M.; Kolstad, K.; Lind, B.; Rudjord, A. L. *et al.* (2002) Temporal trends for ^{99}Tc in Norwegian coastal environments and spatial distribution in the Barents Sea. *Journal of Environmental Radioactivity*, **60**, 49–60.

Bryan, F. (1987). Parameter sensitivity of primitive equation ocean general circulation models. *Journal of Physical Oceanography*, **17**, 970–985.

Burchard, H. (2002) *Applied Turbulence Modelling in Marine Waters*. Springer-Verlag, Berlin, 252 pp.

Burkov, A.; Gerasimenko, A.; Jobson, H.; Novitsky, M.; Voszhennikov, O.; and Zheleznyak, M. (1992). *Methods of Calculating and Forecasting Pollutant Transport in Water Bodies: Hydrological Aspects of Accidental Pollution of Water Bodies*, Ch. 6, Operational

Hydrological Report No. 37, WMO-No. 754. World Meteorological Organization, Geneva, pp. 65–92.

Callender, E.; and Robbins, J. (1993) Transport and accumulation of radionuclides and stable elements in a Missouri River reservoir. *Water Resources Research*, **29**, 1787–1804.

Canuto, V. M.; Howard, A.; Cheng, Y.; and Dubovikov, M. S. (2001) Ocean turbulence. Part I: One-point closure model-momentum and heat vertical diffusivities. *Journal of Physical Oceanography*, **31**, 1413–1426.

Carroll, J.; and Harms, I. H. (1999) Uncertainty analysis of partition coefficients in a radionuclide transport model. *Water Research*, **33**, 2617–2626.

Chang, H. H. (1988) *Fluvial Processes in River Engineering*. John Wiley & Sons, New York, 105 pp.

Chapman, B. M. (1982) Numerical simulation of the transport and speciation of nonconservative chemical reactions in rivers. *Water Resources Research*, **18**, 155–167.

Chemical Processes in the Ecosystems of the Northern Seas (Hydrochemistry, Geochemistry, Oil Pollution) (1997) I. A. Shparkovsky (Ed.). Publ. KNC RAN, Apatity, 404 pp. [in Russian].

Chen, Y. H. (1988) Development of a quasi-nonequilibrium reservoir sedimentation model: REDSED. In: *Twelve Selected Computer Stream Sedimentation Models Developed in the United States*, S-S. Fan (Ed.). Federal Energy Regulatory Commission, Washington, D.C., pp. 491–510.

CHR/KHR (1978) *Das Rheingebiet, Hydrologische Monographie/Lebassin du Rhin, Monographie hydrologique*. Staatsuitgeverij Den Haag, ISBN 90-12-775-0 [in German and French].

Chumichev, V. B. (1995) Sr-90 discharge with main rivers of Russia into the Atlantic Ocean during 1961–1990. In: *Environmental Radioactivity in the Arctic*, P. Strand and A. Cooke (Eds.). Norwegian Radiation Protection, Osteras, Norway, pp. 79–83.

Chumichev, V. B.; and Demyanchenko, Z. A. (1993) Contamination with ^{90}Sr and discharge of radionuclides for the Techa, Karabolka (1974–1990) and Ob′ (1961–1990) Rivers. In: *Environmental Radioactivity in the Arctic and Antarctic: Proc. of the International Conference on Environmental Radioactivity in the Arctic and Antarctic, Kirkenes*, P. Strand and E. Holm (Eds.), pp. 111–114.

Cochrane, T.; Arkin, W.; Norris R.; and Sands, J. (1989) *Soviet Nuclear Weapons: Nuclear Weapons Databook*, Vol. IV. Natural Resources Defense Council/Harper & Row, New York.

Cochrane, T.; Norris, R.; and Bukharin, O. (1995) *Making the Bomb: From Stalin to Yeltsin*. Westview Press, Boulder, CO.

Codell, R. B.; Key, K. T.; and Whelan, G. (1982) *Collection of Mathematical Models for Radionuclide Dispersion in Surface Water and Ground Water*, NUREG0868. U.S. Nuclear Regulatory Commission, Washington, D.C., 271 pp.

Colon, R.; and McMahon, G. F. (1987) BRASS model: Application to savannah river system reserviors. *Journal of Water Resources Planning and Management*, **113**(2), 177–190.

Cooper, L.; Beasley, T.; Aagaard, K.; Kelley, J.; Larsen, I.; and Grebmeier, M. (1999) Distribution of nuclear fuel reprocessing tracers in the Arctic Ocean: Indication of Russian river influence. *Journal of Marine Research*, **57**, 715–738.

Coppe, P. (1993) Estimated models of impact used by Electricité de France within the framework of the French and European regulations. *Radioprotection*, Special Issue. *Proc. of the Joint Seminary from September 15th to 18th, 1992, Fribourg, Switzerland*, pp. 185–189.

Cross-border Environmental Problems Emanating from Defense-related Installations and Activities (1995) Vol. 1: *Radioactive Contamination, Final Report, Phase 1: 1993–1995*, Report No. 204. North Atlantic Treaty Organisation, 306 pp.

Cunge, J. A.; Holly, F. M.; and Verwey, A. (1986) *Practical Aspects of Computational River Hydraulics*. Pitman, London, 250 pp.

Dahle, S.; Matishov, G. G.; Savinova, T. N.; Killie, B.; and Savinov, V. M. (1998) Oil hydrocarbons in bottom sediments of the White Sea. In: *Problems of Study, Rational Use and Protection of Natural Resources of the White Sea*. St. Petersburg, pp. 51–52 [in Russian].

Dahlgaard, H. (1995) Transfer of European coastal pollution to the Arctic: Radioactive tracers. *Marine Pollution Bulletin*, **31**, 3–7.

Dauwalter, V. A. (1988) Heavy metals in bottom sediments of the lake–river system Lake Inari–Paswik River. *Water Resources*, **25**(4), 494–500 [in Russian].

Dauwalter, V. A. (1997) Pollution of bottom sediments of the Paswik River basin with heavy metals. *Geoecology*, **6**, 43–53 [in Russian].

Degteva, M. O.; Kozheurov, V. P.; and Vorob′yova, M. I. (1992) Reconstruction of doses for population caused by radioactive waste discharges in Techa River. *Atomic Energy*, **72**(4), 386–390 [in Russian].

Degteva, M. O.; Kozheurov, V. P.; and Vorob′yova, M. I. (1994) A general approach to dose reconstruction in the population exposed as a result of the release of radioactive wastes into the Techa River. *Science of the Total Environment*, **142**, 49–61.

Delhez, E. J.; and Deleersnijder, E. (2002) The concept of age in marine modelling, II: Concentration distribution function in the English Channel and the North Sea. *Journal of Marine Systems*, **31**, 279–297.

Delhez, E.; Campin, J.; and Deleersnijder, E. (1999) Toward a general theory of the age in ocean modelling. *Ocean Modelling*, **1**, 17–27.

Density of the total precipitation flow at the earth's surface, maps (1996) In: *Atlas of the Water Balance of the Northern Polar Area*, N. N. Bryazgin (Ed.). Gidrometeoizdat, St. Petersburg, pp. 3–14.

De Szoeke, R. A. (2000) Equations of motion using thermodynamic coordinates. *Journal of Physical Oceanography*, **30**, 2814–2829.

Deutsches Gewässerkundliches Jahrbuch des Rheingebiets (1989) Teil 1, 2, 3 [in German].

Di Toro, D. M. (2000) *Sediment Flux Modeling*. John Wiley & Sons, 615 pp.

Dmitriyev, A. A. (1994) *Variability of Atmospheric Processes in the Arctic and Its Account in Long-range Forecasts*. Gidrometeoizdat, St. Petersburg, 208 pp. [in Russian].

Dmitriyev, A. A.; Seltser, P. A.; Kondratiuk, S. I.; and Kuchin, V. A. (1989) *Macroscale Atmospheric Processes and Monthly Mean Weather Forecasts in the Arctic*. Gidrometeoizdat, St. Petersburg, 256 pp. [in Russian].

Doncheva, A. V. (1975) *Impact of Metallurgic Production of PTK*. Nauka, Moscow [in Russian].

Doronin, Yu. P. (1986) *Regional Oceanography*. Gidrometeoizdat, St. Petersburg, 304 pp. [in Russian].

Doronin, Yu. P.; and Ivanov, V. V. (1997) Methods and results of modelling hydrological processes in stratified Arctic estuaries. *Proc. Russian–Norwegian Workshop-95: Nature Conditions of the Kara and Barents Seas*, V. Volkov *et al.* (Eds.), Rapportserie No. 97. Norsk Polarinstitutt, Oslo, pp. 209–213.

Drange, H. (1994) *An Isopycnic Coordinate Carbon Cycle Model for the North Atlantic*, Technical Report. Nansen Environmental and Remote Sensing Center, Bergen, Norway.

Drange, H.; and Simonsen, K. (1996) *Formulation of Air–Sea Fluxes in the ESOP2 Version of MICOM*, Technical Report 125, Nansen Environmental and Remote Sensing Center, Solheimsviken, Norway.

Drange, H.; Dokken, T.; Furevik, T.; Gerdes, R.; and Berger, W. (Eds.) (2005a) *The Nordic Seas: An Integrated Perspective*, Geophyisical Monogr. 158. AGU, Washington, D.C., 366 pp.

Drange, H.; Dokken, T.; Furevik, T.; Gerdes, R.; Berger, W.; Nesje, A.; Orvik, K. A.; Skagseth, Ø.; Skjelvan, I.; and Østerhus, S. (2005b) The Nordic Seas: An overview. In: *The Nordic Seas: An Integrated Perspective*, H. Drange, T. Dokken, T. Furevik, R. Gerdes, and W. Berger (Eds.), Geophysical Monogr. 158. AGU, Washington, D.C., 366 pp.

Drange, H.; Gerdes, R.; Gao, Y.; Karcher, M.; Kauker, F.; and Bentsen, M. (2005) Ocean general circulation modelling of the Nordic Seas. In: *The Nordic Seas: An Integrated Perspective*, H. Drange, T. Dokken, T. Furevik, R. Gerdes, and W. Berger (Eds.), American Geophysical Union, Washington, D.C., pp. 199–220.

Dukowicz, J.; and Baumgardner J. (2000) Incremental remapping as a transport/advection algorithm. *Journal of Computational Physics*, **160**, 318–335.

Dumik, V. P.; and Evseev, V. F. (1993) Ecological safety problems connected at North Range (Island of Novaya Zemlya). In: *Environmental Radioactivity in the Arctic and Antarctic: Proceedings of the International Conference on Environmental Radioactivity in the Arctic and Antarctic, Kirkenes*, P. Strand and E. Holm (Eds.), pp. 357–360.

Dunayev, N. N. (1990) Physical properties of bottom soil of the shelf of the Kara Sea, In: *Modern Sedimentation Processes on the Shelf of the Global Ocean*. Nauka, Moscow, pp. 121–126 [in Russian].

Dunster, H. J.; Howell, H.; and Templeton, W. L. (1958) *District Surveys Following the Windscale Accident, October 1957: Proc. Second Intern. Conf. on the Peaceful Uses of Atomic Energy*, Part 19, pp. 296–308, U.N., Geneva.

EC (1990) *The Radiological Exposure of the Population of the European Community from Radioactivity in North European Marine Waters: Project MARINA*, a report by a group of experts convened by the Commission of the European Communities. European Commission, Luxemburg (EUR 12483 EN).

EC (1996) *Council Directive of 13 May 1996 Laying down Basic Safety Standards for the Health Protection of the General Public and Workers against the Dangers of Ionizing Radiation*, Directive 96/29/Euratom. European Commission, Luxembourg.

Ehrhardt, J. (1997) The RODOS system: Decision support for off-site emergency management in Europe. *Radiation Protection Dosimetry*, **73**, Nos. 1/4, 35–40.

Elder, J. W. (1958) The dispersion of marked fluid in turbulent shear flow. *Journal of Fluid Mechanics*, **8**, 33–40.

Eldevik, T.; Straneo, F.; Sando, A.; and Furevik, T. (2005) Pathways and export of Greenland Sea Water. In: *The Nordic Seas: An Integrated Perspective*, H. Drange, T. Dokken, T. Furevik, R. Gerdes, and W. Berger (Eds.). American Geophysical Union, Washington, D.C., pp. 89–104.

Engelund, F.; and Fredsoe, J. (1976) A sediment transport model for straight alluvial channels. *Nordic Hydrology*, **7**, 293–306.

England, M. (1995) The age of water and ventilation timescales in a global ocean model. *Journal of Physical Oceanography*, **25**, 2756–2777.

Environmental Status of the Varanger–Kola Coastal Area (2003) Report No. APN-414.2127. Akvaplan-niva, Tromsø, Norway, 203 pp.

Eraslan, A. H.; Akin, E. J.; Barton, J. M.; Bledsoe, J. L.; Cross, K. E.; Diament, H.; Fields, D. E.; Fisher, S. K.; Harris, J. L.; Hetrick, D. M.; Holdeman, J. T.; Kim, K. K.; Lietzke,

M. H.; Park, J. E.; Pattersom, M. R.; Sharp, R. D; and Thomas, B. (1977) *Development of a Unified Transport Approach for the Assessment of Power-plant Impact*, ORNL/NUREG/TM-89. Oak Ridge National Laboratory, Oak Ridge, TN, 409 pp.

Evdokimova, Z. A.; Zhilina, N. I.; Kiselev, V. P.; Pechenova, O. I.; Semin, N. N.; Tokarchuk, D. N.; Tokarchuk, A. N.; Yurchenko, A. S.; Dynnik, A. Yu; and Rogachev A. V. (2002) *Structure of the E-maps Archive of the Nuclear Safety Institute (IBRAE) of the Russian Academy of Sciences*, Preprint No. IBRAE-2002-04. Moscow, 60 pp. [in Russian]

FAS (2000a). Available at *http://www.fas.org/man/dod-101/sys/ship/row/rus/627.htm*

FAS (2000b) *Project 949 Granit/Oscar I* and *Project 949A Antey/Oscar II*. Available at *http://www.fas.org/nuke/guide/russia/theater/949.htm*

Fields, D. E. (1976) *CHNSED: Simulation of Sediment and Trace Contaminant Transport with Sediment/Contaminant Interaction*, ORNL/NSF/EATC-19. Oak Ridge National Laboratory, Oak Ridge, TN, 170 pp.

Fields, D. E. (1977) *HOTSET: A One-dimensional, Tidal-transient, Discrete Element Model for Simulating Hydrodynamic Conditions and Absorbed and Dissolved Radioisotope Concentration in Controlled Rivers and Estuaries*, ORNL/CSD-16. Oak Ridge National Laboratory, Oak Ridge, TN, 227 pp.

Fischer, H. B. (1973) *Longitudinal Dispersion and Turbulent Mixing in Open-channel Flow, Annual Review Fluid Mechanics*. University of California.

Flather, R. A. (1976) A tidal model of northwest European continental shelf. *Mémoires de la Société Royale des Sciences de Liège*, **6**, 141–164.

Fletcher, C. A. J. (1988) *Computation Techniques for Fluid Dynamics: Fundamental and General Techniques*. Springer-Verlag, Berlin, 401 pp.

Føyn, L.; and Sværen, L. (1995) *Distribution and Sedimentation of Radionuclides in the Barents Sea*, Report: ICES C. M. Institute of Marine Research, Bergen, Norway.

Frantsen, B.; and Bambulyak, A. (2003) *Oil Transportation from the Russian Part of the Barents Region: The State as of 1 July 2003*. Svanhovd Ecological Center. Available at *www.svanhovd.no*.

Friedrich, H.; and Levitus, S. (1972) An approximation to the equation of state for sea water, suitable for numerical ocean models. *Journal of Physical Oceanography*, **2**, 514–517.

Furevik, T.; Bentsen, M.; Drange, H.; Johannessen, J. A.; and Korablev, A. (2002) Temporal and spatial variability of the sea surface salinity in the Nordic Seas. *Journal of Geophysical Research*, **107**, 8009, doi: 10.1029/2001JC001118.

Furevik, T.; Bentsen, M.; Drange, H.; Kindem, I.; Kvamstø, N.; and Sorteberg, A. (2003) Description and validation of the Bergen Climate Model: ARPEGE coupled with MICOM. *Climate Dynamics*, **21**, 27–51, doi: 10.1007/s00382-003-0317-5.

Galperin, B.; Kantha, L. H.; Hassad, S.; and Rosati, A. (1988) A quasi-equilibrium turbulent energy model for geophysical flows. *Journal of Atmospheric Sciences*, **45**, 55–62.

Gao, Y.; Drange, H.; Bentsen, M.; Johannessen, O. M; and Pettersson, L. H. (2002) *Annual Report of RADARC*, WP5 (0.6), Technical Report No. 210. Nansen Environmental and Remote Sensing Center, Bergen, Norway.

Gao, Y.; Drange, H.; and Bentsen, M. (2003) Effects of diapycnal and isopycnal mixing on the ventilation of CFCs in the North Atlantic in an isopycnic coordinate OGCM. *Tellus*, **55B**, 837–854.

Gao, Y.; Drange, H.; Bentsen, M.; and Johannessen, O. M. (2004) Simulating transport of non-Chernobyl ^{137}Cs and ^{90}Sr in the North AtlanticArctic region. *Journal of Environmental Radioactivity*, **71**, 1–16, doi: 10.1016/S0265-931X(03)00108-5.

Gao, Y.; Drange H.; Bentsen M.; and Johannessen O. M. (2005) Tracer-derived transient time of the eastern waters in the Nordic Seas. *Tellus*, **57B**, 332–340.

Gao, Y.; Drange, H.; Johannessen, O. M.; and Petersson, L. H. (2009) Source and pathways of ^{90}Sr in the North Atlantic/Arctic region: Present day and global warming. *Journal of Environmental Radioactivity*, **100**, 375–395.

Garcon, V. C.; Baumert, H. W.; Schrimpf, W; and Woods, J. D. (1993) Fluctuations: A task package for the physicists. In: *Towards a Model of Ocean Biochemical Processes*, G. T. Evans and M. J. R. Fasham (Eds.). Springer-Verlag, Berlin, pp. 47–70.

Gargett, A. (1984) Vertical eddy in the ocean interior. *Journal of Marine Research*, **42**, 359–393.

Gascard, J. C.; Raisbeck, G.; Sequeira, S.; Yiou, F.; and Mork, K. A. (2004) The Norwegian Atlantic current in the Lofoten Basin inferred from hydrological and tracer data (^{129}I) and its interaction with the Norwegian coastal current. *Geophysical Research Letters*, **31**(1), L01308, doi: 10.1029.

Gaspar, P. (1988) Modeling the seasonal cycle of the upper ocean. *Journal of Physical Oceanography*, **18**, 161–180.

GEBCO (2003) *GEBCO Digital Atlas*, Centenary Edition. British Oceanographic Data Center (CD-ROM).

Gedeonov, L. I. *et al.* (1998) *Radioactive Contamination of the River System Techa–Iset–Tobol–Irtysh–Ob′*, Preprint RI-251. Khlopin Radium Institute, St. Petersburg.

Geoecology of the Shelf and Shores of the Seas of Russia (2001) N. A. Aibulatov (Ed.). Noosfera Publ., Moscow, 428 pp. [in Russian].

Gerland, S.; Lind, B.; Dowdall, M.; Karcher, M.; and Kolstad, A. K. (2003) ^{99}Tc in seawater in the West Spitsbergen Current and adjacent areas. *Journal of Environmental Radioactivity*, **69**, 119–127.

Getzlaff, K.; Boning, C.; and Deng, J. (2006) Lagrangian perspectives on deep water export from the Subpolar North Atlantic. *Geophysical Research Letters*, **33**(L21S08).

Gladkov, G. A; and Sivinzev, Yu. V. (1994) Radiation level in the area of sunken nuclear submarine *Komsomoletz*. *Atomic Energy*, **77**(5), 379–385.

Gofman, D.; Lyashenko, G.; Marinets, A.; Mezhueva, I.; Shepeleva, T.; Tkalich, P.; and Zheleznyak, M. (1996) Implementation of the aquatic radionuclide transport models RIVTOX and COASTOX into the RODOS System. In: *The Radiological Consequences of the Chernobyl Accident: Proceedings of the First International Conference. Minsk, Belarus, March 18–22*, A. Karaoglou, G. Desmet, G. N. Kelly, and H. G. Menzel (Eds.). European Commission. Luxembourg, pp. 1181–1184.

Golubeva, N. I. (2002) Results of the studies of pollution of the atmospheric air by heavy metals in the open areas of the Barents Sea In: *Current Problems of Oceanography of the Shelf Seas of Russia: Abstracts of the International Conference, Rostov-on-Don, June 13–15, Murmansk*, pp. 39–41 [in Russian].

Golubeva, N. I. (2007) Pollution of the Arctic atmosphere by toxic heavy metals. In: *Biology and Oceanography of the Northern Sea Route: The Barents and Kara Seas*. Nauka, Moscow, pp. 173–190 [in Russian].

Great Soviet Encyclopedia (1955) *Yenisei*, Vol. 16. Russian Academy of Sciences, Moscow [in Russian].

Grell, G.; Dudhia, A. J.; and Stauffer, D. R. (1994) *A Description of the Fifth-generation Penn State/NCAR Mesoscale Model (MM5)*, NCAR Technical Note, NCAR/TN-398 + STR.

Guidebook for Hydrological Calculations when Designing Water Reservoirs (1983) Hydrometeoizdat, Leningrad, 283 pp. [in Russian].

Gusev, A.; Mantseva, E.; Shatalov, V.; and Strukov, B. (2005) *Regional Multicompartment Model MSCE-POP EMEP/MSC-E*, Technical Report 5/2005. Available at *http://www.msceast.org/events/review/pop_description.html*

Handbook of Parameter Values for the Prediction of Radionuclide Transfer in Environments (1994) IAEA Technical Report No. 364. Vienna, 284 pp.

Handling of Radioactive Wastes and Spent Nuclear Fuel, Their Utilization and Disposal in 1996–2005 (1995) Russian Federal Program, Approved by the Government of Russian Federation on 23.10.95 [in Russian].

Haney, R. L. (1991) On the pressure gradient force over steep topography in sigma co-ordinate ocean models. *Journal of Physical Oceanography*, **21**, 610–619.

Hansen, B.; and Østerhus, S. (2000) North Atlantic–Nordic Seas exchanges. *Progress in Oceanography*, **45**(2), 109–208.

Harder, M. (1996) Dynamik, Rauhigkeit und Alter des Meereises in der Arktis. Ph.D. thesis, Alfred-Wegner-Institut für Polar- und Meeresforschung, Bremerhaven, Germany.

Harms, I. H.; Hübner, U.; Backhaus, J. O.; Kulakov, M.; Stanovoy, V.; Stepanets, O. L.; Kodina, L.; and Schlitzer, R. (2003) Salt intrusions in Siberian river estuaries: Observations and model experiments in Ob′ and Yenisei. In: *Siberian River Runoff in the Kara Sea: Characterisation, Quantification, Variability, and Environmental Significance*, R. Stein, K. Fahl, D. K. Fütterer, E. Galimov, and O. Stepanets (Eds.). Elsevier, pp. 27–46.

Hatun, H.; Sando, A.; Drange, H.; and Bentsen, M. (2005a) Seasonal to decadal temperature variations in the Faroe–Shetland inflow waters. In: *The Nordic Seas: An Integrated Perspective*, H. Drange, T. Dokken, T. Furevik, R. Gerdes, and W. Berger (Eds.), American Geophysical Union, Washington, D.C., pp. 239–250.

Hatun, H.; Sandø, A.; Drange, H.; Hansen, B., and Valdimarsson, H. (2005b) Influence of the Atlantic subpolar gyre on the thermohaline circulation. *Science*, **309**, 1841–1844.

Havno, K.; Madsen, M. N.; and Dorge, J. (1995) MIKE 11: A generalized river modelling package. In: *Computer Models of Watershed Hydrology*, V. P. Singh (Ed.). Water Resources Publications, Highlands Ranch, CO, pp. 733–782.

Heling, R.; Zheleznyak, M.; Raskob, W.; Popov, A.; Borodin, R.; Gofman, D.; Lyashenko, G.; Marinets, A.; Pokhil, A.; Shepeleva, T.; and Tkalich, P. (1997) Overview of modelling of hydrological pathways in RODOS. *Radiation Protection Dosimetry*, **73**, Nos. 1/4, 67–70.

Hibler, W. D. III (1979) A dynamic thermodynamic sea ice model. *Journal of Physical Oceanography*, **9**, 815–846.

Hibler, W. D. III (1980) Modeling of a variable thickness sea ice cover. *Monthly Weather Review*, **108**, 1943–1973.

Hofer, H.; and Bayer, A. (1993) Calculation of radionuclide dispersion in flowing waters with a dynamic model. *Kerntechnik*, **58**(3), 164–169.

Høibråten, S.; Thresen, P. E.; and Haugan, A. (1997) The sunken submarine *Komsomolets* and its effects on the environment. *Science of the Total Environment*, **202**, 67–78.

Holly, F. M. (1987) Dispersion in rivers and coastal waters. 1: Physical principles and dispersion equations. *Developments in Hydraulic Engineering*, Vol. 3. Elsevier, London.

Holly, F. M.; Yang, J. C.; Schwarz, P.; Schaefer, J.; Hsu, S. H; and Einhellig, R. (1990) *CHARIMA: Numerical Simulation on Unsteady Water and Sediment Movement in Multiply Connected Networks of Mobile-bed channels*, IIHR Report No. 343. Iowa Institute of Hydraulic Research, University of Iowa, Iowa City, IA, 327 pp.

Hu, D. (1997) Global-scale water masses, meridional circulation, and heat transport simulated with a global isopycnal ocean model. *Journal of Physical Oceanography*, **27**, 96–120.

Hubel, K. (1989) *Erstellung eines dynamischen Modells zur Berechnung der Strahlenexposition über den Wasserpfad bei stehenden und fliesenden gewassern.* Jahresbericht Bayersche Landesanstalt für Wasserforschung (St.Sch.1071), Munchen [in German].

Hurrell, W. (1995) Decadal trends in the north Atlantic oscillation: Regional temperatures and precipitation. *Science*, **269**, 676–679.

Hydrologic Engineering Center (1982) *HEC-2 Water-surface Profiles User's Manual*. U.S. Army Corps of Engineering, Davis, CA, 402 pp.

Hydrologic Engineering Center (1977) *HEC-6 Scour and Deposition in Rivers and Reservoirs*. U.S. Army Corps of Engineering, Davis, CA, 295 pp.

IAEA (1985a) *Hydrological Dispersion of Radioactive Material in Relation to Nuclear Power Plant Siting*, Safety Series N50SGS6. International Atomic Energy Agency, Vienna, 116 pp.

IAEA (1985b) *Sediment Kd's and Concentration Factors for Radionuclides in the Marine Environment*, Technical Reports Series No. 247. International Atomic Energy Agency, Vienna.

IAEA (1989) *Evaluation of the Reliability of Predictions Made Using Environmental Transfer Models*, Safety Series No. 100. International Atomic Energy Agency, Vienna.

IAEA (1994) *Handbook of Parameter Values for the Prediction of Radionuclide Transfer in Temperate Environments*, Technical Reports Series No. 364. International Atomic Energy Agency, Vienna.

IAEA (1995) *Sediment Distribution Coefficients and Concentration Factors for Biota in the Marine Environment*, Technical Reports Series No. 422. International Atomic Energy Agency, Vienna..

ICRP (1990) *Recommendations of the International Commission on Radiological Protection*, ICRP Publication 60, Ann. ICRP 21, Nos. 1/3. Pergamon Press, Oxford.

Ilyin, G. V. (2005) Regularities of accumulation and structure of oil hydrocarbons in sediments of the Kara Sea: Theory and practice of complex marine studies in the interest of economy and safety of the Russian North. *Abstracts: International Scientific–Practical Conference, Murmansk, March 15–17*. KNC RAN Publ., Apatity, pp. 60–62 [in Russian].

Ilyin, G. V. (2007) Modern background and regularities of the distribution of metals and microelements in water of the Barents Sea. *Complex Study of the Processes, Characteristics and Resources of the Russian Seas of the North-European Basin*, Issue 2. KNC RAN Publ., Apatity, pp. 517–530 [in Russian].

Ilyin, G. V. (2007a) Composition of hydrocarbons and regularities of their accumulation in sediments of the Kara Sea. In: *Biology and Oceanography of the Northern Sea Route: The Barents and Kara Seas*. Nauka, Moscow, pp. 190–198 [in Russian].

Ilyin, G. V. (2007b) Some peculiarities of accumulation of aromatic hydrocarbons in bottom sediments of the Laptev Sea. In: *Biology and Oceanography of the Northern Sea Route: The Barents and Kara Seas*. Nauka, Moscow, pp. 198–207 [in Russian].

Ilyin, G. V.; and Petrov, V. S. (1994) Studies of the levels of heavy metals in soils. *Ecological Studies of the Zone of Industrial Development of the Shtokman Gas-condensate Field on the Shelf of the Barents Sea*. KNC RAN Publ., Apatity, pp. 44–48 [in Russian].

Ilyin, G. V.; and Shavykin, A. A. (2007) Development of the northern routes for transportation of oil–gas raw materials and the related ecological problems. *Scientific Notes of MGPU: Geographical Science*, Issue 2. Publ. MGPU, Murmansk, pp. 72–91 [in Russian].

Ilyin, G. V.; Shchekaturina, T. L.; and Petrov, V. S. (1996a) Comparative characteristics of the hydrocarbon composition of bottom sediments of the southern Barents Sea. *Okeanologiya*, **36**(5), 787–792 [in Russian].

Ilyin, G. V.; Shchekaturina, T. L.; and Petrov, V. S. (1996b) Oil hydrocarbons in bottom sediments. *Ecosystems, Bioresources and Anthropogenic Pollution of the Pechora Sea*. KNC RAN Publ., Apatity, pp. 107–113 [in Russian].

Ilyin, G. V.; Lucchitta, A.; and Ponitz, P. (1998a) Comparative description of the oxygen regimes in the waters of the continental slopes in the Barents Sea and the Laptev Sea. *Berichte zur Polarforschung*, No. 287, pp. 19–30.

Ilyin, G. V.; Schekaturina, T. L.; and Petrov, V. S. (1998b) Oil hydrocarbons in the Laptev Sea bottom sediments. *Berichte zur Polarforschung*, Bremerhaven, No. 287, 30–40.

Ilyin, G. V.; Matishov, D. G.; and Kasatkina, N. Ye. (2004) Chemical pollution and accumulation of radionuclides in the elements of the ecosystem of the Barents and White Seas. *Complex Study of the Processes, Characteristics and Resources of the Russian Seas of the North-European Basin*. KNC RAN Publ., Apatity, pp. 436–459 [in Russian].

Ilyina, T.; Pohlmann, T.; Lammel, G.; and Sündermann, J. (2006) A fate and transport ocean model for persistent organic pollutants and its application to the North Sea. *Journal of Marine Systems*, **63**, 1–19.

Ivanov, G. I. (2002) *Methodology and the Results of Eco-geochemical Studies of the Barents Sea*, Rotaprint. VNII Okeangeologiya, St. Petersburg, 153 pp. [in Russian].

Ivanov, V. V. (1976) Freshwater balance of the Arctic Ocean. *Proceedings AARI*, **323**, 138–144 [in Russian].

Ivanov, V. V. (1995) The studies of the Kara mouth region, priorities of studies for the modeling of water mass dynamics, thermohaline and ice processes in the Kara mouth region. *NDPE Workshop: Modeling Requirements for Water Mass Dynamics, Ice and River Transports in the Kara Sea, Tjøne, Norway, 26–30 June*, pp. 70–84.

Ivanov, V. V. (1997) Inflow and spreading of river water in the Kara sea. In: *Proceedings of the Scientific Seminar: Nature Conditions of the Kara and Barents Seas, St. Petersburg, 1995*. Oslo, pp. 214–220.

Ivanov, V. V.; and Nikiforov, Ye. G. (1976) Methods for estimating possible changes in the hydrological regime of the Kara Sea as affected by interbasin river runoff removal. *Proceedings AARI*, **314**, pp. 176–182 [in Russian].

Ivanov, V. V.; and Osipova, I. V. (1972) An outflow of the Ob′ waters to the sea and its multi-year variability. *Proceedings AARI*, **297**, 86–91.

Ivanov, V. V.; and Sviyatsky, A. Z. (1987) Numerical simulation of seawater intrusion in a river mouth on the seasonal temporal scale. *Water Resources*, **5**, 116–122.

Izrael, Yu. A. (1998) Radioactive contamination of the Earth surface. *Vestnik RAS*, **68**(10), 15–23 [in Russian].

Janjic, Z. I. (1977) Pressure gradient force and advection scheme used for forecasting with steep and small scale topography. *Beiträge zur Physik der Atmosphäre*, **50**, 186–199.

Jannasch, H. W.; Honeyman, B. D.; Balistrieri, L. S.; and Murray, J. W. (1988) Kinetics of trace element uptake by marine particles. *Geochim. Cosmochim. Acta*, **52**, 567–577.

JNRI (1997) *Sources Contributing to Radioactive Contamination of the Techa River and Areas Surrounding the Majak Production Association, Urals, Russia*. Joint Norwegian–Russian Expert Group, Østerås, Norway.

Jobson, H. E. (1989) *User's Manual for a Flow Model Based on the Diffusion Analogy*. U.S. Geological Survey, Reston, VA, 56 pp.

Johannessen, O. M. (1986) Brief overview of the physical oceanography. In: *The Nordic Seas*, B. G. Hurdle (Ed.). Springer–Verlag, New York.

Johannessen, O. M. (2008) Decreasing Arctic sea ice mirrors increasing CO_2 on decadal time scale. *Atmospheric and Oceanic Science Letters*, **1**, 51–56.

Johannessen, O. M.; Sandven, S.; and Johannessen, J. A. (1991) Eddy-related winter convection in the Boreas Basin. In: *Deep Convection and Deep Water Formation in the Oceans*, J. C. Gascard and P. C. Chu (Eds.), Oceanographic Series. Elsevier, pp. 87–105.

Johannessen, O. M.; Muench, R.; and Overland, J. E. (Eds.) (1994) *The Polar Oceans and Their Role in Shaping the Global Environment*, The Nansen Centennial Volume, Geophysical Monograph 85. American Geophysical Union, Washington, D.C., ISBN 0-87590-042-9.

Johannessen, O. M.; Pettersson, L. H.; Bobylev, L. P.; Neelov, I. A.; Kouraev, A. V.; Shuchman, R. A.; Volkov, V. A.; Donchenko, V. K.; Rastoskuev, V. V; and Hasselman, K. (1996) *Observations and Modelling of Transport and Dilution of Radioactive Waste and Dissolved Pollutants in the Kara Sea*, Final Report Phase I, Technical Report No. 109 (Contract No. INTAS-93-814). Nansen Environmental and Remote Sensing Center, Bergen, Norway, 53 pp.

Johannessen, O. M.; Pettersson, L. H.; Bobylev, L. P.; Neelov, I. A.; Kouraev, A. V.; Volkov, V. A.; Donchenko, V. K.; Rastoskuev, V. V.; Tishkov, V. P.; Stepanov, A. V; and Hasselman, K. (1997) *Observations and Modelling of Transport and Dilution of Radioactive Waste and Dissolved Pollutants in the Kara Sea*, Final Report Phase II, Technical Report No. 144 (Contract No. INTAS-93-814 ext.). Nansen Environmental and Remote Sensing Center, Bergen, Norway, 164 pp.

Johannesen, O. M.; Pettersson, L. H.; Gao, Y.; Nielsen, S. P.; Borghuis, S.; Strand, P.; Reiersen, L. O.; Bobylev, L. P.; Volkov, V.; Neelov, I.; Stepanov, A.; Bobylev, K.; Zheleznyak, M.; and Maderich, V. (2002) Simulation for potential radioactive spreading in the 21 century from rivers and external sources in the Russian arctic coastal zone—RADARC. In: *Fifth Int. Conf. on Environmental Radioactivity in the Arctic and Antarctic*, P. Strand, T. Jølle, and Å. Sand (Eds.). Norway Radiation Protection Authority, , Norway, pp. 53–57.

Johannessen, O. M.; Pettersson, L. H.; Volkov, V. A.; Nielsen, S. P.; Maderich, V. S.; Stepanov, A. V.; Sickel, M.; Gao, Y.; Neelov, I. A.; Bobylev, L. P.; Zheleznyak, M.; Reiersen, L. O.; and Drange, H. (2003) *Simulation Scenarios for Potential Radioactive Spreading in the 21st Century from Rivers and External Sources in the Russian Arctic Coastal Zone—RADARC*, Final Scientific Report, INCO: ICA-CT-2000-10037 (1998—2002). Available at *http://www.nersc.no/RADARC*, 41 pp.

Johannessen, O. M.; Dziuba, N.; Koshebutsky, K.; Maderich, V.; and Zheleznyak, M. (2003a) *Final Scientific Report for Simulation Scenarios for Potential Radioactive Spreading in the 21st Century from Rivers and External Sources in the Russian Arctic Coast Zone*, Tech. Rep. Nansen Environmental and Remote Sensing Center, Bergen, Norway.

Johannessen, O. M.; Gao, Y.; Pettersson, L. H.; and Drange, H. (2003b) *Potential Spreading of Radioactivity from the Sunken Russian Nuclear Submarine K-159 in the Outer Part of the Murmansk*, Special Rep. 74. Nansen Environmental and Remote Sensing Center, Bergen, Norway.

Johannessen, O. M.; Bengtsson, L.; Miles, M. W.; Kuzmina, S. I.; Semenov, V.; Alekseev, G. V.; Nagurny, A. P.; Zakharov, V. F.; Bobylev, L. P.; Pettersson, L. H.; Hasselmann, K.; and Cattle, H. P. (2004) Arctic climate change: Observed and modelled temperature and sea ice. *Tellus*, **56A**(4) 328–341.

Johansen, O. (2000) DeepBlow: A Lagrangian plume model for deep water blowouts. *Spill Science and Technology Bulletin*, **6**, 103–111.

Joint Russian–Norwegian Expert Group for Investigation of Radioactive Contamination in the Northern Seas (1993) *A Survey of Artificial Radionuclides in the Kara Sea*, results from the Russian–Norwegian 1992 Expedition to the Barents and Kara Seas. Norwegian Radiation Protection Authority, Østerås, Norway.

Kabakchi, S. A.; Zagainov, V. A.; Lushnikov, A. A.; Zhuravel, V. I.; Lalushkin, Yu. P.; and Nazin, E. R. (1994) Hypothetical damages of storages of high-active liquid radioactive waste and pulp. *Atomic Energy*, **76**(5), 444–452.

Kalnay, E.; Kanamitsu, M.; Kistler, R.; Collins, W.; Deaven, D. *et al.* (1996) The NCEP/NCAR 40-year reanalysis project. *Bulletin of the American Meteorological Society*, **77**, 437–471.

Karcher, M. J.; Gerland, S.; Harms, I. H.; Iosjpe, M.; Heldal, H. E. *et al.* (2004) The dispersion of ^{99}Tc in the Nordic Seas and the Arctic Ocean: A comparison of model results and observations. *Journal of Environmental Radioactivity*, **74**, 185–198.

Karickhoff, S. W. (1984) Organic pollutant sorption in aquatic systems. *Journal of Hydraulic Engineering*, **110**, 707–735.

Karickhoff, S. W. *et al.* (1979) Sorption of hydrophobic pollutants on natural sediments. *Water Resources*, **13**, 241–248.

Karim, M. F.; and Kennedy, J. F. (1981) *Computer-based Predictors for Sediment Discharge and Friction Factor of Alluvial Streams*, IIHR Report No. 242. Iowa Institute of Hydraulic Research, University of Iowa, Iowa City, IA, 273 pp.

Karim, M. F.; Holley, F. M.; and Yang, J. C. (1987) *IALLUVIAL Numerical Simulation of Mobile-bed Rivers, Part: Theoretical and Numerical Principles*, IIHR Report No. 309. Iowa Institute of Hydraulic Research, University of Iowa, Iowa City, IA, 198 pp.

Katrich, I. Yu.; Nikitin, A. I.; Medinets, V. I.; Lepeshkin, V. I.; Kabanov, A. I.; Semko, N. N; and Bazhanov, V. N. (1992) Dynamics of the radioactive contamination caused by CNPP accident on observed data 1986–1990. In: *Ecological and Hydrophysical Consequences of the Nuclear Accidents*, V. A. Borzilov and I. I. Kryshev (Eds.). Hydrometeorological Publ., Moscow, pp. 57–61.

Kershaw, P.; and Baxter, A. (1995) The transfer of reprocessing wastes from north-west Europe to the Arctic. *Deep Sea Research*, **42**, 1413–1448.

Kershow, P. J.; Gurbott, P. A.; Woodhead, D. S.; Leonard, K. S.; and Rees, J. M. (1997) Estimation of fluxes of ^{137}Cs in northern waters from recent measurements. *Science of the Total Environment*, **202**, 211–223.

Kistler, R. *et al.* (2001) The NCEP–NCAR 50-year reanalysis: Monthly means CD-ROM and documentation. *Bulletin American Meteorological Society*, **82**(2), 247–267.

Konoplev, A. V.; and Golubenkov, A. V. (1991) Modelling of the vertical migration in soil as result of a nuclear accident. *Meteorologia i Gidrologia*, **10**, 62–68.

Konoplev, A. V.; Bulgakov, A. A.; Popov, V. E.; and Bobovnikova, Ts. I. (1992) Behaviour of long-lived Chernobyl radionuclides in a soil–water system. *Analyst*, **117**, 1041.

Koshebutsky, V.; Maderich, V.; Nesterov, O.; and Heling, R. (2004) Modelling of heat dispersion in inland waters and coastal areas of sea. *Applied Hydromechanics*, **6**(78), 34–44.

Kosmakov, E. B. (1996) *Definition of Potential Places of Radioactive Contamination Pollution of the Yenisei River Valley between Atamanovo and Strelka (Based on the Analysis of Hydrological Processes)*. Research Institute of Natural Ecological Systems, Divnogorsk, Russia.

Kostyanitsyn, M. N. (1964) *Hydrology of the Dnipro and Southern Buh Estuary*. Gidrometeoizdat, Moscow, 336 pp.

Kouraev, A. V. (1996) *Sources of Radioactivity in the Kara Sea*, Technical Report No. 4. NIERSC, St. Petersburg, 64 pp.

Koziy, L.; and Maderich ,V. (1997) A simulation of seasonal sea ice dynamics in coastal areas. *Proc. First International Conference: Port, Coast, Environment, Varna, Bulgaria*, Vol. 2, pp. 220–228.

Koziy, L.; Maderich, V.; Margvelashvili, N.; and Zheleznyak, M. (1998) Three-dimensional model of radionuclide dispersion in the estuaries and shelf seas. *Journal of Environmental Modeling and Software*, **13**(5/6), 413–420.

Koziy, L.; Maderich, V.; Margvelashvili, N.; and Zheleznyak, M. (2000) Numerical modelling of seasonal dynamics and radionuclide transport in the Kara Sea: Oceanic fronts and related phenomena. *Proc. Konstantin Fedorov Intl. Memorial Symp*, IOC Workshop Rep. Series No. 159. UNESCO, pp. 296–301.

Krone, R. B. (1962) *Flume Studies of the Transport of Sediment in Estuarial Processes*, Final Report. Hydraulic Engineering Laboratory and Sanitary Engineering Research Laboratory, University of California, Berkeley.

Kruglov, A. K. (1994) *How the Atomic Industry in the USSR was Created.* CNII Atominform, Moscow, 380 pp. [in Russian].

Kryshev, I. I. (Ed.) (1997) *Environmental Risk Analysis for the Ural Radioactive Pattern.* Nuclear Society of Russia, Moscow, 214 pp. [in Russian].

Kryshev, I. I.; Romanov, G. N.; Sazykina, T. G.; Isaieva, L. N.; and Kholina, Yu. B. (1997) *Radioecological Problems of the Southern Urals.* Nuclear Society of Russia, Moscow, 118 pp. [in Russian].

Kryshev, I. I.; Romanov, G. N.; Chumichev, V. B.; Sazykina, T. G.; Isaeva, L. N; and Ivanittskaya, M. V. (1998a) Radioecological consequences of radioactive disharges into the Techa River on the Southern Urals. *J. Environ. Radioactivity*, **38**, 195–209.

Kryshev, I. I.; Romanov, G. N.; Sazykina, T. G.; Isaeva, L. N.; Trabalka, J. R.; and Blaylock, B. G. (1998b) Environmental contamination and assessment of doses from radiation releases in the Southern Urals. *Health Physics*, **74**(6), 687–697.

Kulebakina, L. G.; and Polikarpov, G. G. (1991) Radioecological monitoring of the Black Sea basin following the Chernobyl NPP accident. In: *Proceedings of Seminar on Comparative Assessment of the Environmental Impact of Radionuclides Released during Three Major Nuclear Accidents: Kyshtym, Windscale, Chernobyl, held in Luxembourg*, Report EUR 13574, V.II, pp. 607–648.

Kuzmina, S.; Johannessen, O. M.; Bengtsson, L.; Aniskina, O. G.; and Bobylev, L. P. (2008) High northern latitude surface air temperature: Comparison of existing data and creation of a new gridded data set 1900–2000. *Tellus*, **60**, 289–304.

Kuznetsov, Yu. V.; Nosov N. A.; and Legin, V. K. (1993) Dumping of radioactive waste disposal in the sea environment: Scientific and practical aspects. In: *Environmental Radioactivity in the Arctic and Antarctic: Proceedings of the International Conference on Environmental Radioactivity in the Arctic and Antarctic, Kirkenes*, P. Strand and E. Holm (Eds.), pp. 37–52.

Kuznetsov, Yu. V. *et al.* (2001) *Final Project Technical Report of ISTC 748-97.* Khlopin Radium Institute, St. Petersburg.

Laissaoui, A.; Abril, J. M.; Perianez, R.; García-León, M.; and García-Montano, E. (1998) Determining kinetic transfer coefficients for radionuclides in estuarine waters: Reference values for ^{133}Ba and effects of salinity and suspended load concentrations. *Journal of Radioanalytical Nuclear Chemistry*, **237**, 5–61.

Ledwell, J. R.; Watson, A. J.; and Law, C. S. (1993) Evidence for slow mixing across the pycnocline from an open ocean tracer release experiment. *Nature*, **364**, 701–703.

Levitus, S.; and Boyer, T. P. (1994) *World Ocean Atlas 1994, Vol. 4: Temperature*, NOAA Atlas NESDIS 4, Washington, D.C.

Levitus, S.; Burgett, R.; and Boyer, T. P. (1994) *World Ocean Atlas 1994, Vol. 3: Salinity*, NOAA Atlas NESDIS 3. Washington, D.C.

Liquid Pathway Generic Study (1978) *Impact of Accidental Radioactivity Releases to Hydrosphere from Floating and Land-based Nuclear Power Plants*, USNRC, NUREG-044. U.S. Nuclear Regulatory Commission, 212 pp.

List of Fishery Economic Standards: Maximum Permissible Concentrations (MPC) and Approximately Safe Impact Levels (ASIL) of Harmful Substances for Water of the Water Bodies of Importance for Fishery (1999) Publ. VNIRO, Moscow, 305 pp. [in Russian].

Livingston, H.; Kupferman, S.; Bowen, V.; and Moore, R. (1984) Vertical profile of artificial radionuclide concentrations in the central Arctic Ocean. *Geochimica et Cosmochimica Acta*, **48**, 2195–2203.

Loeng, H.; Ozhigin, V.; Adlandsvik, B.; and Sagen, H. (1993) Current measurements in the northeastern Barents Sea. *Proceedings of the ICES Statutory Meeting*, 22 pp.

Loeng, H.; Ozhigin, V.; and Adlandsvik, B. (1995) Water fluxes through the Barents Sea. *Proceedings of the ICES Statutory Meeting*, 12 pp.

Long-term Observations on Regime and Resources of Surface Waters (1986) Vol. 1 (12), *Yenisei Basin*. Gidrometeoizdat, Leningrad [in Russian].

LTI (2000) Modeling analysis of PCB and sediment transport in support of the Kalamazoo River Remedial Investigation/Feasibility Study. Appendix to Blasland, Bouck & Lee, Inc./ Allied Paper, Inc. (2000) *Portage Creek, Kalamazoo River Superfund Site RI/FS Supplement to the Kalamazoo River RI/RS*. Limno-Tech.

Lyszov, V. N.; Ivanov, A. B.; and Kolyshkin, A. E. (1993) Radioecological issues of Tomsk accident. *Atomic Energy*, **74**(4), 364–367 [in Russian].

Maderich, V.; Dziuba, N.; Koshebutsky, V.; Zheleznyak, M.; and Volkov, V. A. (2005) An assessment of flux of radionuclide contamination through the large Siberian rivers to the Kara Sea. *Radioprotection*, **40**, S413–S419.

Maderich, V.; Heling, R.; Bezhenar, R.; Brovchenko, I.; Jenner, H.; Koshebutsky, A.; Kuschan, A; and Terletska, K. (2008) Development and application of 3D numerical model THREETOX to the prediction of cooling water transport and mixing in inland and coastal waters. *Hydrological Processes*, doi: 10.1002/hyp.6985.

Maksimov, I. V. (1939) On the determination of the ordinal value of the summer water outflow from the Barents to the Kara Sea through Yugorsky Shar Strait. *Problemy Arktiki*, Nos. 7/8 [in Russian].

Marchesiello, P.; McWilliams, J. C.; and Shchepetkin, A. (2001) Open boundary conditions for long-term integration of regional oceanic models. *Ocean Modelling*, **3**, 1–20.

Margvelashvili, N.; Maderich, V.; and Zheleznyak, M. (1997) THREETOX: A computer code to simulate three-dimensional dispersion of radionuclides in stratified water bodies. *Radiation Protection Dosimetry*, **73**, Nos. 1/4, 177–180.

Margvelashvili, N.; Maderich, V.; and Zheleznyak, M. (1999) Simulation of radionuclide flux from Dnieper–Bug Estuary into the Black Sea. *Journal of Environmental Radioactivity*, **43**, 157–171.

Margvelashvili, N.; Maderich, V.; Yuschenko, S.; and Zheleznyak, M. (2002) 3D modelling of the mud and radionuclide transport in Chernobyl cooling pond and Dnieper–Bug Estuary. *Fine Sediments Dynamics in the Marine Environment: Proceedings of INTERCOH-2000*, J. C. Winterwerp and C. Kranenburg (Eds.). Elsevier, pp. 595–610.

Marinets, A.; Gofman, D.; and Zheleznyak, M. (1996) Using GIS for modelling radionuclide transport in complex river-reservoir network. *HydroGIS 96: Application of Geographical Information Systems in Hydrology and Water Resources Management (Proc. Vienna Conf. 1996)*, IAHS Publ. No. 235. International Association of Hydrological Sciences, Christchurch, New Zealand, 325– 330.

Marsov, G. N. (1939) Water balance of Yugorsky Shar Strait. *Problemy Arktiki*, No. 4, 13–22 [in Russian].

Martinova, A. *et al.* (1995) The study of the radionuclides carried to the Northern Seas by the Yenisei River, radioactive contamination of the Yenisei River. *Journal of Radioactivity*, **29**(3), 84.

Matishov, D. G.; and Matishov, G. G. (2001) *Radiational Ecological Oceanology*. Kola Science Center RAS, Apatity, 417 pp. [in Russian].

Matishov, D. G.; and Matishov, G. G. (2004) *Radiational Ecological Oceanology*. Springer/Praxis, Heidelberg, Germany/Chichester, U.K., 450 pp.

Matishov, G. G.; Matishov, D. G.; and Nazimov, V. V. (1994) *Levels and Main Directions of Radionuclide Transport in the Barents and Kara Seas (1990–1993)*, map. Russian Academy of Sciences/Murmansk Marine Biology Institute.

Matishov, G. G.; Volkov, V. A.; and Denisov, V. V. (1998a) Current views on the circulation structure of warm Atlantic water in the northern Barents Sea. *Reports of the Russian Academy of Sciences*, **362**(4), 553–556 [in Russian].

Matishov, G. G.; Golubeva, N. I.; Afanasyev, M. I.; and Burtseva, L. V. (1998b) Levels of pollutants in the snow of the Kara and Pechora Seas. *Doklady of Russian Academy of Sciences*, **361**(5), 715–718 [in Russian].

Matishov, G.; Golubeva, N.; and Burtseva, L. V. (2004a) Assessing the flows of heavy metals from the atmosphere through precipitation in the Barents Sea area: Evolution of marine and terrestrial ecosystems in periglacial zones. *Abstracts of the International Conference, September 6–8, Rostov-on-Don*, pp. 76–80 [in Russian].

Matishov, G.; Golubeva, N.; Titova, G.; Sydnes, A.; and Voegele, B. (2004b) *Global International Water Assessment: Barents Sea GIWA Regional Assessment*, Vol. 11. University of Kalmar, Sweden, 114 pp.

Matuschenko, A. M.; Kaurov, G. A.; Krasilov, G. A.; and Haritonov, K. V. (1994) Nuclear polygon without "top secret" label (dates, events). *Proceedings of the Integrated Marine Arctic Expedition*, Issue IY: *Novaya Zemlya*, Vol. 3, P. V. Boyarsky (Ed.). Moscow, pp. 54–67 [in Russian].

Mauritzen, C.; Hjøllo, S.; and Sandø, A. (2006) Passive tracers and active dynamics: A model study of hydrography and circulation in the northern North Atlantic. *Journal of Geophysical Research*, **111**(C08014), 1–15.

McDougall, T. J. (1987) Thermobaricity, cabbeling and water-mass conversion. *Journal of Geophysical Research*, **92**(C5), 5448–5464.

McDougall, T.; and Dewar, W. (1998) Vertical mixing, cabbeling and thermobaricity in layered models. *Journal of Physical Oceanography*, **28**, 1458–1480.

McDougall, T.; and Jackett, D. R. (2005) An assessment of orthobaric density in the global ocean. *Journal of Physical Oceanography*, **35**, 1458–1480.

Mehta, J.; Hayter, P.; Parker, W.; Krone, R.; and Teeter, A. (1989) Cohesive sediment transport, I: Process description. *Journal of Hydraulic Engineering*, **115**, 1076–1093.

Mellor, G. L. (1991) An equation of state for numerical models of ocean and estuaries. *Journal of Atmospheric and Oceanic Technology*, **8**, 609–611.

Mellor, G. L.; and Kantha, L. (1989) An ice–ocean coupled model. *Journal of Geophysical Research*, **94**, 10837–10954.

Mikhailov, V. N. (1997) *River Deltas of Russia and Neighbouring Countries*. GEOS, 413 pp. [in Russian].

Millero, F. J. (1978) Freezing point of seawater. In: *Eighth Report on the Joint Panel on Oceanographic Tables and Standards, Paris*, UNESCO Tech. Pap. Mar. Sci. No. 28, Annex 6. UNESCO, Paris.

Mining Chemical Combine (2000) *Official Prospect* [in Russian].

Miyazawa, Y.; and Minato, S. (2001) POM and two-way nesting POM study of Kuroshio damping phenomenon caused by a strong wind. *Journal of Oceanography*, **56**, 275–294.

Modelling and Study of the Mechanisms of the Transfer of Radioactive Material from Terrestrial Ecosystems to Water Bodies (1996) Final Report ECP-3 Project. CEC, Brussels (Belarus, Russia, Ukraine program on the radiological consequences of the Chernobyl accident).

Mokrov, Yu. G. (1996) The prediction of ^{90}Sr transport by the Techa River. *Problems of Radiation Safety*, No. 1, 20–27.

Mokrov, Yu. G. (2000) Retrospective estimation of an effluent ^{90}Sr in the Kara Sea with waters of the Ob′ River between 1949 and 1990. *International Conference: Radioactivity after Nuclear Explosions and Accidents, Moscow, April*, Vol. 1. Hydrometeoizdat, St. Petersburg.

Molvaer, J.; Knutsen, J.; Magnusson, J.; Rygg, B.; Skei, J.; and Serensen, J. (1997) Klassifisering av miljekvalitet i fjorder og kystfarvann. *Veiledming, SFT Veiledming*, **97**(03), 36 [in Norwegian].

Monte, L.; Håkanson, L.; Periañez, R.; Laptev, G.; Zheleznyak, M.; Maderich, V.; Angeli, G.; and Koshebutsky, V. (2006) Experiences from a case study of multi-model application to assess the behaviour of pollutants in the Dnieper–Bug Estuary. *Ecological Modelling*, **195**, 247–263.

Morozov, A.; Zheleznyak, M.; Aliev, K.; Bilotkach, U.; and Votsekhovitch, O. (1996) Prediction of radionuclide migration in the Pripyat River and Dnieper Reservoirs and decision support of water protection measures on the basis of mathematical modelling. *Extended Abstracts: International Conference "One Decade after Chernobyl: Summing up the Radiological Consequences of the Accident", Vienna, Austria, April 8–12.*

Mundschenk, H.; Krause, W. J.; Dersch, G.; and Wengler, P. (1994) *Überwachung der Bundeswasserstrassen auf radioaktive Stoffe im Normal- und Ereignisfall: Konzept, Methoden und Ergebnisse*, Bericht BfG-0783. Bundesanstalt für Gewässerkunde, Koblenz [in German].

Mundschenk, H.; and Tolksdorf, W. (1988) Methodische Unterschungen zur Sedimentation mit Hilfe radioactiver Leitstoffe am Beispel eines Kleinhafens am Mittelrhein. *Deutsche Gewasserkundliche Mitteilungen*, **32**, 110–119 [in German].

Murray, F. W. (1967) On the computation of saturation vapor pressure. *Journal of Applied Meteorology*, **6**, 203–204.

NATO Report No. 204 (1995a) *Cross-border Environmental Problems Emanating from Defence-related Installations and Activities*, Vol. 1: *Radioactive Contamination*, Final Report, Phase 1: 1993–1995. North Atlantic Treaty Organization, 306 pp.

NATO Report No. 206 (1995b) *Cross-border Environmental Problems Emanating from Defence-related Installations and Activities*, Summary Final Report, Phase 1: 1993–1995. North Atlantic Treaty Organization, 26 pp.

NATO (1998) *Cross-border Environmental Problems Emanating from Defence-related Installations and Activities, Phase II: 1995–1998*, Vol. 2: *Radioactive Contamination of Rivers and Transport through Rivers, Deltas and Estuaries to the Sea*, Report No. 225. NATO, 136 pp.

Neelov, I. A.; and Kouraev, A. V. (1996) *3D Circulation Model of the Kara Sea*, Technical Report No. 3, Project INTAS-93-0814. NIERSC, St. Petersburg, Russia, 49 pp.

Nemirovskaya, I. A. (2005) Hydrocarbons in the ecosystem of the White Sea. *Okeanologiya*, **45**(5), 678–688 [in Russian].

Neyelov, I. A. (1982) A mathematical model of synoptic eddies in the ocean. *Oceanology*, **22**(6), 872–875.

Nies, N.; Harms, H.; Karcher, M.; Dethleff, D.; Bahe, C.; Kuhlmann, G.; Oberhuber, J.; Backhaus, J.; Kleine, E.; Loewe, P.; Matishov, D.; Stepanov, A.; and Vasiliev, O. (1998) Anthropogenic radioactivity in the Nordic Seas and the Arctic Ocean: Results of a joint project. *Deutsche Hydrographische Zeitschrift*, **50**(4), 313–343.

Nikanorov, A. M.; Ivanov, V. V.; and Bryzgalo, V. A. (2007) *Rivers of the Russian Arctic under Modern Conditions of Anthropogenic Influence*. Russian Academy of Sciences, Rostov-on-Don, 272 pp.

Nikiforov, Ye. G.; Panov, V. V.; and Shpaikher, A. O. (1974) *Kara Sea: Oceanographic Encyclopedia*. Gidrometeoizdat, Leningrad, 632 pp. (Russian edition of *The Encyclopedia of Oceanography* (1966) R. W. Fairbridge (Ed.), Reinhold, New York.)

Nikiforov, Ye. G.; and Shpaikher, A. O. (1980) *Typical Features in the Formation of Large-scale Oscillations of the Hydrological Regime of the Arctic Ocean*. Gidrometeoizdat, Leningrad, 270 pp. [in Russian].

Nikipelov, B. V.; Romanov, G. N.; Buldakov, L. A.; Babaev, N. S.; and Kholina, Yu. B. (1989) Radiation accident in South Urals in 1997. *Atomic Energy*, **67**(2), 74–80.

Nikitin, A. I.; Katrich, I. Yu.; Kabanov, A. I.; Chumichev, V. B.; and Smagin, V. M. (1991) Radioactive contamination of the Arctic Ocean from observations in 1985–1987. *Atomic Energy*, **71**(2), 169–172 [in Russian].

Nilsen, T. (1993) Ny storsatsing på atomkraftverk. *Bellona Magasin*, August/September, 28–32 [in Norwegian].

Nilsen, T. (1993a) Avfall på avveie. *Bellona Magasin*, August/September, 34–47 [in Norwegian].

Nilsen, T. (1993b) Aktuelt fra Russland. *Bellona Magasin*, Desember/January, No. 6, 38–39 [in Norwegian].

Nilsen, T. (1994) I skyggen av Sibirs plutonium fabrikken. *Bellona Magasin*, **6**, December/January, 26–37 [in Norwegian].

Nilsen, T.; Kudrik, I.; and Nikitin, A. (1996) The Russian Northern Fleet: Sources of radioactive contamination. *Bellona Magasin*, **2**, 94–102.

Nilsen, J. E. Ø.; Gao, Y.; Drange, H.; Furevik, T.; and Bentsen, M. (2003) Simulated North AtlanticNordic Seas water mass exchanges in an isopycnic coordinate OGCM. *Geophysical Research Letters*, **30**, doi: 10.10292002GL0657.

Noskov, A. I. *et al.* (1960) *Navigation Charts of the Tom River from Tomsk to the Confluence with the Ob River*, River Bed Exploring Crew #12, M. M. Shumilova (Ed.). Ob′ Basin Management Department, Tomsk, Russia.

Nosov, A. V. (1997) Study of radioactive substance migration mechanisms in the Yenisei floodplain. *Meteorology and Hydrology*, No. 12 [in Russian].

Nosov, A. V.; and Martynova, A. M. (1996) Analysis of radiation situation on Yenisei after decommissioning of the two flow-through reactors of Krasnoyarsk MCC. *Atomic Energy*, **81**(3), 226–232.

Nosov, A. V.; Ashanin, M. V.; Ivanov, A. V.; and Martynova, A. M. (1993) Radioactive pollution of Yenisei stipulated by discharges from Krasnoyarsk Mining and Chemical Combine. *Atomic Energy*, **74**(2), 144–150.

Novitsky, M. A.; Vakulovsky, S. M.; Nikitin, A. A.; and Tereb, L. A. (1999) Evaluation of Kara Sea radioactive contamination as a consequence of potential accidents at the Mayak PA. *Proc. Fourth International Conference on Environmental Radioactivity in the Arctic, Edinburgh, Scotland.* Norwegian Radiation Protection Authority, pp. 36–38.

Nuclear Tests of the U.S.S.R.: General Characteristics, Goals, Nuclear Test Organizations in the U.S.S.R. (1997) RFNC-VNIIEF, Moscow [in Russian].

Oey, L.-Y. (2006) An OGCM with movable land–sea boundaries. *Ocean Modelling*, **13**, 176–195.

Oey, L.-Y; and Chen, P. (1992) A nested-grid ocean model: With application to the simulation of meanders and eddies in the Norwegian coastal current. *Journal of Geophysical Research*, **97**, 20063–20086.

Onishi, Y. (1977) *Mathematical Simulation of Sediment and Radionuclide Transport in the Columbia River*, Rep. BNWL2228. Battelle Pacific Northwest Laboratory, Richland, Washington, D.C.

Onishi, Y. (1981) Sediment and contaminant transport model. *Journal of the Hydraulic Division ASCE*, **107**(HY9), 1089–1107.

Onishi, Y. (1993) *Sediment Transport Models and Their Testing*, Lecture Series. NATO Advanced Studies Institute, Pullman, WA.

Onishi, Y.; Serne, J.; Arnold, E.; Cowan, C.; and Thompson, F. (1981) *Review: Radionuclide Transport, Sediment Transport, Water Quality, Mathematical Modeling and Radionuclide Adsorption/Desorption Mechanism*, NUREG/CR-1322. Pacific Northwest Laboratory, Richland, Washington, D.C., 512 pp.

Onishi, Y.; Wheelan, G.; and Skaggs, R. L. (1982) *Development of a Multimedia Radionuclide Exposure Assessment Methodology for Low-level Waste Management*, PNL3370. Pacific Northwest Laboratory, Richland, Washington, D.C.

Onishi, Y.; Dummuller, D. C.; and Trent, D. S. (1989) *Preliminary Testing of Turbulence and Radionuclide Transport Modeling in Deep Ocean Environment*, PNL-6853. Pacific Northwest Laboratory, Richland, Washington, D.C.

Orlanski, I. (1976) A simple boundary condition for unbounded hyperbolic flows. *Journal of Computational Physics*, **21**, 251–269.

Orlob, G. T. (Ed.) (1983) *Mathematical Modeling of Water Quality: Streams, Lakes, and Reservoirs*, International Series on Applied Systems Analysis. IIASA/Pitman Press, London, 518 pp.

Orre, S.; Gao, Y.; Drange, H.; and Nilsen, J. E. Ø. (2007) A reassessment of the dispersion properties of ^{99}Tc in the North and Norwegian Sea. *Journal of Marine Systems*, **30**, 24–38, doi: 10.1016/j.jmarsys.2006.10.009.

Orvik, K.; and Skagseth, Ø. (2003) The impact of the wind stress curl in the North Atlantic on the Atlantic inflow to the Norwegian Sea toward the Arctic. *Geophysical Research Letters*, **30**, doi:10.1029/2003GL017932.

Østerhus, S.; Turrell, W. R.; Jónsson, S.; and Hansen, B. (2005) Measured volume, heat, and salt fluxes from the Atlantic to the Arctic Mediterranean. *Geophysical Research Letters*, **32**(L07603), doi: 10.1029/2004GL022188.

Ozhigin, V. K. (1997) On the changes in currents in the Barents Sea and assessment of water exchange at its boundaries: Ecological monitoring of western Arctic seas (from concept to practice). *Abstracts: International Conference, Murmansk, October 23–25*, 168 pp. [in Russian].

Paluszkiewicz, T.; Hibler, L. F.; Becker, P.; Mandych, A.; Richmond, M. C.; and Thomas, S. (1997) An assessment of the flux of ^{90}Sr through the Ob′ River and estuary to the Kara Sea. *Science of the Total Environment*, **202**, 43–56.

Paluszkiewicz, T.; Hibler, L. F.; Richmond, M. C.; Bradley, D. B.; and Thomas, S. (2001) Modeling the potential radionuclide transport by the Ob′ and Yenisei rivers. *Marine Pollution Bulletin*, **43**, 111–121.

Parkinson, C. L.; and Washington, W. M. (1979) A large-scale numerical model of sea ice. *Journal of Geophysical Research*, **84**, 311–337.

Partheniades, E. (1965) Erosion and deposition of cohesive soil. *Journal of the Hydraulic Division ASCE*, **91**, 105–139.

Pavlov, V. (1994) *Oceanographical Description of the Kara and Barents Sea*, working material of the International Arctic Seas Assessment Project. International Atomic Energy Agency, Vienna, 134 pp.

Periañez, R. (2004) The dispersion of ^{137}Cs and $^{239,240}Pu$ in the Rhone River plume: A numerical model. *Journal Environmental Radioactivity*, **77**, 301–324.

Periañez, R. (2005) Modelling the transport of suspended particulate matter by the Rhone River plume (France): Implications for pollutant dispersion. *Environmental Pollution*, **133**, 351–364.

Petrov, O. (1993) Environmental and sanitation problems related to radioactive waste dumping into the Arctic seas. In: *Environmental Radioactivity in the Arctic and Antarctic: Proceedings of the International Conference on Environmental Radioactivity in the Arctic and Antarctic, Kirkenes*, P. Strand and E. Holm (Eds.), pp. 93–100.

Plotitsyna, N. F. (2002) Persistent chlorinated hydrocarbons in the Barents Sea water. *Proc. Environmental Pollution of the Arctic: The Second AMAP International Symposium, Rovaniemi, Finland, October*. Arctic Monitoring and Assessment Program (AMAP), Oslo, p. 11.

Plotitsyna, N. F.; Kireyeva, L. I.; Shalyapina, T. N.; Zimoveyskova, T.; and Golubeva, T. A. (2002) State of pollution of the Barents Sea ecosystem. *Abstracts: Conference Oil and Gas of the Arctic Shelf, Murmansk, November 13–15*. CD-ROM, [in Russian].

Popov, A.; Catsaros, N.; Maderich, V.; and Yuschenko, S. (2002) Reconstruction of land-to-sea radionuclide flux by POSEIDON/RODOS model chain: The Thermaikos Gulf case. *Radioprotection*, **37**, CI, 677–682.

QEA (1999) *PCBs in the Upper Hudson River*, Vol. 2: *A Model of PCB Fate, Transport and Bioaccumulation*. Quantitative Environmental Analyses, Albany, New York (prepared for General Electric).

Radioactive contamination of the Black and Azov seas (1998) In: Voitsekhovich, O. V. (Ed.). *Radioecology of Water Objects of the Chernobyl NPP Accident Impact Area*. Chernobylinterinform, Kiev, pp. 127–151.

R-ArcticNET (2001) *A Regional Hydrographic Data Network for the Pan-Arctic Region*, Version 2.1. Water Systems Analysis Group, University of New Hampshire (CD-ROM).

Raisbeck, G.; and Yiou, F. (1999) ^{129}I in the oceans: Origin and applications. *The Science of the Total Environment*, **237**, 31–41.

Raudkivi, A. (1967) *Loose Boundary Hydraulics*. Pergamon Press, New York.

Redi, M. H. (1982) Oceanic isopycnal mixing by coordinate rotation. *Journal of Physical Oceanography*, **12**, 1154–1158.

Reed, M.; Johansen, O.; Brandvik, P. J.; Daling, P.; Lewis, A.; Fiocco, R.; Mackay, D.; and Prentki, R. (1999) Oil spill modelling toward the close of the 20th century: Overview of the state of the art. *Spill Science and Technology Bulletin*, **5**, 3–16.

Reed, R. K. (1977) On estimating insolation over the ocean. *Journal of Physical Oceanography*, **7**, 482–485.

Report on Environmental Protection and Rational Use of Natural Resources of Murmansk Oblast′ in 2004 (2005) KaeM, Apatity, 99 pp. [in Russian].

Report on Environmental Protection and Rational Use of Natural Resources of Murmansk Oblast′ in 2005 (2006) MIP-999, Murmansk, 120 pp. [in Russian].

Report on Environmental Protection and Rational Use of Natural Resources of Murmansk Oblast′ in 2006 (2007) MIP-999, Murmansk, 159 pp. [in Russian].

Report on the Environmental State and Protection of the Murmansk Oblast' in 2002 (2003) KaeM, Apatity, 130 pp. [in Russian].

Review of Environmental Pollution in the Russian Federation for 2000 (2001) Roshydromet, Moscow, 238 pp. [in Russian].

Ritter, L.; Solomon, K. R.; Forget, J.; Stemeroff, M.; and O'Leary, C. (1995) *Persistent Organic Pollutants*, Assessment Report for the International Program on Chemical Safety (IPCS). WHO, Geneva.

Romankevich, Ye. A.; Danyushevskaya, A. I.; Belyayeva, A. N.; and Rusanov, V. P. (1982) *Biogeochemistry of Organic Matter of the Arctic Seas*. Nauka, Moscow, 240 pp. [in Russian].

Rovinsky, F. Ya.; Teplitskaya, T. A; and Alekseyeva, T. A. (1988) *Baseline Monitoring of Polycyclic Aromatic Hydrocarbons*. Gidrometeoizdat, Leningrad, 224 pp. [in Russian].

Russia: River Basins (1999) Aerokosmologiyz, Yekaterinburg, 520 pp. [in Russian].

Santschi, P. H.; and Honeyman, B. D. (1989) Radionuclides in aquatic environments. *Radiation Physics and Chemistry*, **34**, 2213–2407.

Sapozhnikov, V. V.; and Sokolova, S. A. (1994) Distribution of pollutants in water and bottom sediments of the White Sea. In: *Complex Studies of the Ecosystem of the White Sea*. VNIRO, Moscow, pp. 104–108 [in Russian].

Sarmiento, J. L.; Feely, H. W.; Moore, W. S.; Bainbridge, A. E.; and Broecker, W. S. (1976) The relationship between vertical eddy diffusion and buoyancy gradient in the deep sea. *Earth and Planetary Science Letters*, **32**, 357–370.

Savinov, V. M.; Savinova T. N.; Carrol, J.; Matishov, G. G.; Dahle, S; and Naes, K. (2000) Polycyclic aromatic hydrocarbons (PAHs) in sediments of the White Sea, Russia. *Marine Pollution Bulletin*, **40**(10), 807–818.

Savinova, T. N. (1994) *Study of Chlorinated Hydrocarbons in Bottom Sediments: Ecological Studies of the Zone of Industrial Development of the Shtokman Gas-condensate Field on the Shelf of the Barents Sea*, Preprint. Russian Academy of Sciences, Apatity, pp. 50–53 [in Russian].

Sayles, F. L.; Livingston, H. D.; and Panteleev, G. P. (1997) The history and source of particulate ^{137}Cs and $^{239,240}Pu$ deposition in sediments of the Ob River Delta, Siberia. *Science of the Total Environment*, **202**, 25–41.

Scheringer, M.; Salzmann, M.; Stroebe, M.; Wegmann, F.; Fenner, K; and Hungerbühler, K. (2004) Long-range transport and global fractionation of POPs: Insights from multimedia modeling studies. *Environmental Pollution*, **128**, 177–188.

Schnoor, J. L. (1996) *Environmental Modeling: Fate and Transport of Pollutants in Water, Air, and Soil*. John Wiley & Sons, New York, 704 pp.

Schnoor, J. L.; Mossman, D. J.; Borzilov, V. A.; Novitsky, M. A.; Voszhennikov, O. I.; and Gerasimenko, A. K. (1992) Mathematical model for chemical spills and distributed source runoff to large rivers, In: *Fate of Pesticides and Chemicals in the Environment*, J. L. Schnoor (Ed.). John Wiley & Sons, New York, 436 pp.

Schott, F.; and Brandt, P. (2007) Circulation and deep water export of the subpolar North Atlantic during the 1990s. In: *Ocean Circulation: Mechanisms and Impacts*, AGU Geophysical Monograph. American Geophysical Union, Washington, D.C., pp. 33–46.

Schuckler, M.; Kalckbrenner, R.; and Bayer, A. (1976) *Zukunftige radiologische Belastung durch kerntechnische Anlagen im Einzugsgebiet des Oberrheins*, Teil 2: *Belastung über den Wasserweg: Conference Dusseldorf Proceedings*. Eggenstein-Leopoldshafen, Germany [in German].

Scientific–Methodological Approaches to Assessment of the Impact of Oil–Gas Production on the Ecosystems of the Arctic Seas (1997) G. G. Matishov *et al.* (Eds.). Russian Academy of Sciences, Apatity, 393 pp. [in Russian].

Sea Water Quality from Hydrochemical Indicators: Yearbook 2002 (2005) Gidrometeoizdat, St. Petersburg, 126 pp. [in Russian].

Semenov, A. (1993) Impact of the Chernobyl accident on Kola Peninsula radiation conditions. In: *Environmental Radioactivity in the Arctic and Antarctic: Proceedings of the International Conference on Environmental Radioactivity in the Arctic and Antarctic, Kirkenes*, P. Strand and E. Holm (Eds.), pp. 353–356.

Shevchenko, V. P. (2006) *Influence of Aerosols and Marine Sedimentation in the Arctic*. Nauka, Moscow, 226 pp. [in Russian].

Shin, C. S.; and Gloyna, E. F. (1970) Mathematical model for the transport of radionuclides in a stream system. In: *Environmental Surveillance in the Vicinity of a Nuclear Facility*, W. Reinig and C. Charles (Eds.). Thomas Publishing, Springfield, IL, pp. 518–532.

Shine, K. P. (1982) Parameterization of shortwave flux over high albedo surfaces as function of cloud thickness and surface albedo. *Quarterly Journal of the Royal Meteorological Society*, **110**, 747–760.

Simonsen, K.; and Haugan, P. M. (1996) Heat budgets of the Arctic, Mediterranean and sea surface heat flux parameterizations for the Nordic Seas. *Journal of Geophysical Research*, **101**, 6553–6576.

Slavik, O.; Zheleznyak, M.; Dzuba, N.; Marinets, A.; Lyashenko, G.; Papush, L.; Shepeleva, T.; and Mihaly, B. (1997) Implementation of the decision support system for the river–reservoir network affected by releases from the Bohunice NPP, Slovakia. *Radiation Protection Dosimetry*, **73**, Nos. 1/4, 171–175.

Smagorinsky, J. (1963) General circulation experiments with primitive equations, 1: The basic experiment. *Monthly Weather Review*, **91**, 99–164.

Smith, J.; Ellis, K.; and Boyd, T. (1999) Circulation features in the central Arctic Ocean revealed by nuclear fuel reprocessing tracers from Scientific Ice Expeditions 1995 and 1996. *Journal of Geophysical Research*, **104**(C12), 29663–29677.

Smith, J.; Jones, E.; Moran, S.; Smethie, W.; and Kieser, W. (2005) Iodine 129/CFC 11 transit times for Denmark Strait Overflow Water in the Labrador and Irminger Seas. *Journal of Geophysical Research*, **110**, doi: 10.1029/2004JC002516.

Smitz, Y.; and Everbecq, E. (1986) Modeling the behaviour of radionuclides in aquatic ecosystems. Paper presented at *CEC Seminar on Cycling of Long-lived Radionuclides in the Biosphere: Observations and Models, Madrid, September*.

Smolarkiewicz, P. K.; and Clark, T. L. (1986) The multidimensional positive definite advection transport algorithm: Further development and applications. *Journal of Computational Physics*, **67**, 396–438.

Solovieva, Z. S.; and Antonova, L. V. (1995) Spatial–temporal variability of the ice regime elements in the lower reaches and mouth areas of rivers in the Kara Sea basin. In: *Natural Conditions of the Kara and Barents Seas: Proc. of the Russian–Norwegian Workshop-95*, Rapportserie No. 97. Norsk Polarinstitut, Oslo, pp. 257–259.

Sørensen, H. R.; Kjelds, J.; Deckers, F.; and Waardenburg, F. (1996) Application of GIS in hydrological and hydraulic modelling: DLIS and MIKE11-GIS. In: *Proc. HydroGIS 96: Application of Geographic Information Systems in Hydrology and Water Resources Management, Vienna, April*, IAHS Publ. No. 235. International Association of Hydrological Sciences, Christchurch, New Zealand.

Sorteberg, A.; Furevik, T.; Drange, H.; and Kvamsto, N. (2005) Effects of simulated natural variability on Arctic temperature projections. *Geophysical Research Letters*, **32**(L18708), doi: 10.1029/2005GL023404.

Sources contributing to radioactive contamination of the Techa River and areas surrounding the Mayak Production Association, Urals, Russia (1997) *Program on Investigations of Possible Impacts of the Mayak PA Activities on Radioactive Contamination of the Barents and Kara Seas*. Joint Norwegian–Russian Expert Group for Investigation of Radioactive Contamination in the Northern Areas, Norway.

Stanovoy, V. V. (2006) Variability of the thermohaline structure of water in the estuaries of the Kara Sea. *Proceedings AARI*, **447** [in Russian].

State of the Environment of Murmansk Oblast′ in 2000–2001 (2002) MIP-999, Murmansk, 186 pp. [in Russian].

State Water Cadastre (various years) *Annual Observations Data*, Vol. 7: *Ob′ and Yenisei Basins* [in Russian].

Steinberg, L. J. *et al.* (1997) Characterisation of parameters in mechanistic models: A case study of a PCB fate and transport model. *Ecological Modelling*, **97**, 35–46.

Stockholm Convention on Persistent Organic Pollutants (2001) UNEP. Available at *http://chm.pops.int/*

Strand, P. (1998) Radioactivity. In: *AMAP Assessment Report: Arctic Pollution Issues*. Arctic Monitoring and Assessment Program, Oslo, pp. 526–552.

Strand, P.; Aarkrog, A.; Bewers, J.; Tsaturov, Z.; and Magnusson, S. (1996) Radioactive contamination of the Arctic marine environment. In: *Radionuclides in the OceansInputs and Inventories*. Inst. de Protection et de Sûrété Nucléaire, France, pp. 95–119.

Strand, P.; Balonov, M.; Aarkrog, A.; Bewers, M. J.; Howard, B.; Salo, A.; and Tsaturov, Y. S. (1998) Radioactivity. In: *AMAP Assessment Report: Arctic Pollution Issues*. Arctic Monitoring and Assessment Program, Oslo, pp. 859.

Strand, P.; Howard, B.; Aarkrog, A.; Balonov, M.; Tsaturov, Y.; Bewers, J.; Salo, A.; Sickel, M.; Bergman, R.; and Rissanen, K. (2002) Radioactive contamination in the Arctic: Sources, dose assessment and potential risks. *Journal of Environmental Radioactivity*, **60**, 5–21.

Study of Influence of Land-based Sources of Radionuclides on Radioactive Contamination of the Kara Sea through Ob′ and Yenisei River Systems (2002) Tasks and Final Reports, INTAS-97-31278. Nansen Environmental and Remote Sensing Center, Bergen, Norway.

Sun, S.; Bleck, R.; Rooth, C.; Ducowicz, J.; Chassignet, E.; and Killworth, P. (1999) Inclusion of thermobaricity in isopycnic-coordinate ocean models. *Journal of Physical Oceanography*, **29**, 2719–2729.

Surface Water Quality of the Russian Federation (2004) Gidrometeoizdat, St. Petersburg, 427 pp. [in Russian].

TAMS (2000) *Revised Baseline Modelling Report, Hudson River PCBs Reassessment RI/FS: Further Site Characterisation and Analyses*, Vol. 2D. TAMS Consultants, Limno-Tech, Menzie Cura & Associates, Tetra Tech (prepared for the U.S. Environmental Protection Agency (Region 2) and U.S. Army Corps of Engineers, Kansas City District).

The Digital Chart of the World for Use with ARC/INFO (1993) Data Dictionary. Environmental Systems Research Institute, Redlands, CA.

The Soviet Arctic: The Seas and the Islands of the Arctic Ocean (1970) Nauka, Moscow, 528 pp. [in Russian].

Thibodeaux, L. J. (1996) *Environmental Chemodynamics: Movement of Chemicals in Air, Water, and Soil*. John Wiley & Sons, New York, 624 pp.

Thiede, J.; and Johannessen, O. M. (2008) Climate change in the North: Past, present and future. *Episodes—Journal of International Geoscience*, **31**(1), 163–167.

Timofeyev V. T. (1961) *The Water Masses of the Arctic Basin*. Gidrometeoizdat, Leningrad, 190 pp. [in Russian].

Timofeyev, V. T. (1963) Interaction of Arctic Ocean water with Atlantic and Pacific water. *Oceanology*, **3**(4), 569–578 [in Russian].

Tkalich, P.; Zheleznyak, M.; Lyashenko, G.; and Marinets, A. (1994) RIVTOX: Computer code to simulate radionuclide transport in rivers. In: *Computational Methods in Water Resources X*, Vol. 2, A. Peters *et al.* (Eds.). Kluwer Academic Publishers, Dordrecht, The Netherlands, pp. 1173–1180.

Tolman, H. L. (1999) *User Manual and System Documentation of WAVEWATCH-III*, Version 1.18, Technical Note 166, NOAA/NWS/NCEP/OMB. National Oceanic and Atmospheric Administration, Silver Springs, MD.

Toole, J. M.; Polzin, K. L.; and Schmitt, R. W. (1994) Estimates of diapycnal mixing in the abyssal ocean. *Science*, **264**, 1120–1123.

Trapeznikov, A.; Aarkrog, A.; Kulikov, N.; Nielsen, S. P.; Pozolotina, V.; Polikarpov, G.; Trapeznikova, V.; Chebotina, M.; Chukanov, V.; and Yushkov, P. (1993) Radioactive contamination of the Ob river system from the nuclear enterprise Majak in the Urals. In: *Environmental Radioactivity in the Arctic and Antarctic: Proc. International Conference on Environmental Radioactivity in the Arctic and Antarctic, Kirkenes*, P. Strand and E. Holm (Eds.), pp. 135–150.

Treshnikov, A. F.; and Baranov, G. I. (1972) *The Structure of Circulation of the Arctic Basin Waters*. Gidrometeoizdat, Leningrad, 158 pp. [in Russian].

Tretyakov, M. V. (2001) Modelling of the seasonal and synoptical variability of hydrological fields of Yenisei Estuary. Dissertation Cand. Sci., Arctic and Antarctic Research Institute, St. Petersburg.

Tsumune, D.; Aoyama, M.; and Hirose, K. (2003) Numerical simuation of ^{137}Cs and $^{239/240}$Pu concentration by an ocean general circulation model. *Journal of Environmental Radioactivity*, **69**, 61–84.

Turanov, I. (1963) *Water Exchange through the Arctic Straits: A Review of Current Studies*, Report on the OK-9/3, AARI Archives. Arctic and Antarctic Research Institute, St. Petersburg, 100 pp. [in Russian].

UNSCEAR (1982) *Ionizing Radiation: Sources and Biological Effects*, U.N. Scientific Committee on Effects of Atomic Radiation Report to the General Assembly. United Nations, 773 pp.

USNRC (1978) *Liquid Pathway Generic Study*, Rep. NUREG0440. U.S. Regulatory Commission, Washington, D.C.

Vakulovsky, S. M. (1993) Transport of artificial radioactivity by the Ob′ to the Arctic seas. In: *Environmental Radioactivity in the Arctic and Antarctic: Proc. International Conference on Environmental Radioactivity in the Arctic and Antarctic, Kirkenes*, P. Strand and E. Holm (Eds.), pp. 107–110.

Vakulovsky, S. M. (1995) Estimation of the impact of radioactive waste discharge into the Yenisei River on Kara Sea radioactive contamination. *Journal of Environmental Radioactivity*, **29**(3), 88–90.

Vakulovsky, S. M.; Kryshev, I. I.; Nikitin, A. I. *et al.* (1995) Radioactive contamination of the Yenisei River. *Journal of Environmental Radioactivity*, **29**(3), 225–236.

Valsaraj, K. T. *et al.* (1997) A quasi steady state pollutant flux methodology for determining sediment quality criteria. *Environmental Toxicology and Chemistry*, **16**(3), 391–396.

van Leer, B. (1979) Toward the ultimate conservative difference scheme, V: A second-order sequel to Godunov's method. *Journal of Computational Physics*, **32**, 101–136.

van Mazijk, A. (1996) *One-dimensional Approach of Transport Phenomena of Dissolved Matter in Rivers: Communication on Hydraulic and Geotechnical Engineering*, Report No. 96-3. Delft University of Technology, Faculty of Civil Engineering, 310 pp.

van Rijn, L. C. (1981) Model for sedimentation predictions In: *Proc. XIX Congr. International Association for Hydraulic Research Congress, New Delhi, India*, pp. 321–328.

van Rijn, L. C. (1984a) Sediment transport, Part I: Bed load transport. *J. Hydraul. Engineering*, **110**, 1431–1456.

van Rijn, L. C. (1984b) Sediment transport, Part II: Suspended load transport. *J. Hydraul. Engineering*, **110**, 1613–1641.

Vinje, T.; and Volkov, V. (Eds.) (1997) *Russian–Norwegian Oceanographic Program: KAREX–94*, Cruise Report.

Voinov, G. N. (1999) *Tidal Phenomena in the Kara Sea*. Russian Geographical Society, St. Petersburg, 110 pp.

Voitsekhovitch, O. V.; Zheleznyak, M. J.; and Onishi, Y. (1994) *Chernobyl Nuclear Accident Hydrological Analyses and Emergency Evaluation of Radionuclide Distribution in the Dnieper River, Ukraine during the 1993 Summer Flood*, PNL-9980. Battelle Pacific Northwest Laboratories, Richland, Washington, D.C., 96 pp.

Voitsekhovitch, O.; Sansone, U.; Zheleznyak, M.; and Bugai, D. (1996) Water quality management of contaminated areas and its effect on doses from aquatic pathways: The radiological consequences of the Chernobyl accident. *Proceedings of the First International Conference, Minsk, Belarus, March 18–22*, A. Karaoglou, G. Desmet, G. N. Kelly, and H. G. Menzel (Eds.). European Commission. Luxembourg, pp. 401–410.

Volkov, V. A.; Johannessen, O. M.; Borodachev, V. E.; Voinov, G. N.; Pettersson, L. H.; Bobylev, L. P.; and Kouraev, A. V. (2002) *Polar Seas Oceanography: An Integrated Case Study of the Kara Sea*. Springer/Praxis, Heidelberg, Germany/Chichester, U.K., 450 pp.

Vorobyev, V. I. (1959) The coastline length of the USSR Seas. *Geographicheskiy Sbornik*, No. 13 [in Russian].

Waters, R. D.; Kompton, K. L.; Novikov, V.; and Parker, R. L. (1999) *Releases of Radionuclides to Surface Waters at Krasnoyarsk-26 and Tomsk-7*, RR-99-3. International Institute for Applied Systems Analysis, Laxenburg, Austria.

Westall, I. C.; Zachary J. L.; and Morel, F. M. (1976) *MINEQL: A Computer Program for the Calculation of Chemical Equilibrium Composition of Aqueous Systems*, Tech. Note 18. Department of Civil Engineering, Massachusetts Institute of Technology, Cambridge, MA, 102 pp.

Wetmore, J. N.; and Fread, D. L. (1983) *The NWS Simplified Dam Break Flood Forecasting Model for Desk-top and Hand-held Microcomputers*. Hydrologic Research Laboratory, Office of Hydrology, National Weather Service, National Oceanic and Atmospheric Administration, Silver Springs, MD.

White, A.; and Gloyna, E. F. (1969) *Radioactivity Transport in Water: Mathematical Simulation*, EHE-70-04, Technical Report No. 19. University of Texas, Austin, TX, 231 pp. (prepared for the U.S. Atomic Energy Commission).

Won Seo, I.; and Sung Cheong, T. (1998) Predicting the longitudinal dispersion coefficient in natural streams. *J. Hydraulic Eng.: Proc. ASCE*.

Wool, T.; Ambrose, R.; Martin, J.; and Comer, E. (2005) *Water Quality Analysis Simulation Program (WASP)*, Version 6.0: *User's Manual*. Environmental Research Laboratory, U.S. Environmental Protection Agency (Region 4), Athens, GA.

Yablokov, A. V.; Karasev, V. K.; Rumyantsev, V. M.; Kokeev, M. E.; Petrov, O. J.; Lystsov, V. N.; Emelyanenkov, A. F.; and Rubtsov, P. M. (1993) *Facts and Problems Related to Radioactive Waste Disposal in Seas Adjacent to the Territory of the Russian Federation*, Nos. 13, 15, 16. Administration of the President of the Russian Federation, Zeleniy Mir, Moscow [in Russian].

Yapa, P. D.; and Weerasuriya, S. A. (1997) Spreading of oil spilled under floating broken ice. *Journal of Hydraulic Engineering*, **123**, 676–683.

Yapa, P. D.; Zheng, L.; and Nakata, K. (1999) Modeling underwater oil/gas jets and plumes. *Journal of Hydraulic Engineering*, **125**, 481–491.

Yearbook of Sea Water Quality by Hydrochemical Indicators for 1996 (1997) Gidrometeoizdat. St. Petersburg, 117 pp. [in Russian].

Yearbook of Surface Water Quality of the Russian Federation for 1996 (1997) Gidrometeoizdat, St. Petersburg, 737 pp. [in Russian].

Yearbook of Water Quality of the Barents Sea (1991) Murmansk UGMS, Murmansk, 61 pp. [in Russian].

Yearbook of Water Quality of the Barents Sea by Hydrochemical Indicators for 1991 (1992) Murmansk UGMS, Murmansk, 47 pp. [in Russian].

Yearbook of Water Quality of the Barents Sea by Hydrochemical Indicators for 1992 (1993) Murmansk UGMS, Murmansk, 39 pp. [in Russian].

Zalesak, S. (1979) Fully multidimensional flux-corrected transport algorithms for fluids. *Journal of Computational Physics*, **31**, 335–362.

Zapadka, T.; Wozniak, S. B.; and Wozniak, B. (2001) A simple formula for the net long-wave radiation flux in the southern Baltic Sea. *Oceanologia*, **43**, 265–277.

Zatonsky, M. Ya. (1970) Waves. In: *The Soviet Arctic: Seas and Islands of the Arctic Ocean*. Nauka, Moscow, pp. 197–199.

Zhang, D. L., Chang, H. R.; Seaman, N. L.; Warner, N. L.; and Fritsch, J. M. (1986) A two-way interactive nesting procedure with variable terrain resolution. *Monthly Weather Review*, **114**, 1330–1339.

Zheleznyak, M. (1988) On the structure of the near-bottom oscillating turbulent boundary layer. *Hydromechanics*, **58**, Naukova Dumka, Kiev, 1–8.

Zheleznyak, M. (1990) Mathematical models of radionuclide dispersion in a reservoir set. In: *System Analysis and Methods of Mathematical Modeling in Ecology*. V. Glushkov Institute of Cybernetics, Kiev, pp. 48–58 (in Russian).

Zheleznyak, M. (1997) Multiple scale analyses of radioactive contamination for rivers and reservoirs after the Chernobyl accident. *Multiple Scale Analyses and Coupled Physical Systems: Proc. Saint-Venant Symposium, August 28–29*. Presses de l'ecole nationale des ponts et chaussées, Paris, pp. 45–52.

Zheleznyak, M.; and Marinets, A. (1993) Modeling of long-wave propagation in river channels on the basis of the "diffusion wave" equation. In: *Computational Technologies*, Vol. 2, No. 6. Institute of Computational Technologies of Russian Academy of Sciences, Novosibirsk, pp. 147–154 [in Russian].

Zheleznyak, M.; Vojtcekhovich, O.; Demchenko, R.; Kuzmenko, Yu.; Tkalich, P.; and Khursin, S. (1992a) Simulating the effectiveness of measures to reduce the transport of radionuclides in the Pripyat–Dnieper. *Proc. International Seminar on Intervention Levels and Countermeasures for Nuclear Accidents, Cadarache, France, October 7–11, 1991*, Radiation Protection-54, EUR 14469. Commission of the European Communities, pp. 336–362.

Zheleznyak, M.; Demchenko, R.; Khursin, S.; Kuzmenko, Yu.; Tkalich, P.; and Vitjuk, N. (1992b) Mathematical modeling of radionuclide dispersion in the Pripyat–Dnieper aquatic system after the Chernobyl accident. *Science of the Total Environment*, **112**, 89–114.

Zheleznyak, M.; Demchenko, R.; Khursin, S.; Kuzmenko, Y.; Tkalich, P., and Vitjuk, N. (1992) Mathematical modelling of radionuclide dispersion in the Pripyat Dnieper aquatic system after the Chernobyl accident. *The Science of the Total Environment*, **112**, 89–114.

Zheleznyak, M.; Demchenko, R.; Kuzmenko, Yu.; Tkalich, P.; Dzuba, N.; and Mezhueva, I. (1993a) Mathematical modelling of radionuclide dispersion in the water bodies of the Chernobyl NPP zone and in the Dnieper Reservoirs. *Hydrological Impact of Nuclear Power Plants: Proc. UNESCO Workshop-92.* UNESCO, Paris, pp. 173–185.

Zheleznyak, M.; Tkalich, P. V.; Lyashenko, G. B.; and Marinets, A. V. (1993b) Radionuclide aquatic dispersion model: First approach to integration into the decision support system on the basis of post-Chernobyl experience. *Radiation Protection Dosimetry*, **6**, 7–43.

Zheleznyak, M.; Kuzmenko, Yu.; Tkalich, P.; Dzuba, N.; Gofman, D.; Golovanov, I.; Marinets, A.; and Mezhueva, I. (1994) Modelling of radionuclides transport in the set of river reservoirs. In: *Computational Methods in Water Resources X*, Vol. 2, A. Peters *et al.* (Eds.). Kluwer Academic Publishers, Dordrecht, The Netherlands, pp. 1189–1196.

Zheleznyak, M.; Blaylock, G.; Gontier, G.; Konoplev, A. *et al.* (1995) Modeling of radionuclide transfer in rivers and reservoirs: Validation study within the IAEA/CEC VAMP Program. *International Symposium on Environmental Impact of Radioactive Releases, IAEA, Vienna, May 8–12,* IAEA-SM-339. International Atomic Energy Agency, Vienna, pp. 330–331.

Zheleznyak, M.; Heling, R.; Raskob, W.; Popov, A.; Borodin, R.; Gofman, D.; Lyashenko, G.; Marinets, A.; Pokhil, A.; Shepeleva, T.; and Tkalich, P. (1996) Modelling of hydrological pathways in RODOS. *The Radiological Consequences of Chernobyl Accident: Proceedings of the First International Conference, Minsk, Belarus, March 18–22,* A. Karaoglou, G. Desmet, G. N. Kelly, and H. G. Menzel (Eds.). European Commission. Luxembourg, pp. 1139–1148.

Zhizhanov, A. V. (1980) Results of the expedition to and hydrological studies of the Ob′ mouth area in 1977–1979. *Proc. AARI*, **394**, 51–62.

Zhukinsky, V. N. *et al.* (1989) *The Dnieper and Buh Estuary Ecosystem.* Naukova Dumka, Kiev, 239 pp.

Index

MIX
Papier aus verantwortungsvollen Quellen
Paper from responsible sources
FSC® C105338
FSC
www.fsc.org

If you have any concerns about our products,
you can contact us on
ProductSafety@springernature.com

In case Publisher is established outside the EU,
the EU authorized representative is:
Springer Nature Customer Service Center GmbH
Europaplatz 3, 69115 Heidelberg, Germany

Printed by Libri Plureos GmbH
in Hamburg, Germany